Recent Advances in the Scientific Research on Ancient Glass and Glaze

Series on Archaeology and History of Science in China

Series Editor: Fuxi Gan *(Shanghai Inst. of Optics and Fine Mechanics, Chinese Academy of Sciences, China)*

Vol. 1: History of Modern Optics and Optoelectronics Development in China
edited by Fuxi Gan and Shouyun Tian

Vol. 2: Recent Advances in the Scientific Research on Ancient Glass and Glaze
edited by Fuxi Gan, Qinghui Li and Julian Henderson

SERIES ON ARCHAEOLOGY AND HISTORY OF SCIENCE IN CHINA

VOL. 2

Recent Advances in the Scientific Research on Ancient Glass and Glaze

Editor-in-Chief

Fuxi Gan
Chinese Academy of Sciences, China

Co-editors

Qinghui Li
Chinese Academy of Sciences, China

Julian Henderson
The University of Nottingham

World Scientific

World Century

Published by

World Century Publishing Corp.
27 Warren Street
Suite 401-402
Hackensack, NJ 07601
USA

and

World Scientific Publishing Co. Pte. Ltd.

5 Toh Tuck Link, Singapore 596224

USA office: 27 Warren Street, Suite 401-402, Hackensack, NJ 07601

UK office: 57 Shelton Street, Covent Garden, London WC2H 9HE

Library of Congress Cataloging-in-Publication Data
Names: Gan, Fuxi, editor. | Li, Qinghui, 1973– editor. | Henderson, Julian, 1953– editor.
Title: Recent advances in the scientific research on ancient glass and glaze / editor-in-chief: Fuxi Gan (Chinese Academy of Sciences, China) ; co-editors: Qinghui Li (Chinese Academy of Sciences, China), Julian Henderson (The University of Nottingham).
Other titles: Series on archaeology and history of science in China ; v. 2.
Description: Singapore ; Hackensack, NJ : World Scientific, [2016] | Series: Series on archaeology and history of science in China ; vol. 2 | Includes bibliographical references and index.
Identifiers: LCCN 2015031776| ISBN 9789814630276 (hardcover : alk. paper) | ISBN 9814630276 (hardcover : alk. paper) | ISBN 9789814630283 (pbk. : alk. paper) | ISBN 9814630284 (pbk. : alk. paper)
Subjects: LCSH: Glassware, Ancient--Research. | Glassware, Ancient--China. | Glassware, Ancient--Japan. | Glazing (Ceramics)--Research--China. | Glazing (Ceramics) | Glass manufacture--Research. | Glass manufacture--China. | Glass manufacture--Japan.
Classification: LCC TP850 .R43 2016 | DDC 748.20931--dc23
LC record available at http://lccn.loc.gov/2015031776

British Library Cataloguing-in-Publication Data
A catalogue record for this book is available from the British Library.

In-house Editor: Christopher Teo

Typeset by Stallion Press
Email: enquiries@stallionpress.com

Printed in Singapore

Preface

It has been almost ten years since the 2005 Shanghai International Workshop on Archaeology of Glass was held in April 2005, Shanghai. The topic of that symposium was the ancient glass research along the Silk Road. Its proceedings in Chinese edition and English version were published in 2007 and 2009 respectively by the Fudan University Press and the World Scientific Publishing. The readers offered good response and evaluation.

In 12–14 November 2012, the second International Workshop on Archaeology of Glass — International Symposium on Ancient Glass (ISAG'2012), was also held in Jiading District, Shanghai. About 20 more presentations were given at the meeting. At a panel discussion before the meeting close, all the participants suggested to publish the proceedings of the second meeting. After the ISAG'2012, with the organization of Prof. Julian Henderson, we got two chapters from new contributors. Based on this background, a book entitled 'Recent Progress of Scientific Research on Ancient Glass and Glaze' will be published through two years' preparation for collecting manuscripts and edit processing.

The aim of the book is to report the recent research development of ancient glass and glaze technology in the world and the exchange of the East and West along the Silk Road. The contents of this book are dedicated to promote the exchanges between researchers in both social and scientific fields, and to deepen the relative studies in the world.

The scope of this book covered includes the new archaeological findings of ancient glass and faience in the world, relationship of glass-making with glazing technology, development and application of modern technique used for the characterization of ancient glass and glaze, compound colorants/opacifiers among ancient glass, early exchanges of culture and technique between China and elsewhere along the Silk Road, and so on.

This book provides a comprehensive overview of the contributors' significant discoveries in the research field of ancient glass, glaze and related materials, such as the early Byzantine glass from Bulgaria, glasses from Kofun and Byodoin Temple in Japan, the finding of the chimes made of lead-barium glass of the Western Han Dynasty from China, characterization of the famous Islamic glass wares excavated from the Famen Temple's underground palace of the Tang Dynasty, new opinions on the origin and technical development of ancient faience and glaze in China, ancient potash glasses in Asia and tin or antimony-based compound colorants used in ancient soda-lime glasses, and the tracing of early exchanges between West and East though the scientific study of ancient glasses.

This book also focuses on the application of modern non-invasive techniques for characterization of cultural relics, such as the using of portable XRF, portable and mobile laser Raman spectroscopy, Optical Coherence Tomography (OCT) and isotope analysis. The other techniques include SEM-EDS, LA-ICP-AES, FT-IR microscopy, XRD and so on. Basing on the reported results, we may learn how the modern techniques are used to solve specific archaeological questions.

Publication of this book is supported by the Compass Plan Project from the State Administration of Cultural Heritage, China (Grant No. 20110303). The other related projects include the Major State Basic Research Development Program of China (973 Program, Grant No. 2012CB720901, 2012CB720906) and the National Natural Science Foundation of China (Grant No. 51402326, 11374314).

Co-editor Prof. Qinghui Li from the Shanghai Institute of Optics and Fine Mechanics (SIOM), CAS, undertakes the organization and editing work of this book. Co-editor Prof. Julian Henderson from the University of Nottingham pays great effort to go over the most manuscripts contributed by Chinese scholars and to prepare the nice preface. Dr. Hongxia Zhao, Dr. Song Liu and Mr. Junqing Dong from SIOM all take part in the edition of this proceeding. I would like to extend my acknowledgments to them.

Gan Fuxi
Shanghai, 2015

E-mails of the Corresponding Authors

Chapter	Name	E-mail
1	Fuxi Gan	fxgan@mail.shcnc.ac.cn
2	Julian Henderson	julian.henderson@nottingham.ac.uk
3	Brigitte Borell	b_borell_seidel@hotmail.com
4	Izumi Nakai	inakai@rs.kagu.tus.ac.jp
5	Laure Dussubieux	ldussubieux@fieldmuseum.org
6	Qinghui Li	qinghuil@sina.com
7	Akiko Inoue	aurora@ff.iij4u.or.jp
8	Qinghui Li	qinghuil@sina.com
9	Jianfeng Cui	cuijianfeng@pku.edu.cn
10	Qian Cheng	gzcq503@163.com
11	Thomas Smith	acxts3@nottingham.ac.uk
12	Nikolaos Zacharias	zacharias@uop.gr
13	Hongxia Zhao	tulipzhao@163.com
14	James W. Lankton	jamesyuri@gmail.com
15	Fuxi Gan	fxgan@mail.shcnc.ac.cn
16	Maolin Zhang	mlzhang@ustc.edu
17	Weidong Li	liwd@mail.sic.ac.cn
18	Julian Henderson	julian.henderson@nottingham.ac.uk
19	Guoxia Li	liguoxia@zzu.edu.cn
20	Xiaowei Xia	172065140@qq.com
21	Junqing Dong	djqing@siom.ac.cn

(*Continued*)

(*Continued*)

Chapter	Name	E-mail
22	Song Liu	littleprincels@163.com
23	Qinglin Ma	qinglinma226@126.com
24	Xi Chen	ahaqchen@126.com
25	Julin Wang	julinwang@126.com
26	Yimin Yang	yiminyang@ucas.ac.cn
27	Fuxi Gan	fxgan@mail.shcnc.ac.cn

Contents

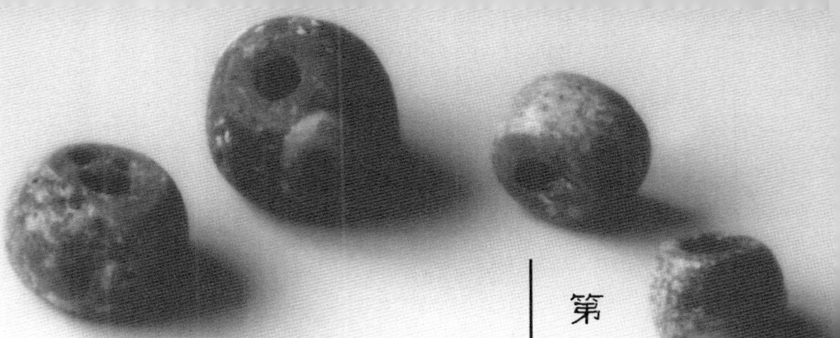

Chapter 1 第一章

The Ancient Glass Road: The Cultural and Technical Exchange of Chinese Glass and Faience with Outside China before the Han Dynasty (200 B.C.)

Fuxi Gan[*,†]

[*] Shanghai Institute of Optics and Fine Mechanics, Chinese Academy of Sciences, Shanghai 201800, China

[†] Fudan University, Shanghai 200433, China

1.1 Introduction

The commonly discussed Silk Road has become synonymous with the main artery connecting Asia with Europe for economic, political, cultural and technological exchanges in ancient times. Zhangqian's travels to the Western Region (~200 B.C.) was a magnificent undertaking. It greatly promoted the economic, cultural and technological exchanges between the four powerful empires at that time — the Hellenistic/Roman Empires in the West, the Han Dynasty in the East, the Kushan Empire in South Asia and the Hellenistic Empire in Central Asia. Therefore, the Silk Road was first recognised owing to Zhangqian's travels to the Western Regions (200 B.C.–200 A.D.). More physical evidence and related literary records related will be discussed below.[1,2]

There are other descriptions of contacts between ancient peoples in China and the outside world dating to before the Silk Road in ancient Chinese legends such as "King Mu travelled to West in Zhou Dynasty" (1500 B.C.). From this, one may conceive faintly that there had been a road linking the Eurasian steppe with China along which ancient people could travel. Some people called this the Northern Steppe Road or the Steppe Silk Road.[3] However, the literary records about Proto-history are few, and written characters had not yet appeared during the Neolithic age. To understand the culture of proto-historic times, we have to depend on archaeological studies. Through field excavation, cultural relics, scientific analysis of archaeological materials, and comparisons in a worldwide context; we can investigate their origins and development, find the traces of the exchanges between China and the outside world, and gain some inspiration.

1.2 Ancient Chinese faience and its origin

Glazed quartz, named faience in the West, appeared *c.* 3500–3000 BC (the middle Neolithic age). Sintered quartz sand contains a small amount of glass so it is not real glass. Faience pre-dated glass and dates to the same period painted pottery appeared in China. Chinese faience has been found in Gansu, Inner Mongolia, Shaanxi, Shanxi and Henan provinces along the Yellow River valley, Hubei and Jiangsu provinces along the Yangtze River valley, dating from the Early Western Zhou Dynasty to the Early Warring States Period (1046–400 B.C.). Chinese faience used quartz sand as its main raw material ($SiO_2 > 90\%$) as in western faience. The fluxing agent used in Chinese faience was plant ash, in which the content of K_2O is higher than that of Na_2O.[4–6] In Egypt, natron was used as the fluxing agent. Recent measurements show that some Chinese faience samples contain higher Na_2O levels relative to K_2O.[7]

The faience of the early Western Zhou Dynasty has also been unearthed at Ejinaqi in north-western Inner Mongolia.[4] This indicates that the early faience in China might come from the west, but most of it was made in China. Table 1.1 lists the chemical composition of glazed tubes and beads found in Pingdingshan and Xichuan in Henan Province.[8–10] In the Early Western Zhou, only few faience

Table 1.1 PIXE analytical results of the faience artifacts dated to the Zhou Dynasties unearthed from the Pingdingshan and Xichuan, Henan Province (wt.%)

Sample No:	Artifact	Date	Place	Na_2O	MgO	Al_2O_3	SiO_2	P_2O_5	SO_3	K_2O	CaO	TiO_2	V_2O_5	Fe_2O_3	CuO	PbO	BaO
HNZZ-81	Blue bead	Early Western Zhou	Pingdingshan	0.55	0.91	3.10	91.29	1.25	n.d.	0.17	0.19	0.23	n.d.	0.59	1.71	0.00	n.d.
HNWKII-91–1	Blue bead			1.20	0.72	1.91	91.56	n.d.	n.d.	0.24	0.68	n.d.	n.d.	0.63	2.95	0.71	n.d.
HNWKII-92–1	Blue bead			1.16	0.16	1.18	91.30	0.64	n.d.	0.23	0.29	0.03	n.d.	0.27	2.08	2.63	n.d.
HNWKII-64	Blue bead	Middle Western Zhou		1.50	0.84	5.71	86.30	n.d.	0.54	0.91	0.35	0.06	n.d.	0.77	2.52	n.d.	n.d.
HNWKII-70	Blue bead			n.d.	0.24	2.08	85.26	0.84	0.55	0.35	0.62	n.d.	n.d.	0.10	9.07	n.d.	n.d.
HNZZ-17	Blue tube	Late Western Zhou		n.d.	0.40	1.71	92.50	1.91	0.28	0.25	0.65	n.d.	n.d.	0.08	1.71	n.d.	n.d.
HNZZ-18	Blue tube			n.d.	0.34	1.94	89.20	3.11	0.51	0.25	2.23	0.04	n.d.	0.26	1.86	n.d.	n.d.
HNWKII-75–1	Blue tube	Early Spring and Autumn Period		n.d.	0.11	1.10	94.99	0.41	n.d.	0.18	0.20	0.03	n.d.	0.21	2.77	n.d.	n.d.
HNWKII-78a	Blue bead			n.d.	0.27	1.79	95.37	0.31	n.d.	0.34	0.10	0.06	n.d.	0.13	1.65	n.d.	n.d.
HNWKII-82	Blue bead	Late Spring and Autumn Period	Ye County, Pingding-shan	2.14	0.67	3.46	85.65	1.33	n.d.	2.39	0.47	0.11	n.d.	0.35	2.37	0.96	0.10
HNWKII-21	Blue bead	Middle Spring and Autumn Period	Xichuan	0.91	0.89	7.26	83.12	1.16	1.75	1.12	0.67	0.19	n.d.	1.12	1.54	n.d.	n.d.
HNWKII-24	Blue bead			n.d.	0.83	3.65	92.13	1.31	0.18	0.43	0.13	0.03	n.d.	0.33	0.96	n.d.	n.d.
HNWKII-30	Blue bead	Late Spring and Autumn Period		0.57	0.39	1.47	94.25	0.44	0.43	0.37	0.39	0.04	n.d.	0.12	1.49	n.d.	n.d.
HNWKII-34	Blue tube			n.d.	0.50	6.05	89.39	0.41	0.07	0.79	0.38	0.20	n.d.	0.38	1.76	n.d.	n.d.
HNWKII-44	Blue bead			1.10	1.02	4.61	76.95	0.23	1.27	1.47	1.00	n.d.	n.d.	0.58	11.19	n.d.	n.d.
HNWKII-58	Blue bead	Middle Warring States Period		3.72	1.72	4.88	50.58	n.d.	1.72	1.45	1.21	n.d.	0.31	0.98	8.61	10.39	12.51
HNWKII-59	Blue bead			3.99	1.43	5.47	51.73	0.60	2.14	1.43	0.92	n.d.	0.12	0.75	7.40	9.99	12.49

n.d.: not detected.

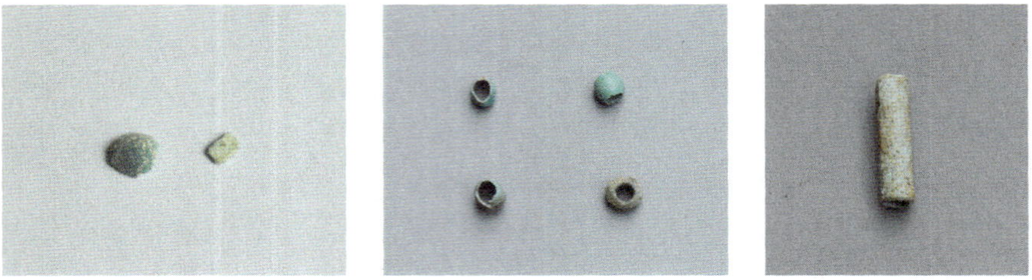

(a) (b) (c)

Fig. 1.1 Faience beads excavated from the cemetery of Ying State in Pingdingshan dated from the early to middle Western Zhou. (a) HNWKII-92, the "Liaozhu"of the Early Western Zhou; (b) HNWKII-69~72, the "Liaozhu"of the Middle Western Zhou; (c) HNWKIII-78, the "Liaoguan"of the Middle Western Zhou

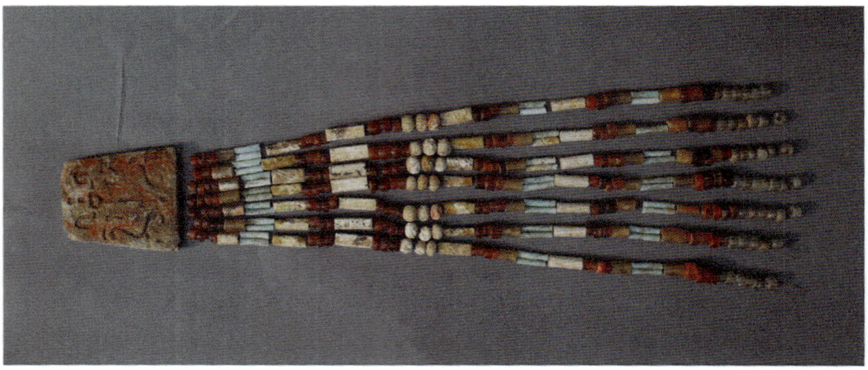

Fig. 1.2 A set of jade pendants of the late Western Zhou excavated from the cemetery of Ying State in Pingdingshan. The green tubes indicated by red arrows are faience tubes

beads, most of which were fragments, have been found. More faience beads and tubes have been found in middle Western Zhou contexts.

Figures 1.1 and 1.2 show the green faience beads of the Early and Middle Western Zhou and an assemblage of faience with a jade pendant of the Late Western Zhou excavated from the cemetery of Ying State in Pingdingshan. By the late Western Zhou, frit artefacts appeared, in which there was a higher proportion of glass. The appearance of Chinese 'frit' occurred about 1000 years later than that in Western Asia.

1.3 Ancient Chinese glass and the cultural and technical exchange between China and the outside — The Ancient Glass Road

Glass making in western Asia and Egypt started from 2500 B.C., the same time as bronze making in China. The earliest glass artefacts from China were found in Baicheng and Tacheng in Xinjiang. The earliest glass artefact in inner China (in the areas of the Yellow River and Yangtze River valleys) dates to the early Warring States Period.

A great number of glass beads were unearthed from Kiziltur Cemetery, Baicheng in Xinjiang. These monochromatic glass beads date to the Spring and Autumn periods (1000–800 B.C.). Their chemical compositions closely resemble that of the ancient glass beads found in Mesopotamia, belonging to the soda lime silicate glass family (Na_2O-CaO-SiO_2) (Table 1.2).[11,12] However, some very different yellow glasses contain elevated levels of PbO and Sb_2O_3 in (e.g. XJ-2B, XJ-2C) . The high PbO and Sb_2O_3 contents are thought to be related to the presence of lead antimonite ($Pb_2Sb_2O_7$), a kind of colourant/opacifier in yellow glass. The antimony-based colourant/ opacifiers — for example, $Pb_2Sb_2O_7$ or $CaSb_2O_6$ — are found in some beads of plant-ash type soda-lime glasses dated to about 1000–500 BC in Xinjiang.[12] It has been reported that lead antimonate yellow was used in Egyptian faience and glass artefacts as early as the XVIIIth Dynasty (c. 1567–1320 B.C.) and lasted until the Third Intermediate Period (c. 1070–713 B.C.), including some from El-Amarna and Malkata.[13,14] To our knowledge, no antimony-based opacifiers were identified among the native PbO–BaO–SiO_2, PbO–SiO_2 and K_2O–CaO–SiO_2 glasses in central China from the 5th century B.C. to the 17th century A.D. The number of tombs in Kiziltur cemetery from which glasses were unearthed constitutes about 25 percent of the total tombs in this cemetery. These Kiziltur tombs belong to people below the noble class, which indicates that glass was relatively popular at that time.

A number of inlaid so called 'eye beads' were unearthed from the Xujialing Tomb in Xichuan (Fig. 1.3a),[15] Jiuxian village in Ye county (Fig. 1.3b) and Luoyang in Henan,[16] Leigudun tomb in Sui County

Table 1.2 Chemical compositions of early glass beads (1000.–600 B.C.) unearthed from Baicheng, Xinjiang, China (wt.%)

Sample No.	SiO_2	Na_2O	CaO	MgO	K_2O	Al_2O_3	PbO	BaO	CuO	Fe_2O_3	TiO_2	ZnO	MnO	B_2O_3	Sb_2O5	P_2O_5
XJ-1A	63.10	18.27	5.88	5.20	2.57	1.12	0.09	0.02	0.79	0.57	0.07	0.05	0.04	0.10	n.d.	1.02
XJ-2A	65.38	11.54	8.88	5.02	1.59	1.99	1.93	0.01	0.01	1.03	0.02	0.02	0.04	0.06	0.72	0.09
XJ-2B	64.31	12.05	4.80	2.67	2.42	1.36	9.01	0.008	0.001	1.10	0.07	0.02	0.02	0.22	1.60	0.34
XJ-3A	65.19	15.27	6.65	3.66	2.93	1.44	0.02	0.02	0.76	0.86	0.13	0.06	0.03	0.10	n.d.	2.54
XJ-4A	66.11	14.29	6.61	4.58	2.19	1.89	0.62	0.01	0.90	1.07	0.17	0.05	0.03	0.03	1.44	n.d.
XJ-30	75.44	9.08	7.74	3.35	1.51	1.43	0.02	0.005	0.56	0.34	0.11	0.11	0.02	0.24	0.03	n.d.
XJ-40	68.88	15.93	3.11	4.03	2.20	0.87	0.02	0.005	1.10	0.56	0.04	0.05	0.08	0.10	0.01	n.d.
XJ-2C-white	65.45	15.93	10.01	4.76	1.06	0.12	0.63	n.d.	n.d.	0.29	n.d.	n.d.	0.04	n.d.	0.71	n.d.
XJ-2C-yellow	65.16	14.79	8.01	4.78	0.81	0.40	4.19	n.d.	n.d.	0.39	n.d.	n.d.	0.05	n.d.	0.42	n.d.

n.d.: not detected.

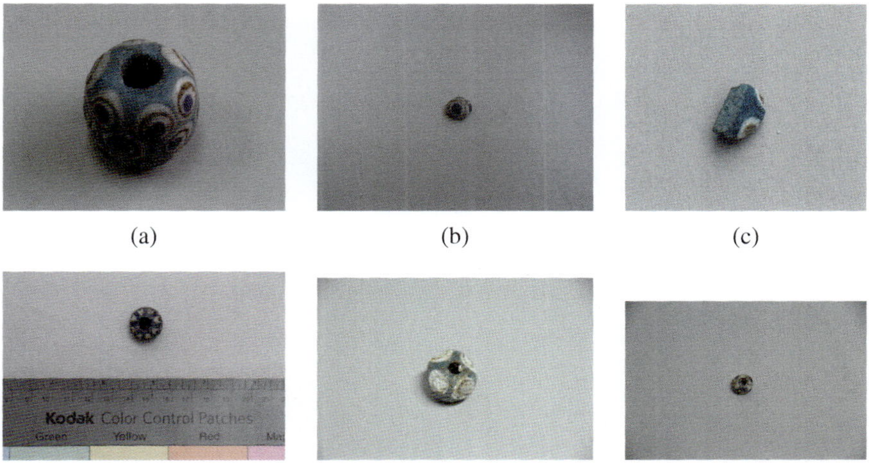

(a) (b) (c)

(d) (e) (f)

Fig. 1.3 Na_2O-CaO-SiO_2 glass eye-beads unearthed from China dating to the Eastern Zhou. (a) No.HNZZ-01 unearthed from Xujialing tomb in Xichuan in Henan province dating to the early Warring States Period; (b) No.HBWKI-18 unearthed from Leigudun tomb in Sui county in Hubei province dating to the early Warring States Period; (c) No.HNWKII-83 unearthed from Jiuxian in Ye county in Henan province dating to the late of Spring and Autumn period; (d) No.5:6055 unearthed from *Chu* tomb in Jingzhou in Hubei province dating to the Warring States Period; (e) No.PM68:5 unearthed from Xiongjiazhong tomb in Jingzhou in Hubei province dating to the Warring States Period; (f) No.PM66:3 unearthed from Xiongjiazhong tomb in Jingzhou in Hubei province dating to the Warring States Period

(Fig. 3c)[17] and the *Chu* tomb in Jingzhou (Figs. 1.3d, 1.3e and 1.3f) in Hubei.[18] These glass beads date to the late Spring and Autumn to Warring States Periods (500–200 B.C.). The shapes and patterns of most beads are similar to that of ancient inlaid Egyptian beads (1300 B.C.), all of which have a blue body, dark blue pupils and brown circles (Fig. 1.4).

Recently, the chemical composition of these eye beads (body, pupils and circles) have been determined by using non-destructive techniques (PIXE, WDXRF and pXRF) and SEM-EDS. The results show that their chemical compositions are all quite similar. Table 1.3

Fig. 1.4 Glass eye-beads of the New Kingdom in ancient Egypt (1567–1055 B.C.)

gives the chemical compositions of the different parts of the eye beads unearthed from central China, such as the Xujialing tomb,[15,19] the Leigudu tomb,[20] the tomb of Marquis Yi of the Zeng State,[19] Jiuxian village and Luoyang,[16] *Chu* tombs at Jingzhou,[18] as well as ancient Egyptian eye beads.

By comparing the analytical results, we can see that the main chemical compositions of the eye beads unearthed from these tombs is quite similar to Egyptian eye beads: all of them are soda lime silicate glass (Na_2O-CaO-SiO_2) without PbO or with a very low PbO content. It is interesting that the blue, brown and white part of some samples (i.e. HB-1, HB-6, HNWKII-83, lgd-1 and PM66:3) have higher Sb_2O_3 contents than others. The colourant/opacifier of $CaSb_2O_6$ was identified by Raman in samples of HB-1, HB-6 and HNWKII-83, and the colourant/opacifier $Ca_2Sb_2O_7$ was also found in HB-6.[19] With reference to other published compositional results for other similar glass beads it is clear that calcium antimonate compounds were used as the colourant/opacifier.

It is worth noting that the pale blue, green and white bodies of the glass beads from Xichuan, Luoyang and Jingzhou etc. contain very low impurities. They were evidently made with pure raw materials. The content of Fe_2O_3 in the body is less than 1%, and CuO ranges from 1.21% to 2.64%, a little higher than that of Fe_2O_3. The body of the eye beads is pale blue or green, caused by the Cu^{2+} ion. Quite low CoO was found (or not detected) in the glass body, but in

Table 1.3 Chemical compositions of glass eye-beads dated to the Warring States Period unearthed from different places of China (wt.%)

Place	Sample No	Measured point	Na_2O	MgO	Al_2O_3	SiO_2	P_2O_5	Cl	K_2O	CaO	TiO_2	MnO	Fe_2O_3	CoO	CuO	ZnO	PbO	BaO	Sb_2O_3
Xujialing Xichuan, Henan	HNZZ-01	blue body	9.53	0.19	3.4	72.8	0.31	0.61	2.15	8.54	n.d.	n.d.	0.58	0.03	1.21	0.03	n.d.	n.d.	0.00
		blue pupil	9.91	0.97	3.5	70.5	0.22	0.68	1.17	8.62	n.d.	n.d.	2.86	0.25	0.48	0.04	n.d.	n.d.	0.00
		red+while circle	7.37	1.17	3.8	75.0	0.23	0.83	1.01	8.73	n.d.	n.d.	0.76	0.04	0.13	0.04	n.d.	n.d.	0.00
Leigudun Sui, Hubei	HB-1*	Blue body	15.04	0.3	1.2	71.98	n.d.	n.d.	0.3	8.24	n.d.	n.d.	0.32	n.d.	1.75	n.d.	n.d.	n.d.	0.88
	lgd-1☆	blue body	4.29	0.33	2.14	75.06	n.d.	n.d.	0.97	11.34	n.d.	n.d.	0.81	n.d.	0.90	n.d.	n.d.	n.d.	2.59
		brown eye circle	4.62	0.32	2.18	73.69	n.d.	n.d.	0.92	11.49	n.d.	n.d.	1.00	n.d.	0.45	n.d.	n.d.	n.d.	3.34
Leigudun Sui, Hubei	HBWKI-18	Light blue body	13.43	5.78	5.33	64.28	0.34	n.d.	2.15	6.45	n.d.	n.d.	0.71	0.02	1.30	0.03	n.d.	n.d.	n.d.
		dark blue eye part	14.58	6.57	4.99	62.41	0.12	n.d.	2.20	6.69	n.d.	n.d.	0.84	n.d.	1.34	0.00	n.d.	n.d.	n.d.
		interlayer	13.58	5.59	5.35	66.41	0.54	n.d.	1.91	5.21	n.d.	n.d.	0.38	n.d.	0.78	0.00	n.d.	n.d.	n.d.
Tomb of Marquis Yi of the Zeng State In Sui county, Hubei Province	HB-6*	Dark blue eye pupil	15.4	0.4	1.74	69.9	n.d.	n.d.	0.4	8.54	n.d.	0.03	1.48	0.08	0.24	n.d.	0.72	n.d.	1.14
		White circle	15.33	0.33	1.49	70.86	n.d.	n.d.	0.37	8.42	n.d.	0.03	1.01	n.d.	0.39	n.d.	0.65	n.d.	1.12

(*Continued*)

Table 1.3 (*Continued*)

Place	Sample No	Measured point	Na_2O	MgO	Al_2O_3	SiO_2	P_2O_5	Cl	K_2O	CaO	TiO_2	MnO	Fe_2O_3	CoO	CuO	ZnO	PbO	BaO	Sb_2O_3
Jiuxian village at Ye county, Henan	HNWKII-83a	blue body	1.89	0.92	3.52	77.01	1.06	0.00	0.86	9.58	0.03	0.07	0.84	n.d.	2.80	n.d.	0.59	n.d.	n.d.
	HNWKII-83b	yellow eyeball	0.00	1.59	3.75	79.32	1.01	0.00	0.44	9.89	0.25	0.00	0.95	n.d.	0.92	n.d.	0.48	n.d.	n.d.
	HNWKII-83c*	fresh blue body	14.99	1.31	3.82	67.76	0.31	n.d.	0.40	7.21	n.d.	n.d.	0.27	n.d.	2.00	n.d.	n.d.	0.19	0.72
	HNWKII-83d*	white eye part	2.82	1.26	2.87	55.43	0.15	n.d.	0.29	5.59	n.d.	n.d.	n.d.	n.d.	0.03	n.d.	n.d.	0.06	4.68
International Trade Building at Luoyang, Henan	HNLY-08a	blue body	7.78	1.01	7.57	71.30	0.56	0.55	0.51	7.91	0.10	0.00	0.69	n.d.	1.06	n.d.	n.d.	n.d.	n.d.
	HNLY-08b	eye part	0.00	1.97	9.19	54.40	2.41	0.53	0.32	22.10	1.32	0.34	2.44	0.67	0.49	n.d.	n.d.	n.d.	n.d.
	HNLY-08c	blue eyeball	9.75	0.72	4.65	69.50	0.33	0.58	1.40	8.21	0.12	n.d.	3.31	0.28	0.36	n.d.	n.d.	n.d.	n.d.
Zhen-zhi-chang at Luoyang, Henan	HNLY-09a	black eyeball	2.35	2.30	11.96	67.94	0.64	n.d.	3.62	4.81	0.39	n.d.	3.39	n.d.	0.33	n.d.	n.d.	n.d.	n.d.
	HNLY-09b	yellow body	0.97	0.86	8.49	77.75	0.59	n.d.	0.96	4.60	0.34	n.d.	2.98	0.03	1.00	n.d.	n.d.	n.d.	n.d.
	HNLY-09c	green body	0.70	1.17	7.38	78.50	0.68	n.d.	1.84	4.04	0.15	n.d.	3.04	n.d.	2.45	n.d.	n.d.	n.d.	n.d.
	HNLY-09d	vitreous body	0.76	0.94	5.76	85.60	0.26	0.06	0.68	2.53	0.09	n.d.	1.69	0.07	1.55	n.d.	n.d.	n.d.	n.d.
	HNLY-09e※	vitreous body	8.27	2.28	3.72	73.35	n.d.	n.d.	3.58	5.67	0.00	0.02	1.14	n.d.	1.70	n.d.	n.d.	n.d.	n.d.
Chu tomb at Jingzhou, Hubei	5:6055*	bule body	6.42	1.93	7.65	64.58	n.d.		4.23	5.52	n.d.	0.08	0.98	0.11	1.59	n.d.	0.04	5.89	0.71
		blue eyeball	3.69	1.41	12.97	63.75	n.d.		2.92	5.09	n.d.	0.73	1.81	0.20	0.44	n.d.	0.26	5.70	0.86
Xiongjiazhong tomb at Jingzhou, Hubei	PM66:3*	blue eyeball	6.13	0.65	4.31	72.71	n.d.		1.26	8.76	0.00	0.04	2.76	0.18	0.27	n.d.	n.d.	n.d.	3.04
		white body	1.44	1.02	12.83	68.40	1.21		1.36	7.73	0.28	0.05	1.29	n.d.	n.d.	n.d.	n.d.	n.d.	4.21
	PM68:5*	green part	8.37	0.72	6.92	67.85	n.d.		2.07	8.85	0.13	0.03	0.88	n.d.	2.64	n.d.	0.19	n.d.	0.70
		white-yellow eye part	1.12	0.76	6.64	76.32	n.d.		n.d.	10.04	0.15	0.05	1.01	n.d.	n.d.	n.d.	n.d.	n.d.	3.67
Amarna , Egypt			18.36	5.2	3.85	58.5			0.6	10.30	1.53	0.25							

☆Determined by WDXRF;*Determined by pXRF; ※Determined by SEM-EDS; All the others measured by PIXE technique; n.d.: not determined.

the dark blue pupil of the eye beads there is 0.1–0.3% CoO. Co^{2+} is a strong blue colorant, which must have been intentionally added by the glass masters. This is the earliest discovery of cobalt blue in ancient glass and ancient porcelain glaze in China (i.e. the use of CoO as the blue colorant). The use of CoO (content > 0.1%) as a colorant in glass in ancient China started in the Eastern Han Dynasty. However, in ancient Mesopotamia and Egypt, it was used from at least the 14th century BC. Based on the chemical compositions and colourants/opacifiers in these glass eye beads from western and central China, we can guess that this batch of glass eye beads was most probably imported from the West (areas such as Egypt, West Asia and Central Asia). However, there is about 5.8% BaO in the sample 5:6055 from the *Chu* tomb in Jingzhou dating to the Warring States period. The glass eye bead was possibly made in China using local raw materials, but the glass making technology was introduced from outside of China earlier, before 800 B.C.

Related research has shown that the chemical compositions of the largest number of ancient glasses developed inside China and it is quite different to that of ancient Western glasses. The glasses made by the ancient Egyptians, Romans and Persians as well as Islamic peoples mainly belong to the soda-lime-silicate glass system. In contrast, the chemical composition of ancient glass developed in China contains more K_2O, PbO and BaO. Therefore, it can easily be distinguished. The earliest glass made in inner China with these characteristics started from the early to middle Warring States Period, its main chemical composition being the K_2O-CaO-SiO_2 system. From the early-middle Warring States to the Han Dynasty, the glass mainly belongs to PbO-BaO-SiO_2 system and was made mostly in Hubei, Hunan, Anhui and Jiangxi, along the Yangtze River valley. The ancient glass of the Na_2O-CaO-SiO_2 system dates to before the Qin Dynasty: it spread from west to east and from north to south. Various kinds of evidence shows that there had been some routes connecting the East and West before the Silk Road formally came into existence dating from the Han Dynasty. I therefore name it 'The Glass Road'; it is mainly the early Northern Steppe Road.

1.4 Formation of the Ancient Glass Road and the movement of tribes along the Eurasian steppe

It can be seen from the cultural and technological origin of ancient glass and faience artifacts in China, that the contact and exchange between China and the outside world was mainly with Western and Central Asia, along the Steppe Road before the Han Dynasty. So the nomadic tribes between Europe and Asia played an important role.

The Ancient Glass Road started and operated from 1500 BC to 500 B.C., i.e. the Shang and Zhou Dynasties. Figure 1.5 shows one part of the Ancient Glass Road. The western glasses and faience entered the Yining area from north Xinjiang along the Ili River. The monochrome glass beads and eye beads dating to the Pre-Han period excavated in Xinjiang were mostly derived from the north. The glass in the Pre-Qin period come from Hami, running eastward along the northwestern steppe road through Juyanhai and Ejinaqi, in Inner Mongolia, then passing Mt.Yinshan to Baotou, then turning southward to Taiyuan, Louyang and Zhengzhou; or to Xi'an via Huaxian in Shaanxi. It was said that King Mu of the Western Zhou Dynasty

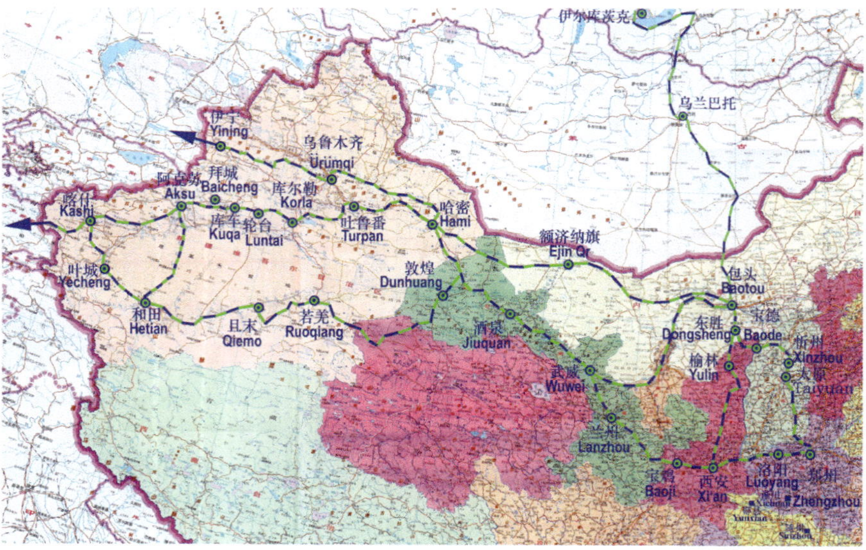

Fig. 1.5 Sketch map showing the routes of ancient Glass Road within modern China

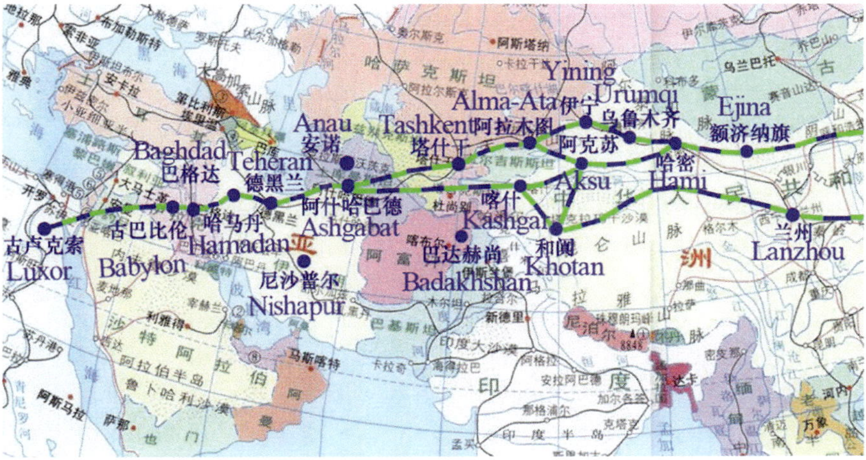

Fig. 1.6 Sketch map showing the westward extending of the ancient Glass Road

returned from his Westward travels and made his way along this route. In comparison, the Tarim Basin was sparsely populated and was blockaded during that period. We also mentioned that early semi-precious stones, such as lapis lazuli, turquoise and nephrite jade came from ancient Afghanistan, Iran and Xinjiang and were transported along this road before 1000 B.C.[21]

Figure 1.6 shows the other section of the glass road. The Glass Road passed the ancient cities of Luxor in Egypt, Babylon in ancient Mesopotamia and Nishapur in Iran. The famous Anau ruin of Turkmenistan is also marked in the figure. The western glasses characterized by the soda-lime silicate system originated from Mesopotamia and Egypt. Alternatively, the trade of ancient glasses started from Bagdad in Iraq, passing through Habadan, Tehran and Mashhad in Iran, Tashkent in Uzbekistan, Almaty in Kazakstan, then to Yining entering China, connecting with the Glass Road inside China.

Starting from the late Neolithic age, the nomadic tribes in the Eurasian steppe moved quite a lot. The earliest tribe organized was of Indo-European people, with an origin north of the Caspian Sea. The Mongolian-Turkish tribe in the East was bounded by the mountains of Altai and Tianshan by the Indo-European tribe. The Indo-European Hittite people entered Asia Minor about 2000 BC, and established the Hittite empire. Later, the Kassite and Hurrian people of Indo-European origins entered Central Asia, and became one of the powerful empires including in Assyria in Mesopotamia.[22] This corresponded to the period of the Shang Dynasty in China. In ca. 1500 BC, the Indo-European Aryan people moved southeastwards into the Indus River, and the Hurrian people most probably moved eastward to the northwest of China.[21]

The Greek word "Skythaio" came from Central Asia and the Asia Minor area. They settled in the Ili and the Altai areas in the 9th century BC and were called the Sakka people in the Chinese history. The Yueh Chih people were called Indo-Scyths in ancient Greek, i.e. the Indo-Scythian origin, they entered the Tarim Basin in 2000 B.C. and moved further eastwards later.

The Hurrians and Scythians all knew the techniques of bronze and iron melting, glass making and possibly jade inlaying. They introduced these techniques to Central Asia and then to the main tribes in the northwest of China (1500–500 B.C.), and later, the techniques were transferred to inner China via the Sakka and Yueh Chih peoples (from 500 B.C.).

1.5 Conclusion

The civilization centers of the ancient world developed independently based on their own modes and specialties, but were not blockaded and isolated. They influenced, infiltrated and fused with each other and promoted the development of human civilization together. It is difficult to ignore that contacts between different civilizations really existed (apart from with some remote Pre-historic and Proto-historic societies). After they emerged the civilization centers expanded outside like waves, and influenced peripheral areas. The exchange and fusion resulted from the movement of nomadic tribes and nationalities. We should study ancient relics and ruins in China, and also we should be concerned with, familiarise ourselves and study those abroad. The cultural and technical exchange between China and the outside world can be found through a comparative study. Based on independent developments combined with absorbing the essence of foreign civilizations, the centers of civilizations developed continually.

References

1. L. Boulnois, *La Route de la Soie*, Arthaud, Paris, France, 1963.
2. Y. P. Zhang, *Story of the Silk Road*, China Intercontinental Press, Beijing, China, 2005.
3. Z. L. Wang, *History of Middle Asia*, Xinjiang People's Press, Wulumuqi, China, 2004.
4. F. X. Gan, The Silk Road and ancient Chinese glass, in *Ancient Glass Research along the Silk Road*, eds. F. X. Gan, R. H. Brill and S. Y. Tian (World Scientific Publishing Co. Pte. Ltd, Singapore, 2009), pp. 41–109.
5. X. F. Fu and F. X. Gan, *Jouranl of the Chinese Ceramic Society* **34**, 427 (2006).
6. Y. Qin, X. Chen, X. L. Li and Q. W. Chen, *Journal of the Chinese Ceramic Society* **40**, 567(2012).
7. Z. G. Zhang and Q. L. Ma, Faience beads of the Western Zhou Dynasty excavated in Gansu Province, China: a technical study, in *Ancient Glass Research along the Silk Road*, eds. F. X. Gan, R. H. Brill and S. Y. Tian (World Scientific Publishing Co. Pte. Ltd, Singapore, 2009), pp. 275–299.
8. F. X. Gan, Y. Q. Hu, J. Q. Dong *et al.*, *Jouranl of the Chinese Ceramic Society* **37**, 1005 (2009).
9. J. Q. Dong, Y. Q. Hu, F. X. Gan *et al.*, Analysis on the "Liaozhu" and "Liaoguan" excavated from Yingguo tomb in Ping Ding Shan of Henan Province, in *Cemetery of the Ying State in Pingdingshan,* eds. Henan Institute of Cultural Relics and Archaelology and Pingdingshan Cultural Bureau (Da Xiang Press, Zhengzhou, 2012), Vol. 2, pp. 847–862.
10. Q. H. Li, J. Q. Dong and F. X. Gan, *Journal of Guangxi University for Nationalities (Natural Science Edition)* **15**, 31(2009).
11. F. X. Gan, Q. H. Li, D. H. Gu *et al.*, *Jouranl of the Chinese Ceramic Society* **31**, 663 (2003).
12. Q. H. Li, S. Liu, H. X. Zhao *et al.*, *Archaeometry* **56**, 601(2014).
13. R. J. H. Clark and P. J. Gibbs, *Journal of Raman Spectroscopy* **28**, 99 (1997).
14. Th. Rehren, *Journal of Archaeological Science* **35**, 1345 (2008).
15. F. X. Gan, H. S. Cheng, Y.Q. Hu *et al.*, *Science in China Series E: Technological Sciences* **52**(4), 922 (2009).
16. J. Q. Dong, Q. H. Li, F. X. Gan *et al.*, *Materials China* **31**, 10 (2012).
17. F. X. Gan, H. X. Zhao, Q. H. Li *et al.*, *Jianghan Archaeology* **115**, 108 (2010).
18. Q. H. Li, J. Q. Dong, B. M. Su *et al.*, *Dunhuang Research* **137**, 92 (2013).

19. H. X. Zhao, Q. H. Li, S. Liu *et al.*, *Journal of Raman Spectroscopy* **44**, 643 (2013).
20. Y. Qin, L. Z. She, X. L. Li *et al.*, *Journal of the Chinese Ceramic Society* **37**, 574 (2009).
21. F. X. Gan, *Journal of the Chinese Ceramic Society (In Chinese)* **41**, 458 (2013).
22. L. S. Stavrianos, *The World since 1500: A Global History*, 3rd edn. (Englewood Cliffs, New Jersey, 1975).

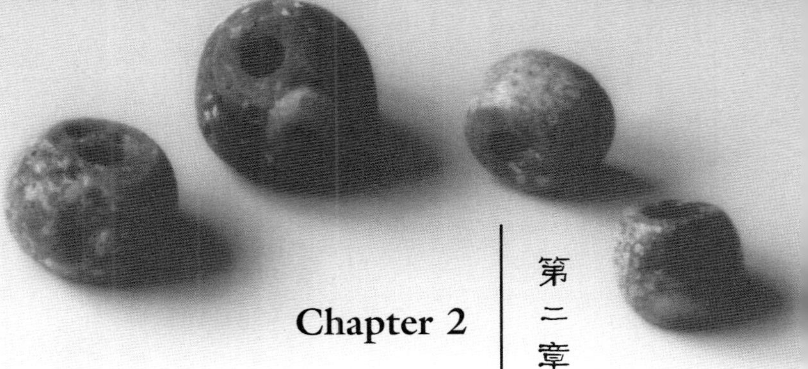

Chapter 2 第二章

Glass Provenance along the Silk Road: The Use of Trace Element Analysis

J. Henderson, S. Chenery,† J. Kröger‡ and E. W. Faber**

*Department of Archaeology, School of Humanities, University of Nottingham, University Park, Nottingham, NG7 2RD, UK

†The British Geological Survey, Keyworth, Nottinghamshire, NG12 5GG, UK

‡Museum für Islamische Kunst, Pergamonmuseum, Geschwister-Scholl-Str. 6, 10117 Berlin, Germany

2.1 Introduction

The Silk Road provided a channel along which humans, religions, diseases, animals, crops, metal, pottery and glass passed from east to west and west to east. In fact the overland 'Silk Road' consisted of a number of different routes, but together ultimately created links between e.g. Scandinavia in the west to the coast of China and to Japan in the east. In the 'western' part of the medieval exchange network the routes bifurcated: one originating in Scandinavia and passing through Bulgaria, the other passing through northern Italy and along the Dalmatian coast. Both eventually met up in Constantinople, and extended to the east through Anatolia, northern Syria, northern Iraq, Iran and into central Asia.

Trade in glass and ceramics along the Silk Road was fed by production centres both close to the Silk Road and at some distance from it. Artisans of different ethnic origins contributed to the production of materials along the Silk Road. Their ethnicities must have had an impact on the technology involved (the raw materials and production processes) and the forms of decoration and glass vessel shapes.

Scientific characterisation of ancient glass provides a means of tracing production and distribution patterns along the Silk Road.[1] It therefore provides a means of identifying and measuring interaction between different nodes of production and trade. This is partly dependant on the identification of primary production centres, preferably as a result of archaeological investigations, but also historical evidence. It is therefore important that appropriate techniques of scientific analysis are used for the charactersiation of glass made at such centres so that the results can been used to distinguish it from glass made at other production centres. Given the wide environmental and geological contrasts to be found along the Silk Road there is a clear potential for this to be successful, some of which has been realised.[2,3]

Natron glass technology was dominant in western Europe and the Mediterranean well before plant ash glass technology was reintroduced fully c. 9th century AD. Butrint in Albania located on the Dalmatian coast was an important Balkan trade centre on the Silk Road.[4] Scientific analysis of 1st–8th century natron glasses found there using electron probe microanalysis shows that a complex range of glass sources were used. At least six sub-types of soda-lime glasses were identified.[5] This work has demonstrated both the complexity of the range of raw materials with varying impurities used to make the glasses and the increasing potential to ascribe sources for different compositional types and therefore to suggest trade in such glass. Analysis of glasses that travelled along the Silk Road from sites such as Lou Lan on the 'southern desert route', Hazar-tam near Kashgar and Qoch Homa at Kucha all in Xinjiang[6] has provided insights into their chemical compositions and the likely range of areas in which the glasses were manufactured, such as central Asia and the Middle East.

Moreover, Liu *et al.*[7] have demonstrated by analysing 65 glass beads from Xinjiang using high resolution portable XRF that their technology changed from — high alumina glasses, some also containing high potassium oxide, mainly deriving from Kashan controlled areas in India, Pakistan and Afghanistan in the 1st to 4th centuries AD — to plant ash glasses considered to have derived from Sasanian areas such as in Iraq; the Sasanian territories extended well outside Iraq and connected the Mediterranean sea with India.

These three studies provide interesting examples of how scientific analysis is beginning to answer questions about glass technology production, provenance and trade in the context of Silk Road research.

Isotopic analysis of natron and plant ash glasses can provide different and related insights into ancient glass technology and provenance. The use of thermal ion mass spectrometry to investigate both glass raw materials and glasses has the potential to provide an independent geological provenance for glass. If combined isotopic and chemical characterisation of glass from production centres can be carried out it provides an index against which glasses of unknown origin can be compared. Neodymium isotopes ($^{143}Nd/^{144}Nd$) provide a means of characterising silica in quartz, sand and glasses. Strontium isotopes ($^{87}Sr/^{86}Sr$) characterise the calcium in shell and bone fragments and in plant ashes (according partly to the geological formation ages of the calcium). The isotopes remain unaltered (there is no fractionation) when the glass is made. Both for plant ash and for natron glasses raw materials from potential and actual locations where raw materials have been drawn from have been isotopically characterised.[8–11] By comparing Nd and Sr isotopes in plant ashes, quartz pebbles and sand samples it has been possible to define clear factory signatures for Islamic plant ash glasses in the Middle East — in Syria, Lebanon and Israel, an area where glass was manufactured for local consumption, short distance trade and for trade along the Silk Road.[9,12–14] For natron glasses it has also been possible to bolster evidence for both Levantine production zones and 'non-Levantine' production zones[15–17] and provide evidence for the mixing of raw glasses.[18]

So although major and minor chemical and isotopic analysis of Middle Eastern glasses provides some intriguing insights, trace element characterization has been used to a lesser extent. This paper will focus on the trace element analysis of Middle Eastern glasses dating to between the 9th and 14th centuries. With a broad geographical provenance of plant ash glasses found in the Levant and Syria provided by major and minor elemental oxide concentrations in mind, this paper addresses several research questions using trace element analysis:

(1) How does trace element analysis contribute to our understanding of plant ash (soda-lime) glass technology?
(2) To what extent does the use of trace element analysis of plant ash glass support the results from using major and minor elemental analysis?
(3) Where the definition of a 'production zone' is unclear using major and minor analysis, does trace element analysis provide a better definition?
(4) How does trace element analysis contribute to the provenancing of plant ash (soda-lime) glass?
(5) Using trace elements is there evidence for mixing/recycling?
(6) Does the use of trace element analysis suggest that multiple production zones existed?

2.2 Scientific analysis of Middle Eastern Islamic plant ash glasses

2.2.1 *Chemical analysis*

The variation in elemental concentrations in plant ash glasses of all periods is wide. This is due to a number of factors, both cultural and technological, which is discussed at length elsewhere.[19,20] The chemical and mineralogical compositions of the raw materials, especially the plant ashes, can vary widely and (proportions of) the impurities in the raw materials are transferred to the glasses made from them. Given that there is such a wide range of factors that determine the chemical compositions of glasses, it is perhaps surprising that chemical analysis has produced definable and broad compositional groups which link both to different chronologies and to production zones. In view of the potential for glass recycling, the chemical analysis of raw furnace glasses (from primary glassmaking sites) forms an important part of these investigations. It is a reasonable assumption that if production occurred on a large scale locally-made glasses were mainly recycled.

The chemical compositions of different glass melts from a primary glassmaking furnace would be expected to vary somewhat depending on the chemical/ isotopic compositions of the raw materials used, but the fact that it is possible to define regional/geographical compositional groups shows that recycling has apparently not disrupted 'local' compositional groupings significantly.

Sayre and Smith defined two distinct Islamic compositional types: plant ash soda-lime glasses and high lead oxide glasses (a mixture of lead and soda-lime glass).[21] By chemically analysing Egyptian Islamic glass weights which can be dated by their impressed inscriptions. Sayre,[22] Matson,[23] and Gratuze and Barrandon[24] showed that before 845 A.D. the glass weights were made with natron and after this plant ashes were used instead. While it is very unusual to be able to date glass objects so closely — or at least when the object was stamped — this is not quite the same as the date when the glass was made. The glass used for these glass weights may have been made somewhat earlier. Moreover, assuming that the glass for the weights was made in Egypt (raw glass may also have been imported), this technological change is a reflection of what was happening in Egypt, which had extensive local sources of natron, and not necessarily in other parts of the Islamic world.

Plots of major and minor element oxide concentrations in plant ash glasses from the Levant and the Middle East have already produced groupings that correlate with glasses found in different regions.[20, 25–27] There are exceptions from these groupings but mainly not for furnace glasses. These can be explained as resulting from trade in raw glass and glass vessels, though this is not an entirely satisfactory explanation.

Thus some glasses found in the Levant can be distinguished from some made in Syria and others found in Iran by their CaO:MgO ratios. Levantine glasses have the highest calcium and lowest magnesium oxide levels and the Iranian ones the lowest calcium oxide and highest magnesium oxide levels.[20] These correlate with probable production areas for glass vessels decorated in specific ways. These include some lustre painted vessels from the Levant and Syria, most scratch decorated from Syria and almost all line and facet cut (colorless) vessels from Iraq/Iran. The fact that not all glasses fall into 'expected' zones is predictable given the trade and mixing of glasses that occurred. For example, coloured glass found at Nishapur, north-east Iran mainly falls into the 'Syrian' area. This clearly needs further scientific investigation. Nevertheless these groupings do provide a broad geographical provenance that reflects broad geological contrasts in the origins of the raw materials used to make the glasses across the area. Both magnesium and calcium are likely to have been introduced in the plant ash.

Compositional results for Mamluk (12th–14th centuries) mosque lamps (Figs. 2.1 and 2.2) dedicated to specific emirs in Cairo (Egypt)[28] do not fit the broad trend (on the assumption that the lamps were made in Egypt) but fall into the 'Syrian' group. This could therefore suggest that the raw glass used in these lamps was manufactured in Syria rather than in the Levant or in Egypt. This is an example of how trace element analysis can be used to investigate the glasses further.

Set against these broad geographical trends there is also evidence for probable experimentation in glass technology in the 9th century. This was a time when the powerful 'Abbasid caliphate unified broad areas of the Middle East and central Asia. The caliphate supported the development of the 'sciences' and the 'arts'. Long treatises describing, for example, how to make lustre decorated

Fig. 2.1 Mosque lamp 7, mid 14th century

Fig. 2.2 Mosque lamp 4; mid 13th-early 14th century

glass vessels and different glass colours were written in Arabic by al-Jabir.[29] These 'creative' industries involved these early men of science in experimentation.

The 8th and 9th centuries saw an important transition in glass technology. Although the Sasanians had manufactured and used plant ash glasses in the area of what is today Iraq and Iran (2nd–6th centuries AD), it was natron glass technology that dominated from about 800 BC until about 800 AD in the Mediterranean and northern Europe. In the 9th century glassmakers who probably included Muslims, Jews and Christians appear to have been responsible for fully reintroducing plant ash glass technology and by the 10th–12th centuries large volumes of plant ash glass were being produced in the massive tank furnaces in the Levantine coastal location of Tyre.[30,31] This was a zone where natron glass had previously been fused.

As noted above, major and minor element analysis of Islamic Middle Eastern glass provides evidence for regional production zones. The evidence is quite convincing but not all results fit neatly into their 'expected' groups and the boundaries of some groupings generally overlap or are not clear. This approach therefore provides a good start and as more major and minor elemental data is published, especially for raw furnace glasses excavated from primary glassmaking sites, the groupings may become more coherent. The majority of cameo decorated vessels in contrasting colours fall into the 'Syrian' compositional group.[20] The provenance of highly coloured glass, such as cobalt blue vessels, and the associated source of an 'Islamic' zinc-rich cobalt-baring mineral, may also be linked to production zones or centres and/or the transport of the colorant to separate production zones or centres.[32]

2.2.2 *Isotopic analysis*

The use of strontium and neodymium isotope analysis of Islamic plant ash glasses may have the potential to gauge their provenance in a more refined way than chemical analysis. As already noted, minor (and trace) elemental concentrations in ancient glasses can be associated with one or more specific raw materials. However, it is not always possible to be certain how such elements are incorporated in the glass melt or whether they are associated with more than one raw material. Aluminium oxide, for example, can be introduced in feldspars that occur in sands, is present in plant ashes (from clays/feldspars in the local soils) and can sometimes enter the glass melt through interaction with the furnace or crucible walls. However, it is often assumed that the alumina only derives from feldspars in sand (which it probably did for natron glasses).

In contrast, strontium is chemically linked to calcium present in plant ashes, shell and bone fragments and other calcium-rich materials that could be used to make glass. Neodymium is chemically linked to silica. So, unlike chemical analysis, there is no question about their associations. Although it is not a direct relationship, the values of $^{87}Sr/^{86}Sr$ and $^{143}Nd/^{144}Nd$ in raw materials (and glasses) are related to the relative geological ages of raw materials. Moreover, these isotopic ratios in the raw materials do not fractionate at the high temperatures at which glass was made.

Isotopic analysis of glasses can therefore provide distinctive groupings based partly on contrasting geological ages of the raw materials used to make them. In some cases strontium is a better discriminator and in others neodymium. However, a combination of both is generally the best. There is, nevertheless, the possibility that the same isotopic signature can be introduced into glass in more than one production location. Therefore each isotopic characterisation of ancient glass needs to be considered according to the archaeological context date and the cultural contexts involved.

In order to place glass isotope results into an environmental context it is important to determine the isotopic characteristics of raw materials selected from a variety of geological environments across the landscape. It is especially important to do this around primary production centres. An index of isotopic signatures across the landscape against which glass isotope results can be compared is produced. An example of this is the isotopic analysis of plants and silica from Syria and the Lebanon.[9,14] There is, for example, a clear contrast between the (higher) Sr isotope signature for plants growing on the limestone steppe of northern Syria and those growing on basaltic soils around Damascus. Therefore glasses made from plants growing on these areas will reflect this contrast. Degryse *et al.* have shown that under somewhat extreme temperate weather conditions (i.e. high rainfall in northwest Scotland) the Sr isotope signature of the plant ash is affected by the hydrology of the immediate area as well as the underlying geology. [12]

However, in areas of low rainfall, such as a semi-desert environment, it is most unlikely that this effect would modify the Sr signature significantly; further research into this effect is therefore necessary. Moreover the somewhat constrained Sr signatures of raw plant ash glasses from the 9th century primary glass making furnaces at al-Raqqa, Syria shows that the plants used to make them were taken from an environment with the same correspondingly constrained Sr isotope signatures based on plants harvested in and near al-Raqqa.

There is a somewhat limited range of silica sources for primary glass production such as coastal and riverine sand and quartz pebbles, though chert is not out of the question.[20] Neodymium isotopes ($^{143}Nd/^{144}Nd$) provide a means of characterising the silica used. There is sufficient contrast found in these isotopes to provide a very useful additional provenancing tool. Most isotopic analyses of sands have been carried out in the west Mediterranean.[10–11] Other neodymium isotope analyses for sands and quartz are from the Levantine coast, riverine sand and quartz pebbles from the Euphrates in northern Syria and inland deposits including some from the evaporite deposit at

Jeboul, south-west of Aleppo in northern Syria.[14] There is a clear contrast between inland and coastal sands, the latter having higher $^{143}Nd/^{144}Nd$ values. The higher Nd signature is found in raw glass from Banias, the Serçe Limani shipwreck, the primary glassmaking centre at Tyre, and vessels from Beirut. The lower Nd signature is found in (inland) primary furnace glass from the glassmaking centre at al-Raqqa and in Damascus glass vessels.

2.3 Materials and methods

2.3.1 *Samples*

The objects sampled are listed in Table 2.1, together with their origins, dates, artefact type and colour.

2.3.2 *Analytical techniques*

Electron probe microanalysis (EPMA)

The electron microprobe analyses were carried out on the glass samples using a JEOL JXA-8200 electron microprobe in the Microanalysis Research Facility in the Archaeology Department, University of Nottingham. The glass samples were then mounted in cold-setting epoxy resin, before being ground and polished using standard sample preparation procedures down to a 0.02 micron final polishing suspension. The samples were then coated with a thin layer of carbon so as to prevent distortion and deflection of the electron beam and so as to optimise imaging capability. BSE images of the samples were then examined in order to select areas of the glassy matrix for chemical analysis. The EPMA-WDS was calibrated against a combination of certified mineral, pure metal and synthetic standards. A multi-element secondary standard, Corning Museum B, was reanalysed at the start and end of each set of analyses, to identify if any instrumental drift during its operation has occurred and so as to establish accuracy and precision. Quantitative compositional analyses were carried out using an analytical set-up of a 20 kV accelerating voltage and a 50nA beam current with a 50 μm defocused beam, and the quantitative results

Table 2.1 The glass samples sampled and analysed using EPMA and LA-ICP-MS

Sample no.	Site	Date	Artefact	Colour
DOH9	Cairo*	c. 1350-65	Mosque lamp, catalogue 7 (Carboni 2003)	Colourless
DOH10	Cairo*	c. 1350-65	Mosque lamp, catalogue 7 (Carboni 2003)	Green
DOH11	Cairo*	c. 1350-65	Mosque lamp, catalogue 7 (Carboni 2003)	Opaque yellow
DOH12	Cairo*	c. 1350-65	Mosque lamp, catalogue 7 (Carboni 2003)	Opaque green
DOH13	Cairo*	c. 1350-65	Mosque lamp, catalogue 7 (Carboni 2003)	Opaque red
DOH14	Cairo*	c. 1300-1340	Mosque lamp, catalogue 4 (Carboni 2003)	Blue
DOH16	Cairo*	c. 1300-1340	Mosque lamp, catalogue 4 (Carboni 2003)	Opaque red
DOH17	Cairo*	c. 1300-1340	Mosque lamp, catalogue 4 (Carboni 2003)	Green
DOH18	Cairo	c.1412-15	Mosque lamp, catalogue 12 (Carboni 2003)	Green
DOH19	Cairo	c. 1412-15	Mosque lamp, catalogue 12 (Carboni 2003)	Blue
DOH20	Cairo	c. 1412-15	Mosque lamp, catalogue 12 (Carboni 2003)	Opaque yellow
DOH21	Cairo	c. 1412-15	Mosque lamp, catalogue 12 (Carboni 2003)	Opaque green
DOH80	Cairo	c. 1300-50	Mosque lamp, catalogue 5 (Carboni 2003)	Opaque green
DAM1	Damascus	12th century	Beaker, poor quality	Green
DAM14	Damascus	12th century	Beaker, poor quality	Colourless
DAM37	Damascus	12th century	Possible Beaker	Pale purple
DAM38	Damascus	12th century	Dimpled beaker	Colourless
DAM43	Damascus	12th–14th centuries	Grenade body	Colourless

(*Continued*)

Table 2.1 (*Continued*)

Sample no.	Site	Date	Artefact	Colour
DAM44	Damascus	12th–14th centuries	Grenade shoulder	Purple
DAM45	Damascus	12th–14th centuries	Grenade base	Purple
DAM46	Damascus	12th–14th centuries	Grenade neck	Purple
BEI48	Beirut	12th–14th centuries	Beaker base	Pale green
BEI49	Beirut	12th–14th centuries	Bowl rim	Purple
BEI51	Beirut	12th–14th centuries	Beaker base, pontil	Colourless
BEI53	Beirut	12th–14th centuries	Beaker base	Pale brown
BEI54	Beirut	12th–14th centuries	Bottle fragment	Colourless
BEI55	Beirut	12th–14th centuries	Bowl fragment	Colourless
BEI108	Beirut	12th–14th centuries	Bowl (?) base pontil	Green
BEI109	Beirut	12th–14th centuries	Bottle rim	Green
KAM1	Khirbat al-Minya	8th century	Window	Purple
KAM2	Khirbat al-Minya	8th century	Window	Purple
KAM3	Khirbat al-Minya	8th century	Window	Brown
KAM6	Khirbat al-Minya	8th century	Flask straight neck (TA150)	Green
KAM7	Khirbat al-Minya	8th century	Flask pushed in base (TA851)	Green
KAM8	Khirbat al-Minya	8th century	Flask neck, white trailed decoration (TA2410)	Purple

(*Continued*)

Table 2.1 (*Continued*)

Sample no.	Site	Date	Artefact	Colour
RAQ34 (15)	Al-Raqqa, TZ	9th century	Raw furnace glass	Purple
RAQ36 (16)	Al-Raqqa, TZ	9th century	Raw furnace glass	Purple
RAQ38 (17)	Al-Raqqa, TZ	9th century	Raw furnace glass	Purple
RAQ41 (18)	Al-Raqqa, TZ	9th century	Vessel rim	Colourless
RAQ42 (19)	Al-Raqqa, TZ	9th century	Scrap	Cobalt blue
RAQ43 (20)	Al-Raqqa, TZ	9th century	Scrap	Opaque red
RAQ44 (21)	Al-Raqqa, TZ	9th century	Raw furnace glass	Purple
RAQ45 (22)	Al-Raqqa, TZ	9th century	Raw furnace glass	Brown
RAQ46 (23)	Al-Raqqa, TZ	9th century	Raw furnace glass	Emerald green
RAQ47 (24)	Al-Raqqa, TZ	9th century	Beaker base	Green
RAQ48 (25)	Al-Raqqa, TZ	9th century	Bottle base	Green
RAQ49 (26)	Al-Raqqa, TZ	9th century	Raw furnace glass	Blue
RAQ50 (27)	Al-Raqqa, TZ	9th century	Bottle rim, weathered	Green
RAQ51 (28)	Al-Raqqa, TZ	9th century	Beaker rim with applied blue string	Pale green
RAQ54 (35)	Al-Raqqa, TZ	9th century	Vessel rim weathered	Colourless
RAQ58 (32)	Al-Raqqa, TZ	9th century	Phial base	Green

(*Continued*)

Table 2.1 (*Continued*)

Sample no.	Site	Date	Artefact	Colour
RAQ60 (34)	Al-Raqqa, TZ	9th century	Raw furnace glass	Colourless
RAQ61 (46)	Al-Raqqa, al-Banat	12th century	Bowl	Green
RAQ63 (36)	Al-Raqqa, TZ	9th century	Bowl rim	Green
RAQ66 (41)	Al-Raqqa, palace	9th century	Window	Green
RAQ67 (42)	Al-Raqqa, palace	9th century	Window	Blue

Notes on samples:

Cairo, Egypt -mosque lamps mainly attributed to a Cairo production:
Catalogue 7 dedicated to sultan al-Nasir Nasir al-Din Hasan (r. 1347-51, 1354-61); Catalogue 4 made for the sultan al-Nasir Muhammad ibn Qalawun (r. 1293-94, 1299-1309, 1310-41);
Catalogue 12 made for the sultan al-Malik al-Mu'ayyad Shaykh (r. 1412-21).

Damascus, Syria vessels from excavations of the Damascus citadel, directed by Dr S. Bertier.

Beirut, Lebanon vessels from BEY 031 excavations in Beirut, directed by Dr. Hans Curvers.

Khirbat al-Minya, Israel samples from the Museum für Islamische Kunst, Berlin; accession number provided where available.

Al-Raqqa, Syria most samples from excavations of primary glass-making site of Tell Zujaj (TZ) directed by Prof. J. Henderson; window glass from excavations of west palace in al-Raqqa directed by Dr. M. Meinecke; vessel sample from Qasr al-Banat, Raqqa from excavations by the Director general of Antiquities and Museums of Syria. Sample numbers in brackets refer to electron microprobe analyses published in [35] in Table 1.

were produced using a PRZ correction routine. The counting times were 30 seconds on the peak and 15 seconds on the background either side of the peak. A defocused beam is used to reduce the effect of the migration of alkalis within the samples. The following relative analytical accuracies were obtained using the Corning B standard as unknown: 3% for Na_2O; 2.5% for SiO_2; 1.5% for K_2O; 1.5% for CaO and 2% for PbO. For minor elements the accuracy was 5% for MgO, 2.5% for Al_2O_3, 15.5% for P_2O_5, 6.5% for FeO, 5% CuO and up to 13% for Cl. For trace elements the levels of detection varied from 75 ppm for Mn and Cr to 455 ppm for Ba. A fuller consideration of analytical errors in the epma analyses of mixed alkali and other ancient glasses is published elsewhere where the conditions used for the analysis of al-Raqqa samples can be found.[33]

Laser-Ablation inductively coupled plasma mass spectrometry (LA-ICP-MS)

To perform the analysis the polished sample discs were placed inside chamber with a window and the laser, NewWave UP193FX 193nm excimer laser system with integrated viewing optics, fired at the sample. The laser operating conditions, fluence C^2 . 4J × cm^2 were chosen to give a crater of 100 μm in diameter and approximately 30 μm deep, removing less than a microgram of material over 30 seconds. Each sample or calibration standard was ablated three times in succession. Ablated material was carried from the chamber by a stream of helium gas (0.80 L × min^{-1}), mixed with an argon make up gas (0.85 L × min^{-1}), to the quadrupole ICP-MS (Agilent 7500 series). The ICP-MS was repeatedly fast jumped through the 47 isotopes of the major/trace elements, returning an integrated response every 5s; data being recorded continuously in time resolved analysis (TRA) fashion. The integrated signal from each ablation was blank subtracted and calibrated to that produced by a standard glass of known concentration (NIST SRM612). To allow for differential ablation rates between samples and sample-standards responses were normalised to the known Si concentration as determined by EPMA. These calculations were performed off-line in an Excel spreadsheet. Calibration and quality control glasses samples (NIST SRM610) were run with each sample block.[34] Quality control data demonstrated across six analytical sessions and n = 84 analyses, a precision for all elements between 2–13 RSD% except P (62%); K (15%); Sn (24%). Whilst accuracy ranged between –9 and +16% for all elements except Mg (+24%); P (+26%); Sb (+24%). All of these were deemed "fit-for-purpose" except possibly P.

2.4 Results and discussion

2.4.1 *Major and minor elements analysis*

An interim discussion of the mosque lamp samples is published elsewhere.[28] Those for Al-Raqqa are already published[35]: the figure in brackets following the RAQ sample number is the analysis number in the publication. Results for the Damascus and Beirut samples are in press.[14]

The translucent glass samples are invariably coloured by a combination of iron and manganese oxides, producing green, brown, blue and purple colours under different oxidising/reducing conditions in the furnace. Manganese appears to have been added deliberately and this is generally the case in Middle Eastern Islamic glass.[25,35] Colouration of Islamic translucent glasses has been discussed elsewhere in some detail so not would not be discussed further here. Opaque yellow enamels (DOH 11 and 20) are opacified and coloured by lead stannate ($Pb_2Sn_2O_7$); opaque green enamel samples (DOH 12, 21, 80) with a combination of copper and lead stannate; opaque red enamel samples (DOH 13 and 16) with iron oxide (no copper was detected). The opacification of Islamic glasses has also been discussed in some detail elsewhere.[36–38]

Figures 2.3 and 2.4 are bi-plots of the samples subjected to both EPMA and LA-ICP compared with 9th and 12th century raw furnace glasses from al-Raqqa and raw glass from 11th century al-Raqqa and raw glass from the early 11th century shipwreck off the coast of Turkey at Serçe Limani.[39,40] The expected elevated levels of magnesia and potassium oxide levels that are found in plant ash glasses and lower levels in natron glasses have been detected. The lower levels of both are also found in some enamel samples containing high lead oxide used in the decoration of the mosque lamps. There is a broad positive correlation between MgO and K_2O but some individual samples do not fit the correlation; nor do the 9th and 12th century raw furnace glasses from al-Raqqa or the Serçe Limani glasses. There is a limited relationship between the geographical origin of the glass and the relative K_2O/ MgO levels.

As discussed above relative levels of MgO and CaO provide a broad geographical provenance for selected groups of glasses, bearing in mind that not all vessel glasses will necessarily have been

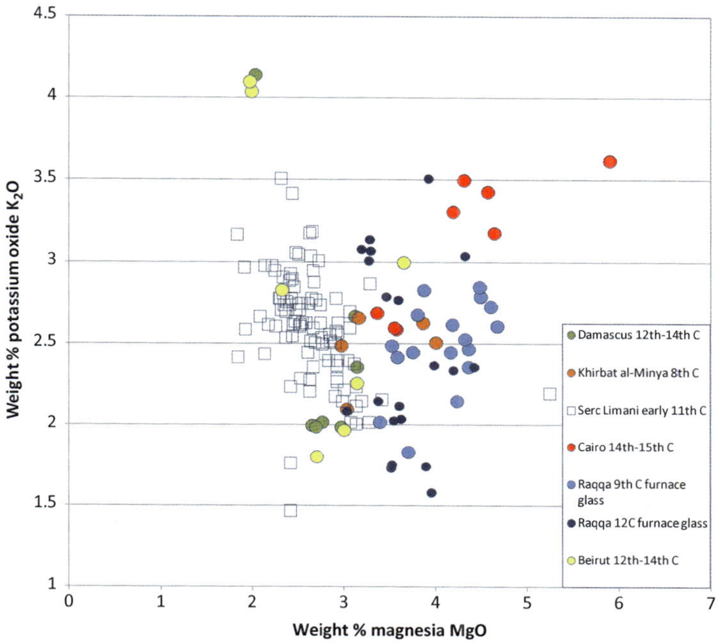

Fig. 2.3 Weight % magnesia versus weight % potassium oxide in Islamic glasses from the Middle East (Kam = Khirbat al-Minya). Samples containing low levels of both are either natron glasses (from Khirbat al-Minya or high lead enamel samples from 'Cairo')

made at or near the location where they were discovered. In Fig. 2.4 some broad distinctions are apparent. All of the raw 9th and 12th century furnace glasses from al-Raqqa and all mosque lamp glasses that contain more than 7% CaO fall on the right hand side of the figure. On the other hand glasses found in the Levant (Khirbat al Minya, most samples from Beirut and the Serçe Limani shipwreck raw glass) and Damascus fall to the left of the line. Natron glasses from Khirbat al Minya containing low calcium oxide might be expected to have an Egyprian origin.[41]

2.4.2 *Trace element analysis*

Glass raw materials are (often) associated with particular raw material types. For example, cesium (Cs), boron (B), beryllium (Be) and

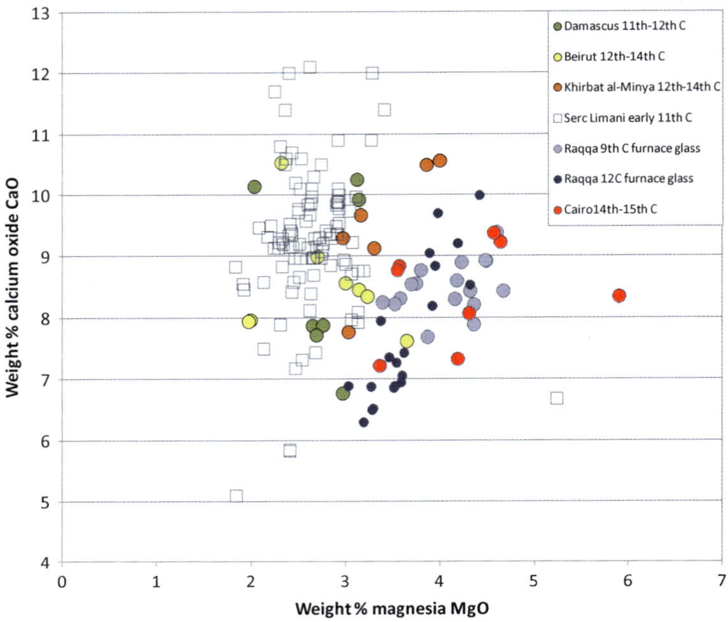

Fig. 2.4 Weight % magnesia versus weight % calcium oxide in Islamic glasses from the Middle East (Kam = Khirbat al-Minya). Samples containing low levels of both are either natron glasses (from Khirbat al-Minya) or high lead enamel samples from 'Cairo'.

lithium (Li) are associated with alkaline raw materials such as potassium (K) and sodium (Na); niobium (Nb), ytterbium (Yb), hafnium (Hf), lanthanum (La), tantalum (Ta) and zirconium (Zr) are associated with minerals in silica sands. Examples are monazites [lanthanum La], rutile [niobium Nb] and zircon [Hafnium Hf]. Molybdenum (Mo) and tungsten (W) are often associated with colorants. Chromium (Cr) and vanadium (V) can be associated with the presence of iron (Fe) and manganese (Mn) colorants; zinc (Zn) and cadmium (Cd) with copper (Cu); silver (Ag) with lead (Pb); nickel (Ni) with iron, cobalt (Co) and sometimes copper. Arsenic (As) is sometimes introduced with cobalt and copper.

Figure 2.5 shows evidence for use of different alkaline plant ashes sources with respect to the geographical origins of the glasses. The mosque lamp glasses possibly made in Cairo mainly have distinctive high Cs/K ratios. Khirbat al-Minya glasses cluster together and have

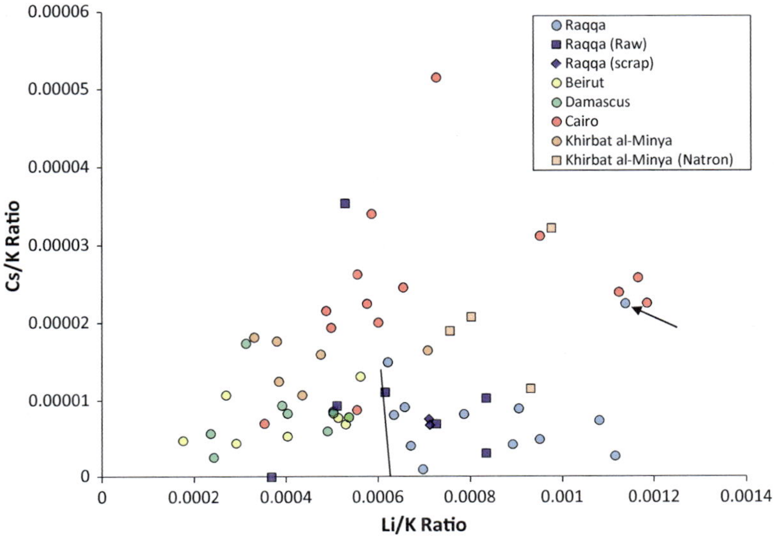

Fig. 2.5 Cs/K ratio versus Li/K ratio in the samples tested

somewhat higher Cs/K values than the glass from Beirut and Damascus; whilst the natron glasses have higher Li/K ratios than the plant ashes but similar Cs/K from Khirbat al-Minya suggesting Li is more dominant in the natron. The plotted points for Beirut and Damascus overlap to some extent.

These results provide a way of distinguishing between the mosque lamps possibly made from glass made in Cairo and al-Raqqa glasses that minor impurities do not provide (Figs. 2.3 and 2.4). As found with the EPMA analyses, the Levantine glasses from Kirbat al-Minya and Beirut and those from Damascus fall into the same field, even though there is some distinction between Khirbat al-Minya and the other glasses using trace elements associated with the plant ashes. The plotted Raqqa glass that falls outside the main group with a Cs/K ratio of just above 0.00002 is a 12th century vessel fragment from the Princess's Palace (Qasr al Banat) — located in Fig. 2.5 with an arrow. It has been made with plant ashes which contain the characteristic Li/K and Cs/K ratios of the mosque lamps. This suggests it was probably not made in al-Raqqa, but perhaps in Cairo.

Al-Raqqa glasses are can mainly be separated from the other glasses having high Li/K and low Cs/K ratios. The raw and scrap glass tending to have lower Li/K than the manufactured vessel glass. This therefore provides a local signature for the glasses. Two Raqqa raw samples have amongst the lowest Li/K values, lying close to the Levantine glasses. Radiogenic isotope results for Raqqa furnace glasses shows that there is a 'core' group of $^{87}Sr/^{86}Sr$ signatures associated with calcium/plant ashes but that there is a far greater variation of $^{144}Nd/^{143}Nd$ isotopic signatures associated with silica indicating that different silica sources were used to make them.[9] One of the raw glasses (RAQ38-purple furnace glass) has a particularly anomalous alkali ratio and is the only raw glass with a Cs/K ratio above 0.00003; this is explained by its heterogenous nature and very close proximity to the base of the furnace implying poor fusion and mixing of raw materials. Variable Nd isotope signatures suggest multiple sources and although the Sr isotopes imply a consistent plant ash source, the trace alkali elements indicate the situation may be more complicated. Therefore, perhaps we should consider a scenario whereby different plants from the same area were used allowing a constant Sr isotope ratio but with different trace alkali concentrations.

Conventional wisdom is that zircons in the silica source provide the REE and in particular the Nd that allows isotope ratio measurements and the contribution from the plant ash or colourants is minimal. Therefore, we might assume that a simple plot of Zr v Yb, another heavy REE assumed to come from zircon, should be highly correlated. However, Fig. 2.6 again suggests the situation is not quite so simple. There is a broad correlation between these elements. However inspection of individual site data shows significant divergences. Beruit material splits into two groups of (i) <50mg/kg Zr and (ii) >100 mg/kg Zr; but there is an uncorrelated range of Yb between 0.2 and 0.6 mg/kg Yb. In contrast, Raqqa raw glass and scrap has a consistent Zr range between 100–140 mg/kg Zr but Yb between 0.3 and 0.9 mg/kg Yb. Yet in the Raqqa vessel glass Zr ranges between <40 and 170 mg/kg, whilst the Yb retains a similar range to raw-scrap material. Again, with Khirbat al-Minya material the natron and plant ash fluxes have different chemistry-origins with majority of samples

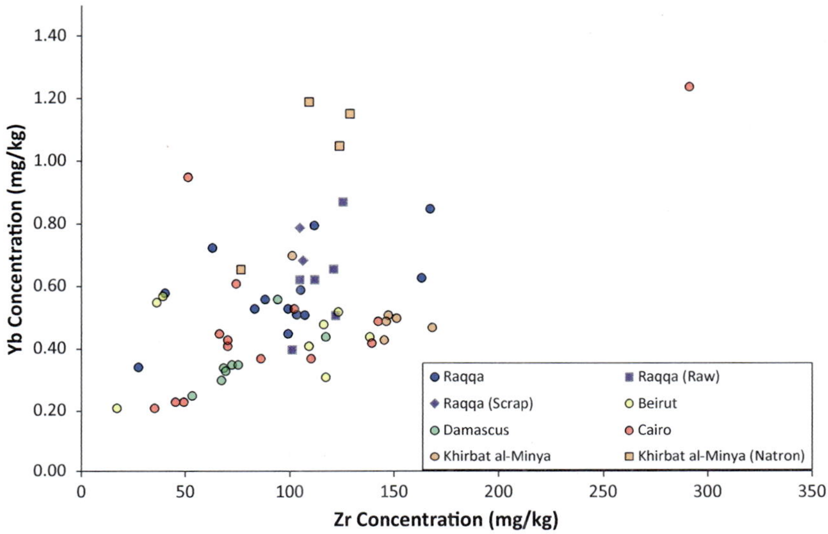

Fig. 2.6 Zr concentration (mg/kg) versus Yb concentration (mg/kg) in the samples tested

having both different Zr and Yb concentrations according to flux type. Clearly, a simple model of REE being associated solely with Zr presumed to be from a zircon mineral does not explain these results.

A number of useful insights are provided by plotting Zr v Ti concentrations of glasses after Shortland *et al.*, 2007 (Fig. 3).[41] Titanium (Ti), like Zr, is assumed to come from detrital heavy minerals such as rutile or sphene; the heavy minerals, and their major elements, are likely to be in different proportions in different source materials. Compared with Shortland *et al.*'s data for much earlier (Late Bronze Age) glasses from a similar area, all the samples from the current study fall outside the chemically pure Mesopotamian material field (assumed to be crushed quartz) and more than half have a greater proportion of Zr-Ti than the Egyptian field indicating the use of less pure raw materials. Helpfully, this plot separates Beirut and Damascus: the alkali ratio data plotted in Fig. 2.5 does not. The Khirbat al-Minya data is distinguished particularly by high Zr-Ti i.e. the most impure raw materials. The Raqqa raw glasses have a relatively constant Zr concentration but, as for Yb, variable Ti contents. Although the Yb

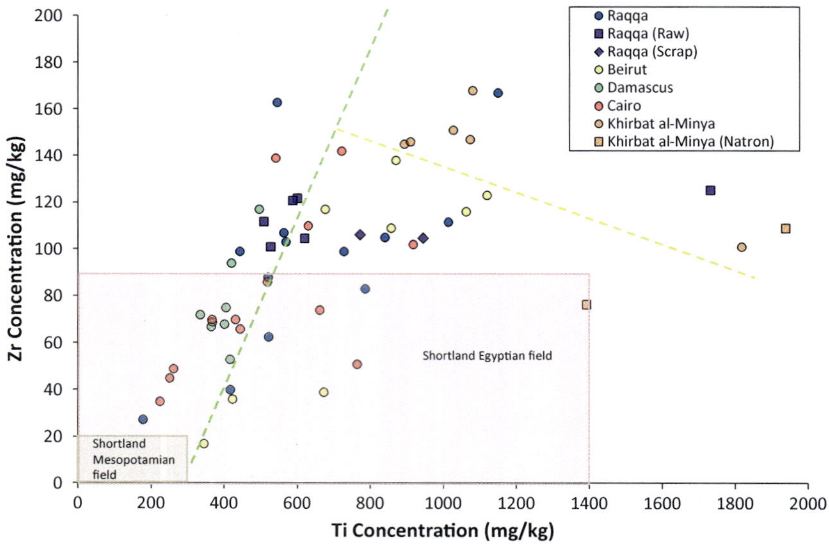

Fig. 2.7 Titanium concentrations (mg/kg) versus Zr concentrations (mg/kg) in the samples tested

levels appear to characterise al-Raqqa raw furnace glasses, at first glance it appears that the Li/K ratio is less discriminatory for them. The Raqqa vessel glass and the Cairo mosque lamps have very variable Zr and Ti. The Cairo mosque lamp enamel samples (DOH-11;DOH-12; DOH-21 & DOH-80) have the highest Pb concentrations and the lowest Zr (<52 mg/kg).

2.5 Conclusions

This set of trace element results produced using LAICPMS for 8th-15th century Islamic glass in the Middle East provide evidence for separate production zones in northern Syria (Raqqa), Beirut, Damascus and Cairo all of which may have produced glass for consumption along the Silk Road. The al-Raqqa samples tested include the only raw furnace glasses that were known to have been made there nevertheless there is a relatively consistent correlation between chemistry and different geographical origins of the samples. Samples from Cairo, Beirut and Damascus are contemporary or partially contemporary. There is, nevertheless, a possibility that different sources of raw materials with somewhat different trace element characteristics were used to make glass in the same place in different periods.

Certain trace elements such as Zr and Ti confirm the distinction between and Beirut and Damascus plant ash glasses found using isotopic analysis.

The association of REEs and other diagnostic elements with sand or with plant ash used for the manufacture of plant ash glass may be too simple and needs to be revisited in relation to the primary source materials, especially plant ashes.

Separable compositional groups from different origins in the current trace element data set suggests that minimal recycling occurred within compositional glass types.

Although the Yb levels appear to characterise al-Raqqa raw furnace glasses, at first glance it appears that the Li/K ratio is less discriminatory for them.

References

1. F. X. Gan, The Silk Road and ancient Chinese glass, in *Ancient Glass Research along the Silk Road*, eds. F. X. Gan, R. H. Brill and S. Y. Tian (World Scientific Publishing Co. Pte. Ltd, Singapore, 2009), pp. 41–108.
2. L. Dussubieux and B. Gratuze, Origine et diffusion du verre dans le monde Indien et en Asie du sud-est: L'importance du dosage des elements-traces, *Revue d'Arch'eometrie* **27**, 67–73 (2003).
3. Qinghui Li, Yongchun Xu, Ping Zhang *et al.*, Chemical composition analyses of early glasses of different historical periods found in Xinjiang, China, in *Ancient Glass Research along the Silk Road*, eds. F. X. Gan, R. H. Brill and S. Y. Tian (World Scientific Publishing Co. Pte. Ltd, Singapore, 2009), pp. 331–357.
4. Richard Hodges, Adriatic sea trade in a European perspective', in *From one sea to another. Trading places in the European and Mediterranean Early Middle Ages*, eds. S. Gelichi and R. Hodges, Proceedings of the International Conference, Comacchio, March 27th–29th, 2012.
5. N. Schibille, Supply routes and the consumption of glass in first millennium CE Butrint (Albania), *Journal of Archaeological Science* **38**, 2939–2948 (2011).
6. R. H. Brill, Opening remarks and setting the stage: lecture at the 2005 Shanghai International Workshop on the archaeology of glass along the Silk Road, in *Ancient Glass Research along the Silk Road*, eds. F. X. Gan, R. H. Brill and

S. Y. Tian (World Scientific Publishing Co. Pte. Ltd, Singapore, 2009), pp. 109–147.

7. S. Liu, Q. H. Li, F. Gan *et al.*, Silk Road glass in Xinjiang, China: chemical compositional analysis and interpretation using a high-resolution portable XRF spectrometer, *Journal of Archaeological Science* **39**, 2128–2142 (2012).
8. J. Henderson, J. A., Evans, H. J. Sloane *et al.*, The use of oxygen, strontium and lead isotopes to provenance ancient glasses in the MiddleEast, *Journal of Archaeological Science* **32**, 665–673 (2005).
9. J. Henderson, J. Evans, Y. Barkoudah, The roots of provenance: glass, plants and isotopes in the Islamic Middle East, *Antiquity* **83**, 414–429 (2009).
10. D. Brems, M. Ganio, K. Latruwe *et al.*, Isotopes on the beach Part 1: strontium isotope ratios as a provenance indicator for lime raw materials used in Roman glass making, *Archaeometry* **55**, 214–234 (2013).
11. D. Brems, M. Ganio, K. Latruwe *et al.*, Isotopes on the beach, Part 2: Neodymium Isotopic Analysis for the provanancing of Roman glass making, *Archaeometry* **55**, 449–464 (2013).
12. P. Degryse, A. Shortland *et al.*, Considerations on the provenance determination of plant ash glasses using strontium isotopes, *Journal of Archaeological Science* **37**, 3129–3135 (2010).
13. P. Degryse, I. Freestone, J. Schneider and J. Jennings, Technology and provenance study of Levantine plant ash glass using Sr-Nd isotope analysis. In *Glass in Byzantium — Production, Usage, Analyses*, eds. J. Drauke and D. Keller (Romisch-Germanischen Zentralmuseums, Mainz, 2010), pp. 83–91.
14. J. Henderson, J. Evans, S. Bertier *et al.*, Glass production and provenance in the Islamic Middle East: chemical and isotopic analysis in an environmental context, 2015, submitted.
15. I. C. Freestone, K. A., Leslie, M., Thirlwell and Y. Gorin-Rosen, Strontium isotopes in the investigation of early glass production: Byzantine and early Islamic glass from the Near East, *Archaeometry* **45**, 19–32 (2003).
16. K. A. Leslie, I. C., Freestone, D., Lowry and M. Thirlwell, The provenance and technology of Near Eastern glasses: Oxygen isotopes by laser fluorination as a complement to strontium, *Archaeometry* **48**, 253–270 (2006).
17. P. Degryse and J. Schneider, Pliny the Elder and Sr-Nd isotopes: tracing the provenance of raw materials for Roman glass production, *Journal of Archaeological Science* **35**, 1993–2000 (2008).
18. P. Degryse, J. Schneider, U. Haack *et al.*, Evidence for glass 'recycling' using Pb and Sr isotopic ratios and Sr-mixing lines: the case of Early Byzantine Sagalassos, *Journal of Archaeological Science* **33**, 494–501 (2006).
19. C. M. Jackson, C. A. Booth and J. W. Smedley, Glass by design? Raw materials, recipes and compositional data, *Archaeometry* **47**, 781–795 (2005).

20. J. Henderson, Ancient Glass, an interdisciplinary exploration, Cambridge University Press, New York and Cambridge, 2013.
21. E. V. Sayre and R. W. Smith 1961, Compositional categories of ancient glass, *Science* **133**, 1824–26 (1961).
22. E. V. Sayre, Summary of the Brookhaven Program of analysis of ancient glass, in *Applications of Science in examinations of works of Art*, Proceedings of the Seminar held at the Boston Museum of Fine Arts, Boston, Mass., 7th–16th September 1965, Museum of Fine Arts, Boston, pp. 145–154.
23. F. R. Matson, The manufacture of eighth-century Egyptian glass weights and stamps, in *Early Arabic Glass Weights and Stamps*, eds. C. Miles (Numismatic Notes and Monographs, no. 111, The American Numismatic Society, New York, 1948) 31–69.
24. B. Gratuze and J.-N. Barrandon, Islamic glass weights and stamps, *Archaeometry* **32**, 155–162 (1990).
25. R. H. Brill, Some Thoughts on the Chemistry and Technology of Islamic Glass, in *Glass of the Sultans*, The Metropolitan Museum of Art in association with The Corning Museum of Glass, New York, Benaki Museum, and Yale University Press, pp. 25–45 (2001).
26. J. Henderson, Glass trade and chemical analysis: A possible model for Islamic glass production', in *Échanges et commerce du verre dans le monde antique*, eds. D. Foy and M.-D. Nenna (Actes du colloque de l'Association Francaise pour l'Archéologie du Verre, Aix-en-Provence et Marseille, 7–9 juin 2001, Montagnac: Éditions Monique Mergoil, 2003), pp. 109–123.
27. N. Kato, I. Nakai and Y. Shindo, Transitions in Islamic plant ash glass vessels: on-site chemical analyses conducted at the Raya/al Tur area on the Sinai Peninsula, Egypt, *Journal of Archaeological Science* **37**, 1381–1395 (2010).
28. J. Henderson, Technological analysis of Mamluk glass, in *Mamluk enamelled and Gilded glass in the Museum of Islamic Art, Qatar*, eds. S. Carboni (The Museum of Islamic Art, Doha, 2003), pp. 28–31.
29. A. Y. Al-Hassan, An eighth century Arabic treatise on the colouring of glass Kitāb al-durra al-maknūna (the book of the hidden pearl) of Jābir ibn Hayyān (c. 721–c. 815), *Arabic Sciences and Philosophy* **19**, 121–156 (2009).
30. F. Aldsworth, G. Haggerty, S. Jennings *et al.*, Medieval Glassmaking at Tyre, Lebanon, *Journal of Glass Studies* **44**, 49–66 (2002).
31. I. C. Freestone, Composition and affinities of glass from the furnaces on the island site', Tyre, *Journal of Glass Studies* **44**, 67–77 (2002).
32. J. Henderson, Islamic Glass Production and the possible evidence for trade in a cobalt-rich colorant, in *Gilded and enamelled glass from the Middle East*, ed. R. Ward (British Museum Press, 1998), pp. 116–121.
33. J. Henderson, Electron probe microanalysis of mixed-alkali glasses, *Archaeometry* **30**, 77–91 (1988).

34. S. Chenery and J. M. Cook, Determination of rare earth elements in single mineral grains by laser ablation microprobe–inductively coupled plasma mass spectrometry — preliminary study, *Journal of Analytical Atomic Spectrometry* **8**, 299–303 (1993).
35. J. Henderson, S. McCloughlin and D. McPhail, Radical changes in Islamic glass technology: evidence for conservatism, and experimentation with new glass recipes, *Archaeometry* **46**, 439–468 (2004).
36. I. C. Freestone and C. P. Stapleton, Composition and technology of Islamic enameled glass of the thirteenth and fourteenth centuries', in *Gilded and enamelled glass from the Middle East*, ed. R. Ward (The British Museum Press, London, 1998), pp. 122–128.
37. M. Verità, Analytical investigation of European enameled beakers of the 13th and 14th centuries, *Journal of Glass Studies* **37**, 83–98 (1995).
38. J. Henderson, J. A. Allan, Enamels on Ayyubid and Mamluk glass fragments, *Archaeomaterials* **4**, 167–183 (1990).
39. G. F. Bass, B. Lledó, S. Matthews and R. H. Brill (eds.), *Serçe Limani: Glass of an eleventh century shipwreck*, Volume 2, Rachel Foundation Nautical Archaeology Series, College Station, Texas: Texas A & M University Press, 2009.
40. R. H. Brill, Chemical analyses, in *Serçe Limani: Glass of an eleventh century shipwreck*, eds. G. F. Bass, R. H. Brill, B. Lled´o and S. D. Matthews (Rachel Foundation Nautical Archaeology Series, Volume 2, College Station, Texas: Texas A & M University Press, 2009), pp. 459–496.
41. M.-D. Nenna, M. Picon and M. Vichy, Ateliers primaries et secondaires en Égypte á l'époque gréco-romaine, in *La Route du verre: Ateliers primaires et secondaires du second millénaire av. J.-C. au Moyen Age*, ed. M.-D. Nenna (Lyon: Maison de l'Orient Méditerraneéan-Jean Pouilloux, 2000), pp. 97–112.
42. A. J. Shortland, N. Rogers and K. Eremin, Trace element discriminants between Egyptian and Mesopotamian Late Bronze Age glasses, *Journal of Archaeological Science* **34**, 781–789 (2007).

Chapter 3 第三章

Travels of Glass Vessels along the Maritime Silk Road

Brigitte Borell
Independent Scholar,
c/o Institute for Classical Archaeology, University of Heidelberg,
Marstallhof 4, 69117 Heidelberg, Germany

3.1 Introduction

The aim of this paper is to raise some questions about the 'social life' of glass vessels by way of adopting ideas put forward by Arjun Appadurai.[1] Particular emphasis is laid on their role in trade and as prestige objects. Through its history, the social role of glass as a man-made material is highly dependent on its technological development. The invention and development of technologies for producing vitreous materials has been a slow process extending over a very long period of time in the ancient Near East and in Egypt: from glazed stones to faience, occasional small glass objects and finally, in the mid-second millenium BCE — together with a glass production on a larger scale — the first small glass vessels formed over a core. Early glass production in Egypt took place under royal patronage, if not as a royal monopoly. Glass was an exclusive material, a luxury object, in value and function comparable to that of precious stones and gold. Glass vessels continued to be rare and luxurious objects for a long part of their history.

Before the introduction of glass-blowing, forming a hollow glass object was a special challenge for the ancient craftsman. In the course of time, several different techniques were developed, not all of them fully understood. The invention of glass-blowing in the Levant in the mid-first century BCE and its further development during the first century CE mark a dramatic change in the manufacture of glass vessels. Glass-blowing did not only allow fast production, in particular, for closed shapes like small bottles, but also the making of very thin-walled glass vessels which needed only a small amount of glass material. From the amount of glass previously needed for a typical mould-formed bowl, a number of small glass bottles could be blown in a comparatively short time. It is important to keep this aspect of the history of glass in mind, because the 'social life' of glass vessels is closely connected to these technological developments. With regard to glass vessels, we see a very gradual change lasting for more than one and a half thousand years, from being exclusive accessories of splendour for kings and royalty to wider circles of wealthy élites, and finally, to objects of regular use for affluent classes of society.

In particular, this paper intends to look into the roles of glass vessels transported along the maritime routes of the so-called Silk Road across the Sea. The first part deals with glass vessels, mainly of Mediterranean origin, which were exported as far as China. The second part is devoted to a special group of glass vessels found predominantly in the Gulf of Tonking region and probably produced there.

3.2 Glass vessels from the Mediterranean along the maritime routes in the first century CE

A good example to use for following the paths of Mediterranean glass vessels are ribbed bowls. They are not only frequently encountered, owing to their large-scale production in the late first century BCE and the first century CE, but, because of their characteristic shape and decoration, they can be recognised, even when only a small fragment is preserved. Judging from the finds, they seem to have played a major role in the maritime trade of the period. They might not only have been used as a commodity, but also might have been used as presents for local rulers.

They are found in the Red Sea ports Myos Hormos (present-day Quseir al-Qadim) and Berenike, along the north Somali coast, and in southeastern Arabia, along the Straits of Hormuz. They were also brought to India.[2] Fragments of at least two, one blue and one green, were found in Pattanam, a recently discovered site identified as the famous port called Muziris.[3] Ribbed bowls were also found in Taxila and in Begram. Although the date and the nature of the so-called "Treasure House" at Begram are still debated, there is general agreement among the scholars that the glass vessels arrived there via the Red Sea trade across the Indian Ocean to Barbarikon, a port in the Indus Delta, from where they travelled further, first up-river, and then over land routes. The two storage rooms at Begram also contained other rare objects, among them lacquer ware from China, variously dated between the first century BCE and the mid-first century CE.[4–6] These might have come along overland routes of the Silk Road, although combinations of land and sea routes to India may not be excluded.[7] Ribbed bowls are also known from sites along the southeastern coast of India, in particular from Arikamedu: fragments of at least six ribbed bowls have been reported from the site.[2]

The most spectacular find is, of course, the discovery of such a ribbed bowl in a burial in the Yangzi delta, at Ganquan, Hanjiang (Jiangsu), about 80 km northeast of Nanjing.[8,9] Due to the previous looting of the tomb, only fragments were recovered during excavation (Table 3.1 analysis no. 1). The burial has been identified as that of Liu Jing, one of the younger sons of the Emperor Guangwu (r. 25–57 CE), who was nominated King of Guangling in 58 CE and committed suicide in 67 CE. The glass bowl is not one of the common monochrome examples, but it is made of mosaic glass, which involved several more labour-intensive steps enhancing its prestigious value. As Marianne Stern has pointed out "Mosaic glass vessels were among the finest luxury ware of the early Roman Empire." The marbled pattern was intended to reproduce the appearance of natural banded stone, and was "particularly favoured for ribbed bowls because the more the hot glass was deformed, the more interesting the marbling became".[10]

Table 3.1 Chemical analyses (wt%) of glass vessels of Mediterranean origin found in tombs in Jiangsu from the Eastern Han period (25–220 CE) and Eastern Jin period (265–420 CE)

No.	1	2	3	4a	4b	5a	5b	6
Date	E.Han	E.Han	E.Jin	E.Jin	E.Jin	E.Jin	E.Jin	E.Jin
SiO_2	64.79	68.37	69.39	67.70	68.41	69.15	69.95	72.04
Na_2O	18.18	16.01	19.60	19.23	18.82	15.84	16.10	14.76
CaO	7.66	4.62	6.81	6.05	5.80	5.79	5.72	6.29
K_2O	0.88	0.48	0.49	0.45	0.38	0.33	0.27	0.32
MgO	0.61	0.79	0.27	0.94	1.00	0.55	0.63	0.61
Al_2O_3	3.44	1.76	1.89	3.43	2.20	2.09	2.09	2.30
Fe_2O_3	1.30	0.61		0.58	0.63	1.22	1.17	1.34
TiO_2					0.080		0.099	0.09
Sb_2O_5			0.29					
MnO	2.45	0.58		1.63	1.34	0.03	0.002	0.03
CuO	0.03	2.88		0.02		0.27	0.22	0.25
CoO				traces				0.12
SnO_2								0.003
PbO							0.24	0.21
P_2O					0.09		0.07	0.05

Notes on samples:
No. 1: Tomb 2 at Ganquan, Hanjiang, Jiangsu.[9]
No. 2: Tomb at Laohudun, Jiangsu.[13]
No. 3: Tomb at Nanjing Daxue Beiyuan, Nanjing, Jiangsu.[60]
No. 4a and 4b: Great tomb at Beigushan, Nanjing, Jiangsu (pale yellowish cup fragment).[34]
No. 5a and 5b: Great tomb at Beigushan, Nanjing, Jiangsu (fragment of thin-walled dark blue glass vessel).[34]
No. 6: Tomb at Qinglongshan, Jiangning, near Nanjing, Jiangsu.[22]

In this case, the colour range of the mosaic glass, translucent reddish purple (coloured with manganese) and opaque creamy-white glass, points to the imitation of one of the costliest luxury objects in antiquity, the famous and extremely expensive murrhine vessels made of a rare natural stone. The mineral of which the *vasa murrina* were made has been convincingly identified as a certain variety of fluorspar which occurs typically in the colours ranging from purple to blue and from white to yellowish. Pliny (*Natural History* 37.18–22) reports that the stone for the murrhine vessels came from Parthia, in present-day Iran.[11,12] The first murrhine vessels ranked among the most costly luxuries; the Emperor Nero (54–68 CE) paid the fabulous sum of one million sesterces for one cup. The real thing was affordable for only a few of the top élite. Such a glass imitation is to be regarded as a product with a high prestige value. In maritime trade contacts, such an item of luxury glassware would have been used as a present to the élites in foreign lands.

Another Mediterranean glass vessel, a monochrome cup made of transparent blue glass (Fig. 3.1), was found in the same area, at Laohudun, Ganquan, in a tomb probably dating from the early Eastern Han period (25–220 CE).[13] The owner of the tomb was certainly a person of high social status. The chemical analysis of the glass shows that it is a mineral soda glass of Mediterranean type (Table 3.1 analysis no. 2). It has very thin walls, and is certainly a blown vessel. Unfortunately, most of it was badly crushed into tiny

Fig. 3.1 Eastern Han tomb at Laohudun, Hanjiang, Jiangsu. Fragmented blown glass cup, Roman, 1st century CE. Estimated rim diameter 7.8 cm. Photo: Wang Zhigao

fragments, and it is difficult to say anything definite about its shape. However, it seems likely that it was one of the early blown cups of Isings form 12, a very characteristic and popular Roman glass vessel shape in the earlier part of the first century CE.[14] The excavation report speaks about a diameter of 7.8 cm — that would be in good agreement with this class of cups. It further mentions a preserved height of 9.2 cm, hence it might have been a cup of the high variety (variant of Isings form 29).[15–17] A closer examination of the fragments might reveal remains of cut grooves, since the Roman cups of this class were frequently decorated with such horizontal grooves on the exterior.

In all likelihood, such glass vessels were shipped along the maritime routes from the southern seas, probably used as prestigious gifts as part of trade transactions, whereas other commodities — many of them of a perishable nature — played the dominant role in maritime trade.

In the centuries around the turn of the millenium, the networks connecting different maritime routes reached to the east as far as China, and to the west — via the Red Sea — ultimately to the Mediterranean. The Malay Peninsula is like a barrier between the Indian Ocean and the South China Sea, and it appears that here, in the region of the Isthmus of Kra, the different maritime routes were connected by land routes across the peninsula. Such a land crossing between months of sailing is mentioned in a Chinese source (*Han shu*, chapter 28B). It describes a Han period sea voyage from the Gulf of Tonking to a place called Huangzhi, which is usually sought in southeastern India, and is possibly to be identified with Kanchipuram.

In recent years, sites, on the eastern and western coasts of the Isthmus of Kra, have increasingly yielded archaeological evidence for such transpeninsular crossings. Finds include Indian pottery, Roman intaglios, and Mediterranean glass fragments, as well as fragments of Han period storage jars. At Phu Khao Thong, a site on the western coast of the peninsula, a few fragments of mosaic glass vessels came to light.[18] One of them (Fig. 3.2) appears to stem from the curved side of a hemispherical moulded bowl. The pattern, with spirals of opaque yellow glass in a matrix of translucent bluish-green glass, points to a

Fig. 3.2 From Phu Khao Thong, Ranong province, Thailand. Fragment of a late Hellenistic or early Roman mosaic glass bowl, mould-formed, 1st century BCE or early 1st century CE. Photo: Boonyarit Chaisuwan

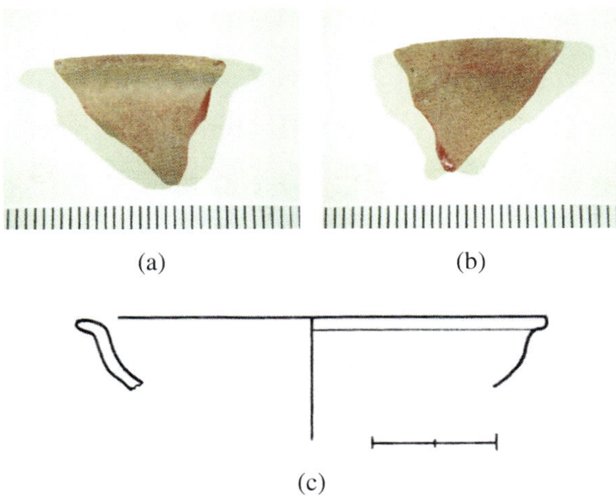

Fig. 3.3 From Ta Chana, Suratthani province, Thailand. Rim fragment of a mould-formed cup of red opaque glass, Roman, first half of 1st century CE. Estimated rim diameter 7.6 cm. Photos and drawing: Brigitte Borell

date possibly still in the first century BCE, or in the early first century CE.[19,20] A site on the eastern coast, Ta Chana, yielded a rim fragment of opaque bright red glass (Fig. 3.3). The shape as well as the colour of the glass is typical for a certain class of small moulded bowls of early Roman date produced in the late first century BCE and the first half of the first century CE.[20,21] The chemical analyses of both fragments, carried out by James Lankton, confirm the Mediterranean origin of the glass.

3.3 Glass vessels from the Mediterranean found along the maritime routes in the Eastern Jin period (317–420 CE)

In China, we find significant archaeological evidence for glass vessels of Mediterranean and Near Eastern origin in tombs of the Eastern Jin period (317–420 CE).[22] Most of them were found in tombs in the area around Nanjing, then the captial, Jiankang, of the Eastern Jin.

Among the best-known are two beakers (Figs. 3.4 and 3.5) with cut decoration from one of the tombs of the Wang family at Xiangshan (tomb 7), possibly the burial of Wang Yi, who died in 322 CE.[23] The owner of the tomb was buried between his two wives. One glass beaker of almost clear colourless glass was found at the head of the man's coffin (M7:23; H 10.4 cm, D 9.4 cm). The fragments of the other beaker of the same type and size, but of slightly yellowish glass, were found underneath a bronze wine warmer (M7:31) at the coffin-head-end of the woman buried south of him, that is, to his right. This was a far richer burial compared to the modest burial of the woman

Fig. 3.4 Eastern Jin tomb 7 at Xiangshan, Nanjing. One of two glass beakers with facet cutting, probably Roman, late 3rd or early 4th century CE. H 10.4–10.9 cm. Photos: Nanjing Municipal Museum

Fig. 3.5 Eastern Jin tomb 7 at Xiangshan, Nanjing. The other one of the two glass beakers with facet cutting, probably Roman, late 3rd or early 4th century CE. H 10.4–10.9 cm. Photos: Nanjing Municipal Museum

on the northern side. Both beakers have thick walls, measuring between 0.5 and 0.9 cm, and have slightly everted rounded rims. They are richly embellished with facet cutting.[24] They may be attributed to the Roman tradition of facet cutting, which flourished in the third and fourth centuries CE. The peculiar way in which the large oval facets are surrounded by a cut groove is already known from glass vessels found at Dura Europos in Syria, which predate 256 CE.[10,25] They have also been compared to a glass beaker said to come from Iran in the British Museum, which has been labelled as early Sasanian.[26] The glass of the two beakers from tomb 7 Xiangshan has apparently not yet been subjected to chemical analysis, which might resolve the question of whether it belongs to the glassmaking tradition characteristic of the Levant, which used mineral soda, or to that of the Middle East, which used a soda derived from plant-ash and has elevated levels of K_2O and MgO.[27–30]

The Wang were a powerful family from Shandong, who preferred emigration to the south in the troubled years before the fall of Luoyang to peoples from the north and the end of the Western Jin in 311 CE. Like other émigré families of the northern élite, they assisted in re-establishing the dynasty as the Eastern Jin in the south. The Wang family had close ties to the court. Wang Yi had a reputation as a painter and calligrapher. He was an instructor of Sima Shao, the later Emperor Ming (r. 323–325 CE), who honoured him on the occasion of his funeral like a member of the imperial family.[31,32] Besides the glass beakers, Wang Yi's tomb contained another exceptional object: a gold ring with a diamond was found in the tomb in a position suggesting that it had been placed on the chest of the body as a gift of condolence. In the tomb of another member of the Wang family, Xingzhi, who died in 340 CE, a vessel made of a nautilus shell mounted in gilded bronze was found. The nautilus, which lives only in the tropical waters of the Indo-Pacific, could only be obtained via the southern seas. Such exceptional and luxurious goods, including the glass beakers, were probably tributes brought by envoys from foreign countries, "gifts that then found their way into a very limited section of the upper classes".[31]

Another fragmented vessel was recovered from the tomb at Nanjing University, which is considered to be an imperial tomb of the Eastern Jin. The glass has been analysed and proved to be of Mediterranean origin (Table 3.1 analysis no. 3). The fragments are from a thin-walled cup (wall thickness is given as 0.1 cm) with an everted rim (10 cm in diameter) and a decoration of widely set shallow oval facets (Figs. 3.6a and 3.6b). The walls are curved, hence the

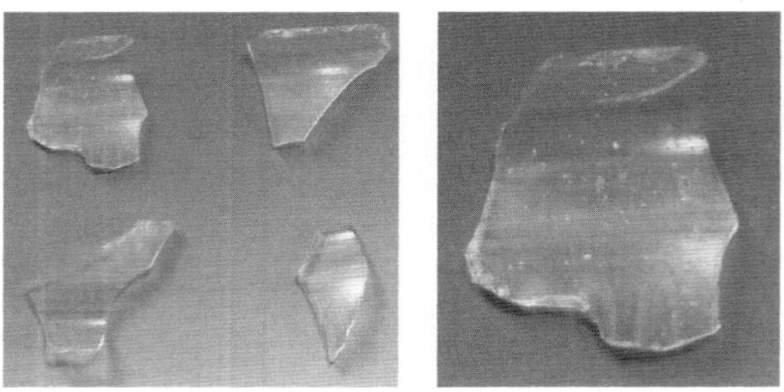

(a) (b)

Fig. 3.6 Eastern Jin tomb at Nanjing University. Fragments of a glass cup with facet cutting, Roman, 3rd or 4th century CE. Estimated rim diameter 10 cm. Photo: Wang Zhigao cup was of a more globular shape. Such late Roman thin-walled cups decorated with shallow facets arranged in different patterns are well-known from contexts of the third and fourth centuries CE.[10,33]

More fragments of Mediterranean glass vessels were found in the great tomb at Beigushan, which is also considered to be an imperial tomb. According to the brief report of 1983, it contained the fragments of three glass vessels, two made of clear glass with a yellowish tinge, and one made of dark blue glass.[34] Two fragments — one of clear glass, the other of blue glass — have been analysed by two different institutions, the Nanjing Geological and Mining Research Institute and the Beijing Building Materials Research Institute. The chemical analyses of both glasses indicate that they are of a natron 'Mediterranean' type. The blue glass (Table 3.1 analyses 5a.b) stems apparently from a thin-walled glass vessel; the report gives its wall thickness as between 0.1 and 0.3 cm. The 25 fragments of clear glass (Table 3.1 analyses 4a.b) have cut decoration, and the report compares them in shape, decoration and general appearance with the two glass beakers from the Wang tomb 7 at Xiangshan. However, judging from the photograph (Fig. 3.7) these fragments might come from two different types of glass vessels. According to the report, their thickness varies within a range of 0.2 and 0.5 cm. Some of them seem indeed to be similar to the beakers in the Wang tomb: with a row of

Fig. 3.7 Great Eastern Jin tomb at Beigushan, Nanjing. Fragments of one or two glass cups with facet cutting. Roman, 3rd or 4th century CE. Photo: Wang Zhigao

closely-set short ovals, horizontal grooves, and possibly large ovals below. These fragments might also have the thicker walls around 0.5 cm, like the Xiangshan beakers (between 0.5 and 0.9 cm). If the published analysis is that of such a fragment, this would indirectly support the Mediterranean origin of the Wang beakers. Other fragments, however, seem to resemble the cup from the tomb at Nanjing University: with a thin convex wall, everted rim, and a decoration of shallow, elongated ovals, which are widely spaced.

In a tomb of the Gao family burial site at Xianheguan, in the northeastern outskirts of Nanjing, a glass cup which could be restored from fragments was found (Figs. 3.8a and 3.8b). Gao Song, a high official of the Eastern Jin dynasty who died in the year 366 CE, is buried in Xianheguan tomb 2, together with his wife, who died before him in 356 CE.[35,36] Tomb 6, which contained the glass cup, is considered as the burial with the highest status among this group of tombs, and is dated to the early Eastern Jin period. It has been proposed to be the tomb of Gao Song's parents. The glass cup was found in the wife's coffin, and was possibly placed on a lacquer stand. The glass has not yet been analysed, but the vessel-type is very similar to that of the cup from the Nanjing University tomb. In shape and style, they both compare to late Roman glass cups decorated with a

(a)

(b)

Fig. 3.8 Eastern Jin Tomb 6 at Xianheguan, Nanjing. Glass cup with facet cutting, probably Roman, 3rd or 4th century CE. H 7 cm, rim diameter 9.1 cm. Photos: Brigitte Borell variety of patterns of zoned facet cutting, which are well represented in third and fourth century contexts.[10,37]

The finds in tomb 4 at Fuguishan, presumably also the burial of a member of the imperial family, included — placed in a silver bowl — a glass bowl of globular shape with everted rim.[24,38] To date, the glass has not yet been analysed, but the bowl is probably of Middle Eastern origin. Its shape compares to one of the two glass bowls from the Western Jin (265–316 CE) tomb 4021 at Wulidun in Echeng, Hubei (actually two similar bowls, 4021.3 and 4021:4, had been found, the latter only in fragments)[8,39] and to the bowl from an Eastern Jin tomb at Pingshigang in Zhaoqing, Guangdong, dated to 325 CE.[40,41] These bowls have been analysed and proved to be made of a plant-ash glass characteristic of a Middle Eastern origin. The Eastern Jin tomb 5 at Shidaigang in Xiatang in Guangzhou also yielded fragments of a Middle Eastern vessel made of dark blue glass.[22,41] The glass vessel of Fuguishan tomb 4 was carefully set into a shallow silver bowl, indicating the high esteem in which the glass vessel was held.

In 2008, fragments of another vessel of dark blue glass were found in an Eastern Jin tomb at Qinglongshan in Jiangning.[22] The fragments are very small and give only few clues to the original shape. The glass vessel had convex walls, its size would be in the range of the other glass vessels — with a diameter between 7 and 9 cm. The chemical analysis carried out by James Lankton and Bernard Gratuze (Table 3.1 analysis 6) shows that the glass belongs to a 'Mediterranean' glass composition.

The burial-place of high-ranking members of the Wen family, another of the prominent families of the time, was located at Guojiashan in Nanjing.[22,42] Tomb 13, dating to the late Eastern Jin or early Southern Dynasties, was presumably the tomb of Wen Songzhi, the grandson of Wen Jiao.[43] It contained numerous glass vessel fragments, which appear to stem from three different glass vessels: one of them made of clear glass, one of dark blue glass, and the third one made of clear glass with streaks of blue-green glass. Unfortunately, they survived only in tiny fragments, therefore — without chemical analyses — nothing more precise can be said about their origin, whether they are made of 'Mediterranean' or Middle Eastern glass. Like the other tombs of the Wen family, tomb 13 at Guojiashan was richly furnished, the remains of the three glass vessels further attest to the high status of the tomb and its owner.

The picture emerges that the imported glass vessels from the Nanjing tombs share certain characteristic features. As far as analysis is concerned, the glass belongs to the Mediterranean glass-making tradition. The glass vessels made of dark blue glass appear to have no further decoration, those of clear glass are decorated with facet cutting, an embellishment which greatly increased the value of the glass vessels. Some of the tombs containing glass vessels are considered to be tombs of members of the imperial Sima family, possibly of the emperor himself, as has been suggested for the tomb at Nanjing University and the great tomb at Beigushan, the others were burials of members of the great aristocratic families with close connections to the court. This supports the view of Annette Kieser that the possession of such exotic luxury objects was limited to the court of the Eastern Jin emperor and to aristocratic circles very close to him. They might have been brought as tributary gifts from foreign countries to the emperor and were then, as special favours, passed on to selected high-ranking persons.

3.4 Chinese glass vessels made of lead-barium glass

In the past, a view prevailed that the production of ancient glass vessels was a monopoly of Western Asia, Egypt, and the Mediterranean. This is not the case: discoveries in the last fifty years have changed this view. As is now well-known, already in the Han period glass vessels were manufactured in China, made of characteristic Chinese lead-barium glass, and in characteristic Chinese shapes. Such glass vessels have been discovered in two Western Han tombs dating from the second century BCE. Two eared cups and a circular tray were discovered in the tomb of Liu Sheng, King of Zhongshan, at Mancheng, Hebei.[44] Liu Sheng was a son of the Han Emperor Jing (r. 157–141 BCE), and he died in 113 BCE. In the tomb at Beidongshan in Xuzhou, Jiangsu, remains of sixteen cylindrical cups made of lead-barium glass were found.[45,46] The tomb also dates from the second century BCE; however, owing to several phases of looting before excavation took place, the owner of this tomb, certainly a King of Chu and a member of the imperial family, has not yet been identified. Several candidates have been suggested, among them, Liu Jiao, the younger brother of Emperor Gao (r. 206–195 BCE), who died in 179 BCE, and Liu Dao, who died in 128 BCE, thus limiting the dating of the tomb to a period covering the second and third quarters of the second century BCE.[46]

Still, finds of glass vessels made of lead-barium glass are very rare. With their turbid, slightly opaque appearance and their colour-range from light green to white, they were clearly intended to imitate jade. In fact, initially, the excavators at Beidongshan mistook the cylindrical glass cups for jade.[8,47] It seems the glass vessels were specifically manufactured for persons of the highest social level: to date at least, they have been found only in tombs of high-ranking members of the imperial family.

3.5 Han period glass vessels from the Gulf of Tonking region made of potash glass

A different class of glass vessels has been discovered in Han period tombs in regions adjoining the Gulf of Tonking area. They were found predominantly in Guangxi, and were studied mainly by Huang Qishan, who devoted several contributions to this group, during the early period of his research. The majority of these glass vessels are small cups without foot or handles, measuring 4–7 cm in height and 7–9 cm in diameter. Their base is usually convex, sometimes curving softly inward, thus forming a concave dimple in the centre. A characteristic of theses cups are the horizontal ribs, usually three, on the exterior located about two-thirds up their height (Fig. 3.9). A variant of this cup has a stemmed foot. Another shape is a saucer or tray with flaring walls and a flat base. One such saucer was found in a context forming a set with a cup on a stemmed foot (Fig. 3.10). In addition, there are shallow bowls with a rounded base. The vessel walls are between three and four millimeters thick and sometimes even thicker. The wall often thickens towards the bottom.

Fig. 3.9 Western Han tomb 70 at Wenchangta, Hepu, Guangxi. Mould-formed glass cup. H 5.2 cm, rim diameter 7.4 cm. Photo: Brigitte Borell

Fig. 3.10 Eastern Han tomb at Nandoucun, Guixian (Guigang), Guangxi. Set of glass cup on stemmed foot and glass tray, mould-formed. H of cup 8.2 cm, rim diameter 6.4 cm. Photo: Brigitte Borell

Huang Qishan lists sixteen glass vessels of this group found in Han period tombs in Hepu, near the coast, and in Guixian, further inland.[48,49] They come from eleven tombs, which contained either one or two glass vessels. The preliminary chronology of the tombs suggests a time-span for the production of the glass vessels from the first century BCE to the first and possibly second centuries CE. In the tomb at Muzhuling, which contained a shallow bowl, some *da quan wu shi* coins were also found. Therefore this tomb cannot be dated earlier than the year 7 CE, when these coins were introduced, it might be dated contemporaneous with the coins or later. A more detailed chronology is to be expected when all these burial contexts are fully published.

The glass vessels were certainly highly-valued objects, they are found in wealthy burials of high social status. In at least two cases the occupant of the tomb is known. Tomb 34 at Honglingtou, probably dating from the late Western Han period, contained a bronze seal with the name Wang Qing.[48,49] The occupant of tomb 1 at Huangnigang in Hepu, a very wealthy high status burial, probably dating from the early first century CE, has been identified as Chen Bao, who was county magistrate of Xuwen in the Hepu commandery (pers.comm. Xiong Zhaoming, March 2009). The cup was found in the rear chamber of the tomb, which indicates the value attached to the glass vessel, apparently intended for the personal use of the deceased. In other tombs, the glass vessels were carefully placed inside a bronze dish or bronze basin; in Wang Qing's tomb it had been set inside a lacquer container.[48,49]

Two more glass vessels of this group were found in the northern part of neighbouring Vietnam. A shallow bowl (Figs. 3.11a and 3.11b) was found in Lao Cai in the upper Red River valley in a burial complex of a first century BCE to first century CE date.[50] Several such find groups were discovered there during construction work, and were brought to the local museum, but their archaeological context had already been destroyed. Judging from the wealth of these find groups, in particular, the presence of nineteen Heger I bronze drums and other prestige objects, these must have been burials of a locally-ruling élite. The group which contained the glass dish was particularly rich, and

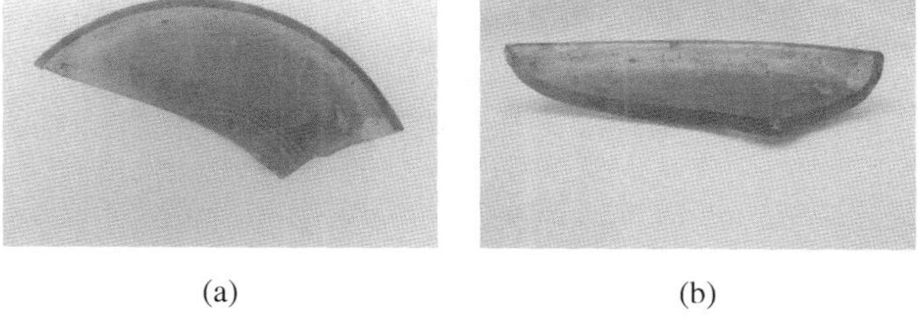

(a) (b)

Fig. 3.11 From Lao Cai, Lao Cai province, Vietnam. Fragment of a mould-formed shallow glass dish. 1st century BCE to 1st century CE. Estimated rim diameter 13.5 cm. Photos: Brigitte Borell might represent either an exceptionally wealthy élite burial or, alternatively, more than one burial of high status. The other one is a cup which is still unpublished. It was found near a brick tomb in Bac Ninh Province near Hanoi, but without a clearly datable context. According to information received by Yuko Hirano (pers.comm. Dec. 2009), it is made of a translucent glass with a greenish tint, and is similar in type to the cup from Wenchangta, Hepu (Fig. 3.9).

Recently, a cup of this type made of dark blue glass has been found in a Han period burial in Hunan (Q. H. Li, pers.comm. Nov. 2012). In addition, eight more vessels may be assigned to this group: a cup formerly in a private collection in San Francisco (Fig. 3.12), a cup and a shallow bowl in the Musée Guimet, Paris,[9] a cup in the Freer Gallery of Art, Washington, and two cups and two saucers in a private collection in Hong Kong (the latter five examples still unpublished).

When, in the 1950s, the first of these glass vessels were excavated from Eastern Han tombs in the Guixian area, they came as a great surprise. At that time, glass vessels found in China in contexts of the early periods were held to be imports from the West, and the glass vessels from Guangxi came to be regarded as imported via the southern seas. However, there is no closely corresponding shape to be found in the well-known inventory of Mediterranean or Western Asiatic glass vessels.[51] In the 1980s, Shi Meiguang and his colleagues identified the glass of two of the Guangxi vessels as well as that of other objects of characteristic Chinese shape found in Guangxi as a potash glass.[52] This is the same result as that produced by Robert Brill for a cup formerly in the Shorenstein collection.[53] The view that

Fig. 3.12 Formerly in the Shorenstein collection. Mould-formed glass cup. Rim diameter 7.7 cm. Photo: The Corning Museum of Glass

these potash glass objects might have been locally made in southern China opened up new approaches for research.

Since then, many more chemical analyses of glasses found in Asia have been carried out, and potash glass is now well established as a compositional family of Asian glass known from finds in South Asia, Southeast Asia, and East Asia. Several subgroups of Asian potash glasses are now recognised. The cups and bowls discussed here (Table 3.2 analyses 1–9) belong to the subgroup with low lime, which has a

Table 3.2 Chemical analyses (wt%) of glass vessels dating from the Han period (206 BCE–220 CE) found in tombs in Guangxi, Guangdong, and Henan

No.	1	2	3	4	5	6	7	8	9	10	11
Date	Han	W.Han	W.Han	W.Han	W.Han	W.Han	W.Han	E.Han	E.Han	W.Han	W.Han
SiO_2	76.78	79.69	73.69	74.62	78.29	76.28	74.94	77.70	xxx	xx	82.12
Na_2O	0.68		1.58	1.56		0.27	0.16				0.66
CaO	0.94	0.41		0.68	0.12	0.54	0.03		xxx	xx	1.91
K_2O	14.94	16.22	16.53	16.01	17.28	15.43	15.99	16.80	xx	xx	12.18
MgO	0.37	0.01	0.66	0.41		0.47	0.15				0.41
Al_2O_3	3.41	2.14	5.68	5.36	1.99	3.28	4.16	3.17			2.14
Fe_2O_3	0.92	1.36	0.73	0.71	0.56	0.47	0.60	0.78	x	x	0.34
TiO_2	0.27	0.22									0.15
Sb_2O_5	0.11										
MnO	0.15		0.69	0.64			0.01				0.05
CuO	0.01		0.25		1.67	0.01	1.24	1.62		x	0.00
SnO_2	0.09										
PbO	0.40									x	
P_2O	0.12										
BaO									x	x	0.02

Notes on samples:

Samples from Guangxi:
No. 1: Formerly Shorenstein Coll[53]; **No. 2**: Wenchangta T.70; No. **3:** Honglingtou T.11: 22; **No. 4**: Honglingtou T.11: 21; **No. 5:** Huangnigang T.1; **No. 6**: Fengliuling T.2: stemmed cup; **No. 7:** Fengliuling T.2: cup; **No. 8:** Guixian, Qiche Road T.5; **No. 9:** Guixian, Railway Station (analyses nos.2-9[60,61]).

Sample from Guangdong:
No. 10: Hengzhigang T.2061.[9,51]

Sample from Henan:
No. 11: Chenpeng T.68.[63]

distinct distribution in northern Vietnam and southwestern China, in particular, in Guangxi.[54–57]

The supposition of regional primary production of the glass, as well as of the manufacture of the glass vessels, finds support in a text passage in the *Baopuzi neipian* compiled by Ge Hong towards the end of the Western Jin period (265–316 CE): "The 'crystal' bowls, which are made in foreign countries, are in fact prepared by compounding five sorts of [mineral] ashes. Today, among the people of Jiao and Guang many obtained this method and make them". Most likely, this reflects the production of the potash glass vessels of our group, as has been pointed out already by others. The two regions "Jiao" and "Guang" comprise the three former Han period commanderies in present-day Vietnam, and the four in present-day Guangxi and Guangdong.[58]

With good reason, therefore, the potash glass, as well as the finished glass vessels, may be considered to be regional products. Possibly, a number of workshops were involved in the manufacture of the glass vessels. Where these were located is still unknown. A possible candidate might be the area of Hepu. A likely alternative might be the area of the commandery of Jiaozhi, comprising the delta area of the Red River and part of its valley. According to the late Western Han census of the years 1–2 CE, Jiaozhi was by far the most densely populated commandery in the region. Hopefully, future excavations might yield evidence for such glass workshops.

Noteworthy is the large percentage among the surviving vessels of at least about ten or eleven examples made from almost clear glass, otherwise rarely used (in contrast to the strongly coloured blue or greenish-blue glasses). These glass cups and bowls made of translucent glass of natural aqua colour, with a bluish or greenish tint, might have been intended to imitate the foreign 'crystal' bowls mentioned by Ge Hong.

Of interest is also the term *zhu* used in the text to describe the making of the glass bowls. It is taken from the vocabulary of foundry work for casting metal in moulds. The glass cups were clearly made with the help of moulds. It is also interesting to note that vessels of almost identical shape have closely similar measurements. One series of cups has diameters of 7.7 cm, another of 9.2 cm; the saucers have flat bases of 9 cm in diameter. This might point to a prototype, from which the moulds could repeatedly be formed. The moulds themselves would probably perish in the process of manufacturing the glass vessels. One possible method would be chip-casting, a method by which small glass chips are filled in the hollow space between the moulds, then put in a kiln to be melted.[59] Alternatively, the pressing of hot glass in a mould has been suggested for the Gulf of Tonking cups.[60–62] A certain amount of hot glass — in a state of viscosity comparable to that of honey — would be pressed into an outer mould with the help of an inner mould or plunger. Techniques of mould-pressing hot glass appear to have been widely applied in Chinese glass-working, for instance to make *bi* disks and other ornaments. Such a process of hot-working the glass would also account for some irregularities and distortions which could not be explained by a manufacture with the method of chip-casting. Such irregularities are particularly noticeable on the trays with flaring walls of uneven height.[62]

A small glass bowl of a different type was found in a late Western Han tomb at Chenpeng in Henan.[9,63] The bowl with a rim diameter of 6.2 cm has convex walls, a flat base, and two horizontal grooves cut on the exterior below the rim. It is made of a clear glass with a slightly yellowish tinge. The glass has been analysed, and proved to be a potash glass, but with slightly higher lime content than the typical Guangxi glass (Table 3.2 analysis 11). The small bowl is certainly the product of a different workshop. However, as the excavation report suggests, it might possibly have been made somewhere in the southern regions, and carried northwards to Nanyang.

This present state of knowledge suggests a reconsideration of the three mould-made bowls found in the mid-Western Han tomb 2061 in Guangzhou dated to the early first century BCE.[8,9,64] Their shape, with convex walls, has some resemblance to the bowl from Chenpeng. In size, they are larger, with diameters of about 10.6 cm, and they are made of a dark blue glass. The glass of one of the bowls has been analysed by X-ray fluorescence (Table 3.2 analysis 10). However, even though this analysis remains inconclusive, it does not exclude a potash glass.[9,58] In that case, the three glass bowls from this tomb might be regarded as forerunners of the later production of potash glass vessels. Future investigations might provide more insight into this matter.

To return to the question about the 'social life' of these locally produced glass vessels, it seems that they were available to the upper social strata which included local élites and high officials, but possibly also other social groups of a certain wealth. This would be quite understandable, considering their proposed local production. A more detailed picture is to be expected after full publication of the tomb contexts. The few data available on the placement of the glass vessels within the tombs leads to the conclusion that the glass vessels were certainly considered to be very valuable objects, and were held in high esteem, even though they were not such exotic items as the glass vessels brought from the Mediterranean region and western Asia. In any case, the number of surviving examples of these potash glass vessels is surprisingly high, in view of the apparently limited time-span of their production. In addition, glass vessels of this Gulf of Tonking group had a wider geographical distribution. The glass cup recently found in the north, in Hunan, has already been mentioned.

The rim fragment of a glass cup strikingly similar to the Guangxi cups was found in Arikamedu, on the east coast of South India, about 120 km south of Kanchipuram, which is often identified with Huangzhi mentioned in the *Han shu*. The fragment was found during Mortimer Wheeler's excavation in 1945. Unfortunately, the present location of the fragment is unknown: it is possibly lost. It is described as being made of bluish-green glass with three horizontal ribs. Although classified in the initial publication as a Roman glass, it is in shape, appearance, and measurements very close to the cup series from Guangxi with diameters of 7.7 cm.[9,51] It was found in a layer dated to the first century BCE, possibly extending into the early first century CE. This is also in good agreement with the date of the glass vessels from the Tonking Gulf region. The cup fragment would have reached Arikamedu along the maritime routes, together with other luxury goods like Chinese silks.

As has already been mentioned, according to the *Han shu*, such a voyage from the Gulf of Tonking to Huangzhi in India included a land crossing commonly thought to refer to the Isthmus of Kra region. Recently, several fragments of glass vessels of our group have been identified from two sites south of the Kra Isthmus. Most of them come from Ta Chana on the east coast, and one fragment from Bang Kluay on the west coast. Although the fragments are small, their shapes, with the three horizontal ribs, closely parallel the Guangxi cups. The composition of their glass has been analysed by James Lankton, it is a low-lime potash glass like that of the Guangxi vessels.[65] The fragments made mainly of blue, green, and clear glass point to at least eight individual cups of the type known from the Guangxi tombs. There has been as yet no archaeological excavations of these two sites, but they are considered to be way stations and trading stations connected to the ends of transpeninsular routes. In addition, at Ta Chana, some specialised industrial activities, like the making of beads in glass and semi-precious stone, appear to have taken place, as indicated by the waste material discovered. The context in which the vessel glass fragments survived might have been that of glass-working, where they were melted down to be used for making beads, for example. A vessel fragment distorted by heat and apparently manipulated with a tool seems to support such an interpretation.[50]

3.6 Conclusion

As is not uncommon in archaeology, the tentative conclusions offered here can not be other than provisional in nature. Full publication of the tomb contexts and new finds may add to the picture, or even answer some of the questions posed. Possibly, more questions have been raised than answers could be given.

However, in summarising the different aspects of the role of glass vessels in Chinese contexts, their status as precious objects remains beyond doubt. The Han period glass vessels made from Chinese lead-barium glass are still very rare. Although remains of nineteen glass vessels survived, they come from only two tomb contexts of the second century BCE. Their owners were persons of the highest social rank. It has been pointed out that these glass vessels were meant to simulate jade. This is also the case for numerous other types of funerary glass objects, which were usually crafted from jade, but were also made of lead-barium glass. Even though the glass was used as a substitute for jade, and the manufacture of such glass objects with the help of moulds could be done in a relatively short time compared to the labourious carving of the stone, it would be erroneous to think of glass as a commonplace material in this period. Quite the contrary: although less highly prized than jade in the hierarchy of materials, glass should still be regarded as a precious material.

Documented finds of glass vessels imported from the Mediterranean region are also very rare in the Han period. They could have reached China via the overland routes of the Silk Road. However, written sources and archaeological evidence attest to the glass vessels being one of the commodities transported along the maritime routes from the Red Sea ports to India and beyond. It seems likely that the glass vessels found in tombs of the Eastern Han period in the area of Yangzhou came along the maritime routes. The ribbed bowl formed over a mould and made of mosaic glass imitating veined natural stone, as well as the fragile blown glass cup made of thin-walled transparent blue glass were certainly exotic and highly treasured objects in the possession of persons of the highest social class. The situation is similar in the Eastern Jin tombs of the fourth century CE, where finds of imported glass vessels of Mediterranean and Middle Eastern origin are limited mainly to tombs of the imperial family, or of high-ranking members of the aristocracy closely connected to the court. Particularly notable is the number of glass vessels made of clear glass decorated with facet cutting. These transparent glass vessels and the play of light in the cut decoration might have been a special attraction.

In contrast, the picture is different for the Tonking Gulf group of mould-formed glass vessels made of potash glass. They occur predominantly in Guangxi and northern Vietnam, and probably originated in the area of the Gulf of Tonking. They seem to have been available not only to high-ranking officials and local élites, but also to a wider circle of persons of the upper social level. As far as can be deduced at present, they were produced for a certain time during the later Western and Eastern Han periods, then their production seems to have been discontinued. They survived in surprisingly large numbers: to date, thirty-six individual examples are known, and it appears that they were produced in quantities sufficient even to be exported.

Acknowledgments

For fruitful discussions, generous assistance in obtaining photographs, and other help — not least with translations of Chinese texts — the author wishes to thank An Jiayao (Beijing), Robert Brill (Corning), Boonyarit Chaisuwan (Muang Sing), Bernard Gratuze (Orleans), Sabine Hieronymus (Heidelberg), Huang Qishan (Nanning), Annette Kieser (Marburg), James Lankton (Seoul), the Lao Cai Provincial Museum (Lao Cai), Li Qinghui (Shanghai), the Museum of the Guangxi Zhuang Autonomous Region (Nanning), the Nanjing Municipal Museum (Nanjing), Pham Minh Huyen (Hanoi), Bunchar Pongpanich (Bangkok), Marianne Stern (Hilversum), Wang Zhigao (Nanjing), Xiong Zhaoming (Nanning).

References

1. A. Appadurai, Introduction: commodities and the politics of value, in *The Social Life of Things. Commodities in Cultural Perspective*, ed. A. Appadurai (Cambridge University Press, Cambridge, 1986), pp. 3–63.
2. E.M. Stern, Early Roman export glass in India, in *Rome and India. The Ancient Sea Trade*, ed. V. Begley, and R.D. De Puma (University of Wisconsin, Madison, 1991), pp. 113–124.
3. P.J. Cherian, V. Selvakumar, and K.P. Shajan, The Muziris heritage project: excavations at Pattanam — 2007, *Journal of Indian Ocean Archaeology* **4**, 1–10 (2007).
4. M. Pirazzoli-t'Serstevens, Les laques chinois de Begram, *Topoi* **11(2)** 473–484 (2001).
5. S. Luo, The date and nature of the Begram treasure, *Kaogu* **2**, 68–80 (2011).
6. L.R. Zhang, Chinese lacquerwares from Begram: date and provenance, *International Journal of Asian Studies* **8**, 1–24 (2011) .
7. R. Mairs, Glassware from Roman Egypt at Begram (Afghanistan) and the Red Sea trade, *British Museum Studies in Ancient Egypt and Sudan* **18**, 61–74 (2012). http://www.britishmuseum.org/research/publications/online_journals/bmsaes/issue_18/mairs.aspx

8. J.Y. An, The art of glass along the Silk Road, in *China — Dawn of a Golden Age 200–750 AD*, ed. J.C.Y. Watt (Metropolitan Museum of Art, New York, 2004) pp. 57–65.
9. B. Borell, Trade and glass vessels along the maritime Silk Road, in *Glass along the Silk Road, from 200 BC to AD 1000*, ed. B. Zorn and A. Hilgner (Römisch-Germanisches Zentralmuseum Mainz, Mainz., 2010), pp. 127–142.
10. G.D. Weinberg and E.M. Stern, *Vessel Glass*. The Athenian Agora vol. 34 (American School of Classical Studies at Athens, Princeton, 2009), pp. 36–38; 94–96.
11. D. Williams, Crystalline Matter. *British Museum Magazine* **48**, 47 (2004).
12. A. Trussaud, and M. Vickers, Ancient murrhine ware and its glass evocations, *Journal of Glass Studies* **49**, 143–152 (2007).
13. Yangzhou bowuguan, Jiangsu Hanjiangxian Ganquanxiang Laohudun Han mu (Han tomb at Laohudun in Ganquanxiang of the county of Hanjiang, Jiangsu), *Wenwu* **10**, 62–70 (1991).
14. C. Isings, *Roman Glass from Dated Finds* (J.B.Wolter, Groningen, 1957).
15. D. Foy and M.-D. Nenna, *Tout feu tout sable. Mille ans de verre antique dans le Midi de la France* (Musées de Marseille, Aix-en-Provence, 2001), pp. 26–27 fig. 9; p. 107 no. 131.5.
16. D. Foy and M.-D. Nenna, Productions et importations de verre antique dans la vallée du Rhône et le Midi méditerranéen de la France (Ier-IIIe siècles), in *Échanges et commerce du verre dans le monde antique*, ed. D. Foy and M.-D. Nenna (Mergoil, Montagnac, 2003), pp. 227–296, esp. p. 242 fig. 53–54.
17. V. Arveiller-Dulong and M.-D. Nenna, *Les verres antiques. Vol.2. Vaisselle et contenants du Ier siècle au début du VIIe siècle après J.-C.* (Musée du Louvre, Paris, 2005), nos. 11, 14–16.
18. B. Chaisuwan, Early contacts between India and the Andaman Coast in Thailand from the second century BCE to eleventh century CE, in *Interactions between South and Southeast Asia*, ed. P.-Y. Manguin, A. Mani, and G. Wade (Institute of Southeast Asian Studies, Singapore, 2011), pp. 83–112.
19. D. Foy, Une production des bols moulés à Beyrouth à la fin de l'époque hellénistique et le commerce de ces verres en Méditerranée occidentale, *Journal of Glass Studies* **47**, 11–55 (2005).
20. F.D. Grose, *The Toledo Museum of Art: Early Ancient Glass. Core-formed, Rod-formed and Cast Vessels and Objects from the Late Bronze Age to the Early Roman Empire, 1600 B.C. to A.D. 50*, (Hudson Hills Press, New York, 1989), pp. 189–191; 256–261.
21. S.M. Goldstein, *Pre-Roman and Early Roman Glass in the Corning Museum of Glass* (The Corning Museum of Glass, Corning, N.Y., 1979), no. 298.
22. Z.G. Wang, Liu Chao mu zang chutu boli rongqi mantan/On the glass containers unearthed from the Six Dynasties tombs, *Nanjing Bowuguan* **12**, 55–63 (2012).

23. Nanjing shi bowuguan, Nanjing Xiangshan 5 hao 6 hao 7 hao mu qingli jianbao/A brief report on the excavation of tombs nos. 5, 6 and 7 at Xiangshan, Nanjing, *Wenwu* **11**, 23–41 (1972).
24. J.Y. An, Glass vessels and ornaments of the Wei, Jin and Northern and Southern Dynasties periods, In *Chinese Glass*, ed. C. Braghin (Leo S. Olschki, Florence, 2002), pp. 45–70.
25. C.W. Clairmont, *The Glass Vessels*. The Excavations at Dura-Europos Final Report IV, Part V, (Dura-Europos Publications, New Haven, 1963), pp. 70–72, group f.
26. British Museum Collection Database. "1966,1212.1" www.britishmuseum.org/collection, British Museum. Online. Accessed 18/04/2013
27. R.H. Brill, Chemical analyses of some Sasanian glasses from Iraq, in: *Sasanian and Post-Sasanian Glass*, ed. D. Whitehouse (The Corning Museum of Glass, Corning, N.Y., 2005), pp. 65–88.
28. I. C. Freestone, Glass production in Late Antiquity and the Early Islamic period: a geochemical perspective, in *Geomaterials in Cultural Heritage*, ed. M.Maggetti and B.Messiga (The Geological Society, London, 2006), pp. 201–216.
29. P. Mirti, M. Pace, M.M. Negro Ponzi, and M. Aceto, ICP-MS analysis of glass frgaments of Parthian and Sasanian epoch from Seleucia and Veh Ardasir (Central Iraq), *Archaeometry* **50**(3), 429–450 (2008) .
30. P. Mirti, M. Pace, M. Malandrino, and M.M. Negro Ponzi, Sasanian glass vessels from Veh Ardasir: new evidence by ICP-MS analysis, *Journal of Archaeological Science* **36**, 1061–1069 (2009).
31. A. Kieser, Northern influence in tombs in southern China after 317 CE? — A re-evaluation, In *Han Tang zhi jian wenhua yishu de hudong yu jiaorong/Between Han and Tang. Cultural and artistic interaction in a transformative period*, ed. Wu Hung (Wenwu chubanshe, Beijing, 2001), pp. 231–272, esp. p. 253.
32. A. Kieser, „Laid to rest there among the mountains he loved so well"? In search of Wang Xizhi's tomb, *Early Medieval China* **17**, 74–94 (2011).
33. J.-P. Brun, Le verre dans le Désert Oriental d'Ègypte: contextes datés, In *Échanges et commerce du verre dans le monde antique*, ed. D. Foy and M.-.D. Nenna (Mergoil, Montagnac, 2003), pp. 377–387.
34. Nanjing shi bowuguan, Nanjing beijiao Dong Jin mu fajue jianbao/A brief report on the excavation of an Eastern Jin tomb in the northern suburb of Nanjing, *Kaogu* **4**, 315–322 (1983).
35. Nanjing shi bowuguan, Jiangsu Nanjing Xianheguan Dong Jin mu/The Eastern Jin tombs at the Xianhe temple in the outskirts of Nanjing, *Wenwu* **3**, 4–40 (2001).
36. Nanjing Municipal Museum, The Eastern Jin tombs at Xianheguan in Nanjing, Jiangsu, *Chinese Archaeology* **2**, 242–250 (2002).
37. E.M. Stern, *Roman, Byzantine, and Early Medieval Glass: 10 BCE — 700 CE* (Hatje, Ostfildern, 2001), esp. no. 57.

38. Nanjing shi bowuguan *et al.*, Jiangsu Nanjing shi Fugui shan Liu Chao mudi fajue jianbao/Preliminary report of the excavation of a Six Dynasties cemetery at Fugui Hill in Nanjing city, Jiangsu, *Kaogu* **8**, 27–47 (1998).
39. Nanjing daxue *et al.*, *Echeng Liu Chao mu/Tombs of the Six Dynasties at Echeng* (Ke xue chubanshe, Beijing, 2007), pp. 303–305 fig. 219,3 pl. 15.14.
40. Q.H. Li, The tables of chemical analysis of Chinese ancient glass artifacts, in *Zhongguo gu dai boli ji shu de fa zhan/Development of ancient Chinese glass*, (Shanghai ke xue ji shu chubanshe, Shanghai, 2005), pp. 281–326, esp. 296 analysis 199.
41. F.X. Gan, The Silk Road and ancient Chinese glass, in *Ancient Glass Research along the Silk Road*, ed. F.X. Gan, R.H. Brill, S.Y. Tian (World Scientific, Singapore, 2009), pp. 41–108.
42. Nanjing shi bowuguan, Nanjing shi Guojiashan Dong Jin Wen shi jia zu mu/ Wen kin tombs of the Eastern Jin period on Guojia Hill in Nanjing City, *Kaogu* **6**, 3–25 (2008).
43. A. Kieser, New insight on émigré tombs of the Eastern Jin in Jiankang, in *The Yields of Transition: Literature, Art and Philosophy in Early Medieval China*, ed. J.A. Rošker and N.V. Suhadolnik (Cambridge Scholars Publishing, Newcastle upon Tyne, 2011), pp. 53–73.
44. B.D. Yang (ed.), *Zhongguo jin yin boli falang qi quan ji*, vol.4: *Boliqi* (Hebei meishu chubanshe, Shijiazhuang, 2004), pl. 46 and 47.
45. M.G. Shi, Y.D. Li, and F.Z. Zhou, Some new glass finds in China, *Journal of Glass Studies* **34**, 23–26 (1992).
46. J.C.S. Lin (ed.), *The Search for Immortality. Tomb Treasures of Han China* (Fitzwilliam Museum, Cambridge, 2012), pp. 49–57; pp. 168–169 no. 59.
47. C. Braghin, Polychrome and monochrome glass of the Warring States and Han periods, In *Chinese Glass*, ed. C.Braghin (Leo S. Olschki, Florence, 2002), pp. 3–43, esp. p. 32 footnote 105.
48. Q.S. Huang, Guangxi Han dai boli yu hai shang si chou zhi lu, in *Hai shang si chou zhi lu yan jiu*, ed. C.J. Wu and Y.S. Lü (Ke xue chu ban she, Beijing, 2006), pp. 154–163.
49. Q.S. Huang, Guangxi Han dai boli y Yue ren hai shang si chou zhi lu de tan suo, In *Ou Luo yi cui: Guangxi Baiyue wen hua wen wu jing pin ji*, ed. Zhongguo guo jia bo wu guan, Guangxi Zhuangzu Zizhiqu bo wu guan bian ji (Zhongguo she hui ke xue chu ban she, Beijing, 2006), pp. 268–281.
50. B. Borell, The Han period glass dish from Lao Cai, northern Vietnam, *Journal of Indo-Pacific Archaeology* **32**, 70–77 (2012).
51. B. Borell, The glass vessels from Guangxi and the maritime Silk Road in the Han Period (206 BCE to 220 CE), in *Unearthing Southeast Asia's Past. Selected Papers from the 12th International Conference of the European Association of Southeast Asian Archaeologists* vol.1, ed. M.J. Klokke and V. Degroot (National University of Singapore, Singapore, 2013), pp. 141–154.
52. M.G. Shi, O.L. He, and F.Z. Zhou, Investigation on some Chinese potash glasses excavated in Han Dynasty tombs, in *Archaeometry of Glass. Proceedings of the*

Archaeometry Session of the XIV International Congress on Glass 1986 New Delhi India, ed. H.C. Bhardwaj (Indian Ceramic Society, Calcutta, 1987), pp. 15–20.

53. R.H. Brill, *Chemical Analyses of Early Glasses*, 2 vols. (The Corning Museum of Glass, Corning, N.Y., 1999), vol. 2, p. 358 no. 5770.
54. F.X. Gan, H.S. Cheng, Q.H. Li, Origin of Chinese ancient glasses — study on the earliest Chinese ancient glasses, *Science in China Series E: Technological Sciences* **49**(6), 701–713 (2006) .
55. F.X. Gan, H.S. Cheng, Q.H. Li, Zhongguo gu dai boli de qi yuan — Zhongguo zui zao de gu dai boli yan jiu, *Zhongguo Kexue (China Science)* **37**(3), 382–391 (2007).
56. H.X. Zhao, Q.H. Li, F.X.Gan, and H.C. Cheng, Guangxi Hepu diqu chutu Han dai gu boli de zhizi ji fa X-yingguang fenxi, *He Ji Shu/Nuclear Techniques* 30(1), 27–33 (2007) .
57. Q.H. Li, W.Z. Wang, Z.M. Xiong, F.X. Gan, and H.S. Cheng, PIXE study on the ancient glasses of the Han Dynasty unearthed from Hepu county, Guangxi, In *Ancient Glass Research along the Silk Road*, eds. F.X. Gan, R.H. Brill, and S.Y. Tian (World Scientific, Singapore, 2009), pp. 397–412.
58. B. Borell, Han Period glass vessels in the early Tongking Gulf region, In *The Tongking Gulf Through History*, eds. N. Cooke, T.N. Li, and J.A. Anderson (University of Pennsylvania Press, Philadelphia, 2011), pp. 53–66.
59. E.M. Stern, and B. Schlick-Nolte, *Early Glass of the Ancient World 1600 B.C. – 50 A.D.* (Hatje, Stuttgart, 1994), pp.49–53.
60. J.Y. An, *Early Chinese Glassware*, translation by M. Henderson of J.Y. An Jiayao, *Kaogu Xuebao* **4**: 413–448 (1984) (Oriental Ceramic Society Translations No.12, 1987).
61. Q.S. Huang, Zhongguo nanfang he xinan de gudai boli jishu/Ancient glass technology in the South and South-West of China, In *Zhongguo gudai boli jishu de fazhan/Development of Chinese ancient glass*, ed. F.X. Gan (Shanghai kexue jishu chubanshe, Shanghai, 2005), pp. 182–199.
62. B. Borell, Han period glass vessels from the Gulf of Tonking region: aspects of their technology, In *Annales du 18e Congrès de l'Association Internationale pour l'Histoire du Verre, Thessaloniki September 2009*, ed. D. Ignatiadou and A. Antonaras (ZITI Publishing, Thessaloniki, 2012), pp. 491–496.
63. Nanyang Municipal Institute of Cultural Relics and Archaeology (Henan Province), Henan Nanyang shi Chenpeng cun 68 hao Han mu/Tomb no. 68 of the Han period at Chenpeng village in Nanyang city, Henan, *Kaogu* **10**, 33–39 (2008).
64. Guangzhou shi bowuguan *et al.*, *Guangzhou Han mu/Excavation of the Han tombs at Guangzhou* (Wenwu chubanshe, Beijing, 1981), pp. 238–239 fig. 143,4 pl. 67,1.
65. L. Lankton, B. Pongpanich, B. Chaisuwan, and B. Gratuze, Chinese Han period glass cup fragments in peninsular Thailand, paper presented at the 19th Indo-Pacific Prehistory Association Congress, Hanoi, Vietnam, November 29–December 5, 2009.

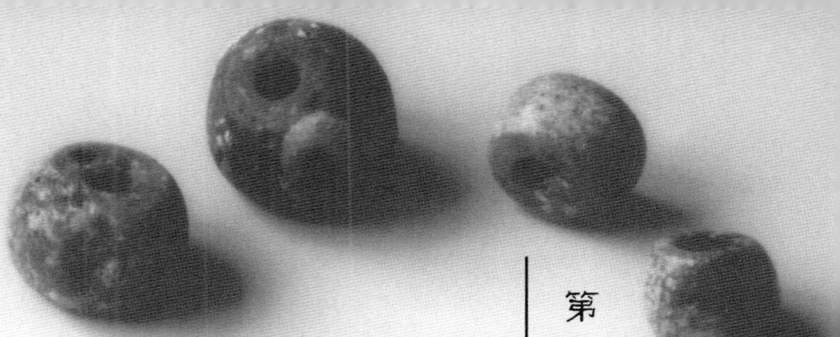

Chapter 4

第四章

Chemical Composition of Glass Beads Excavated from Kofun (ca. AD 2nd to 7th c.) in Western Japan by Portable XRF Showing Glass Trade among Asian Countries

Izumi Nakai and Junko Shirataki
Faculty of Science, Tokyo University of Science,
Kagurazaka, Shinjuku, Tokyo 162–8601, Japan

4.1 Introduction

It is known that the use of glass in Japan began at the end of 3rd century BC.[1–3] Most of the glass were made into beads used for accessories and their occurrence as grave goods varied in Kofun in burial mounds or tombs. On the other hand, local domestic production of glass from natural resources must wait till 7th century AD, when the production of lead silicate glass began in Asukaike, Nara prefecture, Japan. Recently, Kumamoto suggested that lead isotope analysis of yellowish green glass beads excavated from Odara Kofun (mid 5th century AD), showed that lead from South East Asia was used.[4] The chemical compositions of these glasses have not been published yet. The author (I.N.) has been studying the glass in West Asia and trade between West Asia and South East Asia including Japan.[5,6] We have been therefore interested in knowing the origin of glass beads at the begining of the first millennium AD in Japan. Archaeological study and our preliminary analytical study on the Kumamoto's glass disclosed that the glass beads excavated from Kofun were well preserved and were made in a rich variety of glass types.[7,8] In this study, we have analyzed the glass beads excavated from 10 diffrent Kofun in Kumamoto and Okayama prefectures. We know that all of the glass beads analyzed in this study were imported from somewhere outside of Japan. The object of the present study is to reveal the origin of the glass beads distributed across Japan, especially in Kumamoto and Okayama areas dated from between 2nd to 7th century AD. Another purpose is to study colorants of the glass beads.

T. Koezuka *et al.* are pioneers of the archaeometric study of early glass in Japan. They classified the Japanese glass in Yayoi and Kofun periods into 6 types according to their chemical compositions as shown in Fig. 4.1.[1–3] They analyzed the glass based on destructive analysis techniques such as using an X-ray microprobe. So, the number of the analyzed glass is limited and a large number of glass in Japan remain without scientific analysis. Since 2003, the present authors have developed a portable XRF suitable for nondestructive analysis of glass objects.[9–12] This instrument allows us to analyze light elements such as Na, which is a major component of glass, and Mg, a key element in revealing alkali sources in glass: i.e., natron or plant ash. Thus, we have started a comprehensive analysis of ancient glass excavated all over Japan using our portable XRF. The glass objects

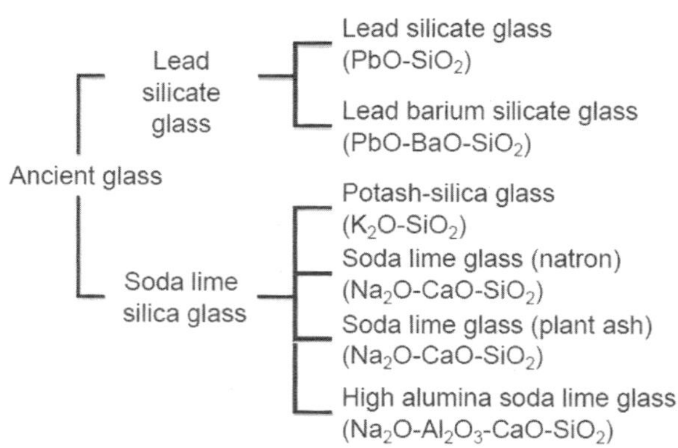

Fig. 4.1 Glass types in ancient Japan

analyzed in this study have never been analyzed before. The present paper will also show how much information can be obtained from truly nondestructive quantitative analysis of glass. This is very important point in the analysis of cultural heritage, which often remains unanalyzed because any sampling or polishing is strictly prohibited.

4.2 Experimental

4.2.1 *Samples*

Table 4.1 A list of analyzed samples

Site	Date	Number of samples
Kumamoto		
Odara Kofun	mid 5th c. A.D.	20
Etafuneyama Kofun	late 5th–early 6th c. A.D.	11
Kunigoshi Kofun	late 5th or early 6th c. A.D.	16
Kitaoka yokoanagun	late 6th c. A.D.	18
Daibo Kofun	early 6th c. A.D.	7
Umade Kofun	mid–6th c. A.D.	21
Syoji Kofun	6th c. A.D	4
Osugi	2nd–3rd c. A.D	8
Okayama		
Tsukadan Kofun	mid 6th–early 7th c. A.D.	15
	early 6th–late 6th c. A.D.	11

Table 4.1 shows a list of samples quantitatively analyzed in this study. A total of 131 glass beads excavated from 10 different Kofuns in Kumamoto and Okayama prefecture, Japan have been analyzed. These Kofuns were built from the 2nd to the 7th century AD. The samples were subjected to the nondestructive XRF analysis and some were analyzed by μ-Raman spectroscopy.

4.2.2 *XRF analysis*

A portable XRF spectrometer, OURSTEX 100FA II-L (OURSTEX) was brought to the museum facilities in Kumamoto and Okayama.

The XRF instrument was specially designed for glass analysis. The analytical procedures and conditions are described elsewhere.[5,6] The measurement time was 200 sec (live time) and tube current (mA) was adjusted so that the dead time does not exceed 30%. Standard glass samples (NIST SRM610, 612, 621, 1412, 1830, 1831) and 15 synthetic glasses that had been analyzed using inductively coupled plasma-atomic emission spectrometry (ICP-AES) by the authors' research group, were used to generate calibration curves for quantitative analyses. XRF spectra for the light elements from Na to Si were measured using white X-ray mode with an X-ray tube voltage of 6 kV and beam current of 1.5 mA. For heavier elements from Ti to Zr, monochromatic X-ray mode operated at 40 kV and 1.0 mA was used. For the analysis of Ca, K, Sn and Sb, white X-ray mode was used at 40 kV 0.25 mA. The measured net XRF intensity was normalized to that of the Rayleigh (elastic) scattering peak of Pd Lα (2.84 keV) line for the 6 kV data and the Compton scattering peak of Pd Kα (21.125 keV) line for the 40 kV data.

A multivariate statistical analysis using StatPartner ver. 2.0 was carried out to characterize the chemical composition of the glass samples and to distinguish between glasses with similar compositions. Principal component analysis (PCA) was applied so as to classify the analyzed glass. The following 10 elements were used for the diagnostics: Mg, Al, K, Ca, Ti, Mn, Fe, Cu, Sr, and Pb. For cluster analyses of the samples, Ward's method was applied and the following elements were used as index elements: Mg, Al, K, Ca, Ti, Fe, Rb, and Zr. The dendrogram was obtained using the Ward's method.

4.2.3 *Structure analysis*

Raman spectra were measured using MiniRam, B&W TEK Inc. (U.S.A.). The instrument has a light weight (2 kg) and is suitable for *in situ* analysis with a compact lithium battery; the instrument is also capable of μ-Raman analysis using microscopic video system (weight: 2 kg). The following measurement conditions have been applied: laser wavelength of 785 ±1 nm, laser power of maximum 300 mW, laser size of 90 μm in diameter with 20× objective lens (microprobe mode). The laser power of the instrument was limited to less than 30% to avoid a thermal damage of the artifacts. The instrument has a 2048 pixel CCD detector. The measurement region is 100 to 3150 cm^{-1} (Raman shift), and the resolution is about 10 cm^{-1} FWHM.

Gandolfi camera was used to obtain powder diffraction data of a small crystalline sample. The diffraction pattern was measured with Cu Kα X-ray (Ni filter) and recorded on an Imaging Plate.

4.3 Results

4.3.1 *Chemical compositional classification*

Figure 4.2 shows the result of the cluster analysis showing a dendrogram of the glass samples based on the chemical compositions of the analyzed glass. The glasses clearly separate into 4 compositional types: high alumina potash-silica glass K_2O-Al_2O_3-SiO_2, high alumina soda glass (Na_2O-Al_2O_3-CaO-SiO_2), potash-silica glass (K_2O-SiO_2) and soda lime glass (Na_2O-CaO-SiO_2). Lead barium silicate glass (PbO-BaO-SiO_2) is

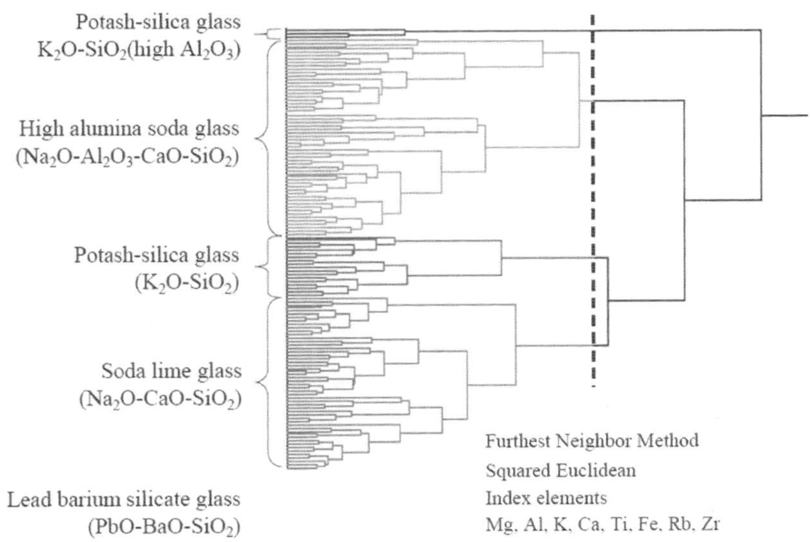

Fig. 4.2 Cluster analysis: dendrogram of the glass samples obtained on the basis of chemical composition

the only glass which contained high level of Pb, and is obviously of a different type from the others. Therefore, the data were not included into the cluster analysis. These results are consistent with classification of Japanese ancient glass proposed by Koezuka *et al* shown in Fig. 4.1.

Table 4.2 Average bulk composition in wt% for the different types of glass

Type	Number	Na_2O	MgO	Al_2O_3	SiO_2	K_2O	CaO
Na_2O-CaO-SiO_2	52	4.27	1.43	4.06	71.05	2.57	6.13
Na_2O-Al_2O_3-CaO-SiO_2	57	8.26	0.37	9.54	67.34	2.74	2.92
K_2O-SiO_2	18	1.43	0.27	3.40	71.74	8.79	1.24
K_2O-SiO_2 (high Al_2O_3)	3	1.48	0.36	6.74	72.19	11.53	0.23

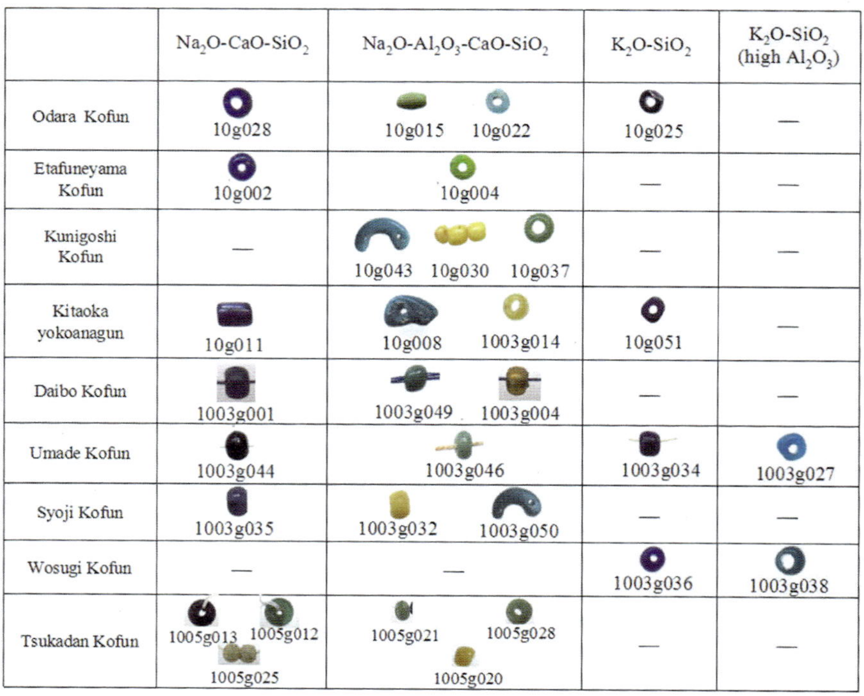

Fig. 4.3 Photographs of the representative samples with their compositional types and name of Kofun

It is remarkable to see that almost all glass types found in Japan except for lead silicate glass are observed in glass excavated from a limited region (Kumamoto and Okayama prefectures) and occur within a limited time period: spanning between the 2nd and 7th century AD. The average chemical composition of each type of glass was calculated and is shown in Table 4.2. A summary of glass types found in each Kofun and their representative samples with sample No. and photos is given in Fig. 4.3. Their complete analytical data are listed in Table 4.3.

4.3.2 *Origin of yellow colors*

We have studied the origin of the yellow color of the yellowish green beads from Odara Kofun. Several compositional types of yellow pigments are known to be used in glass in ancient times: the Pb-Sn type ($PbSnO_3$, Pb_2SnO_4), the Pb-Sb type ($Pb_2Sb_2O_7$) and the Pb-Sb-Sn type ($Pb_2SbSnO_{6.5}$).[13] XRF spectrum of the green bead is given in Fig. 4.4. The glass contained significant amounts of Cu, Pb, and Sn. Cu could have given blue color to the soda lime glass. It is expected that a combination of blue and yellow should give green as is shown later. The Pb-Sn type colorant should give a yellow color. Raman spectroscopy was applied to identify the yellow pigment, which was fortunately found as an inclusion of yellow powder crystals into the glass matrix. The μ-Raman spectrum of the yellow crystals in the green bead (sample number: 10g016) from Odara Kofun is shown in Fig. 4.5, which also shows the reference spectra of synthetic $PbSnO_3$ and Pb_2SnO_4 for comparison.[12] The results clearly show that the yellow colorant is $PbSnO_3$ and not Pb_2SnO_4. Similar Raman spectra were obtained for other three beads (10g014, 10g015, 10g017).

Fortunately we also found an inclusion exposed on the outside (see Fig. 4.6) of the glass bead from Kunigoshi, Kofun. This allows us to sample a small amount of the yellow powder. Thus, we collected the crystals, which was subjected to the X-ray powder

Table 4.3 A complete analytical data for the samples shown in Fig. 4.3*

Site	File No.	Type	Main component / wt%												Minor component / ppm						Dark blue color-ants	Flux
			Na_2O	MgO	Al_2O_3	SiO_2	K_2O	CaO	MnO	Fe_2O_3	CuO	PbO	SnO_2	Total	TiO_2	CoO	NiO	Rb_2O	SrO	ZrO_2		
Etafuney-amaKofun	10g002	Na_2O-CaO-SiO_2	10.1	3.7	3.3	73.4	3.18	5.55	0.20	1.11	0.09	0.19	nd	101.1	1470	560	100	20	510	160	Cu	Plant ash
	10g004	Na_2O-Al_2O_3-CaO-SiO_2	9.4	0.4	9.1	70.8	2.58	2.57	0.13	1.91	0.75	2.20	0.52	101.7	12550	70	70	100	370	810		
Kitaoka yokoa-nagun	10g011	Na_2O-CaO-SiO_2	4.5	2.6	4.2	73.2	3.27	6.90	0.13	1.31	0.13	0.25	nd	97.0	2550	500	130	20	480	160	Cu	Plant ash
	10g051	K_2O-SiO_2	1.2	0.3	3.2	70.6	7.69	1.64	1.24	1.39	0.01	nd	nd	87.4	1260	290	110	240	130	80	Mn	
	10g008	Na_2O-Al_2O_3-CaO-SiO_2	3.5	1.0	9.3	71.8	3.21	3.48	0.14	1.24	0.63	0.41	0.11	95.5	5360	50	130	80	420	660		
	1003g014	Na_2O-Al_2O_3-CaO-SiO_2	9.9	0.9	9.5	70.6	1.87	1.94	0.13	2.47	0.01	1.22	0.26	99.6	6530	70	90	60	280	370		
Odara Kofun	10g025	K_2O-SiO_2	1.9	0.2	2.6	69.3	13.65	1.73	1.58	1.21	0.01	nd	nd	92.4	1460	340	130	240	150	90	Mn	
	10g028	Na_2O-CaO-SiO_2	12.3	nd	3.7	73.3	1.40	5.78	1.39	0.87	0.02	0.12	nd	99.4	2310	860	230	20	510	180	Mn	Mineral soda
	10g015	Na_2O-Al_2O_3-CaO-SiO_2	7.7	0.0	9.3	68.8	3.23	2.46	0.19	2.32	0.93	3.24	0.82	100.1	8620	70	130	90	370	770		
	10g022	Na_2O-Al_2O_3-CaO-SiO_2	4.3	0.3	11.7	70.8	5.75	1.32	0.07	0.46	0.33	0.18	0.03	95.6	2740	nd	80	90	270	560		

*nd: not detected

(Continued)

Table 4.3 (*Continued*)

Site	File No.	Type	Main component / wt%												Minor component / ppm						Dark blue colorants	Flux
			Na_2O	MgO	Al_2O_3	SiO_2	K_2O	CaO	MnO	Fe_2O_3	CuO	PbO	SnO_2	Total	TiO_2	CoO	NiO	Rb_2O	SrO	ZrO_2		
Kunigoshi Kofun	10g042	Na_2O-Al_2O_3-CaO-SiO_2	9.4	nd	11.5	71.2	4.40	2.20	0.10	0.87	0.60	0.15	0.04	100.9	3940	nd	90	80	460	730		
	10g030	Na_2O-Al_2O_3-CaO-SiO_2	11.5	nd	12.1	69.5	3.81	1.50	0.11	1.26	0.02	3.47	0.58	104.6	6020	90	80	130	530	760		
	10g037	Na_2O-Al_2O_3-CaO-SiO_2	12.0	0.2	9.7	69.9	2.38	1.96	0.12	1.89	0.54	0.92	0.19	100.7	7970	60	150	60	360	630		
Daibo Kofun	1003g001	Na_2O-CaO-SiO_2	5.3	2.8	5.2	71.6	3.62	6.51	0.23	1.08	0.06	0.15	nd	96.8	1790	300	90	20	560	100	Cu	Plant ash
	1003g049	Na_2O-Al_2O_3-CaO-SiO_2	11.9	0.1	8.7	67.6	2.64	4.24	0.09	1.37	0.93	0.77	0.19	99.2	5540	nd	110	60	400	440		
	1003g004	Na_2O-Al_2O_3-CaO-SiO_2	3.8	1.2	12.3	63.7	2.05	3.40	0.13	2.52	nd	nd	nd	89.7	4670	80	90	10	590	550		
Umade Kofun	1003g027	K_2O-SiO_2(high Al_2O_3)	1.7	0.6	6.2	73.4	9.14	0.21	0.08	0.57	1.47	0.44	0.17	94.4	3910	nd	110	830	30	240		
	1003g034	K_2O-SiO_2	1.6	nd	4.1	73.3	10.51	1.11	1.39	1.13	0.01	nd	nd	93.5	1720	450	120	230	130	90	Mn	
	1003g044	Na_2O-CaO-SiO_2	6.2	2.6	4.1	72.7	3.50	6.65	0.19	1.17	0.06	0.15	nd	97.7	2400	330	110	30	540	160	Cu	Plant ash
	1003g046	Na_2O-Al_2O_3-CaO-SiO_2	7.8	0.5	8.1	65.1	2.30	4.50	0.13	1.44	0.62	1.26	0.25	92.2	1170	670	210	20	520	110		

*nd: not detected

(*Continued*)

Table 4.3 (*Continued*)

Site	File No.	Type	Main component / wt%												Minor component / ppm						Dark blue color-ants	Flux
			Na_2O	MgO	Al_2O_3	SiO_2	K_2O	CaO	MnO	Fe_2O_3	CuO	PbO	SnO_2	Total	TiO_2	CoO	NiO	Rb_2O	SrO	ZrO_2		
Syoji Kofun	1003g035	Na_2O-CaO-SiO_2	4.5	2.7	4.0	72.8	3.35	5.81	0.12	1.00	0.04	0.15	nd	95.1	5590	60	80	100	480	770	Cu	Plant ash
	1003g032	Na_2O-Al_2O_3-CaO-SiO_2	9.7	0.5	10.8	67.8	3.50	1.75	0.11	1.09	0.03	2.07	0.47	98.3	4380	30	100	120	480	760		
	1003g050	Na_2O-Al_2O_3-CaO-SiO_2	4.3	0.1	10.1	65.9	3.37	4.75	0.12	0.96	0.52	0.25	0.06	91.0	5410	40	100	70	430	610		
Wosugi	1003g042	K_2O-SiO_2	1.4	nd	3.5	73.0	12.33	0.91	1.65	1.57	0.02	nd	nd	94.6	1170	250	120	140	70	70	Mn	
	1003g038	K_2O-SiO_2 (high Al_2O_3)	1.2	0.2	6.5	71.8	13.82	0.15	0.20	1.13	2.10	0.44	0.24	98.1	1870	510	130	210	80	90		
Tsukadan Kofin	1005g012	Na_2O-CaO-SiO_2	10.3	1.9	3.7	70.1	2.88	9.50	0.10	0.78	0.00	0.10	nd	99.6	1310	30	90	10	350	120		
	1005g027	Na_2O-CaO-SiO_2	6.9	1.5	4.4	68.0	3.44	6.88	0.27	1.15	0.04	0.14	nd	93.0	1450	370	110	30	470	120	Cu	Plant ash
	1005g020	Na_2O-Al_2O_3-CaO-SiO_2	11.5	0.8	7.7	59.1	1.51	3.20	0.15	2.75	0.01	1.23	0.26	88.9	6440	90	130	70	330	380		
	1005g025	Na_2O-CaO-SiO_2	2.3	0.8	7.4	69.7	3.46	4.55	0.11	0.75	0.01	0.10	nd	89.5	2100	60	80	60	650	140		
	1005g028	Na_2O-Al_2O_3-CaO-SiO_2	4.2	2.2	9.6	66.4	2.09	2.69	0.12	1.82	0.01	0.94	0.27	91.6	10760	60	130	90	470	1150		
	1005g021	Na_2O-Al_2O_3-CaO-SiO_2	10.7	0.4	9.7	66.9	2.23	2.85	0.10	1.32	0.01	0.55	0.11	95.2	3570	60	110	60	440	400		

*nd: not detected

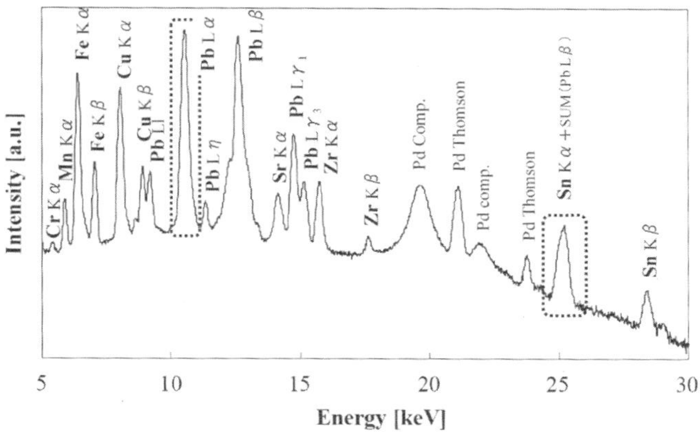

Fig. 4.4 XRF spectrum of yellow green glass beads from Odara Kofun

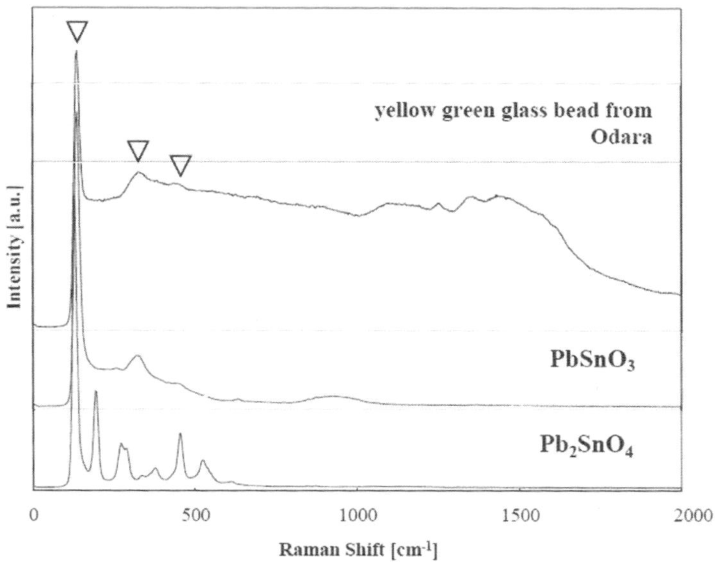

Fig. 4.5 Raman spectrum of a yellow pigment in green glass bead from Odara Kofun

diffraction analysis using a Gandolfi camera. The obtained diffraction pattern on the IP is shown in Fig. 4.6 and the diffraction data are listed in Table 4.4. The reference powder diffraction data of $PbSnO_3$ (PDF 17–0607) are also shown in Table 4.4 for comparison. The two data agreed well with each other and thus, the yellow pigment was clearly identified as $PbSnO_3$. This was the first direct and solid identification of $PbSnO_3$ in ancient Asian glass beads.[8]

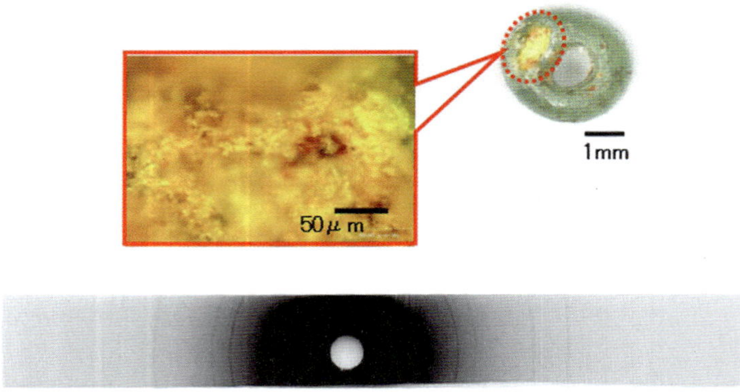

Fig. 4.6 Photo of the yellow powder found in green bead from Kunigoshi and its powder diffraction photograph recorded on IP

Table 4.4 A comparison of diffraction data of yellow pigment from Kunigoshi with reference data

Yellow pigment		$PbSnO_3$ (PDF17–0607)		
d(Å)	intensity	d(Å)	intensity	hkl
3.417	m			
3.220	w	3.220	vw	311
3.091	vs	3.090	vs	222
2.972	w			
2.682	m	2.674	m	400
2.454	vw	2.453	vw	331
2.106	m			
1.895	s	1.890	s	440
1.786	m			
1.617	s	1.612	s	622
1.544	m	1.544	vw	444

4.4 Discussion

4.4.1 *Limitation and possibilities of nondestructive analysis*

The present XRF analyses were carried out truly nondestructively. The samples were simply wiped with a clean soft paper with ethyl alcohol prior to the XRF measurement. Because the samples are part of our precious cultural heritage, the polishing of the samples with sand paper was not allowed by the museums. It is apparent from Table 4.2 that an average Na_2O content of 4.27 wt% is too low for it to be soda lime glass. Na_2O contents of typical soda lime glass are 15 to 20 wt%. This is undoubtedly due to the weathering of the glass surface, which was caused by the long term underground burial conditions. Liu *et al.* studied the effect of surface weathering onto the XRF data and reported the effect of polishing for Na, K, etc.[14] This is the limitation of the nondestructive analysis of archaeological glass by XRF technique. However, our experience suggests that though contents of Na and Si are affected seriously by weathering, it is not so much for the rest of the elements as long as the glass has shiny surface.[5,15] This is supported by the fact that Fig. 4.2 was obtained based on nondestructive XRF data still giving clear information as to the basic classification of glass. This paper will show how much information we can extract from truly nondestructive energy dispersive XRF analysis of glass, although analytical data of Na_2O and SiO_2 could not be used because of the weathering.

4.4.2 *Classification based on the chemical composition*

To characterize the major components of glass explicitly, a ternary diagram using the following three variables was calculated.

$$(K_2O+MgO)/(K_2O+MgO+Al_2O_3/3+CaO/3)$$

$$(Al_2O_3/3)/(K_2O+MgO+Al_2O_3/3+CaO/3)$$

$$(CaO/3)/(K_2O+MgO+Al_2O_3/3+CaO/3)$$

Here, K_2O, MgO, Al_2O_3, and CaO are concentrations (wt%) of each glass sample. The result is given in Fig. 4.7. The ternary diagram

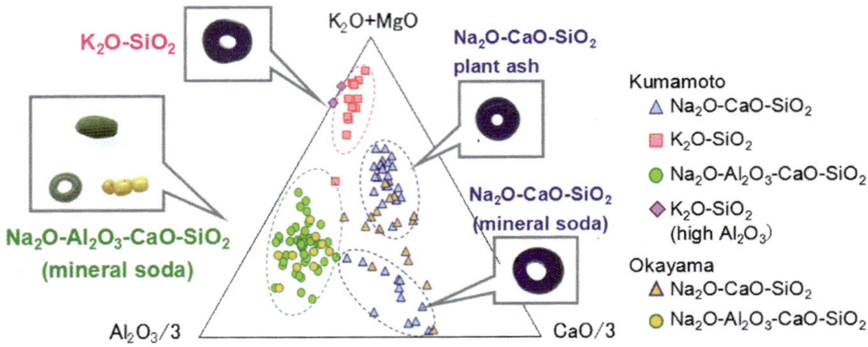

Fig. 4.7 Ternary plot of chemical composition showing the difference in origin

exhibits four distinct clusters in accordance with the classification of the four compositional types revealed by the cluster analysis as shown in Fig. 4.2. Each type of the glass is going to be explained below.

Potash-silica glass (K_2O-SiO_2)

Potash silica glass shows distinctively high K_2O+MgO values forming a cluster close to the top apex of the triangle in Fig. 4.7. A typical example is blue glass bead (10g025) from Odara Kofun (Fig. 4.3): K_2O 13.65 wt%, Al_2O_3 2.6 wt%. The concentration of MnO is high at 1.58 wt%. It is known that potash silica glass and lead barium silicate glass were found across China from the Warring States period to the Han Dynasty (400 BC–200 AD). This suggests that the present glass may have been imported from China. It is known that these glasses were widely distribute from India to South East Asia and Korea.[16,17]

On the other hand, there are several potash silica glasses with high levels of Al_2O_3 and low levels of CaO. Light blue glass (1003g027) from Umade Kofun (Al_2O_3 6.2 wt%, K_2O 9.14 wt%, CaO 0.21 wt%, CuO 1.47 wt%, and CoO not detected) and greenish blue glass (1003g38) from Wosugi Kofun (Al_2O_3 6.5 wt%, K_2O 13.82 wt%, CaO 0.15 wt%, CuO 2.10 wt% and CoO 510 ppm) belong to this type. It is known that these type of glass was distributed from South China to middle Vietnam.[16,17]

Soda lime glass (Na_2O-CaO-SiO_2)

The soda lime glass forms clusters in two regions in Fig. 4.7. This seems to reflect the difference in alkaline fluxing agents used in glass production. It is well known that two types of alkaline flux were used to produce glass in ancient times: plant ash and mineral soda. The former is literally ash of certain types of plants. The latter is well known as natron, which is a natural mineral with chemical formula $Na_2CO_3 \cdot 10H_2O$, but often used as a general term referring to minerals of sodium carbonates used in ancient West Asia.

Figure 4.8 shows the plot of the potassium oxide (K_2O) content vs. magnesium oxide (MgO) content for the soda lime glass with deep blue color including Korea glass reported by Lankton *et al.*[18,19] It is well known that soda-lime glass with both K_2O and MgO levels being greater than about 1.5% were made with certain types of plant ashes as their source of soda. These glasses are known as having Sasanian or Islamic glass composition. On the other hand, Roman and Hellenistic glasses were soda lime glass with significantly lower levels of K_2O and MgO, having been made with natron. Figure 4.8 shows that both types were observed in the present sites studied. Typical natron glass is dark blue glass (10g028) from Odara Kofun, with MgO not detected and K_2O 1.4 wt%; while typical plant ash dark blue glass (1003g001) from Daibo Kofun has MgO 2.8 wt% and K_2O at

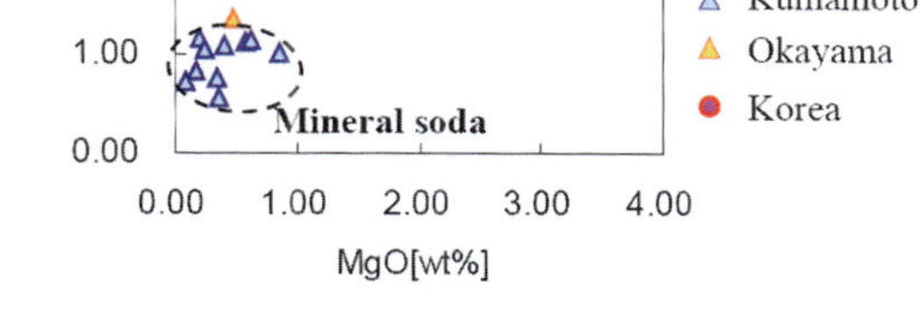

Fig. 4.8 MgO vs. K_2O plot of deep blue Na_2O-CaO-SiO_2 glass

3.62wt%. However, the occurrence of natron glass is rather rare in Japan.[20] Soda lime glass is thus believed to be of West Asia origin but detailed production places have not been clarified yet.

High alumina soda glass(Na_2O-Al_2O_3-CaO-SiO_2)

The fourth group is high alumina soda glass (Na_2O-Al_2O_3-CaO-SiO_2). This cluster is clearly seen in the ternary diagram (Fig. 4.7). It is characterized by its high alumina content, 9.54 wt% as shown in Table 4.2. These glass beads tend to be colorful, such as yellow, green, and light blue. These colored glasses could be classified as Indo Pacific beads, which are found in South India, South East Asia, and East Asia.

The concentrations of TiO_2, Fe_2O_3, and ZrO_2 reflect the geological characteristics of sands used as the silica source of the glass production. Thus another ternary diagram using the following three variables was calculated based on their % weights.

$$(TiO_2/5)/(TiO_2/5+Fe_2O_3/50+ZrO_2)$$

$$(Fe_2O_3/50)/(TiO_2/5+Fe_2O_3/50+ZrO_2)$$

$$ZrO_2/(TiO_2/5+Fe_2O_3/50+ZrO_2)$$

Among them, TiO_2, Fe_2O_3, and ZrO_2 are concentrations (wt%) of each glass sample. The result is shown in Fig. 4.9. As can be seen from Fig. 4.9, the plots of the high alumina soda glass form a marked cluster, which suggests that the sands with similar geological characteristics were used for the production of the glass of this type. The cluster formation for these glasses is also clear in the ternary diagram of Fig. 4.7. These results may suggest that these beads were produced in a limited area using natural resources with a similar geological origin.

The yellow green beads from Odara Kofun belong to this glass type. XRF spectra of green, yellowish green and yellow beads are compared in Fig. 4.10. The major difference is only the peak height of Cu Kα line. The yellow sample shows the lowest Cu Kα peak, then the yellowish green one (CuO/PbO=0.3), followed lastly by the green one which shows the highest (CuO/PbO=0.6). The greenish color was produced

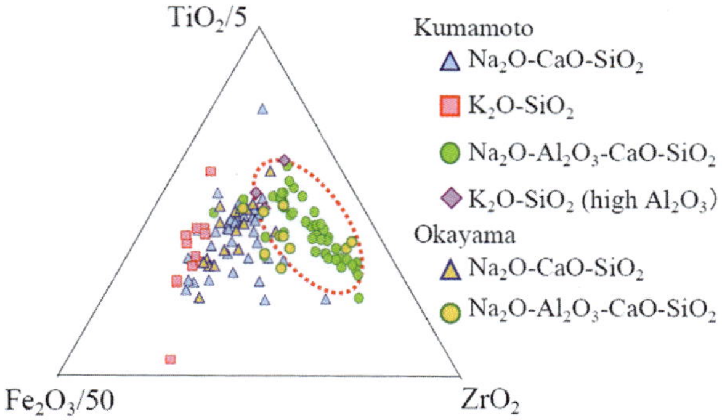

Fig. 4.9 Ternary plot showing compositional characteristics of the high alumina soda lime glass

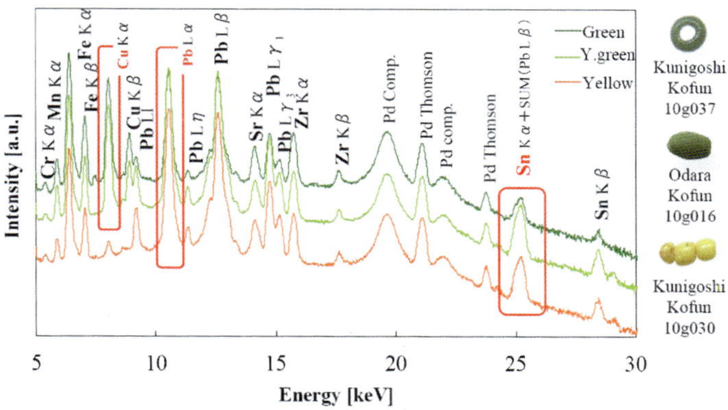

Fig. 4.10 A comparison of XRF spectra of yellow, yellowish green and green glass beads

by adding Cu^{2+} ion into the yellow glass colored with $PbSnO_3$. Namely, when Cu concentration is high, the glass becomes green in color. Figure 4.10 indicates that other components show no significant difference even between two different Kofuns: Kunigoshi and Odara.

We have also examined the compositional uniformity of the colorants. Figure 4.11 shows the SnO_2 vs PbO plot for the yellow to green glass. It is found from Fig. 4.11 that the data are located along a straight line indicating that these colored beads were produced using the yellow colorants with an uniform recipe. It is estimated that the

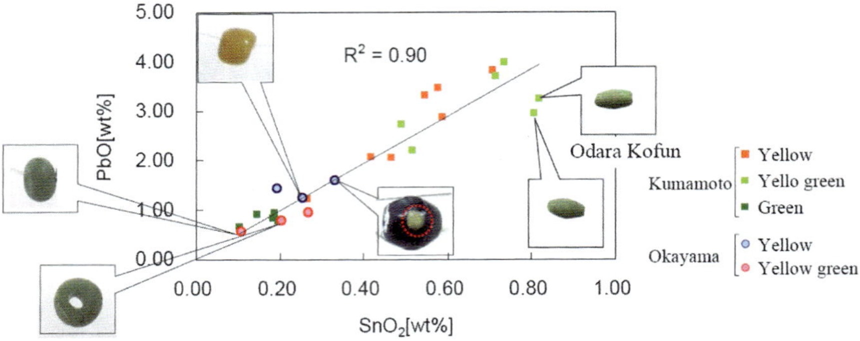

Fig. 4.11 SnO_2 vs. PbO plot for yellow-green glass beads

reported lead isotope analysis of Odara green beads determined that the measured lead component of the yellow colorant was $PbSnO_3$. The possible lead source is South East Asia, probably the mine is estimated to be in Thailand.[4] The observed uniformity of colorant, recipe, sands, flux agent suggest that the origin of these high alumina soda lime glass may be somewhere around the Malay Peninsula and there may have been a large production center for this particular type of glass beads as well as a fixed source where the colorants originate from.

Lead barium silicate glass (PbO-BaO-SiO$_2$)

Figure 4.12 shows an XRF spectrum of blue Magatama (a comma-shaped bead) and the photograph. Intensity of X-ray was expressed in log scale. The spectrum shows strong Pb L lines together with the characteristic Ba Kα peak indicating that the glass is lead barium glass, which is typical Chinese glass in the Han dynasty. The blue colorant is Cu. This glass was originally imported from China to Japan, but the glass was reformed probably in Japan into this (Magatama) form.

Cobalt blue glass

Colorant of the dark blue glass is the Co ion. However, several Co-rich colorants were used to produce dark blue glass. Koezuka reported that there are two typical types, Mn type and Cu type.[19] The glass of the former type contains more than 1% of MnO, while

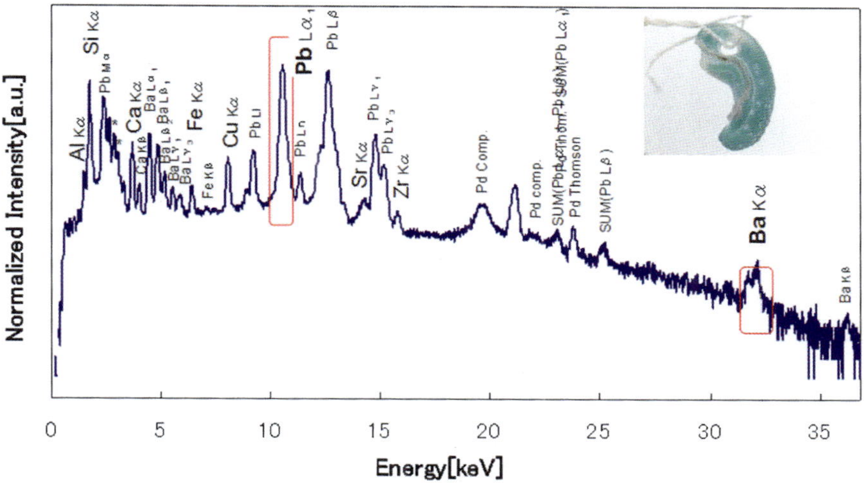

Fig. 4.12 XRF spectrum of lead barium silicate glass (PbO-BaO-SiO_2) from Tsukadan Kofun

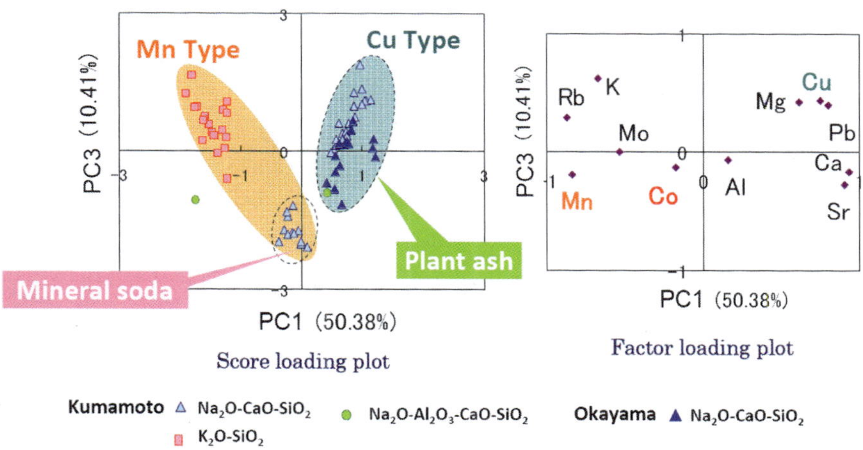

Fig. 4.13 Classification of dark blue colorants based on principal component analysis

that of the latter type contains CuO around 0.1% with MnO less than 0.5%. Principal components analysis was carried out using analytical data of the present dark blue glass and the result is shown in Fig. 4.13. It can clearly be seen from Fig. 4.13 that all plant ash soda lime glass colorants belong to the Cu type, while potash glass to the Mn type. In contrast, mineral soda (natron) glass colorants belong to the

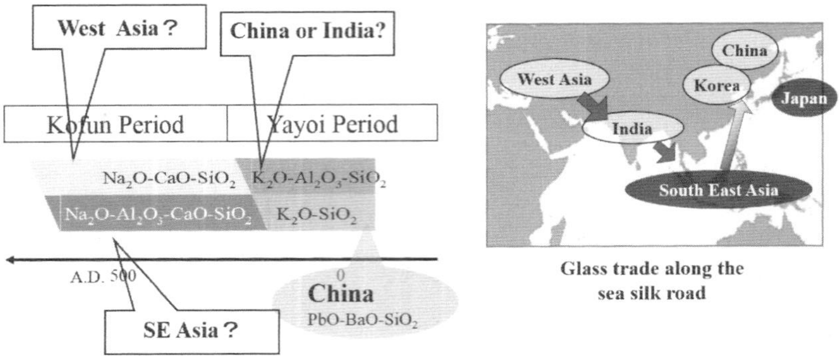

Fig. 4.14 Summary of glass trade between Asian country and Japan

Mn type and occupy a distinct region in Fig. 4.13. This difference may have arisen from the difference in Co source. High alumina soda lime glasses were split into the two regions.

4.5 Conclusion

Origin of the analyzed glass could be summarized in Fig. 4.14. The present analytical study revealed lively tradings of glass among Japan and Asian countries during 2nd to 7th century AD. It is remarkable that the flow of glass started from as far west as India or West Asia and ended in the Far East country, Japan, as is well reflected on glass beads excavated from a small region in Japan. The Sea Silk Road shown in Fig. 4.14 seems to play an important role in the exchange of these glass and other goods.[17] The present study also demonstrated that the nondestructive energy dispersive XRF analysis can offer rich information as to the characteristic chemical composition of glass, though the sodium content may largely decrease because of the surface weathering which impacts and affects the results of some other oxides.

Acknowledgements

We are grateful to Mr. T. Ikeda, K. Sakaguchi, K. Goto and T. Araki of Kumamoto Prefectural Office and M. Yasukawa of Okayama Municipal Office for offering their precious samples for our study.

The authors are grateful to Dr. K. Tantrakarn of TUS for his kind help in preparation of this manuscript and XRF analysis of some of the glass. Thanks are also due to Dr. Y. Abe of TUS for his kind help in Raman analysis. The present work was partly supported by a Grant-in-Aid for Scientific Research ((B): 20404005) from the Ministry of Education, Science, Sports, and Culture, Japan.

References

1. T. Koezuka, *Glass* (in Japanese) **53**, 3–9 (2009).
2. T. Koezuka, *Cultura Antiqua* (in Japanese) **48**, 47–57 (1996).
3. T. Koezuka, *Scientific Study in Historical Transition of Japanese ancient glass* (in Japanese), Doctoral Thesis, 7–8 (1997).
4. S. Uu, S. Shimomura, Y. Hirao, T. Ikeda, *The 26th conference of Japan Society for Scientific Studies on Cultural Properties abstract* (in Japanese), 272 (2009).
5. N. Kato, I. Nakai, Y. Shindo, *J. Archaeol Sci.* **36**, 1698–1707 (2009).
6. A. Hokura, T. Sawada, I. Nakai *et al.*, Chemical analysis of the glass vessel in Toshodaiji Temple designated a national treasure through a portable X-ray fluorescence spectrometer — Where did the glass vessel come from?, in *Ancient Glass Research along the Silk Road*, eds. F. X. Gan, R. H. Brill and S. Y. Tian (World Scientific Publishing Co. Pte. Ltd, Singapore, 2009), pp. 231–241.
7. Kumamoto education committee (eds.), *Report on investigation of decorated tomb in Kumamoto prefecture* (1984) (in Japanese).
8. J. Shirataki, Y. Abe, K. Tantrakarn *et al.*, *Archaeology and Natural Science*, (in Japanese) **63**, 30–52 (2012).
9. I. Nakai, S. Yamada, A. Hokura *et al.*, X-Ray Spectrom. **34**, 46–51 (2005).
10. K. Kikugawa, Y. Abe, T. Sanada, I. Nakai, Adv. *X-Ray. Chem. Anal.* (in Japanese) **40**, 325–337 (2009).
11. K.Tantrakarn, N. Kato, A. Hokura,*et al., X-Ray Spectrom.* **38**, 121–127 (2009).
12. Y. Abe, I. Nakai, K. Takahashi *et al.*, *Analytical and Bioanalytical Chemistry* **395**, 1987–1996 (2009).
13. D. Hradil, T. Grygar, J. Hradilova *et al., Journal of Cultural Heritage* **8**, 377–386 (2007).
14. S. Liu, Q. H. Li, F. X. Gan *et al.*, *J. Archaeol Sci.* **39**, 2128–2142 (2012).
15. N. Kato, I. Nakai, Y. Shindo, *J. Archaeol Sci.* **37**,1381–1395 (2010).
16. F. X. Gan, H. S. Cheng, Q. H. Li, *Science in China Series E: Technological Sciences* **49**, 701–713(2006).
17. F. X. Gan, R. H. Brill, T. Shouyun (eds.), *Ancient Glass Research along the Silk Road, 1st edn.* (World Scientific Publishing Co. Pte. Ltd, Singapore, 2009), pp. *1–108.*

18. J. W. Lankton, L. Dussubieux, *Journal of Glass Studies* **48**, 121–144 (2006).
19. J. W. Lankton, B. Gratuze, G. H. Kim, L. Dussubieux,Silk Road glass in ancient Korea: the contribution of chemical composition analysis, in *Glass along the Silk Road from 200 BC to AD 1000*, eds. B. Zorn, A. Hilgner (*Verlag des Römisch*-Germanischen Zentralmuseums, Mainz, 2010), pp. 221–236.
20. T. Kezuka, T. Tamura, K. Oga, *Bunkazai* (in Japanese), **566**, 13–26 (2010).

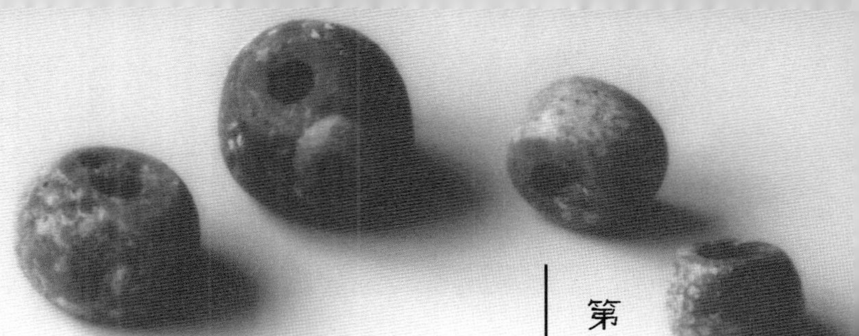

Chapter 5 第五章

Potash Glass: A View from South and Southeast Asia

Laure Dussubieux
The Field Museum, 1400 South Lake Shore Drive,
Chicago, Illinois, 60605, USA

5.1 Introduction

In South and Southeast Asia, between the 4th c. BC and the 3rd c. AD two main types of glass dominate: a mineral soda — alumina glass, also called m-Na-Al glass, and a potash glass. Other glass types exist, but those two glasses are by far the most common, and can be found in the forms of beads or small personal ornaments at almost every site from this period. Based on distinctive trace element patterns and archaeological evidence, it was established that the m-Na-Al glass was manufactured at different locations in South Asia and at different periods.[1,2] This glass was traded as raw material[3] or finished objects through South and Southeast Asia. While the m-Na-Al glass and its different sub-types were discussed in great detail in several publications,[4,5] the potash glass did not benefit from the same attention despite being of an equal importance. In an attempt to provide a better definition of potash glass and its sub-groups, using data obtained from four sites, Ban Don Ta Phet, Khao Sam Kaeo, Phu Khao Thong in Thailand and Arikamedu in India, that yielded large amounts of potash glass, questions related to the circulation and provenance of this glass type will be investigated.

5.2 Origin of the data

5.2.1 *Elemental analysis*

Most of the data discussed here was obtained by the author using laser ablation — inductively coupled plasma — mass spectrometry (LA-ICP-MS). A few samples from Arikamedu were measured using fast neutron activation analysis (FNAA). Data for this site obtained by other researchers will also be used despite the fact that they might be incomplete, lacking trace elements and in some instances some minor elements. LA-ICP-MS is particularly well suited to the study of glass as it can provide the concentrations of 50 to 60 elements within a few minutes with limits of detection in the range of the ppm or below. The use of laser ablation (LA) as a sampling technique allows for the removal of a small quantity of material without visible damage to the artifacts. This technique, first used in the mid-1990s is gaining popularity for the study of ancient glass and has led to important discoveries in terms of provenance and glass trade using trace element concentrations (e.g. [4], [6], [7]).

The compositional data for the sites of Ban Don Ta Phet, Khao Sam Kaeo and Phu Khao Thong were obtained at the Elemental Analysis Facility at the Field Museum between 2006 and 2009. A Bruker inductively coupled plasma — mass spectrometer (ICP-MS), connected to a New Wave UP213 laser for direct introduction of solid samples, was used to generate elemental compositions that include major, minor and trace elements. A detailed account of the technique and its performance can be found in Dussubieux *et al.* (2009).[8] The results for BDTP were discussed, partially, in Lankton *et al.* (2008).[9] The compositions of the KSK samples are available in Dussubieux and Bellina-Pryce (forthcoming).[3] PKT glass compositions were interpreted in Dussubieux *et al.* (2010).[10]

Compositions for the Arikamedu glass were measured between 1997 and 2000 at the Centre Ernest Babelon, IRAMAT, CNRS, Orléans, France and are reported in Dussubieux (2001).[11] Descriptions of the LA-ICP-MS instrumentation, the analytical protocol and the data calibration method are available in Gratuze (1999).[12] FNAA was carried out on a few samples at the Centre d'Etudes et de Recherches par Irradiation, CNRS, Orléans, France. Details about this analytical technique and the analytical parameters can be found in Gratuze and Barrandon (1990).[13]

Concentrations will be reported in weight percent of oxide or part per million (ppm) of elements.

5.2.2 *Sites*

While the distribution of potash glass is extremely wide, encompassing South, Southeast and East Asia, including China, a few sites deserve a special attention for their early dating or for the quantity and quality of the artifacts made from potash glass. One of these sites is Ban Don Ta Phet (BDTP). This site is an Iron Age (4th c. BC) cemetery located in the Kanchanaburi Province of West Central Thailand. It yielded a large quantity of glass ornaments including several thousands of beads and a few bracelets. The glass beads are available in a wide range of shapes; however the most common beads are small opaque red beads made by the drawn technique. A few of those beads were analyzed and belong to the m-Na-Al 3 glass type that was quite likely manufactured in Northeastern India (see [1], [10] for more details about the m-Na-Al 3 glass). More interesting are transparent beads available in a range of colors, including different shades of blue and green along with colourless. Some of them were manufactured using the drawn technique, others were shaped using the same lapidary techniques developed to produce stone ornaments.[14] All transparent beads have a potash-rich composition.

Ornaments and glass bracelets similar to those found at Ban Don Ta Phet were recovered at Khao Sam Kaeo (KSK), on the east coast of the Thailand peninsular. This early urban site, located on the east side of the narrowest part of the Thai-Malay Peninsula close to the Tha Taphao River, dates from the 4th c. –2nd c. BC. It yielded evidence of ornament manufacturing in the form of knapped and hot-worked waste and unfinished or mis-shapen ornaments.[15] Interestingly, the glass found at Khao Sam Kaeo was mostly translucent green and two main compositions were identified: potash and m-Na-Al 3.[3,9] The combined presence of these two glass types provides a signature for the earliest glass found in Southeast Asia.[5] Glass ornaments similar in shape and compositions were also found outside Thailand in Vietnam at the site of Giong Ca Vo dating from the 4th c. BC.[9] The material recovered from Khao Sam Kaeo is not compatible with the South Asian-centered view that prevailed until recently for glass artifact production in the early periods. Indeed, as early as the 4th to 2nd c. B.C. production centres for beads and bracelets were operating in Southeast Asia and they may have traded over relatively long distances in Southeast Asia.

The two other sites that will be considered in this paper are Arikamedu in India and Phu Khao Thong in Thailand. Both sites are later than BDTP and KSK; their occupation started around the 2nd c. BC, and both yielded large proportions of potash glass. Arikamedu is located on the southeastern coast of India. It was excavated several times;[16–18] it was a trading port as well as an industrial center where glass beads (among other goods) were manufactured. Peter Francis Jr gave this site the central role for the production of drawn beads in South India, suggesting that Arikamedu was the main drawn bead procurement center for South and Southeast Asia until the 2nd century AD.[19] Beads and glass fragments from Arikamedu were analyzed with different analytical techniques[11,20–26] and a large proportion of potash glass was identified.

Phu Khao Thong (PKT) is part of a trading complex including several other sites located in the Sukhsamran district, Ranong Province, on the Andaman coast. Like Arikamedu, PKT yielded a range of artifacts suggesting that those sites were part of a trading network extending over long distances. The presence of chunks of raw glass and unfinished beads suggests that monochrome drawn beads were manufactured at PKT.[27,28] Table 5.1 gives the number of compositions that will be discussed below.

Table 5.1 Number of potash compositions that will be used in the discussion below

Site	Number of potash compositions
BDTP	35
KSK	82
PKT	32
Arikamedu	22 from Dussubieux (2001) and 22 from other sources.

5.3 Potash glasses at BDTP, KSK, PKT and Arikamedu

5.3.1 *Different potash glasses*

First, it is necessary to define what it is referred to as 'potash glass' here. A potash glass is a glass type with potash (K_2O) as the major constituent after silica. Generally, constituents such as magnesia are fairly low (<1%) suggesting a mineral origin for the potash. Saltpeter or potassium nitrate (KNO_3), a salt that exists naturally in some soils, is a possible mineral source of potash. As indicated by a wide range of concentrations of lime (CaO) and alumina (Al_2O_3), sands with a wide range of compositions were mixed with saltpeter. Based on the concentrations of these two oxides (CaO and Al_2O_3), three groups of potash glass were defined (Table 5.2).[9]

The lime and alumina concentrations for the potash glass samples found at BDTP, KSK, PKT and Arikamedu are shown in Fig. 5.1. The three types of compositions can be found at these sites; however, an important overlap exists. With alumina concentrations generally below 3% and lime concentrations below 4%, most of the glass samples from Arikamedu belong to the moderate lime and alumina (m-K-Ca-Al) type. A large proportion of the glass from KSK and PKT falls within the same group. Some samples from these two sites, with alumina concentrations as high as 6%, belong to the m-K-Al glass group. Very few samples have low alumina and relatively high lime concentrations (m-K-Ca glass type). All the samples with such a composition are from BDTP. A few samples from KSK and PKT have higher lime and alumina concentrations. They also contain higher magnesia concentrations. Most of those samples are red and contain copper. It is not uncommon in red glass to find higher concentrations

Table 5.2 Definition of three potash glass sub-types

Potash glass	Abbreviation	CaO	Al_2O_3
high lime — low alumina	m-K-Ca	3%<CaO<7%	<3%
low lime — high alumina	m-K-Al	<3%	3%<Al_2O_3<7%
moderate lime and alumina	m-K-Ca-Al	<3%	<3%

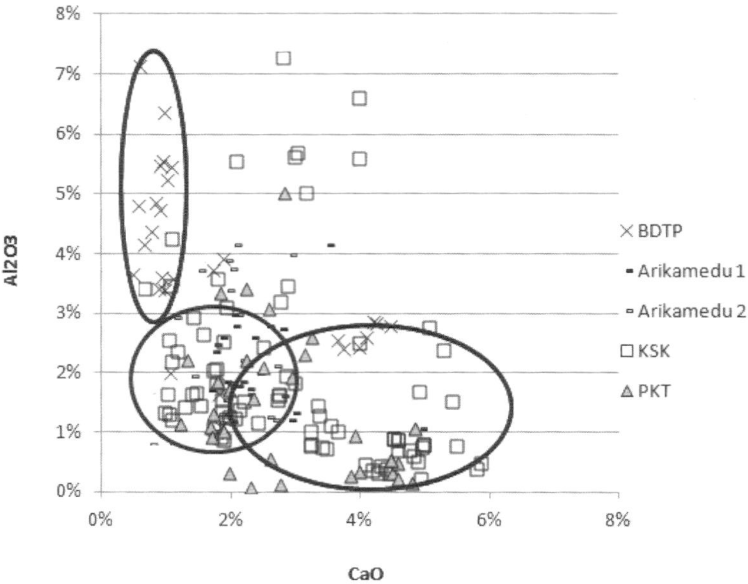

Fig. 5.1 Al_2O_3-CaO biplots for potash samples from BDTP, KSK, PTK and Arikamedu. Arikamedu 1 data are from Dussubieux (2001). Arikamedu 2 data are from Subramanian 1950; Lal 1952; Tornati and van der Sleen 1960; Lamb 1966; Glover and Henderson 1995; Singh 1989; Brill 1999; Dussubieux 2001; Hancock 2014

of constituents such as magnesia, potash, lime or phosphorus suggesting that a reducing ingredient containing these different oxides might have been added to the glass to facilitate the red color.[29]

Because of the overlaps between the three groups as defined using only alumina and lime, trace elements were also investigated to refine the grouping and understand better what kind of raw materials were used in the different potash glasses. Liu *et al.*[30] used rubidium and strontium for that purpose. They identified two main groups of glass: one of them has higher strontium and low rubidium (group I) and the other has higher rubidium and lower strontium concentrations (group II). In group II, rubidium concentrations were correlated to potash and alumina concentrations and this correlation was interpreted as the presence of minerals such as potash-rich feldspars in the sand. In Liu's work, no correlation was observed between strontium and lime concentrations for either group I nor II.

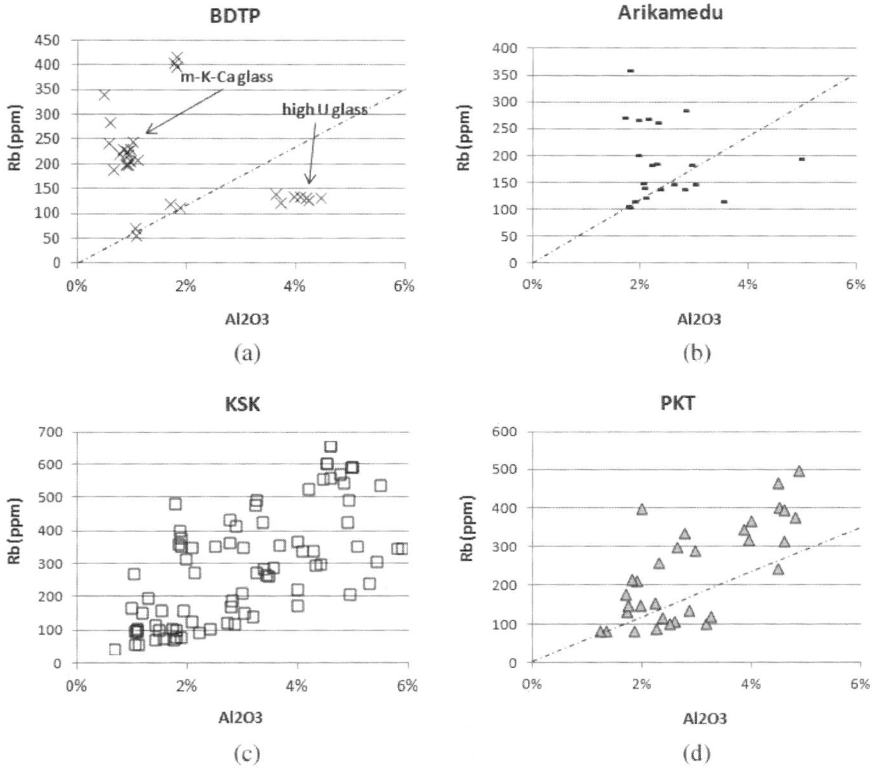

Fig. 5.2 a, b, c, d. Al_2O_3-Rb biplots for BDTP, KSK, PKT and Arikamedu. The dotted line on each graph is a curve with same parameters to visualize more easily how similar is the data distribution for each site

Looking at rubidium and alumina concentrations for the potash glasses from BDTP, KSK, PKT and Arikamedu (Fig. 5.2a, b, c, d), it is possible to observe for some sites (KSK and PKT) a correlation between the alumina and rubidium concentrations. It is important to note that, in some cases, higher concentrations of cobalt are also correlated to higher concentrations of rubidium[11] meaning that rubidium may have several sources: sand with potassium-rich feldspars as suggested by Liu *et al.*[30] but also cobalt ores including rubidium as an impurity. In contrast, for BDTP, with the exception of a few samples that seems to exhibit a similar alumina-rubidium correlation, no clear pattern could be identified for most of the potash artifacts. A few samples have rather high rubidium concentrations despite low alumina (2% or lower). These samples make up the m-K-Ca group shown on Fig. 5.1. Other samples have fairly low rubidium concentrations (<150 ppm) despite alumina concentrations of approximately 4%. All the samples in this group exhibit higher soda (Na_2O) and uranium concentrations (>19 ppm) along with higher concentrations of trace elements such as the rare earth elements. Those high uranium samples are part of the m-K-Al group.

In the glass from the BDTP, KSK, PKT and Arikamedu, strontium and lime are systematically correlated; however, two different correlations were identified as shown on Figs. 5.3a, b, c, d. Arikamedu

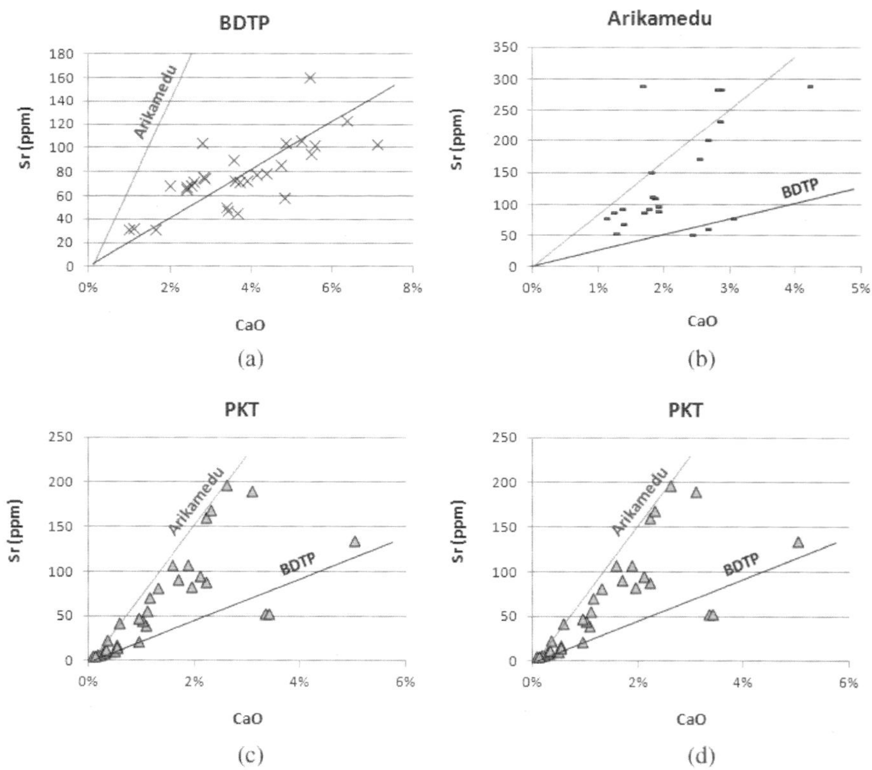

Fig. 5.3 a, b, c, d. CaO-Sr biplots for BDTP, KSK, PKT and Arikamedu. The two dotted lines on each graph are indicating the trends for the BDTP and Arikamedu data to show that they are dissimilar from each other but similar respectively to KSK and PKT

and Phu Khao Thong have similar variations for their strontium and lime concentrations, which are different from KSK and BDTP suggesting the use of different raw ingredients.

5.3.2 *Colors and colorants*

It is important to emphasize that for each site, the samples considered here may not be representative of the range of color (and shapes) found at each site. Despite this fact, a few comments can be made based on the color distribution in the analyzed samples (Table 5.3). At each site a large proportion of the potash glass samples are transparent. At BDTP, all of them are transparent and 68% of the analyzed samples are very lightly colored, being Colourless or bluish or greenish. Dark blue is a very popular color in general. Whereas it only represents 17% of the analyzed samples at BDTP and 27% at KSK,

Table 5.3 Color distribution of the analyzed samples for BDTP, KSK, PKT and Arikamedu. "o" is for opaque and "t" for transparent or translucent. *include analysis by the author and others. See the Arikamedu site description for the list of references

	BDTP (%)	KSK (%)	PKT (%)	Arikamedu* (%)
Black	0	6	0	5
o. green	0	8	3	18
o. red	0	2	7	7
o. yellow	0	0	10	5
Colourless	34	5	10	0
t. bluish or bluish-greenish	34	0	7	9
t. dark blue	17	27	47	52
t. green	0	13	7	0
t. purple	0	9	0	5
t. turquoise blue	14	31	10	0

almost half of the analyzed artifacts from PKT (47%) and Arikamedu (52%) were dark blue.

Colourless glass does not contain any decolorizing agent such as manganese or antimony that was used for example around the Mediterranean area at the same period.[31] Iron, a natural colorant often present in the sands used to manufacture glass, has fairly low concentrations (0.7% or lower), suggesting that a careful raw material selection along with a control of the furnace atmosphere were likely to have been responsible for the absence of colour in the glass.

Transparent glass samples with colors ranging from green to light blue were either coloured by the presence of copper or the presence of iron or by a combination of the two elements.

Purple samples contain manganese with concentrations as high as 5.2%. Dark blue glass may also contain fairly high amounts of manganese brought unintentionally with the cobalt that is the main colorant for these artifacts. Indeed, cobalt is often associated with a number of elements specific for the ore source. Two types of ores were used to color dark blue potash glass. In most of the potash dark blue artifacts, cobalt is associated with higher concentrations of manganese and other elements such as barium (Co-Mn-Ba). A second type of cobalt is associated with higher arsenic concentrations (Co-As). No specific correlation was found between either of the three potash groups defined earlier and a type of cobalt. Glass samples containing cobalt and arsenic were found at Ban Don Ta Phet, and Khao Sam Kaeo, which were the earliest sites. High manganese cobalt is also present in KSK glasses. No high arsenic cobalt ore was identified in Arikamedu and PKY glasses; dark blue glass was only produced with the addition of Co-Mn-Ba ore. It seems therefore that for the 4th to 2nd c. BC period the two ores were used that only the Co-Mn-Ba ore was available later even if it is not possible to explain why.

Opaque glass samples, whether black, red, yellow or green, are not as abundant as transparent or translucent glass. ‘Black’ glass samples are present at KSK and Arikamedu. Black can be produced by a range of elements such as iron or manganese or cobalt if present at high concentrations. At KSK, black samples are in reality dark purple as they contain manganese with concentrations ranging from 3.2 to

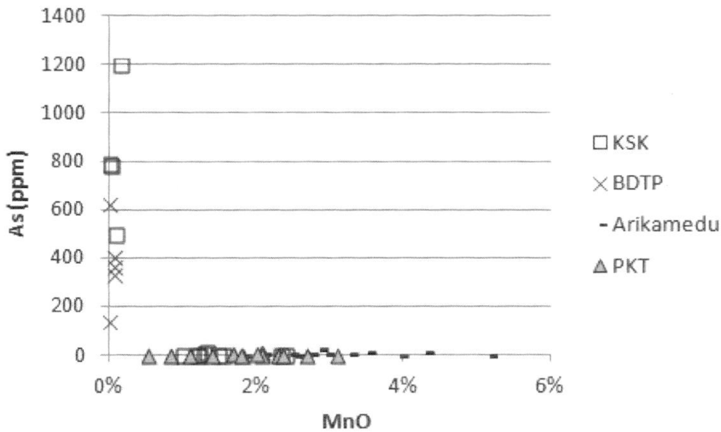

Fig. 5.4 Arsenic and manganese concentrations in the dark blue glass, containing cobalt, from BDTP, KSK, PKT and Arikamedu

5%. At Arikamedu, one sample contains 6.5% of manganese.[26] In the other black samples from that site, this constituent is less than 1%.

Red potash glass appears at KSK, PKT and Arikamedu and contains copper oxide with concentrations ranging from 1.4 to 3.4%. At the three sites, red glass samples sometimes contain more magnesia and lime compared to the other glass samples. This excess of magnesia and lime was observed in other types of red glass[1] and was certainly due to the addition of an ingredient that would have improved the conditions necessary to obtain the opaque red color.

Yellow and green glass samples generally contain a yellow opacifier that in the presence of copper in the glass matrix confers a green color to the glass. Yellow glass samples appear only at PKT and Arikamedu and they contain lead and tin concentrations ranging respectively from 26 to 32% and from 4.3 to 8.4%. Green samples are also present in KSK. They all contain tin and lead but in much lower concentrations: 12 to 20% for lead and 2 to 3% for tin at KSK and Arikamedu. At PKT, one yellow glass was analyzed and its tin concentration was approximately 7%. For all three sites, copper oxide concentrations range from 0.8 to 2%.

5.4 Discussion

Potash glass is extremely common through South and Southeast Asia. The four sites selected here are particularly important for the abundance of glass artifacts they yielded, the diversity of the material and the potential for three of the sites to have been manufacturing workshops for ornaments (KSK, PKT and Arikamedu). It is certain that this glass was traded as raw material and also as finished objects. At a given site, potash glass is often present along with other types of glass (e.g. the m-Na-Al glasses) and it seems that procuring glass from different places was a way to obtain a wide range of colors and transparencies. m-Na-Al glasses are in general opaque and available in a range of vivid colors including yellow, red, orange and green. For example, the m-Na-Al 1 glass type manufactured in South India and or Sri Lanka at periods when potash glass was produced is rarely transparent with the exception of turquoise blue glass.[4,5] With green, dark blue and other transparent blues or colourless glass, the potash glass was complementing the palette offered to the consumers having access to the m-Na-Al 1 glass.

5.4.1 *Sub-groups*

The composition of potash glass is very variable and discrete groups are difficult to identify since elements such as alumina, lime, strontium and rubidium, that are usually used to defined sub-groups vary widely, and concentrations for the sub-groups overlap. Two small groups identified at BDTP stand out. The m-K-Ca group has, in addition to the low alumina and high lime, fairly high rubidium concentrations. A high uranium potash glass was also identified as a subgroup of the m-K-Al group. BDTP is characterized by its very early dating suggesting that maybe at this period several small workshops were using similar recipes but were relying on different raw material sources. Very few other similar compositions were found elsewhere. One exception is Myanmar where both low m-K-Ca and high uranium m-K-Al appeared in an assemblage of glass beads from the Samon Valley dating from the Iron Age.[3] High uranium potash glass appears at other sites in South Asia such as at Arikamedu and Kodumanal and in Southeast Asia at Gîông Cá Vô.

Comparing on the one hand, Arikamedu and PKT, the two later sites, and on the other hand BDTP and KSK, the two earliest sites, it appears that potash glass used in early periods may be different from potash glass as later periods. This is particularly obvious for the lime-strontium correlations for the two later sites compared to the two earlier sites. It also seems that during early periods some potash glass types existed but then disappeared. This idea that the industry of potash glass changed over time is reinforced by the use of different types of cobalt ores during early and later periods. Indeed during the early period, two types of cobalt ores were used: the Co-Mn-Ba and the Co-As ores but the Co-As then disappeared. Of course at this stage, it is impossible to say if the Co-As disappeared because the workshop that used it disappeared of if the source became inaccessible or if a more accessible source suddenly became available.

5.4.2 *Origin and circulation*

The origin of potash glass is still quite uncertain even if several elements are pointing toward a Chinese origin. To this day, no primary workshop for the manufacture of this type of glass has been identified. The provenance of the potash glass was explored using isotope analysis.[32,33] Brill and Fullagar[32] compared the strontium isotope ratios for potash glass found at different locations: China, India, Thailand, Vietnam, Korea and Japan. The strontium isotope results were interpreted independently of the elemental composition of the glass and it is not indicated whether the sample belong to any of the groups defined above for potash glasses. At least three different strontium isotope signatures were identified. Therefore it was assumed that potash glass could have been manufactured in at least three locations.

Li[33] considered the lead isotope signatures of some potash glass samples found in the province of Guangxi, dated from the Han period and compared them with the six different lead isotope signatures that Zhu[34] identified in China. Samples belonged either to the high alumina-low lime potash glass or to the moderate alumina and lime potash glass but had similar lead isotopes signatures. Only the 'Southern China Block' region had a lead isotope signature matching the signatures of the samples. As the province of Guangxi is included in the 'Southern China Block,' Li[33] assumed that the potash glass was a local production; however, no comparison was made with the lead isotope signature of the regions of Southeast Asia that border the Southern China Block. They may have identical lead isotope signatures. More recently the study of faience beads from Shaanxi is providing new evidence pointing in direction of a Chinese origin for the potash glass. Indeed, Hu Yan *et al.*[35] analyzed faience beads from Shaanxi, Middle Western Zhou (10th c. BC) that had a potash composition with low soda and very low calcium and from the later Western Zhou (9th to 8th c. BC) that also had a potash composition, but this time with high calcium and low soda.

If the exact locations where the different potash glass were manufactured still remains uncertain, it seems that the sites of Arikamedu and PKT give us a clue about the circulation of the glass. These two sites, when studied in larger contexts, stand out compared to other sites located in the same regions and dated to similar periods because they present a very different glass assemblage with much higher proportions of potash glass and lower proportions of glass from other origins, including India (see [5] for more details about glass type distribution in Southeast Asia). Arikamedu and PKT being ports, it seems logical to imagine that they were points of entry for the potash glass that would have travelled by sea in the forms of raw glass or recyclable material (e.g. waste).[36] It cannot be excluded that other routes and more especially terrestrial routes were used as well.

5.5 Conclusion

In the absence of clearly identified glass primary glass workshops for the production of the potash glass, one needs to rely heavily on chemical compositions to understand better the circulation and provenance of this glass type. It seems that at very early period (4th c. BC) several potash glasses were manufactured and traded as testified by the variety of compositions at the site of BDTP and KSK. Cobalt from a

Co-As ore seems to characterize early potash glass and could be used as a chronological marker in the future. Later, after the 2nd c. BC, with the sites of PKT and Arikamedu, potash glass seems to have a more uniform composition with the disappearance of the m-K-Ca glass type; however the composition of the potash glass samples at these sites vary in a wide range revealing the use of very heterogeneous raw materials sources. At this period, it seems that the potash glass was circulated by sea, possibly as raw glass, and maybe also as finished products to ports such as Arikamedu and PKT where glass objects were made from it and the objects were redistributed from other sites in the same area.

Acknowledgements

The author would like to thank in particular: Berenice Bellina-Pryce, Bernard Gratuze, Boonyarit Chaisuwan and James Lankton. The samples from Arikamedu were provided by the late P. Francis Jr. The LA-ICP-MS laboratory at the Field Museum of Natural History was built with funds from the National Science Foundation (BCS-0320903), an anonymous donation, and the Anthropology Alliance.

References

1. L. Dussubieux and A. Kanungo, Trace Element Analysis of Glass from Kopia, in Glass in Ancient India: *Excavations at Kopia*, eds. Kanungo, A., (KCHR, Triruvananthapuram, 2013), pp. 360–366.
2. B. Gratuze, L. Dussubieux and O. Bopearachchi, Etude de perles de verre trouvées au Sri Lanka, IIIe s. av. — IIe s. ap. JC., *Annales du 14e congrès de l'AIHV*, Venezia-Milano, Italia, 1998, pp. 46–50.
3. L. Dussubieux and B. Bellina-Pryce, Forthcoming Glass from Early Southeast Asian Producing and Trading Centre. In Bellina, B. (ed.), Khao Sam Kaeo: An Early Industrial Port-city between the Indian Ocean and the South China Sea. Paris.
4. L. Dussubieux, C.M. Kusimba, V. Gogte *et al.*, *Archaeometry* **50**, 797 (2008).
5. L. Dussubieux, B. Gratuze and M. Blet-Lemarquand, *Journal of Archaeological Science* **37**, 1645 (2010).
6. A. Shortland, N. Roger and K. Eremin, *Journal of Archaeological Science* **34**, 781 (2007).

7. M. Walton, A. Shortland, S. Kirk *et al.*, *Journal of Archaeological Science* **36**, 496 (2009).
8. L. Dussubieux, P. Robertshaw and M.D. Glascock, *International Journal of Mass Spectrometry* **284**, 152 (2009).
9. J. Lankton, L. Dussubieux and B. Gratuze, *Bulletin de l'Ecole Française d'Extrême-Orient*, **93**, 317 (2006).
10. L. Dussubieux and B. Gratuze, Glass in Southeast Asia, in *50 years of Archaeology in Southeast Asia: Essays in honour of Ian Glover*, eds. Bellina, B., Bacus, E.A., Pryce, T.O. and Wisseman, C.J., (River Books, London, 2010), pp. 247–259.
11. L. Dussubieux, L'Apport de l'ablation laser couplée à l'ICP-MS à l'étude du verre archéologique de l'Océan Indien, Unpublished PhD Thesis, Université d'Orléans (2001).
12. B. Gratuze, *Journal of Archaeological Science* **26**, 869 (1999).
13. B. Gratuze and J.-N. Barrandon, Islamic weights and stamps: analysis using nuclear techniques, *Archaeometry* **32**, 155 (1990).
14. I.C. Glover, 1990, Ban Don Ta Phet: the 1984–85 excavation, in *Southeast Asian Archaeology*, eds. I. Glover and E. Glover, (Oxford: BAR International Series 561, 1986), pp. 139–83.
15. B. Bellina-Pryce and P. Silapanth, *Bulletin de l'Ecole Française d'Extrême-Orient* **93**, 257 (2006).
16. R.E.M. Wheeler, Ghosh and K. Deva, *Ancient India* **2**, 1 (1946).
17. J.M. Casal, *Fouilles de Virampatnam-Arikamedu.* Paris: Publication de la Commission des fouilles archéologiques, Fouilles de l'Inde (1949).
18. V. Begley, The Ancient Port of Arikamedu, in *New Excavations and Researches 1989–1992.* Vol. 1 (Paris, Bangkok : Centre d'histoire et d'archéologie, École Française d'Extrême-Orient, 1996).
19. P. Jr. Francis, Beadmaking at Arikamedu and beyond, *World Archaeology* **23**(1), 28–43 (1991).
20. R. Subramanian, *Current Science* **19**, 19–20 (1950).
21. B.B. Lal, *Ancient India* **8**, 17–27 (1952).
22. M. Tornati and W.G.N. van der Sleen, *Vetro e silicate* **4**(23), 19–24 (1960).
23. A. Lamb, *Journal of Glass Studies* **8**, 80–94 (1966).
24. I. Glover and J. Henderson, Early Glass in South and South East Asia and China. In: *South East Asia and China: Art, Interaction and Commerce.* London: University of London, 1995, eds. R. Scott and J. Guy, pp. 141–70.
25. R.N. Singh, *Ancient Indian glass: Archaeology and technology* (Delhi: Parimal Publications, 1989).
26. R. Hancock, http://thebeadsite.com/mg11-1e1.html [Accessed 22 August 2014], 2014.
27. B. Chaisuwan, *Early contacts between India and the Andaman Coast in Thailand from the 2nd c. B.C. to 11th c. A.D.* Singapore: Institute of Southeast Asian Studies (2007).

28. B. Chaisuwan, R. Naiyawat, *Thung Tuk: A settlement linking together the Maritime Silk Route*. Thailand: Trio Creation (2009).
29. L. Dussubieux, M. Blet-Lemarquand and B. Gratuze, Innovation dans les techniques de coloration: les verres rouges et orange en Asie du Sud, in: *Les Innovations verrières et leur devenir*, Les Cahiers de Verre & Histoire 2, Actes du deuxième colloque international de l'Association Verre & Histoire, Nancy, 26–28 Mars 2009, eds. S. Lagabrielle and C. Maitte, (Paris: Verre & Histoire, 2013), pp. 11–20.
30. S. Liu, Q.H. Li, Q. Fu *et al.*, *X-Ray Spectrometry* **42**(6), 470–479 (2013).
31. E.V. Sayre, Summary of the Brookhaven program of analysis of ancient glass, *Application of Science in Examination of Works of Art — Proceeding of the Seminar*, Boston, Massachusetts, 7–16 September, 1965, (Research Laboratory, Museum of Fine Arts, 1965), pp. 145–153.
32. R.H. Brill and P.D. Fullagar, Strontium-isotope studies of historical glasses and related materials: a progress report, *Annales du 14e Congrès de l'AIHV*, Antwerp 2006, pp. 552–557.
33. F. Li, *Elemental and isotopic analysis of ancient Chinese glass*. Master Dissertation, University of Hong Kong (1999).
34. B. Zhu, *Journal of Geochemical exploration* **55**, 171–181 (1995).
35. Y. Hu, K. Eremin, R. Newman, A. Shortland, 2014, Investigation of early Chinese Faience, poster presented in: *the 40th International Symposium on Archaeometry*, (ISA, 2014).
36. L. Dussubieux, J. Lankton, B. Bellina-Pryce and B. Chaisuwan, 2012, Early Glass Trade in South and Southeast Asia: New insights from two coastal sites, Phu Khao Thong in Thailand and Arikamedu in South India, in: *Crossing Borders in Southeast Asian Archaeology*. Selected papers from the 13th International Conference of the European Association of Southeast Asian Archaeologists, Berlin, 2010, vol. 2 (Singapore, NUS Press, 2010) pp. 307–328.

Chapter 6

第六章

Archaeological and Technical Study of Western Han Dynasty Lead Barium Glass Chimes (Bian Qing) Unearthed from the Jiangdu King's Mausoleum

Qinghui Li, Zebin Li† and Donghong Gu**

*Shanghai Institute of Optics and Fine Mechanics, CAS, Shanghai 201800, China

†Institute of Archaeology, Nanjing Museum, Nanjing 210016, China

6.1 Introduction

Located in Yunshan Village, Maba Town, Xuyi County, Jiangsu Province, the Jiangdu King's Mausoleum stands 73.6 m high on the top of Dayun Hill (Dayun-shan), which is 30 km to the west of Xuyi County and 1km to the east of Site of Dongyang State Capital. It is adjacent to tombs of Han Dynasty nobles on Qingdun Hill and Xiaoyun Hill (Xiaoyun-shan) to the southwest. After more than two years of excavation, from 2009 to 2011, the graveyard layout of the Jiangdu King's Mausoleum is now generally clear to us. It is a rectangle and consists of three main graves, 11 attendant tombs, two chariot and horse pits, two weapon storage pits, two building foundation remains and a path. The main grave is located in an orderly manner in the south of the cemetery and the satellite tombs to the north. The pits and the building foundations are located separately in a strictly controlled order.[1]

Based on the analysis of historical literature and excavated artifacts, it has been confirmed that the occupant of tomb No. 1, M1, was Liu Fei (*c.* 168–127 BC), the first feudal king of the Jiangdu Kingdom of the Western Han Dynasty and the mausoleum at the Dayun Hill was built for him. Within the mausoleum precinct, more than ten thousand artifacts such as lacquer wares, bronzes, gold and silver wares, a jade coffin and jade suits sewn with gold wires were found.

The present study focuses on a set of glass chimes (Bian Qing) from M1, which is a kind of Chinese musical instrument which could be hung on a frame to play complicated tunes. Accompanying these glass chimes, there was also another set of practical bronze bells (Bian Zhong). These chimes and bells are important artifacts which reflect the ritual regulations of Zhong Qing He Ming (bells and chimes respond sympathetically to each other). To our surprise, these glass chimes are of lead-barium-silicate (PbO-BaO-SiO_2) glasses. They are the earliest known glass musical instruments ever found in China. Their discovery makes it necessary to reconsider the technical development of ancient Chinese glassmaking.

6.2 Archaeological observation of the glass chimes

A total of 22 glass chimes were unearthed from the Jiangdu King's Mausoleum. Figure 6.1 shows the condition of these chimes when excavated. It can be seen that these glass chimes were seriously weathered: most surfaces were covered with thick layers of white weathering products. By contrast, in the cross section we can still see clearly the translucent milk-white or light greenish glass and air bubbles in the inner parts of the chimes (Fig. 6.1c). The weight of these chimes varies from about 0.38 kg to 17 kg, depending on their size. The length of them ranges from 20 cm to 70 cm with thicknesses of 1.1–4.4 cm and perforation diameters of 1.6–1.9 cm.

These glass chimes were hung on a shelf with elegant accessory components such as silver discs (Bi) and mythical creatures (Fig. 6.2). It seems that special emphasis was laid on this set of glass Bian Qing.

(a)

(b)

(c)

Fig. 6.1 The glass chimes following excavation (a) and (b), cross section of one chime showing the inner translucent glass (c)

(a)

(b)

Fig. 6.2 Silver disc inlaid in the shelf used to hang the chimes (a) the base in the shape of mythical animal with wings (b)

6.3 Methods used to analyze the glass chimes

6.3.1 *Confocal Raman spectroscopy*

Raman spectra of the samples were obtained non-invasively with a LabRAM XploRa confocal Ramanscope spectrometer, which was provided by the Horiba (China) Trading Co., Ltd. This spectrometer is equipped with two air-cooled laser sources (532 nm and 785 nm) and a multi-channel air-cooled CCD detector. It is convenient to visualize the images of the samples with the external top USB colour camera and coupling optics. At the beginning of each experiment, the system is calibrated using the 520.7 cm^{-1} silicon line from a single crystal silicon (Si) wafer.[2]

Usually, the Raman spectrum was obtained with 532 nm laser beam, a long-working distance objective (50×) and a diffraction grating of 1800 gr/mm. The laser power of the focused spot (beam diameter, ~1 μm) on the artifact changed from about 2 mW to 15 mW, as measured by LabSpec software through the motorized filter wheel with five neutral density filters. The multi-windows mode was used for optimized spectral resolution (better than 2 cm^{-1}/pixel) and scanning time. The Spectra-ID software permits the creation of spectra libraries which can be searched for identification.

6.3.2 *Hand-held portable X-ray fluorescence (HHpXRF) spectrometer*

A Niton™ XL3t 950 GOLDD+ type HHpXRF spectrometer from Thermo Fisher Scientific Inc. was used to determine the bulk chemical composition of these ancient glass chimes. This analyzer uses a silver excitation tube with voltage of 6–50 kV and maximum current of 200 μA. A geometrically optimized large area drift detector is used and the energy resolution is better than 185 eV. The X-ray beam spot focused on the sample is about 3.0 mm in diameter. Up to 30 elements from magnesium (Mg) to uranium (U) can be detected using this analyzer without helium purging or vacuum pumps.

We used the integrated CCD camera of this XRF analyzer for locating and storing images of areas of interest. The quantitative analysis was done using the software package from Thermo Fisher Scientific Inc. with Mining Cu/Zn Modes. More information about handheld XRF analysis can be found elsewhere.[3]

6.3.3 *Lead isotope analysis*

Lead isotope ratios were measured using a multi-collector inductively coupled plasma mass spectrometer (MC-ICP-MS) of the VG Elemental type at the School of Earth and Space, Peking University. A fragment weighing about 10 mg from each selected glass chime was dissolved in pure nitric acid in a 50 ml glass beaker and the clear solution was leached and diluted in a 100 ml flask using deionized water to determine the lead content using ICP-AES. Depending on the lead content, the solution was then diluted down to a lead content of about 1 μg/l. During the analysis, the thallium (Tl) standard SRM997 was added in the solution as an internal standard. A detailed description of the sample preparation procedure is described elsewhere.[4]

Based on six repeated analysis of SRM981, it was determined that the overall analytical error (2σ) for all the data obtained was less than 0.04%. The standard deviation of mean values for the 20 measurements of each of the five ratios of one sample was less than 0.04%.[5]

6.4 Results and discussion

6.4.1 *Analytical results using Raman spectroscopy*

The purpose of Raman analysis was mainly to identify the crystalline phases, either the weathering products or residual raw materials relating to the analyzed samples. In addition, Raman spectroscopy was also used to determine the characteristics of the glassy phases in ancient glasses and glazes. We used mainly the Spectra-ID software and the Horiba JY Raman Libraries to assign the Raman bands obtained from the samples analyzed.

In Fig. 6.3a, it can be seen that the glassy phases in several samples are characterized by the Raman envelopes at around 500 and 1000 cm^{-1}, which are attributed to the Si-O bending and stretching

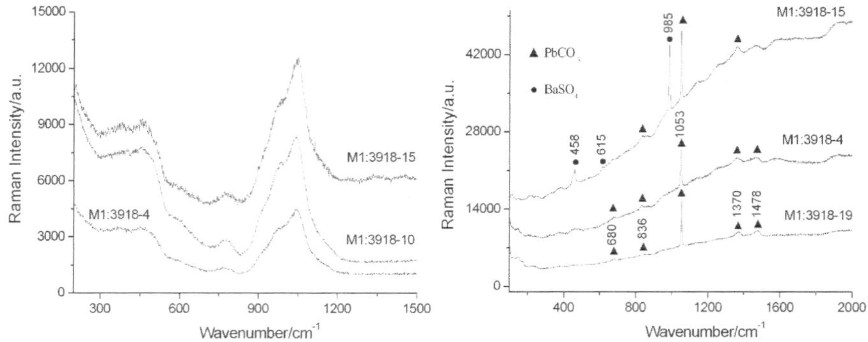

(a) (b)

Fig. 6.3 Selected Raman spectra of different phases identified from different samples. Raman spectra of the detected phases: (a) the glassy matrices; (b) the $PbCO_3$ and $BaSO_4$ crystals

modes respectively.[6] In the white weathered layers, it is very easy to identify cerusite ($PbCO_3$) particles. The Raman bands observed at 680, 836, 1053, 1370 and 1478 cm^{-1} in Fig. 6.3b were attributed to the cerusite particles. Besides cerusite, another weathering product related to lead, mendipite ($Pb_3Cl_2O_2$) with Raman bands at 135 and 273 cm^{-1} was also detected (not given here). Occasionally barite ($BaSO_4$) crystals, which are characterized by the typical Raman bands at 458, 615 and 985 cm^{-1}, were also identified.

6.4.2 *Bulk composition determined by the HHpXRF spectrometer*

The chemical compositions of all the ancient glass chimes analysed are PbO-BaO-SiO_2 glass. Table 6.1 shows the bulk chemical composition obtained including seriously weathered surfaces by the HHpXRF spectrometer. The chemical compositions of less weathered parts are listed in Table 6.2. It can clearly be seen that weathering resulted in an obvious loss of BaO with a reduction of 0.60% to 3.75%.

The levels of PbO detected seem more complex. A comparison between Tables 6.1 and 6.2 shows that PbO can either be increased or reduced by weathering. Combining these results with those of Raman

Table 6.1 Bulk chemical compositions determined by the HHpXRF spectrometer from the seriously weathered parts of the ancient glass chimes (wt%). N.d. denotes not detected

Serial no.	Sample no.	SiO_2	Al_2O_3	PbO	BaO	P_2O_5	Fe_2O_3	CuO	CaO	MnO	ZnO	SnO_2	NiO	Cl
1	M1:3918-3	70.94	0.85	18.99	6.39	0.23	0.18	0.19	1.52	0.13	N.d.	N.d.	0.02	0.55
2	M1:3918-12	65.34	0.98	31.87	0.79	0.40	0.10	0.09	N.d.	0.10	0.02	0.02	0.02	0.27
3	M1:3918-13	74.22	1.31	20.43	0.80	N.d.	0.26	0.02	2.74	0.08	0.01	0.01	0.02	0.10
4	M1:3918-14	77.28	1.13	19.02	1.68	N.d.	0.48	0.04	0.17	0.08	0.01	N.d.	N.d.	0.12
5	M1:3918-2	49.83	1.33	40.28	0.60	0.34	0.29	0.63	6.14	0.12	0.03	0.02	0.03	0.35
6	M1:3918-22	66.89	1.23	26.19	0.61	2.30	1.18	0.13	0.59	0.11	0.02	0.02	N.d.	0.61
7	M1:3918-17	74.67	1.02	23.22	0.61	N.d.	0.14	0.02	0.08	0.07	0.02	0.02	0.01	0.11
8	M1:3918-18	41.51	2.13	51.14	2.61	1.21	0.15	0.03	0.25	0.17	0.04	0.05	0.05	0.66
8	M1:3918-18	51.19	3.49	42.60	0.49	0.85	0.33	0.04	0.17	0.13	0.03	0.03	0.04	0.50
9	M1:3918-4	87.19	0.64	9.78	2.04	N.d.	0.17	0.04	N.d.	0.05	N.d.	N.d.	N.d.	0.08
10	M1:3918-5	61.99	2.15	31.22	3.75	0.09	0.19	0.10	0.10	0.11	0.02	0.02	0.03	0.22
11	M1:3918-1	75.03	1.15	21.34	1.25	N.d.	0.62	0.05	0.19	0.11	0.02	0.01	N.d.	0.18
11	M1:3918-1	75.25	2.03	19.94	1.35	0.28	0.38	0.07	0.21	0.12	0.02	0.01	0.01	0.23
12	M1:3918-7	86.14	1.14	8.65	2.37	0.21	0.89	0.07	0.29	0.09	N.d.	N.d.	0.01	0.14

(*Continued*)

Table 6.1 (*Continued*)

Serial no.	Sample no.	SiO_2	Al_2O_3	PbO	BaO	P_2O_5	Fe_2O_3	CuO	CaO	MnO	ZnO	SnO_2	NiO	Cl
13	M1:3918-6	70.40	1.37	25.82	1.98	N.d.	0.16	0.03	N.d.	0.10	0.02	0.02	0.02	0.09
14	M1:3918-11	90.57	1.10	5.84	1.60	N.d.	0.32	0.09	0.28	0.06	N.d.	N.d.	N.d.	0.13
14	M1:3918-11	32.15	1.37	57.72	5.92	1.60	0.10	0.05	N.d.	0.17	0.05	0.04	0.06	0.78
15	M1:3918-20	52.15	1.41	41.20	3.45	0.71	0.26	0.03	0.12	0.12	0.03	0.03	0.04	0.45
16	M1:3918-15	73.79	2.16	21.63	1.10	0.21	0.61	0.06	0.14	0.07	0.02	0.01	0.01	0.16
17	M1:3918-16	79.43	1.30	18.06	0.78	N.d.	0.17	0.02	0.09	0.06	0.01	0.01	0.01	0.05
18	M1:3918-19	36.66	2.34	55.21	2.42	1.57	0.12	0.02	0.44	0.30	0.05	0.03	0.05	0.80
19	M1:3918-9	75.77	1.39	15.47	6.58	N.d.	0.21	0.04	0.11	0.17	0.01	0.01	0.02	0.20
20	M1:3918-8	26.25	2.57	63.30	4.37	1.99	0.15	0.02	N.d.	0.18	0.05	0.04	0.06	1.03
21	M1:3918-21	42.02	2.26	52.17	0.86	1.45	0.18	0.08	N.d.	0.14	0.03	0.04	0.04	0.73
21	M1:3918-21	56.65	2.72	37.07	1.98	0.37	0.14	0.06	0.51	0.15	0.03	0.02	N.d.	0.29
22	M1:3918-10	83.73	1.66	10.50	2.56	0.30	0.43	0.05	0.47	0.11	N.d.	0.01	N.d.	0.19

Table 6.2 Bulk composition of the ancient glass chimes determined by the HHpXRF spectrometer from the less weathered parts (wt %)

Serial no.	Sample no.	SiO_2	Al_2O_3	PbO	BaO	P_2O_5	Fe_2O_3	CuO	CaO	MnO	ZnO	SrO	SnO_2	NiO	Cl
1	M1:3918-3	51.43	3.02	28.23	13.90	0.51	0.11	0.06	0.55	0.38	0.02	0.04	N.d.	0.05	1.71
2	M1:3918-12	40.01	1.67	33.70	15.71	0.58	0.07	0.03	5.73	0.44	0.02	0.04	N.d.	0.09	1.91
3	M1:3918-13	47.11	3.48	29.55	13.53	0.56	0.28	0.04	4.19	0.38	0.02	0.03	0.01	0.05	0.76
4	M1:3918-14	42.51	2.20	31.40	14.94	0.50	0.15	0.03	6.05	0.43	0.02	0.04	N.d.	0.05	1.68
5	M1:3918-2	46.12	1.81	30.18	14.15	0.52	0.10	0.03	4.71	0.42	0.02	0.04	0.01	0.06	1.83
6	M1:3918-22	35.43	1.77	43.44	10.98	1.32	0.40	0.04	3.95	0.39	0.03	0.06	0.02	0.04	2.11
7	M1:3918-17	45.16	2.24	31.82	14.31	0.50	0.14	0.03	3.78	0.43	0.02	0.04	N.d.	0.04	1.48
8	M1:3918-18	44.13	3.61	35.51	13.81	0.75	0.42	0.38	0.54	0.34	0.03	0.04	N.d.	0.03	0.41
9	M1:3918-4	55.35	2.92	26.86	12.71	0.40	0.16	0.05	0.27	0.34	0.02	0.03	N.d.	0.03	0.85
10	M1:3918-5	47.73	2.17	30.15	15.34	0.66	0.11	0.04	1.39	0.42	0.03	0.04	N.d.	0.04	1.88
12	M1:3918-7	44.35	1.70	28.75	15.34	0.64	0.12	0.03	6.69	0.37	0.01	0.04	0.02	0.03	1.91
13	M1:3918-6	53.06	2.34	28.90	13.00	0.82	0.25	0.03	N.d.	0.37	0.02	0.04	N.d.	0.05	1.12
14	M1:3918-11	51.41	1.30	29.23	13.81	0.61	0.12	0.03	1.12	0.39	0.02	0.04	N.d.	0.04	1.88
15	M1:3918-20	55.05	3.29	27.10	12.46	0.44	0.17	0.05	N.d.	0.35	0.02	0.03	N.d.	0.05	0.98

(*Continued*)

Table 6.2 (*Continued*)

Serial no.	Sample no.	SiO_2	Al_2O_3	PbO	BaO	P_2O_5	Fe_2O_3	CuO	CaO	MnO	ZnO	SrO	SnO_2	NiO	Cl
16	M1:3918-15	55.38	2.18	28.03	12.26	0.53	0.08	0.01	N.d.	0.32	0.02	0.03	0.02	0.04	1.11
16	M1:3918-15	48.21	3.28	30.56	13.94	0.69	0.26	0.08	1.70	0.33	0.04	0.03	N.d.	0.05	0.85
17	M1:3918-16	42.79	1.54	32.27	14.77	0.58	0.10	0.01	5.54	0.44	0.02	0.04	N.d.	0.06	1.83
18	M1:3918-19	34.27	2.59	45.29	15.46	0.88	0.23	0.15	0.27	0.34	0.04	0.02	N.d.	0.03	0.43
19	M1:3918-9	44.79	1.28	32.53	14.83	0.73	0.12	0.03	3.15	0.41	0.02	0.04	N.d.	0.05	2.01
19	M1:3918-9	56.86	1.82	26.48	12.06	0.48	0.06	0.02	N.d.	0.32	0.02	0.04	N.d.	0.04	1.81
20	M1:3918-8	41.66	1.18	33.74	15.88	0.73	0.05	0.04	3.88	0.47	0.02	0.04	0.02	0.05	2.23
21	M1:3918-21	52.63	2.93	28.87	13.14	0.46	0.28	0.07	0.40	0.32	0.02	0.03	N.d.	0.04	0.81
22	M1:3918-10	63.60	1.88	22.26	8.39	1.84	0.50	0.03	0.53	0.20	0.02	0.01	N.d.	0.04	0.68
Average		47.78	2.27	31.08	13.68	0.68	0.19	0.06	2.37	0.37	0.02	0.04	0.00	0.05	1.40

analysis, we could deduce that — where the cerusite compound is deposited — in the weathered surfaces the content of PbO would increase. For example the content of PbO in samples of M1:3918–8, M1:3918–11, M1:3918–18, M1:3918–19 and M1:3918–21 ranges from 51.14% to 63.30% and this is much higher than the average PbO content of 31.08% detected in the less weathered parts. If both PbO and BaO were lost from the weathered layers, the content of SiO_2 can be over 70% and reaches 90.57% in sample M1:3918–11.

Before the present study, fragments taken from the inner parts of three ancient chimes, M1:3918–9, M1:3918–10 and M1:3918–11, were analyzed using a JSM-6490 scanning electron microscope equipped with energy dispersive spectrometer (SEM-EDS). They were also analysed with a Thermo Fisher X Series II type inductively coupled plasma -mass spectrometer (ICP-MS). The average weight percentage for SiO_2, PbO and BaO obtained was about 48%, 34% and 14%, respectively (unpublished data from the Nanjing Museum). This coincides with the results shown in Table 6.2. We can confirm that ancient craftsmen used the same batch materials to make the glass chimes of different sizes.

6.4.3 *Results of the lead isotope analysis*

The results of the lead isotope analysis of 19 of these ancient chimes are listed in Table 6.3. Figure 6.4 shows the scatter diagrams of the lead isotope ratios obtained.

It can be seen that the analyzed samples have very similar isotope ratios. The ratios of $^{207}Pb/^{206}Pb$ and $^{208}Pb/^{206}Pb$ are in the ranges of 0.8856–0.8861 and 2.1716–2.1736 respectively; they fall in the higher isotope range reported by other scholars.[7] The reported results have shown that the ratios of $^{207}Pb/^{206}Pb$ and $^{208}Pb/^{206}Pb$ for the Chinese lead-rich minerals from deposits in northern China fall in the ranges of 0.8507–0.9605 and 2.0834–2.3002 respectively.[8,9] It suggests that the lead-containing materials used in these ancient PbO-BaO-SiO_2 glass chimes found at Dayun Hill were derived from deposits in north of China. However, the exact locations of the deposits used needs to be determined.

Table 6.3 Lead isotope ratios for the ancient glass chimes from Dayun Hill, Xuyi County

Sample no.	$^{207}Pb/^{206}Pb$	$^{208}Pb/^{206}Pb$	$^{206}Pb/^{204}Pb$	$^{207}Pb/^{204}Pb$	$^{208}Pb/^{204}Pb$
M1:3918-2	0.8858	2.1736	17.508	15.508	38.055
M1:3918-3	0.8858	2.1735	17.499	15.500	38.035
M1:3918-4	0.8858	2.1734	17.500	15.502	38.034
M1:3918-5	0.8857	2.1732	17.508	15.508	38.049
M1:3918-7	0.8859	2.1735	17.499	15.501	38.033
M1:3918-8	0.8861	2.1747	17.516	15.522	38.094
M1:3918-9	0.8856	2.1716	17.499	15.498	38.001
M1:3918-11	0.8858	2.1721	17.497	15.499	38.006
M1:3918-12	0.8858	2.1730	17.500	15.501	38.027
M1:3918-10	0.8856	2.1724	17.497	15.497	38.010
M1:3918-13	0.8857	2.1729	17.498	15.499	38.021
M1:3918-14	0.8856	2.1721	17.491	15.489	37.993
M1:3918-15	0.8858	2.1730	17.495	15.496	38.016
M1:3918-16	0.8856	2.1720	17.488	15.487	37.985
M1:3918-17	0.8858	2.1730	17.500	15.502	38.028
M1:3918-18	0.8856	2.1721	17.493	15.492	37.997
M1:3918-19	0.8858	2.1731	17.500	15.502	38.031
M1:3918-20	0.8857	2.1725	17.501	15.500	38.018
M1:3918-21	0.8856	2.1726	17.502	15.500	38.024

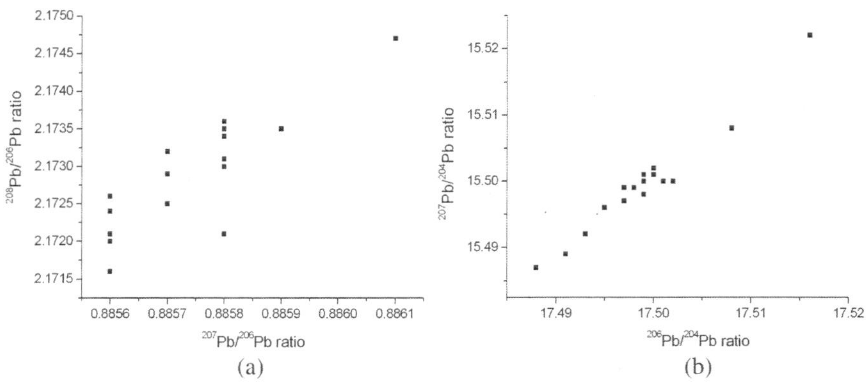

Fig. 6.4. Scatter diagrams of the lead isotope ratios obtained. (a) $^{208}Pb/^{206}Pb$ *vs.* $^{207}Pb/^{206}Pb$; (b) $^{207}Pb/^{204}Pb$ *vs.* $^{206}Pb/^{204}Pb$

6.4.4 *Further remarks*

Up to now, a large number of locally made PbO-BaO-SiO_2 glass artifacts, which date from the middle of the Warring States Period (475–221 BC) to the Western Han Dynasty (202 BC–8 AD), have been unearthed from many places in China.[10–12] Most of these artifacts were used as ornaments (e.g. eye beads and tubes), ritual artifacts such as discs (*Bi*) or for decoration of swords (e.g. sword panels and scabbards). Small flat plaques of Western Han Dynasty date were used to make suits for dead people.[13,14] Occasionally, some PbO-BaO-SiO_2 glass vessels and large animal-shaped artifacts of the Western Han Dynasty date were found in Xuzhou city, Jiangsu Province.[15] The heaviest artifact reported previously is the PbO-BaO-SiO_2 glass disc unearthed from the Mausoleum of the Wudi, Liu Che, of the Western Han Dynasty: it weighs about 1.9 kg.[10] By contrast, 18 of the glass chimes from Dayun Hill are over 2.6 kg. The total weight of the 22 chimes is about 131 kg, and the making of them was time-consuming and needed high quality raw materials. According to this phenomenon, we can say that during the Western Han Dynasty the making of PbO-BaO-SiO_2 glass reached its peak.

A thick flat PbO-BaO-SiO_2 glass piece dating to the Han Dynasty with a length of 32.5 cm, a width of 14.8 cm, a thickness of 3.5 cm and a weight of 5.25 kg was excavated from Jimo county, Shandong Province.[16] It has a hole with a diameter of 1.8 cm, and its weight is similar with that of the glass chime M1:3918–8 from the Dayun Hill. But the position of the hole is in a top corner of the artifact, so it is different from the chimes found at Dayun Hill. Although we still don't know whether the flat glass from Jimo county is a chime or not, it seems necessary to trace its possible relationship with those from Dayun Hill.

Just as discs, cicadas and plaques made from PbO-BaO-SiO_2 glasses were made to imitate stone ones, jade or stone chimes are thought to be one of the reasons why ancient glass chimes found in the Jiangdu King's Mausoleum were made. These ancient glass chimes were thought to be formed by the mould-casting method. Since high lead glass can cause severe corrosion of crucibles during the manufacturing process, enough refractory crucibles were needed to melt the glass before such a large number of glass chimes were cast. To avoid the fracture, the ancient craftsmen probably annealed the chimes after casting.

6.5 Conclusion

In brief, the present study has provided some important information about lead barium glass Bian Qing of the Western Han Dynasty unearthed from the Jiangdu King's Mausoleum. It has been found that these ancient glass chimes of different sizes have very similar chemical compositions, with average contents of about 31.08% PbO, 13.68% BaO and 47.78% SiO_2. Their surfaces were seriously weathered and crystalline $PbCO_3$ particles were easy identified. The 19 samples analyzed have very similar lead isotope ratios. The ratios of $^{207}Pb/^{206}Pb$ and $^{208}Pb/^{206}Pb$ are in the ranges of 0.8856–0.8861 and 2.1716–2.1736, respectively. We would like to suggest that all of the 22 lead barium glass chimes were made by using the same recipe of batch raw materials. The lead-containing minerals used were possibly from deposits located in northern China.

The discovery of lead barium glass Bian Qing from the Jiangdu King's Mausoleum indicates that the making of ancient PbO-BaO-SiO_2 glasses in China reached to its peak during the Western Han Dynasty. Their appearance also provides new evidence for tracing the development of the handicraft in ancient China, and infers that some cultural and social factors would have been essential for the development of ancient PbO-BaO-SiO_2 glasses.

Some researchers have made duplicate ancient glass chimes based on their chemical compositions. The primary simulation experiment using these duplications demonstrates that the lead-barium-silicate glass chimes could generate certain dulcet tones. But we still do not know whether these ancient glass chimes were used to play music or were only artifacts used in ceremonies and funerals. Further research will reveal more secrets related to these ancient glass chimes.

Acknowledgments

This research was supported by the "973" Program (Grant No. 2012CB720906) from the Chinese Ministry of Science and Technology, the Shanghai R&D Public Service Platform Construction Project (Grant No. 13DZ2295800) and the National Natural Science Foundation of China (Grant No. 11374314).

References

1. Nanjing Museum and Bureau for Culture and Broadcasting and News, Xuyi County, *Archaeology (Kaogu)* 7, 53 (2012).
2. H. X. Zhao, Q. H. Li and S. Liu, *Journal of Raman Spectroscopy* **44**, 643 (2013).
3. E. Frahm and R. C. P. Doonan, *Journal of Archaeological Science* **40**, 425 (2013).
4. J. F. Cui, Y. Lei and Z. B. Jin, *Archaeometry* **52**, 597 (2010).
5. J. F. Cui, X. H. Wu and B. L. Huang, *Journal of Archaeological Science* **38**, 1671 (2011).
6. P. Colomban, A. Tournie and L. B. Gurlet, *Journal of Raman Spectroscopy* **37**, 841 (2006).
7. R. H. Brill and H. Shirahata, The second Kazuo Yamasaki TC-17 lectures on Asian glass: recent lead-isotope analyses of some Asian glasses with remarks on strontium-isotope analyses, in *Ancient Glass Research along the Silk Road*, eds. F. X. Gan, R. H. Brill and S. Y. Tian (World Scientific Publishing Co. Pte. Ltd, Singapore, 2009), pp. 149–164.
8. Z. H. Jiang and Q. Y. Zhang, *Journal of the Chinese Ceramic Society* **26(**1), 109–113 (1998).
9. Y. Chen, C. Mao and B. Zhu, *GeoChimica* **3**, 215 (1980).
10. F. X. Gan *et al.* (eds.), *Development of Chinese Ancient Glass*, 1st edn. (Shanghai science and Technology Press, Shanghai, 2005), pp. 252–269.
11. F. X. Gan, R. H. Brill and S. Y. Tian (eds.), Ancient Glass *Research along the Silk Road*, 1st edn. (World Scientific Publishing Co. Pte. Ltd., Singapore, 2009), pp. 1–40.
12. J. Y. Fu and K. Q. Xu, Shanghai Wenbo **2**, 27 (2010).
13. Q. H. Li, D. H. Gu and F. X. Gan, *Nuclear Techniques* **26**, 922 (2003).
14. Q. M. Zhang, *Relics and Museology (Wenbo)* **1**, 59 (2004).
15. Y. D. Li, *Southeast Culture (Dong Nan Wen Hua)* **Z1**, 109 (1990).
16. M. G. Shi and F. Z. Zhou, Investigation on ancient flat glass unearthed from China, in *Proceedings of XVI International Congress on Glass*, Vol. 7 (Madrid, Spain, 1992), pp. 161–166.

Chapter 7 第七章

Study of Byodoin Temple Glass: Discoveries in Japan of Small, Free-Blown Glass Vessels with Lids from the Byodoin Temple and Other Sites from the Heian Period (794–1192)

Akiko Inoue, Junko Shirataki† and Izumi Nakai†*

*Department of the Arts,
Tokai University, 4-1-1 Kitakaname,
Hiratsuka-shi, Kanagawa-ken 259-1292, Japan

†Department of Applied Chemistry,
Tokyo University of Science,
1-3 Kagurazaka, Shinjuku-ku, Tokyo 162-8601, Japan

7.1 Introduction

In the late 7th and 8th centuries, significant numbers of glass beads were fabricated in Japan using domestically produced PbO-SiO_2 glass with high PbO content.[1] Many of these may be found among the treasures of the Shosoin Repository, a collection of artifacts that derives mainly from the Emperor Shomu period of the 8th century, and from the Todaiji Temple in Nara, which was then the capital city of Japan. However, glass findings from the 9th century and later have been fewer in number, prompting many to believe that the use of glass became less popular and production had mostly ceased during the Heian period (794–1192 AD).

The recent discovery of hundreds of glass beads and fragments of small vessels coupled with lids at the Byodoin Temple in Kyoto, which became Japan's new capital in the Heian period, has changed this long-held view. On-site chemical analyses of the objects by our group (Tokyo University of Science) revealed the presence of two compositional types: $PbO\text{-}K_2O\text{-}SiO_2$ (designated as potash lead silica glass) and $PbO\text{-}SiO_2$ (lead silica glass). Both types are widely recognized in China which suggests that the glass objects had been imported from China or their production techniques had entered Japan.

Studies of production methods and shapes by one of the authors (A.I.) have shown close similarities of the Byodoin fragments to a group of vessels and fragments that were discovered in widely separated sites in western Japan, dating from the 12th and the beginning of the 13th centuries. Both the Byodoin and artifacts from other sites are fragments of small vessels and lids made by a free-blowing technique. As they bear close resemblance, it is reasonable to believe that quite a few of these objects were distributed in western Japan at that time. Because of the likelihood that they are closely related to the Chinese glasses, we present here the results of chemical analyses of the Byodoin pieces, with reference to the characteristics of the vessels and their supposed production methods.

7.2 Byodoin glasses

Byodoin Hoo-do (Phoenix Hall) is a splendid temple built by the powerful and aristocratic Fujiwara family in Uji, a suburb of Kyoto, in 1053 AD. The glass artifacts were found in 2004, along with various other objects, in the pedestal of the Amida Nyorai (*Amitābha*) statue, the temple's principal image, during a repair process. We reported our scientific studies of the glasses in 2011 and 2012.[2–4]

The discoveries amounted to more than 240 glass beads and fragments, as well as ~200 glass beads woven into Yoraku chains (*ratnamālā*, an ornament fashioned from a metal chain into which glass beads, bells, and shells are interlaced, and used to decorate Buddha sculptures or pedestals), and 84 glass vessel fragments. It is not known why they had been stored in the pedestal, but because of the luxuriousness and beauty of the Yoraku chains, it is believed that the objects were used to decorate the hall at the time of its foundation, when the Fujiwara family was at the pinnacle of their power.

The chemical composition of a glass reflects its source material and sometimes its place of production, and hence, its origin. Therefore, chemical analysis plays a very important role in the study of historical glass of unknown provenance. We were interested in the origin of the glasses discovered in the Byodoin Temple. Therefore, we chemically characterized the glass objects, using a non-destructive X-ray fluorescence (XRF) technique. We compared the chemical compositions of the Byodoin glass with known historical lead glasses found in Japan and China. In Section 7.4, we will compare the typology of the vessel glass with specimens previously reported from Japan to reveal their historical and technical backgrounds.

7.3 Chemical characterization of Byodoin glass

7.3.1 *Analyzed samples*

Tables 7.1 and 7.2 list the samples quantitatively investigated in this study. The analyzed specimens include 33 vessel fragments and 35 glass beads found in Byodoin. Details of the samples are described in the other sections. Photographs of representative samples of vessel fragments and beads analyzed by non-destructive XRF are shown in Figs. 7.1(a) and 7.1(b), respectively.

7.3.2 *Method of analysis*

A portable Ourstex 100FA IV XRF spectrometer (Ourstex Co.), brought to Byodoin, was used to analyze the samples. The XRF instrument was specially designed for the analysis of glass.[5–8] The analytical procedures and conditions are described elsewhere.[8] The measurement period was 200 s (live time) and the tube current (mA) was adjusted such that the dead time did not exceed 30%. Standard

Table 7.1 Chemical composition of vessel fragments from the Byodoin Temple

Sample No.	Color*	Density/ g·cm^{-3}	Production Technique**	Chemical Composition***/wt% PbO	Na_2O	MgO	Al_2O_3	SiO_2	P_2O_5	K_2O	CaO	TiO_2	Cr_2O_3	MnO_2	Fe_2O_3	CuO	ZnO	Rb_2O	SrO	ZrO_2	As_2O_3†	SnO†
A-01	blue	3.96	f.b.	51.6	n.d.	0.6	1.3	40.3	1.3	3.5	tr.	0.05	0.05	tr.	0.20	0.71	tr.	tr.	n.d.	n.d.		
A-02	blue	3.97	f.b.	51.1	0.4	0.8	2.2	38.0	2.2	3.6	0.4	0.05	0.04	tr.	0.26	0.85	tr.	tr.	n.d.	n.d.		
A-08	blue	4.18	f.b.	54.0	tr.	0.5	1.7	40.0	0.9	1.7	tr.	0.05	tr.	n.d.	0.23	0.61	tr.	tr.	n.d.	n.d.		
A-15	blue	3.97	f.b.	50.7	0.4	tr.	2.1	36.8	3.4	4.0	1.5	0.06	tr.	0.04	0.45	0.41	tr.	n.d.	n.d.	n.d.		
A-18	blue	4.03	f.b.	51.3	0.3	tr.	1.7	39.3	2.4	3.0	1.1	0.06	tr.	0.03	0.25	0.41	tr.	n.d.	n.d.	n.d.		
A-19	blue	3.82	mold	50.9	tr.	tr.	0.8	43.5	1.1	2.7	tr.	0.03	0.05	tr.	0.12	0.25	tr.	tr.	n.d.	n.d.		
A-37	blue	3.58	f.b.	44.7	n.d.	0.8	0.8	46.4	0.7	5.9	n.d.	0.03	tr.	tr.	0.16	0.48	tr.	n.d.	n.d.	n.d.		
B-01	blue	3.87	f.b.	49.7	tr.	0.5	2.0	38.2	2.4	3.4	0.5	tr.	0.03	0.05	1.39	1.44	tr.	tr.	n.d.	n.d.		
B-03	blue	4.02	f.b.	51.4	tr.	tr.	1.7	40.5	2.0	2.8	0.3	0.06	tr.	tr.	0.23	0.70	tr.	n.d.	n.d.	n.d.		
B-05	blue	4.03	f.b.	51.1	tr.	tr.	1.6	39.1	2.5	3.3	0.7	0.06	tr.	tr.	0.25	0.88	tr.	tr.	n.d.	n.d.		
B-07	blue	4.10	winding or working	52.3	tr.	tr.	1.0	38.4	0.9	5.6	n.d.	0.03	tr.	n.d.	0.09	0.99	0.04	tr.	n.d.	n.d.		
B-10	blue	4.09	f.b. + winding	51.9	0.3	0.6	1.8	36.9	3.4	3.2	0.7	0.06	tr.	tr.	0.23	0.75	tr.	n.d.	n.d.	n.d.	○	
B-11	blue	3.92	f.b.	49.3	0.5	tr.	1.8	37.1	5.2	3.2	1.3	tr.	tr.	0.06	0.42	0.75	tr.	n.d.	n.d.	n.d.		
B-12	blue	4.08	f.b. + welding	51.9	tr.	0.6	1.5	37.4	3.1	4.1	0.4	tr.	0.04	tr.	0.21	0.42	0.02	tr.	n.d.	n.d.		
B-14	blue	4.11	f.b. + welding	51.6	tr.	0.5	2.3	36.2	2.1	5.6	0.3	0.05	0.04	0.03	0.21	0.87	0.04	n.d.	n.d.	n.d.		○
B-15	blue	3.84	f.b. + winding	51.0	n.d.	0.4	0.8	42.8	0.3	3.6	0.1	0.03	tr.	tr.	0.24	0.56	0.03	tr.	n.d.	n.d.		○

B-16	blue	4.08	f.b.	53.1	tr.	tr.	0.7	41.8	0.7	2.6	tr.	0.03	0.04	tr.	0.14	0.29	tr.	tr.	n.d.	n.d.		
C-01	green	4.10	f.b.	51.3	tr.	0.7	1.4	33.2	2.6	6.0	0.9	0.04	0.03	tr.	0.37	3.06	0.10	n.d.	n.d.	n.d.		
C-02	blue	3.77	mold	48.8	tr.	1.0	1.4	37.5	4.3	3.9	1.5	0.06	0.04	0.08	0.45	0.71	tr.	n.d.	n.d.	n.d.		
C-03	blue	3.77	mold	48.8	tr.	0.6	2.0	38.3	3.5	4.1	1.3	0.05	tr.	0.04	0.35	0.73	tr.	tr.	n.d.	n.d.		
C-04	blue	3.61	winding + working	46.2	tr.	2.5	1.1	39.2	3.9	5.3	0.5	0.03	tr.	0.04	0.22	0.81	0.04	tr.	n.d.	n.d.		
C-06	blue	3.86	mold	48.7	0.4	1.5	1.8	32.5	7.5	4.7	1.7	0.03	tr.	0.07	0.24	0.70	0.09	n.d.	n.d.	tr.		
C-07	blue	3.81	mold	49.0	0.3	0.5	1.5	36.0	6.2	4.0	1.3	0.05	tr.	0.07	0.35	0.64	0.07	n.d.	n.d.	n.d.	○	
D-01	c.l.	4.09	f.b. + winding	50.3	n.d.	1.5	5.0	29.4	6.0	5.1	1.6	0.07	tr.	0.04	0.48	0.44	tr.	n.d.	n.d.	n.d.		
D-02	m.w.	4.03	f.b.	51.1	n.d.	0.6	1.6	37.6	1.8	5.9	0.9	0.19	tr.	0.04	0.17	0.26	tr.	n.d.	n.d.	n.d.		○
D-03	c.l.	3.95	f.b.	51.3	tr.	0.6	1.5	39.2	1.7	4.8	0.5	0.04	0.03	n.d.	0.18	0.06	tr.	tr.	n.d.	n.d.		○
D-04	c.l.	3.98	f.b. + welding	52.4	n.d.	0.8	1.7	43.6	1.6	2.7	0.9	0.09	tr.	tr.	0.19	0.09	tr.	tr.	tr.	n.d.		
D-07	c.l.	4.01	f.b.	51.2	tr.	0.4	2.2	39.3	1.8	4.3	tr.	0.06	tr.	tr.	0.19	0.09	tr.	n.d.	n.d.	n.d.		
D-09	c.l.	4.02	f.b.	51.7	n.d.	0.6	1.0	41.1	1.5	3.5	tr.	tr.	0.05	tr.	0.14	0.06	tr.	tr.	n.d.	n.d.		
E-01	c.l.	4.00	f.b.+working or winding	51.6	n.d.	0.4	0.7	40.5	0.6	5.7	n.d.	0.05	0.03	tr.	0.16	0.07	tr.	tr.	n.d.	n.d.		
E-02	c.l.	3.84	f.b.	51.3	tr.	0.6	0.8	43.8	0.5	2.5	n.d.	0.04	0.04	tr.	0.13	0.05	tr.	n.d.	n.d.	n.d.	○	
E-03	green	4.22	mold	55.2	n.d.	tr.	1.1	36.4	0.9	4.6	tr.	0.05	tr.	tr.	0.45	0.78	0.05	n.d.	n.d.	n.d.		
E-04	blue	3.86	f.b.	50.7	n.d.	0.5	1.0	41.0	0.7	4.1	tr.	0.03	0.04	0.04	0.12	1.42	tr.	tr.	n.d.	n.d.		

* c.l. : colorless; m.w.: milky white; ** f.b.: free-blown; ***tr.: trace, n.d. not detected; † O: detected

Table 7.2 Chemical composition of glass beads from the Byodoin Temple.

Sample No.	Color	Type	Density/ gcm^{-3}	Chemical Composition/wt%																
				PbO	Na_2O	MgO	Al_2O_3	SiO_2	P_2O_5	K_2O	CaO	TiO_2	Cr_2O_3	MnO	Fe_2O_3	CuO	ZnO	Rb_2O	As_2O_3	SnO_2
E-05	colorless	$PbO-K_2O-SiO_2$	4.00	52	0.3	0.2	0.8	42.4	0.8	3.1	0.1	0.04	0.03	0.01	0.1	0.1	0.02	0.02		
E-07	beige	$PbO-SiO_2$	5.63	74	0.3	0.6	1.2	19.0	2.1	1.2	0.3	0.08	0.03	0.01	0.5	0.4	0.03	0.02	○	
E-08	beige	$PbO-SiO_2$	5.49	72	0.3	0.5	2.2	16.4	4.8	2.1	1.0	0.04	0.02	0.05	0.5	0.4	0.04	0.03		
E-09	colorless	$PbO-K_2O-SiO_2$	3.82	52	0.3	0.3	1.0	42.5	0.5	3.0	0.1	0.02	0.02	0.01	0.1	0.1	0.03	0.02	○	
E-10	dark brown	$PbO-SiO_2$	5.21	69	0.1	0.7	0.7	26.1	1.7	0.4	0.2	0.05	0.05	0.03	1.1	0.1	0.03	0.01		
E-11	green	$PbO-SiO_2$	5.29	70	0.1	0.3	0.7	26.0	1.6	0.4	0.3	0.05	0.04	0.01	0.1	0.4	0.03	n.d.	○	
E-12	dark green	$PbO-SiO_2$	5.23	68	0.4	0.7	1.2	23.6	2.7	0.6	0.5	0.04	0.01	0.04	1.4	1.0	0.02	0.01		
E-13	yellow green	$PbO-SiO_2$	5.43	73	0.2	0.2	1.4	22.4	0.5	1.6	0.1	0.05	0.05	0.02	0.3	0.3	0.02	0.02		
E-14	tea green	$PbO-K_2O-SiO_2$	3.56	50	0.3	0.5	1.4	34.2	5.3	3.7	1.0	0.06	0.03	0.06	0.5	3.0	0.02	0.02		○
E-15	tea green	$PbO-K_2O-SiO_2$	3.64	51	n.d.	0.8	2.1	35.2	1.9	4.7	0.5	0.05	0.03	0.08	0.3	3.6	0.04	0.02		
E-16	tea green	$PbO-K_2O-SiO_2$	3.53	50	0.3	0.7	1.6	33.8	5.5	4.5	1.3	0.07	0.03	0.07	0.5	1.7	0.02	0.02		○
E-18	blue (milky)	$PbO-K_2O-SiO_2$	3.93	52	0.1	0.6	0.8	40.5	1.9	3.2	0.2	0.03	0.03	0.01	0.2	0.8	0.03	0.01		○

E-19	blue	PbO-K_2O-SiO_2	3.86	52	0.1	0.5	1.3	41.4	0.9	3.0	0.2	0.03	0.04	0.02	0.3	0.3	0.02	0.02	
E-20	blue	PbO-K_2O-SiO_2	4.04	51	0.1	0.3	1.0	40.0	0.7	5.3	0.1	0.03	0.03	0.02	0.3	0.8	0.02	0.02	
E-21	blue (milky)	PbO-K_2O-SiO_2	3.81	51	0.3	0.4	1.3	39.1	1.7	4.2	0.2	0.03	0.03	0.03	0.3	1.1	0.03	0.02	○
F-01	colorless	PbO-K_2O-SiO_2	4.00	52	0.2	0.3	1.0	42.8	0.7	1.9	0.2	0.01	0.03	0.03	0.15	0.4	0.02	0.01	
F-02	colorless	PbO-K_2O-SiO_2	3.99	52	0.3	0.2	2.0	42.5	0.4	2.1	0.1	0.04	0.04	0.03	0.3	0.1	0.02	0.02	
F-03	colorless	PbO-K_2O-SiO_2	3.99	52	0.3	0.2	1.2	42.5	0.2	3.3	0.02	0.02	0.04	0.02	0.2	0.1	0.02	0.02	
F-04	colorless	PbO-K_2O-SiO_2	4.04	52	0.2	0.3	0.9	42.8	0.3	2.8	0.05	0.02	0.05	0.02	0.2	0.2	0.03	0.03	
F-05	colorless	PbO-K_2O-SiO_2	3.99	52	0.2	0.3	1.2	43.2	0.3	1.9	0.03	0.04	0.04	0.05	0.2	0.3	0.03	0.02	
F-09	blue (milky)	PbO-K_2O-SiO_2	4.02	52	0.2	0.2	1.4	41.4	0.5	2.2	0.2	0.04	0.03	0.06	0.3	1.5	0.09	0.03	○
F-10	blue	PbO-K_2O-SiO_2	4.04	52	0.2	0.2	0.8	40.6	0.5	4.9	0.1	0.03	0.03	0.02	0.1	0.7	0.05	0.03	
F-26	colorless	PbO-K_2O-SiO_2	3.95	52	0.2	0.4	1.0	41.8	1.6	2.2	0.3	0.02	0.04	0.05	0.27	0.4	0.02	0.02	
H-18	blue (milky)	PbO-K_2O-SiO_2	3.93	52	0.1	1.0	0.9	41.4	0.5	3.0	0.1	0.05	0.04	0.01	0.2	0.4	0.02	0.01	○
H-19	yellow green	PbO-SiO_2	5.32	71	0.2	0.3	1.6	24.6	0.8	0.9	0.2	0.04	0.04	0.03	0.2	0.2	0.04	0.02	

(*Continued*)

Table 7.2 (*Continued*)

Sample No.	Color	Type	Density/ gcm^{-3}	Chemical Composition/wt%																
				PbO	Na_2O	MgO	Al_2O_3	SiO_2	P_2O_5	K_2O	CaO	TiO_2	Cr_2O_3	MnO	Fe_2O_3	CuO	ZnO	Rb_2O	As_2O_3	SnO_2
H-20	beige	$PbO-K_2O-SiO_2$	4.04	52	0.3	0.4	1.2	41.1	2.0	2.8	0.2	0.02	0.02	0.01	0.14	0.1	0.03	0.02	○	
H-23	beige	$PbO-K_2O-SiO_2$	3.99	50	0.5	0.4	2.4	36.4	5.7	2.2	1.3	0.03	0.02	0.05	0.4	0.4	0.03	0.02		
H-26	blue (milky)	$PbO-K_2O-SiO_2$	3.89	51	0.3	0.1	1.8	40.1	2.2	2.9	0.2	0.04	0.05	0.03	0.3	0.6	0.03	0.02		○
H-31	greenish blue	$PbO-K_2O-SiO_2$	4.09	51	0.2	n.d.	0.9	40.9	1.3	4.2	0.1	0.05	0.02	0.01	0.1	0.5	0.03	0.02		
J-03	blue	$PbO-K_2O-SiO_2$	3.72	51	0.2	0.3	1.7	39.2	1.3	3.8	0.8	0.02	0.03	0.03	0.3	0.7	0.02	0.02		
J-08	beige	$PbO-SiO_2$	5.25	69	0.2	0.6	1.2	21.2	4.8	1.0	0.7	0.03	0.03	0.02	0.4	0.5	0.03	0.02		
J-13	beige	$PbO-SiO_2$	5.32	71	0.4	0.7	1.0	23.1	1.5	2.0	0.1	0.10	0.04	0.01	0.22	0.2	0.02	0.01	○	
T-01	dark green	$PbO-SiO_2$	5.25	70	n.d.	0.9	0.9	22.2	2.9	0.3	0.3	0.01	0.02	0.02	0.3	1.8	0.03	0.01		
T-02	pale blue	$PbO-K_2O-SiO_2$	3.93	52	0.2	0.8	1.1	41.0	2.8	1.4	0.4	0.02	0.02	0.02	0.1	0.4	0.02	0.01	○	
T-03	dark green	$PbO-SiO_2$	5.08	67	0.1	0.9	1.0	26.4	1.0	0.6	0.1	0.04	0.05	0.03	1.1	2.0	0.03	n.d.		

(b)

Fig. 7.1 Photos of representative glass vessels fragments (a), and glass beads (b) from the Byodoin Temple

glass samples (NIST SRM 610, 612, 621, 1412, 1830, and 1831) and 15 synthetic glasses that had been previously analyzed by inductively coupled plasma–atomic emission spectrometry (ICP–AES) by the author's group were used to generate calibration curves for quantitative analysis. The measured net XRF intensity was normalized to the Compton scattering peak of the Pd Kα line (21.125 keV).

The density of each sample was measured by the floating method. Carbon tetrachloride was used as a heavy liquid. The weights of the

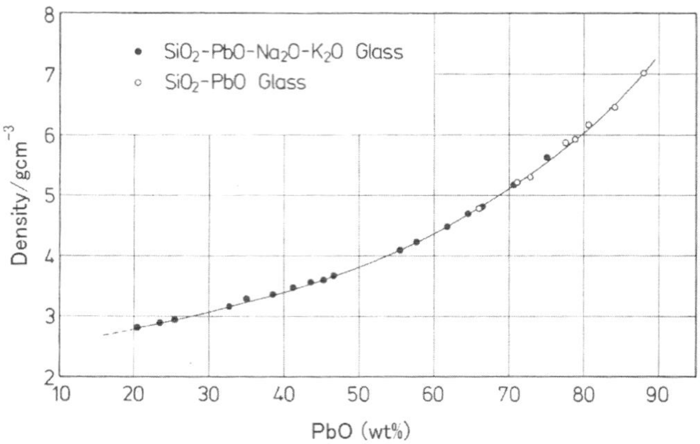

Fig. 7.2 Relationship between the PbO content and density for lead silicate glass

samples in air and after immersion in the heavy liquid were measured, and the volume of the sample was obtained; then the density of the glass was determined. It is well known that the density of lead glass is a function of the Pb concentration in the (potash) lead silicate glass, as shown in Fig. 7.2.[9] Because of its high mass absorption factor, it is difficult to determine accurate concentrations of Pb from XRF analysis only. The PbO content of a sample can be estimated from its measured density by using the calibration graph shown in Fig. 7.2, and the concentrations of the remaining elements can be quantified by the calibration curve method.[10,11]

7.3.3 *Limitations of non-destructive analysis*

The present XRF analyses were carried out non-destructively. Since the XRF technique is more or less a surface analysis, the determination of light elements is affected by the surface conditions.[12] As a result, concentrations of sodium (Na) and potassium (K) tend to decrease due to weathering. In this study, the surface concentrations of K_2O and SiO_2 tended to be lower and higher, respectively, than those of the bulk composition due to the weathering of the glass surface, which was caused by the long-term weathering conditions. This is the limitation of the non-destructive analysis of archaeological glass by the XRF technique.

7.3.4 *Chemical compositions of the Byodoin glass*

The XRF analysis results for the vessel fragments and glass beads are given in Tables 7.1 and 7.2, respectively. The basic chemical type of the glass is (potash) lead silicate glass. Potassium was added to lead silicate glass to lower the melting temperature or reduce the amount of lead used in glass production.

a. *Base glass composition*

As can be seen from Tables 7.1 and 7.2, the Byodoin glasses can be classified into two types on the basis of their chemical compositions: lead silicate glass (the PbO-SiO_2 system) and potash lead silicate glass (the PbO-K_2O-SiO_2 system). All the vessel glasses were classified as potash lead silicate glass. On the other hand, 10% of the glass beads were lead silicate glass without or less potassium and the remaining 90% were potash lead silicate glass. Both types of glasses can be clearly distinguished by density. Figure 7.3 shows a plot of the K_2O content versus density. The density of potash lead silicate glass falls into the range 3.7–4.2 g·cm^{-3}, whereas that of lead silicate glass varies from 5.1 to 5.6 g·cm^{-3}. The latter samples contain low levels of potassium.

The colors of most of the analyzed glasses varied: pale blue, pale greenish blue, greenish blue, milky blue, yellowish green, dark green,

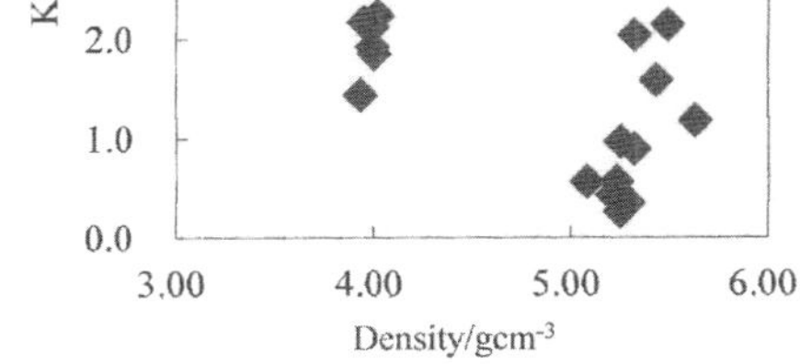

Fig. 7.3 K_2O content versus density

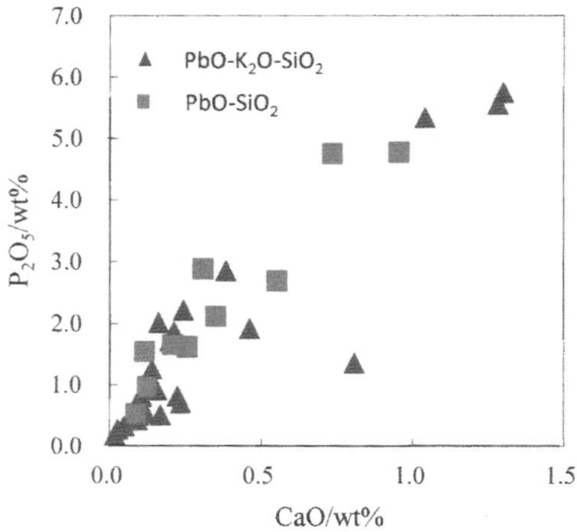

Fig. 7.4 CaO content versus P_2O_5 content

and colorless. The greenish and bluish colors are due to the presence of copper, which produces a green color in lead silicate glass and blue in potash lead silicate glass. The levels of milkiness seem to relate to the SnO_2 contents. Iron is responsible for the greenish and beige colors depending on the oxidation and reduction conditions during glass formation. Characteristically, the CaO content is lower than 2 wt%. From Fig. 7.4, a positive correlation can be observed between the P_2O_5 and CaO levels in the Byodoin glass, implying the addition of calcium phosphate (possibility as bone).

b. *Characteristics of the Byodoin beads*

After analysis, approximately 90% of the 218 pieces of beads were classified as having the PbO-K_2O-SiO_2 composition. Almost all these beads were made by the winding method. The predominant color was blue with a faint greenish tint (E-20 and H-18 in Fig. 7.1(b)), and many were similar in size (6–7 mm in diameter). About 12% of the PbO-K_2O-SiO_2 beads were colorless or had a beige tint, and were 8–9 mm in diameter.

In contrast, the PbO-SiO_2 beads, which accounted for only 10% of the analyzed beads, differed from the PbO-K_2O-SiO_2-type beads in color and shape. A dark green glass that appeared almost black and a yellowish green glass were used (see E-11, 12, and 13 in Fig. 7.1(b)). Pale and deep-beige colored glasses were also observed (E-07 and 08 in Fig. 7.1(b)). The bead sizes varied; some were larger than 10 mm and others were 4–5 mm in diameter. They were made by the winding method, and some were intricately decorated (E-11 and 12 in Fig. 7.1(b)). Decorative beads that are very similar to these Byodoin specimens are also held in the Shosoin Repository collection.[13] The PbO-SiO_2 beads from the Byodoin are thought to be made in the Nara period, and descended to the Fujiwara family, who had a close relationship by marriage with the royal family.

c. *Comparison with glass from the Shosoin Repository*

Because of the apparent similarities between the Byodoin and Shosoin glass beads, as mentioned above, it is interesting to compare their chemical compositions of the same colored beads. Table 7.3 shows the chemical compositions of green glass beads from Shosoin (sample numbers 4, 8, and 9), including other colored beads and Byodoin (E-11 and E-12).

The densities of the three green Shosoin glasses were reported as 5.3, 4.2, and 5.2 g·cm^{-3}. The second value is too low to allow identification as a high lead silicate glass and is probably due to the effects of air bubbles, as mentioned in reference.[13] The remaining two values (5.3 and 5.2 g·cm^{-3}) are similar to those of the Byodoin samples (E-11, 5.29 g·cm^{-3}, and E-12, 5.23 g·cm^{-3}). Thus, the glass beads from both sources are high lead silicate glasses. Their chemical compositions resemble each other: the PbO content of the Shosoin glass is 70–72 wt% while that of the Byodoin glass is 68–70 wt%. Similarly, the SiO_2 contents are 23–25 wt% and 24–26 wt%, respectively. Both are colored by CuO and Fe_2O_3. Their green color glasses contain similar amounts of Na_2O, MgO, Al_2O_3, K_2O, and CaO, as shown in Table 7.3. This shows that both glasses were produced by the same recipe, probably from the same workshop. This interpretation suggests that when the Phoenix Hall was constructed in 1053 AD, the precious antique glass beads of the Nara period were donated by the Fujiwara family and used to decorate the Hall.

Table 7.3 Comparison of the chemical compositions (wt%) of green glass beads from the Shosoin and Byodoin Temple

Site	Sample	Color	Density/ $g \cdot cm^{-3}$	Chemical Composition/wt%																	
				PbO	SiO_2	Na_2O	MgO	Al_2O_3	P_2O_5	K_2O	CaO	Fe_2O_3	CuO	TiO_2	Cr_2O_3	MnO	ZnO	Rb_2O	ZrO_2	As_2O_3	SnO_2
Byodoin	E-11	yellowish green	5.29	70.0	26.0	0.1	0.3	0.7	1.6	0.4	0.3	0.1	0.4	0.05	0.04	0.01	0.03	n.d.		○	
	E-12	dark green + white	5.23	67.9	23.6	0.4	0.7	1.2	2.7	0.6	0.5	1.4	1.0	0.04	0.01	0.04	0.02	0.01			
Shosoin	1	yellow	5.2	73.6	24.1	0.22	0.27	0.06		0.08	0.39	0.14	0.16								
	2	brown	5.3	72.6	23.7	0.1	0.57	0.13		0.18	0.26	0.97	0.19								
	3	r. brown	5.3	73.6	24.8	0.11	0.36	0.3		0.1	0.26	0.69	0.55 (Cu_2O)								
	4	dark green	5.3	71.9	24.7	0.12	0.06	0.31		0.09	0.71	1.08	1.44								
	5	dark brown	5.3	73.3	23.9	0.07	0.14	0.16		0.07	0.38	2.02	0.39								
	6	green part	5.3	72.6	23.5	0.15	0.15	0.15		0.13	0.34	1.15	1.36								
		white part		67.4	31.0	0.2	0.53	0.24		0.12	0.15	0.17	0.42								
	7	white	4.4	66.0	32.8	0.19	tr.	0.19		0.19	tr.	tr.	0.35								
	8	green	4.2	72.4	23.0	0.28	0.21	0.47		0.25	tr.	0.73	1.99								
	9	green	5.2	70.5	24.9	0.49	0.37	0.66		0.32	0.39	0.27	1.58								
	10	pale green	4.6	68.5	30.6	0.11	0.08	0.12		0.24	0.05	0.08	0.15								

Notes: n.d.: not detected; tr.: trace; r.: reddish.

d. *Comparison with other lead silicate glasses from Japan*

The authors (J.S. and I.N.) had the opportunity to analyze the ancient (potash) lead silicate glass from the Oda collection using XRF technique.[11,14] The analytical data for the ancient Japanese lead glasses from the Oda collection are shown in Table 7.4, which also includes the average chemical compositions of lead silica glass (n = 11) and potash lead silicate glass (n = 24) from Byodoin for comparison.

Clearly, the lead glasses from the 6th–8th centuries are of the PbO-SiO_2 type, and those from the 11th to 13th centuries are of the PbO-K_2O-SiO_2 type. The PbO content of the former is 62–70 wt%, whereas that of the latter is 47–51 wt %. Similarly, their density ranges are 4.6–5.3 and 3.7–3.9 g·cm^{-3}, respectively. The SiO_2 contents of the lead silica glass are 22–34 wt%, whereas those of the potash lead silica glass are 40–46 wt%. The K_2O contents of the potash lead silicate glass exceed 3 wt%, as determined by non-destructive XRF analysis. It is thought that these values are a few percent smaller than the bulk K_2O content because of the surface weathering. The chemical composition of a similar blue glass from the Chusonji Temple, Iwate Prefecture, analyzed by Oda via wet chemical analysis, was PbO 47.45%; SiO_2 39.32%; Na_2O 3.38%; and K_2O 7.89%, with a density of 3.76 g·cm^{-3}.[15] All the glasses in Table 7.4 contain MgO, Al_2O_3, CaO, and TiO_2 as minor components at less than 1%, and contain MnO, ZnO, and Rb_2O at ppm levels. The blue to green glasses contain Cu as a colorant from 3000 to 6600 ppm, with Fe_2O_3 from 2000 to 35000 ppm. Some glasses contain Al_2O_3 at levels higher than 1%. The P_2O_5 content of the Byodoin glass is distinctly higher than 1.5%; SrO and the ZrO_2 contents are below the detection limits of the method. Although these distinct compositional differences suggest several different sources, the overall compositions are rather uniform, indicating that they were produced by using a fixed method.

It seems that the two lead glass compositional groups, the PbO-SiO_2 and PbO-K_2O-SiO_2 types, are restricted by chronological differences, namely, the Nara (8c AD) and Heian periods (9–12c AD). This may suggest that there were only a few centers producing glass from

Table 7.4 Chemical composition of ancient lead silicate glass beads from the Oda collection determined by XRF analysis

							Chemical Composition/ wt%									Chemical Composition / ppm						
Sample No.	Site	Era (AD)/ century	Sample	Color	Type	Density/ $g \cdot cm^{-3}$	PbO	Na_2O	MgO	Al_2O_3	SiO_2	P_2O_5	K_2O	CaO	TiO_2	MnO	Fe_2O_3	CuO	ZnO	Rb_2O	SrO	ZrO_2
007	Hakusan-kofun	mid 7 c.	glass bead	green	$PbO-SiO_2$	4.86	63.52	n.d.	0.46	0.62	33.78	0.57	0.32	0.06	0.061	190	2370	3310	300	100	n.d.	n.d.
045	Tatsutago-boyama #3	mid 7 c.to end	glass tube	green	$PbO-SiO_2$	5.11	68.24	n.d.	0.46	4.90	23.78	0.11	0.75	0.75	0.064	130	5000	3790	240	100	n.d.	n.d.
049	Miyajidake shrine	6–7 c.	glass fragment	green	$PbO-SiO_2$	5.04	67.63	n.d.	0.31	0.45	30.34	0.23	0.25	n.d.	0.046	270	2420	4010	230	80	n.d.	n.d.
052	Shosoin	mid 8 c.	glass bead	green	$PbO-SiO_2$	4.59	62.36	0.09	0.68	1.11	33.42	0.27	0.83	0.08	0.059	250	3850	6290	180	50	n.d.	n.d.
059	Horyuji	7 c.	glass bead	blue	$PbO-SiO_2$	5.12	68.15	0.04	0.56	0.54	25.66	0.20	0.59	0.08	0.076	360	35220	4920	130	10	n.d.	n.d.
n = 11	Byodoin	mid 11c.	glass bead	yellow green	$PbO-SiO_2$	5.32	70.28	0.21	0.58	1.18	22.82	2.21	1.00	0.36	0.049	250	5560	6550	290	140	n.d.	n.d.
063	Chusonji	12 c.	glass bead	light blue	$PbO-K_2O-SiO_2$	3.75	46.80	0.09	0.33	1.33	44.86	1.08	3.44	1.09	0.059	180	1990	6420	280	80	n.d.	n.d.
046	Kanazawa-bunko	13 c.	glass rod	colorless	$PbO-K_2O-SiO_2$	3.74	47.38	0.66	1.22	0.57	45.76	0.61	3.30	0.26	0.022	300	830	590	140	190	n.d.	n.d.
n=24	Byodoin	mid 11c.	glass bead	light blue	$PbO-K_2O-SiO_2$	3.91	51.51	0.23	0.42	1.28	40.28	1.65	3.17	0.32	0.033	320	2400	7670	290	190	n.d.	n.d.

Notes: n.d.: not detected

raw materials. At the Asukaike site, Nara, there is evidence for the primary production of lead silicate glass at the end of the 7th century in Japan.[16] We also observed the use of antique glass in Byodoin, because antique glass may have had a special meaning as a treasure in the past. Without knowledge of the isotopic Pb content, it is difficult to further discuss whether there is a possibility of domestic production of these lead glasses. However, the present study of this large quantity of Byodoin glass affords a precious contribution towards an understanding of the glass of the Heian period.

7.3.5 *Origins of lead silicate and potash lead silicate glasses*

The characteristic chemical compositions of the (potash) lead silicate glasses excavated in Japan resemble those of early Chinese high lead silicate glasses and potash lead silicate glasses.[17] The high lead silicate glasses are thought to originate from China. It has been reported that most of the lead silicate glasses date from the Warring State period, and some from the late Spring and Autumn period.[17] The glass gradually evolved into potash lead silicate glass in China, and was produced in large quantities during the Six Dynasties (220–589 AD) and the Sui Dynasty (581–618 AD). High lead silicate glass is found in the late Yayoi period in Japan.[18,19] Glass bracelets made of high lead silicate glass from the 2nd century have been excavated from three sites. Lead silicate glass from the 3rd and 6th centuries has not been excavated in Japan. From the 7th century, a significant production of lead silicate glass is observed.[18,19] As mentioned above, the primary production of lead silica glass started in Asukaike, Nara, at the end of the 7th century. A lot of glass melting pots, galena, quartz, stibnite and charcoal were found at the site.

It should be noted that the lead isotope analysis showed that the Asukaike glasses were produced using galena (PbS) from Japan.[19] As for the potash lead silicate glass, the glass from the 12th century Chusonji Temple (Table 7.4) had been considered to be the earliest examples found in Japan. However, our analyses reveal that the Byodoin glass could belong to an earlier period. Because of the lack of Pb isotope analysis of our samples, we cannot determine the origin of the Pb, but at least the method for preparing the potash lead silicate glass came from China. After the Heian period, Japanese glass was mostly composed of potash lead silicate glass, and this composition continued to be employed, such as for Satsuma Cut Glass, through the end of the Edo period (mid-19th century). Accordingly, information regarding chemical composition is insufficient to clearly characterize lead glass. Further information will be obtained through the Pb isotope analysis of the samples.[20]

7.4 Study of small free-blown glass vessels with lids from the Byodoin Temple

7.4.1 *Characteristics of the Byodoin fragments*

Most of the fragments are small free-blown glasses, and their predominant color (80%) is blue. Many of the pieces closely resemble vessels excavated at Hakata in western Japan, as well as those now held in the Nara National Museum and the Denkoji Temple, which will be referred to in detail in later sections.

Specimen L-01 (Fig. 7.1(a), 33 mm in height) is the lid of a vessel made of dark green glass, and is the piece most indicative of the original shape. It was probably made by blowing a small glass ball, or bubble, which was wound with a thread of glass to fashion a rim; then, a small knob was added at the top. A hole in the bottom (L-01 bottom view, in Fig. 7.1(a)) indicates where a blowing pipe was attached to blow air.

There are several other fragments that seem to have been produced in the same way (B-16 in Fig. 7.1(a)). As shown in the cross-sectional drawing of B-16 (Fig. 7.5), the piece has an uneven

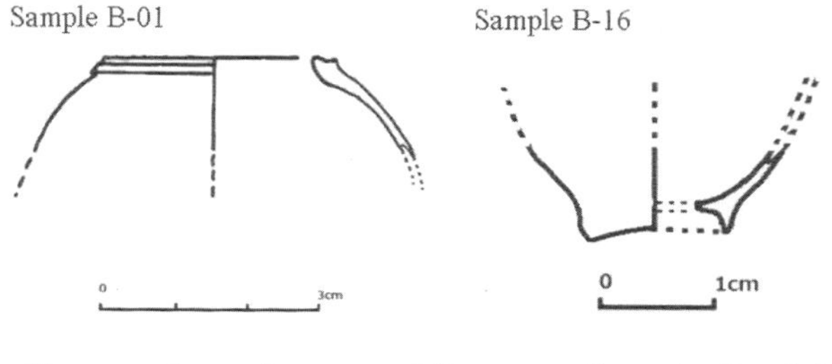

Fig. 7.5 Cross sections of two vessel fragments from the Byodoin Temple

section at the bottom, possibly made when a blow pipe was removed from the glass. This reveals an important piece of information on the shape of the pipe, as well as its diameter (8–9 mm) and thickness (~1 mm). B-10 (Fig. 7.1(a)) is another lid fragment of the same type, with a wound rim. Some lid fragments are decorated with gold-leaf threads; these were chemically confirmed by the current XRF analysis.

B-01 and A-15 (Fig. 7.1(a)) are parts of vessel body rims. The cross section of B-01 (Fig. 7.5) shows that the wall is thick near the rim but becomes thinner toward the bottom. The fragment A-15 shows two noteworthy characteristics: it has an elongated bubble along the line of the rim, and the edge inside the rim is pointed and sharp. Specimen C-01 (Fig. 7.1(a)) is the bottom part of a vessel, whose lid is probably L-01. Its wall is eggshell-thin, and notably, no pontil mark appears at the concave bottom.

7.4.2 *Vessels and fragments in other parts of Japan with similar characteristics to the Byodoin types*

a. *Glass finds at Hakata excavation sites*

Situated at the northern tip of Kyushu, old Hakata and its surrounding areas, including its ports, served as a center of trade with the Chinese continent and Korean peninsula throughout ancient and medieval times. Hakata, in particular, prospered and developed into a metropolitan city at the end of the 11th century, wherein a large number of Chinese were engaged in trade. Large volumes of Chinese-made porcelain and pottery pieces have been unearthed from excavation sites.[21]

A glass vessel and its lid, which was found in 1993 during the 79th excavation of Hakata site by Fukuoka City (Fig. 7.6),[22] is representative of the Hakata findings. It is dated from the first half of the 12th century. The wall of the vessel, now heavily weathered, is extremely thin. The original color is considered to have been green. The Pb isotope analysis results show that the lead in the glass originated from China.[23]

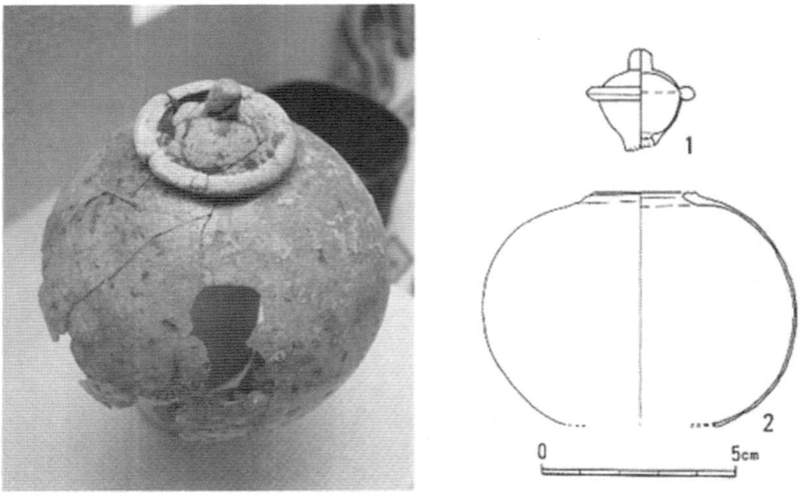

Fig. 7.6 Glass vessel found in 1993 at Hakata site; in the collection of the Fukuoka City Archaeology Center

Hisa reported in 2008 that many fragments of small crucibles, probably used to remelt glass, were also found along with numerous glass objects such as beads and vessel fragments during archaeological excavations from several places in and around the city. Many were originally produced as pots or pitchers in China and converted for use as crucibles. These crucibles were dated from the 11th to the 13th centuries, with the largest portion dated to the late 12th century.[24]

Three chemical analyses of glasses adhering to the crucibles showed that they are of the PbO-K_2O-SiO_2 type. Lead isotope analyses revealed that the ratios matched that of the Tsushima Island mine in Japan.[25,27]

b. *Vessels and lids in the Nara National Museum*

Three glass vessels of the same type are in the collection of the Nara National Museum. No. 865 (Fig. 7.7, 65 mm in belly diameter) is a reliquary vessel for the *sharira* (Buddha's holy ashes). It is placed in a wooden shrine decorated with beads. Though heavily weathered, the original color was green. Nos. 818 and 1132 are set atop the lids of

Fig. 7.7 Glass vessels in the collection of the Nara National Museum

copper-made kyozutsu (*sutra* scroll containers). Only a lid, with a frilled rim, was used in No. 818. A fragment of a similar frilled rim is also preserved in the Byodoin Temple.

The three pieces, which were not archeologically excavated, are said to have been uncovered in Kyushu or areas near Hakata, and have been attributed to the 12th century.[28,29] They are of heavy glass, although no chemical analysis has been carried out.

c. *Vessel for sharira in the Denkoji Temple, Nara*

Figure 7.8 shows a *sharira* vessel, 51 mm in belly diameter, found within the Jizo Bosatsu (*Ksitigarbha*) sculpture of the Denkoji Temple in Nara. According to the document enclosed in the sculpture, an 83-year old woman had the sculpture made in 1228 AD on behalf of her deceased parents. A tip of the knob has been lost but the artifact retains its clear blue color and beautiful luster. Its density

Fig. 7.8 *Sharira* vessel in the collection of the Denkoji Temple, Nara

of 3.8 $g.cm^{-3}$ and beta ray back scattering suggest a PbO content of ~50 wt%.[30]

d. *Kyozuka (sutra mound) finds*

As in China, the practice of burying *sutra* in the ground was seen as a good religious deed in Japan and was actively performed, particularly after the late Heian period. As glass vessels were buried along with the *sutra* in some cases, examples of such items have been excavated in Hiroshima and Osaka prefectures. According to a report, the Pb isotope ratio of a vessel found at the 12th century No. 2 Kyozuka mound of the Makiosan Sefukuji Temple in Osaka was close to that of the Tsushima mine.[31]

e. *Experiments to reproduce vessels and lids*

To understand the manufacturing methods for the glass vessels and lids and to clarify the genesis of their characteristic shapes, one of us (A.I.) attempted to reproduce the pieces in cooperation with N. Fujiwara and H. Kaito of the Tokyo University of the Arts.[4] The glass used had a $PbO\text{-}K_2O\text{-}SiO_2$ composition that approximated to those of the Byodoin and Seiryoji fragments [since the K_2O content on the surface of glass decreases with time, the chemical composition of Sharira vessel in Seiryoji Temple, which had been analyzed by Dr.

K. Yamazaki using a wet method, was used as a model for the experiment. The vessels were brought from China into Japan by a monk in 985.]. Since seamless metal tubes were not available centuries ago, a copper plate, 0.8 mm in thickness, was rolled and blazed to produce a tube with a diameter of 7 mm for use as a blow-pipe. The pipe's size was close to that observed in fragment B-16 from Byodoin.

After several trials, we determined that a lid similar to those among the artifacts could be reasonably produced through the following five steps. (1) Gather a small amount of molten glass at the end of the blow-pipe and blow a small bubble (paraison); (2) Wind a glass thread around the bubble to make a rim; (3) Put a small amount of molten glass at the top to make a knob; (4) Cool the pipe near the glass with a wet cloth to allow it to shrink slightly. Detach the glass by tapping the pipe; (5) Anneal the glass.

We found that the bottom of the reproduced lid becomes uneven (Fig. 7.9a) — as observed in the Byodoin and other pieces — when the glass is gently "removed" from the blow-pipe, rather than cracked at the neck as employed in the western technique. To obtain the desired result, the end of the pipe should not be dipped deeply into the molten glass. This kind of simple free-blowing could be a unique technique prevalent in ancient East Asia. It has been reported that simple small glass bubbles with the same types of bottoms are also held in the Shosoin Repository.[32]

The following describes the manufacturing method for a vessel body. To produce an object with a wide mouth, the glass needs to be

(a) (b) (c)

Fig. 7.9 Experimental pieces made by Fujiwara and Kaito, Tokyo University of the Arts blown in two stages if a thin pipe is used. The making procedure includes six steps as follows below.

(1) Blow a small bubble (first paraison), as for the lid; (2) When the bubble has cooled and solidified, gather additional glass and blow a bigger bubble (second paraison) on top of the first; (3) Make a "kick" at the bottom by slightly pushing in the center of the bottom so that it can stand stably on a flat surface; (4) Cool the pipe near the glass with a wet cloth to allow it to slightly shrink. Detach the glass by tapping the pipe. A gourd-shaped bottle is thus produced (Fig. 7.9b); (5) Anneal the glass; (6) Separate the two paraisons along an engraved line made with a file (Fig. 9c).

The experiments showed that the noteworthy characteristics of the rims observed in the artifacts appear when such processes are followed (Fig. 7.9c). Elongated bubbles around the rim appeared in many of the newly produced pieces, probably because air was sucked in when the molten glass was gathered and blown on the first paraison. Edges inside the rim were pointed and sharp, as was also observed in the A-15 piece from Byodoin, an indication that the jagged rim was left unpolished after the second paraison was cut away. The rim of the vessel, where the two paraisons were combined, was thick.

7.5 Conclusions

As stated earlier, the small vessels and lids dating from the Heian period showed common characteristics, probably because the same production techniques were used. All the analyzed vessel pieces were composed of potash lead silicate (PbO-K_2O-SiO_2) glass. All the vessels were small; the belly size of the largest recovered vessel was 80 mm. The artifacts were quite simple, free-blown glasses that must have been made using only small-scale equipment. Given the similarities in compositions and manufacturing methods, it is reasonable to think the vessels are of the same technological school.

Regarding the origin of the technology, the most likely possibility is China, which had a long history of lead glass production. Furthermore, PbO-K_2O-SiO_2 glass and free-blowing techniques had already been developed by the Song period.[33,34] Many domestically produced free-blown glasses have been discovered in the pagodas of China. Judging from materials such as photos and reports available to the authors, it can be assumed that a greenish blue glass vessel with a belly diameter of 52 mm found in the Tian Feng Ta Pagoda Imperial Tomb, Ningbo City, China,[35,36] for example, and the glass vessels found in Japan were made using the same technology. An exception is a knob on the lid of the Chinese artifact, which seems to have been formed in a different way.

Ningbo and Hakata were then major ports representing East Asia, both vigorously engaged in trade. The possibility is high that both glasses fabricated in China and the technology that produced them was imported to Japan. Glasses thus brought to Japan were treated as objects of high value. In an essay, "*Shin Sarugaku ki*"[37] written by a scholar near the middle of the 11th century — the same period as the construction of the Byodoin temple — "Ruri no Tsubo (glass vessel)" and "fukidama (glass bead or small blown glass)" were mentioned among a list of imports, along with perfumes, medicines, etc. The nobles and the populace were devoted to Buddhism at that time, and the fact that Ruri (glass) was among the treasures used to decorate the Buddha's Paradise was also a reason that glass was highly esteemed.

Because the vessel excavated from Hakata in 1993 (Fig. 7.6) used Chinese lead, its creation in China is highly possible. Alternatively, it could have been made in Hakata using imported metal. On the other hand, there is another fact that lead isotope analysis of potash lead silicate glass adhered to crucibles excavated from 59th, 62nd and 71st survey of Hakata showed that lead from the Tsushima mine, Nagasaki Prefecture was used.[25–27] Although clear evidence that glass blowing was practiced in Hakata has not yet been found, the method could well have transferred to Hakata as the demand for glassware grew in Japan.

Being inclined to believe that the production method we examined actually derives from China, we would like to delve more thoroughly into the materials and artifacts in Japan, China, the Korean peninsula and East Asia, so as to learn more about how the interchanges of glass technology and products occurred.

Acknowledgments

The authors are grateful to the Byodoin Temple for permission to research the precious glasses and to use their pictures in this paper. The Fukuoka City Archaeology Center, Nara National Museum, and Denkoji Temple also kindly granted permission to display the photos of their possessions. Chemical analyses of the Byodoin glasses were carried out with the help of Y. Abe, A. Nakamura, S. Suzuki, and H. Gondai of the Tokyo University of Science, and are greatly appreciated.

References

1. T. Koezuka, Y. Hirao, S. Kawagoe and T. Nishiguchi, Studies on Lead Glass — from Findings at Asukaike Site, *Abstracts of the 10th meeting of Japan Society for Scientific Studies on Cultural Properties* (in Japanese), 100–101 (1993).
2. J. Shirataki, I. Nakai and A. Inoue, Research on Glass Beads found from the Pedestal of the Amida Nyorai Statue in Byodoin Temple, *Hosho Gakuso* (in Japanese) 7, 149–170 (Byodoin, Kyoto, 2011).
3. I. Nakai, J. Shirataki and A. Inoue, Chemical Analyses of Glass Fragments found from the Pedestal of the Amida Nyorai Statue in Byodoin Temple, *Hosho Gakuso* (in Japanese) **8**, 222–232 (Byodoin, Kyoto, 2012).
4. N. Fujiwara, H. Kaito and A. Inoue, Study on Glass vessel Fragments found from the Pedestal of the Amida Nyorai in Byodoin Temple and Experiment on Reproduction of the vessels, *Hosho Gakuso* (in Japanese) **8**, 190–220 (Byodoin, Kyoto, 2012).
5. I. Nakai, S. Yamada, Y. Terada, Y. Shindo and T. Utaka, Development of a portable X-ray fluorescence spectrometer equipped with two monochromatic X-ray sources and silicon drift detector and field analysis of Islamic glasses at an excavation site in Egypt, *X-ray Spectrometry* **34**, 46–51 (2005).
6. K. Tantrakarn, N. Kato, A. Hokura, I. Nakai, Y. Fujii and S. Gluščević, Archaeological analysis of Roman glass excavated from Zadar, Croatia, by a newly developed portable XRF spectrometer for glass, *X-Ray Spectrometry* **38**, 121–127 (2009).
7. N. Kato, I. Nakai and Y. Shindo, Change in chemical composition of early Islamic glass excavated in Raya, Sinai Peninsula, Egypt: on-site analyses using a portable X-ray fluorescence spectrometer, *Journal of Archaeological Science* **36**,1698–1707 (2009).
8. Y. Abe, R. Harimoto, T. Kikugawa, K. Yazawa, A. Nishisaka, N. Kawai, S. Yoshimura, and I. Nakai, Transition in the use of cobalt-blue colorant in the New Kingdom of Egypt, *Journal of Archaeological Science* **39**(6), 1793–1808 (2012).

9. M. Kariya, Specific gravity of Glass, *Glass* (in Japanese) **10**, 23–24 (Association for Glass Art Studies, Japan, 1981).
10. H. Gondai, Y. Abe and I. Nakai, Application of calibration curve method to lead glass/lead-based glaze by using X-ray fluorescence analysis, *Adv. X-ray Chem. Anal. Japan* **42**, 325–340 (2011).
11. J. Shirataki and I. Nakai, Archaeometric analysis of ancient Japanese glass collected by Sachiko Oda, *Bull. Natl. Mus. Nat. Sci. Ser. E* **34** (in Japanese), 61–71 (2011).
12. S. Liu, Q. H. Li, F. Gan, P. Zhang, and J. W. Lankton, Silk Road glass in Xinjiang, China: chemical compositional analysis and interpretation using a high-resolution portable XRF spectrometer, *Journal of Archaeological Science* **39**, 2128–2142 (2012).
13. Shosoin Office (ed.), *Shosoin Glass* (Nihon Keizai Shinbunsha, 1975).
14. S. Oda, Glass in Nara period — Analysis of glasses from Kofukuji, Gankoji, and Horyuji Temples, *Museum* (in Japanese), **192**, 22–28 (Tokyo National Museum, 1967).
15. S. Asahina, G. Aida and S. Oda, Antique glass from Chuson-ji Temple and a general consideration on ancient Japanese glass, *Scientific Papers on Japanese Antiques and Art Crafts* **5**, 1–6 (1953).
16. T. Koezuka, Study on excavated materials from the Asukaike site, *Annural report* (in Japanese, Nara National Research Institute for Cultural Properties, 1992), pp. 56–57.
17. F. X. Gan, The Silk Road and Ancient Chinese Glass, in *Ancient Glass Research along the Silk Road*, eds. F. X. Gan, R. H. Brill and S. Y. Tian (World Scientific Publishing Co. Pte. Ltd, Singapore, 2009), pp. 42–108.
18. T. Koezuka and K. Yamazaki, Chemical composition of ancient glasses found in Japan — A historical survey, *Proceedings of 17th International Congress on Glass*, Beijing, China, 9–14 October 1995, Vol. 6, pp. 469–474.
19. T. Koezuka, Scientific study of evolution of ancient glasses found in Japan (in Japanese), Ph.D. Thesis, Tokyo University of the Arts (1997).
20. R. H. Brill, K. Yamasaki, I. L. Barnes, K. J. R Rosman and M.Diaz, Lead Isotopes in Some Japanese and Chinese Glasses, *Ars Orientalis*, 87–109 (1979).
21. K. Tanaka and I. Sato, Transition of trade ceramics and potteries, in *Excavating Hakata, a Medieval City* (in Japanese), eds. Oba *et al.* (Kaichosha, Fukuoka, 2008), pp. 112–131.
22. Fukuoka City Board of Education (ed.), *Hakata 50* (in Japanese, Fukuoka City, 1996), pp. 92–93.
23. K. Yamazaki and T. Koezuka, Analyses of chemical composition and lead Isotope of a green glass vessel unearthed during the 79th Excavation of Hakata Site, in *Hakata 50* (in Japanese), ed. Fukuoka City Board of Education (Fukuoka City, 1996), pp. 129–131.

24. Y. Hisa, Glass Findings, in *Excavating Hakata, a Medieval City* (in Japanese), eds. Oba *et al.* (Kaichosha, Fukuoka, 2008), pp. 207–210.
25. K. Yamazaki, T. Koezuka and H. Shirahata, Analyses of chemical composition and lead isotope of green glass adhered to an unglazed pot unearthed during the 62th excavation of Hakata Site, in *Hakata 48* (in Japanese), ed. Fukuoka City Board of Education (Fukuoka City, 1995), pp. 239–244.
26. K. Yamazaki and T. Koezuka, Chemical analysis of green glass adhered to an unglazed pot unearthed from Pit 0018 during the 59th excavation of Hakata Site, in *Hakata 36* (in Japanese), ed. Fukuoka City Board of Education (Fukuoka City, 1993), pp. 87–88.
27. K. Yamazaki, T. Koezuka and H. Shirahata, Analyses of chemical composition and lead isotope of a white porcelain fragment and green glass adhered to it unearthed during the 71th excavation of Hakata Site, in *Hakata 53* (in Japanese), ed. Fukuoka City Board of Education (Fukuoka City, 1996), pp. 79–82.
28. Nara National Museum (ed.), *Illustrated Catalogue of Buddhism Arts and Crafts in Nara National Museum* (in Japanese, Nara National Museum, 1992), p. 53.
29. Nara National Museum (ed.), *Illustrated Catalogue of Kyozuka Remains in Nara National Museum*, (in Japanese, Nara National Museum, 1991), p. 68, 88.
30. T. Asahina, F. Yamazaki, I. Otuka *et al.*, On the ancient glass of Sarira colourless glass bottle from Toshodaiji Temple and Sarira blue glass bowl from Denkoji Temple — determination of lead content in glass by means of Beta-ray back scattering, *Scientific Papers on Japanese Antiques and Art Crafts* **6**, 14–18 (1953).
31. H. Mabuchi, *Provenance and Chemical composition of the Ancient Glasses Excavated in Japan* (in Japanese,)Tokyo, 1986–88) pp. 14–15.
32. I. Oga *et al.*, Research on Materials used for Treasures in Shosoin Repository, 1953–55, *Bulletin of Archives and Mausolea Department* (in Japanese) **8**, 80 (Imperial Household Agency, 1957).
33. F. Gan, Evolution of the chemical composition and the origin of making technology of Chinese ancient glass, in *Development of Chinese Ancient Glass* (in Chinese), eds. F. X. Gan *et al.* (Shanghai Science and Technology Press, Shanghai 2005), pp. 230–232.
34. Z. Huang, Ancient Glass Technology in the Sui, Tang and Song Dynasties, in *Development of Chinese Ancient Glass (in Chinese)*, eds. F. X. Gan *et al.* (Shanghai Science and Technology Press, Shanghai 2005), pp. 128–139.
35. Y. Dong (eds.), *A collection of cultural Relics in Ningbo,* 1st edn. (Hua Xia Publishing House, Peking, 1996), pp. 212–213, color plate 112.
36. S. M. Lin, Excavation of the underground palace of Tianfeng Pagoda in Ningbo, Zhejiang Province, *Cultural Relics (Wen Wu)* **6**, 24–25 (1991).
37. A. Fujiwara, *Shin Sarugaku Ki, annotated translation by H. Kawaguchi* (in Japanese), Heibonsha, Tokyo, 1983), pp. 314–315.

Chapter 8 第八章

Chemical Analysis of Tang Dynasty Glass Vessels Unearthed from the Underground Palace of the Famen Temple Using a Portable XRF Spectrometer

Qinghui Li, Jie Jiang,† Xinling Li,† Song Liu,* Donghong Gu,* Julian Henderson‡ and Junqing Dong**

*Center of Sci-tech Archaeology,
Shanghai Institute of Optics and Fine Mechanics,
CAS, Shanghai 201800, China

†Famen Temple Museum,
Fufeng County 722201, Shannxi, China

‡Department of Archaeology, University of Nottingham,
University Park, Nottingham NG7 2RD, UK

8.1 Introduction

The Famen Temple is located in Famen town, Fufeng County, 120 km west of Xi'an City, Shaanxi province. Established during the Eastern Han Dynasty, the temple is renowned as the home of Sakyamuni Buddha relics. In the 1980s, an underground palace was discovered in the temple. Many cultural relics including Mi'se porcelain, gold and silver ceramics, silk, and so on were found in the underground palace. A batch of intact well preserved glass vessels of Islamic style was also found. These glass vessels have attracted wide attention. Some scholars have already studied the glass vessels, but the focus has been on the shapes and patterns.[1–3] Until now, only a very small number of the glass vessels have been analyzed scientifically.[4–5]

In this paper, we discuss the chemical compositions of some glass vessels resulting from the use of a portable XRF spectrometer (pXRF). The results obtained are also compared to those for glass samples from the contemporaneous Islamic world so as to be able to suggest where the glass vessels were made. Thereby the glass vessels are classified according to their chemical compositions. This work provides important information and useful clues about the interconnections between the Islamic world and ancient China along the Silk Road.

8.2 Experimental

8.2.1 *Equipment and method*

An OURSTEX 100FA portable energy dispersive XRF (*p*XRF) spectrometer made by OURSTEX Co., Ltd. was used for the chemical compositional analysis. A palladium (Pd) excitation tube was employed, and the max voltage and the max power were 40 kV and 50W, respectively. The X-ray beam spot focused on the sample is about 2.5 mm in diameter. Three work models, the direct model, the monochromatic model and the filter model, are included here, and their parameters are listed in Table 8.1. The detector is a silicon drifted detector (SDD) with an active area of 5 mm^2 and cooled by Peltier element[6] to –28°C. Two different detection units were employed: one was a low vacuum

Table 8.1 Parameters for the three work models

Parameters	Direct model	Monochromatic model	Filter model
Voltage (kV)	15	40	40
Current(mA)	1	0.5	1
X-ray models	White X-ray	Monochromatic X-ray	White X-ray
Zr filter	OFF	OFF	ON
Time (s)	100	100	100
Elemental range (Z*)	11 ~ 30(Na~Zn)	20 ~ 42(Ca ~ Mo), 74 ~ 92(W ~ U)	47 ~ 56(Ag ~ Ba)

Table 8.2 Parameters for the different detection units

Parameters	Air Unit	Low Vacuum Unit
X-ray source	Pd	
X-ray spot(diameter)	2.5 ~ 3 mm	
Voltage	≤ 40 kV	
Current	0.05–1.5 mA	
Power	≤ 50 W	
Detector	SDD	SDD, developed for light elements
Cooling	Peltier effects, –28°C	
Window materials	Be	MOXTEC AP3.3 organic film
Element range	13 ~ 92 (Al ~ U)	11 ~ 92 (Na ~ U)
Measurement Pressure	10^5 Pa	Low Vacuum (400 ~ 600 Pa)
Chamber	No Chamber	Diameter, 15 cm, height, 10 cm and can extended to 30 cm
Time	100 ~ 300 s	100 ~ 300 s
CCD image system	no	Yes

unit and the other was an air unit. The detector in the vacuum unit is specially developed for light elements, such as Na, Mg, and a polymer window (MOXTEK AP 3.3 film) which can provide superior X-ray transmission in the low-energy range down to Be-Kα is also used as the X-ray window material. A sample chamber is also used. For the air unit, a Be X-ray window is used — no sample chamber is involved. The parameters for the two different detection units are listed in Table 8.2.

A total of twenty-two standard glass samples distributed by Breitlander Eichproben und Labomaterial GmbH, Germany were used to make calibration curves for a number of major and trace elements in silicate glasses. Several reference glass samples, including NIST 1411 and Corning C, were measured so as to evaluate the performance of the method, and the results show that the method has high precision and accuracy. The method, including a discussion of its applicability, has been published elsewhere for the analysis of ancient glass samples from Xinjiang, Guangxi and Hubei provinces.[7–10]

8.2.2 Samples

Eighteen glass vessels excavated from the Tang dynasty underground palace of the Famen Temple were analyzed. Fourteen of these vessels were analyzed using the low vacuum detection unit, the other four were analyzed by the air detection unit because of their large size. In addition, the yellow and black glazes of glass dish with a poppy pattern (1988-0132) were also analyzed using the air detection unit. The flat bottoms or sides were selected for measurement in order to optimize the analytical geometry for quantitative analysis. Information on the glass vessels analyzed is given in Table 8.3.

Table 8.3 Information on the glass vessels found in the underground palace in Famen Temple

Number	Photo	Name	Location analysed	Detection Unit
1988–0124		Blue scratch-decorated glass dish with cross patterns	base	Low vacuum
1988–0125		Undecorated blue glass dish	base	Low vacuum
1988–0126		Undecorated blue glass dish	base	Low vacuum

(*Continued*)

Table 8.3 (*Continued*)

Number	Photo	Name	Location analysed	Detection Unit
1988–0127		Undecorated blue saucer with circular foot	Circular foot	Low vacuum
1988–0129		Gold painted blue glass dish with eight scratch decorated floral patterns	base	Low vacuum
1988–0131		Blue glass dish with string patterns	base	Low vacuum
1988–0132		Bluish-green glass dish with lustre poppy pattern	Base (bluish-green)	Low vacuum
1988–0133		Dark blue glass dish with scratch-decorated maple leaves painted in gold	base	Low vacuum

(*Continued*)

Table 8.3 (*Continued*)

Number	Photo	Name	Location analysed	Detection Unit
1988–0134		Undecorated pale yellow glass saucer	base	Low vacuum
1988–0135		Undecorated pale yellow glass tea cup	base	Low vacuum
1988–0137		Colourless glass cup with rhombus and double circles patterns	base & side	Low vacuum
1988–0138	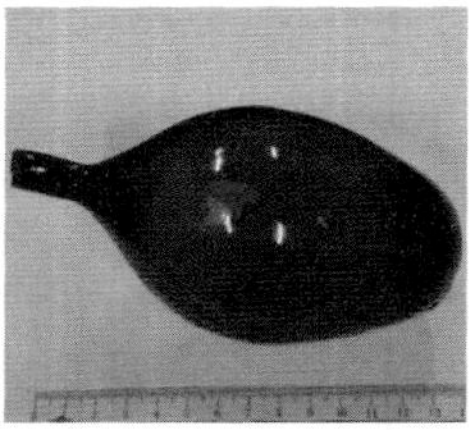	Undecorated blue narrow necked glass bottle	side	Low vacuum
1988–0139	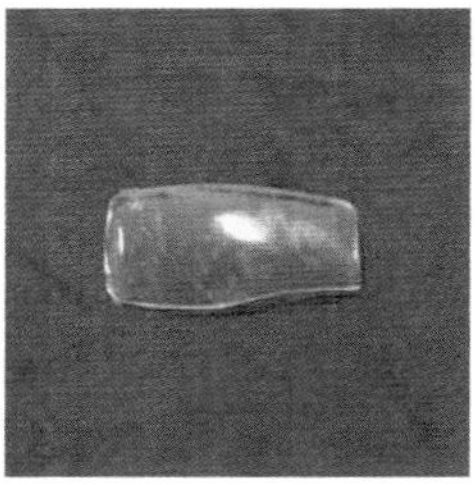	Fragment of cup	side	Low vacuum

(*Continued*)

Table 8.3 (*Continued*)

Number	Photo	Name	Location analysed	Detection Unit
1988–0141		Blue glass dish with scratched maple leaf pattern	base	Low vacuum
1988–0123		Pale yellow bottle with a narrow neck and dish-shaped mouth	Side and brown button	Air
1988–0128		Blue glass dish with eight scratch-decorated floral patterns	base	Air
1988–0132		glass dish with lustre poppy patterns	Yellow and black surface	Air

(*Continued*)

Table 8.3 (*Continued*)

Number	Photo	Name	Location analysed	Detection Unit
1988–0136		Blue glass dish with four scratch-decorated floral patterns	base	Air
1988–0142	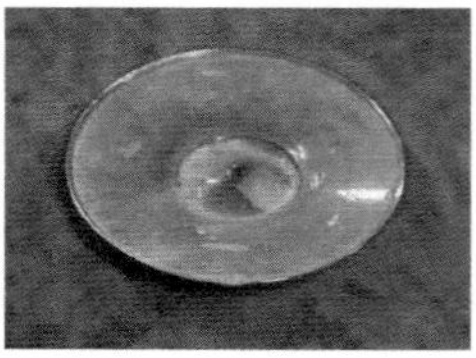	Undecorated light blue glass dish	base	Air

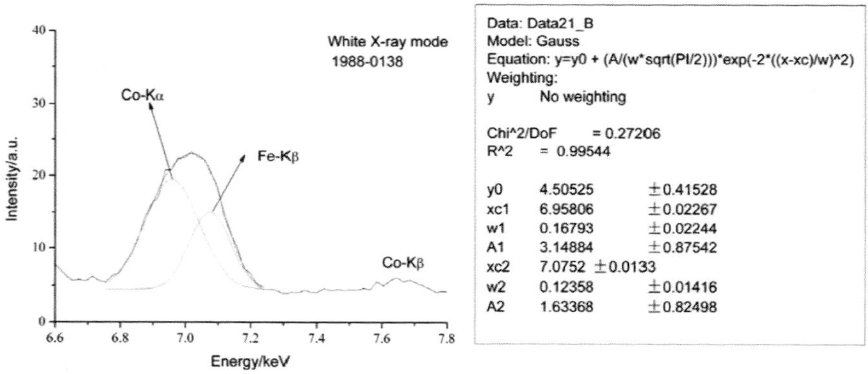

Fig. 8.1 The peak fitting for the overlap peaks of Fe Kß and Co Kα

It is worth noting that the resolution of the SDD X-ray detector is 145 eV (Mn Kα line), thus, the detector can not distinguish between two peaks when the energy difference is lower than 145 eV, and when overlaps between the peaks occur. Taking Fe Kß peak (7.058 keV) and Co Kα peak (6.925 keV), for example, the energy difference for the Fe Kß peak and the Co Kα peak is only 133 eV which is lower than 145 eV, and an overlap occurs. However the Co Kß peak instead of the Co Kα line can be used instead because no overlap occurs with the Co Kß peak. Peak fitting is needed for quantitative analysis of Co.

Figure. 8.1 is an example of peak fitting of Fe Kß peak and Co Kα peak using Origin software (Version 7.5).

8.3 Results and discussion

8.3.1 *Qualitative spectra*

Figure 8.2 shows qualitative spectra of some typical glass vessels obtained by the low vacuum detection unit. An asterisk (*) denotes

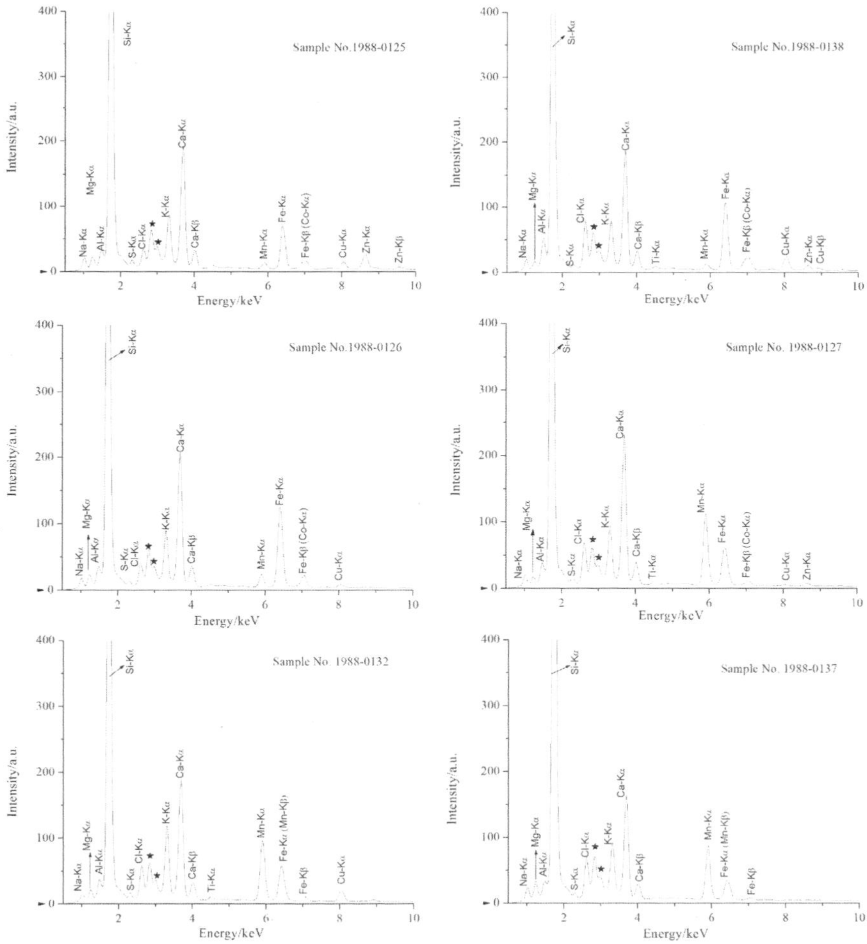

Fig. 8.2 Typical qualitative spectra obtained by low vacuum detection unit of pXRF

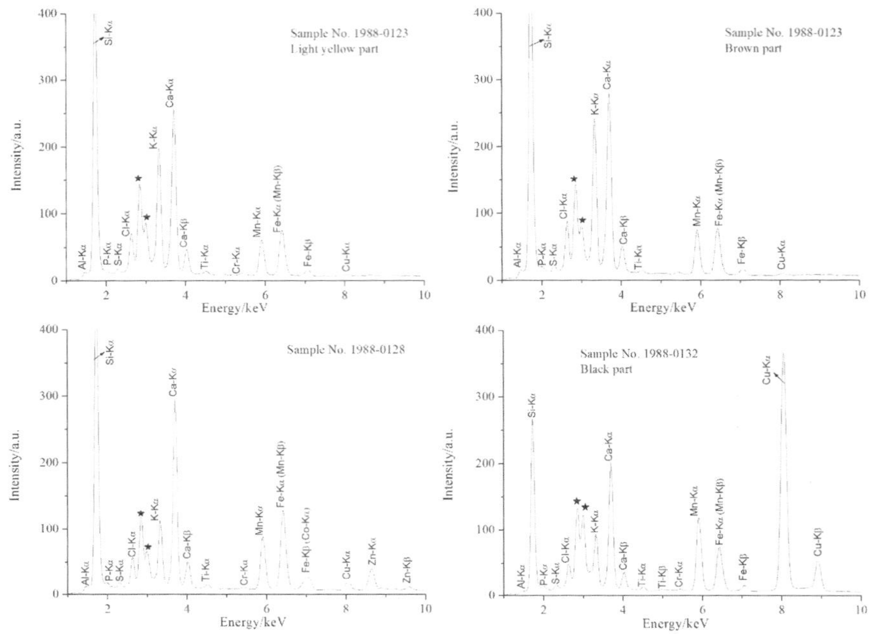

Fig. 8.3 Typical qualitative spectra obtained by air detection unit of pXRF

where the peaks derive from the X-ray target (Pd Lα). As shown in Fig. 2, the characteristic peaks for Na, Si and Ca are obvious and show up for each sample and that of K, Mn, Fe, Co, Cu, Zn also exist; the intensities vary for each sample. The qualitative spectra for the typical samples obtained by the air detection unit are shown in Fig. 8.3. The characteristic peaks of Na and Mg did not show up because of their absorption in air. The characteristic peaks for the other elements are similar to the spectra obtained by the low vacuum detection unit.

8.3.2 *Quantitative results*

The precision and accuracy for the quantitative analysis method can be found elsewhere, where the detection limits for the element oxides of interest, such as those of Na, Mg, Al, Si, K, Ca, Fe, Co and so on, are also given.[7]

Quantitative results obtained by using the low vacuum and air detection units, respectively are given in Tables 8.4 and 8.5. It is

Table 8.4 Quantitative results of glass samples obtained by low vacuum detection unit of pXRF

	wt.%													*ppm*							
编号	Na_2O	MgO	Al_2O_3	SiO_2	P_2O_5	SO_3	Cl	K_2O	CaO	TiO_2	MnO	Fe_2O_3	CoO	Cr	Cu	Zn	Rb	Sr	Pb	Sn	MnO/ CoO	Fe_2O_3/ CoO
1988–0124	12.20	3.96	1.20	69.05	n.d.	0.21	0.60	2.83	7.10	0.04	0.06	2.40	0.27	135	271	104	8	560	167	49	0.24	8.77
1988–0125	12.27	5.44	1.30	67.59	n.d.	0.23	0.27	2.74	6.97	0.05	0.25	1.48	0.33	126	1109	3051	n.d.	680	2300	64	0.76	4.47
1988–0126	10.53	6.18	2.08	65.51	n.d.	0.15	0.33	3.08	8.21	0.08	0.47	2.88	0.27	214	625	120	6	800	828	80	1.71	10.51
1988–0127	11.91	3.14	2.45	64.00	0.45	0.22	0.70	3.23	9.10	0.18	2.73	1.51	0.14	294	272	866	9	880	24	31	19.52	10.77
1988–0129	12.08	5.46	1.74	65.93	0.18	0.16	0.49	2.95	7.94	0.07	0.11	2.48	0.27	241	363	156	19	810	326	61	0.41	9.27
1988–0131	12.25	6.00	1.82	64.29	n.d.	0.18	0.49	2.90	8.22	0.06	0.27	3.02	0.30	158	587	92	9	760	607	70	0.91	10.02
1988–0132	8.74	5.16	2.21	66.79	n.d.	0.20	0.55	4.39	7.52	0.13	2.39	1.37	n.d.	199	2243	125	7	700	318	26		
1988–0133	12.53	4.98	1.45	67.14	n.d.	0.17	0.55	2.68	7.54	0.05	0.10	2.39	0.27	181	291	109	7	650	282	75	0.38	8.98
1988–0134	8.10	5.59	1.41	69.90	0.62	0.39	1.03	3.36	8.16	0.03	0.79	0.53	n.d.	194	41	121	14	710	60	40		
1988–0135	10.65	6.89	0.90	70.19	n.d.	0.18	0.56	2.32	7.12	0.02	0.68	0.42	n.d.	135	5	74	10	510	52	35		
1988–0137	12.52	7.69	1.37	67.05	0.08	0.14	0.59	2.67	5.28	0.01	1.81	0.66	n.d.	176	19	87	8	990	24	29		
1988–0138	11.96	3.40	4.17	64.73	n.d.	0.12	0.85	2.71	7.70	0.14	0.26	2.58	0.54	189	3254	1404	16	220	61	n.d.	0.47	4.75
1988–0139	9.75	6.03	1.88	69.72	n.d.	0.19	0.80	3.60	4.88	0.04	2.21	0.78	n.d.	132	57	142	14	790	25	n.d.		
1988–0141	13.78	4.80	1.91	62.22	n.d.	0.30	3.01	2.44	8.20	0.03	0.36	2.50	0.25	123	477	129	7	800	1362	50	1.42	9.99
Ave.	11.38	5.34	1.85	66.72				2.99	7.42													

"n.d." denotes not detected.

Table 8.5 Quantitative results of glass samples obtained by air detection unit of pXRF

编号	部位	wt.%											ppm								MnO/ CoO	Fe_2O_3/ CoO
		Al_2O_3	SiO_2	P_2O_5	SO_3	Cl	K_2O	CaO	TiO_2	MnO	Fe_2O_3	CoO	Cr	Cu	Zn	Rb	Sr	Pb	Sn		
1988–0123	body	0.69	66.87	0.30	0.17	0.74	5.75	6.33	0.08	0.97	0.84	n.d.	48	80	32	41	800	60	47		
	button	0.63	67.07	0.06	0.15	0.67	5.38	6.71	0.09	0.93	0.95	n.d.	52	76	22	30	570	37	39		
1988–0128	bottom	0.58	66.07	0.32	0.15	0.54	2.72	7.90	0.15	1.37	1.84	0.32	64	145	4187	20	730	310	98	4.28	5.75
	yellow	0.36	58.33	CuO 7.73	0.17	0.50	3.40	7.58	0.14	2.73	1.44	n.d.	4		234	28	520	90	50		
1988–0132	black	0.39	59.02	CuO 8.31	0.21	0.59	2.95	6.99	0.15	2.54	1.37	n.d.	47		295	28	460	140	53		
1988–0136	bottom	0.49	68.06	0.20	0.13	0.66	2.96	7.48	0.10	0.16	2.00	0.22	162	316	66	34	750	151	77	0.73	9.09
1988–0142	bottom	0.58	66.47	0.29	0.18	0.92	1.91	8.56	0.20	2.30	1.14	0.14	55	77	412	29	590	46	35	16.43	8.14

worth noting that the results in Table 8.4 are normalized to 100%. But for the air detection unit, Na_2O and MgO can not be detected, and the results of Table 8.5 are normalized to 83% according to the contents of Na_2O and MgO obtained in Table 8.4.

As can be seen in Table 8.4, the levels of Na_2O and CaO range from 8.10 to 13.78 wt% and 5.28 to 9.10 wt%, respectively. The contents of MgO and K_2O range from 3.14 to 7.69 wt% and 2.32 to 4.39 wt% respectively. The chemical results indicate that the glass samples belong to plant ash soda-lime silica glass type. The contents of Al_2O_3 range from 0.90~2.45 wt%; only one sample has an Al_2O_3 level that reaches as high as 4.17 wt% which is distinctive and not considered initially in the discussion below.

8.3.3 *Classification by transition metals*

According to the contents of MnO, CoO and Fe_2O_3, the glass samples analyzed in this paper can be divided into four 'groups', which is shown in Figs. 8.4a and 8.4b ('group' 4 consists on one sample). Table 8.6 shows the different groups and the sample numbers belonging to each group.

Group I: the samples in group I are colored by the Co^{2+} ion, and the levels of Fe_2O_3 are relative high, while the MnO levels are low. The blue glass hues are likely to be due to the CoO contents. The mineral-rich raw material containing Co used in the samples in this group has is characterised by high levels of Fe_2O_3 and low levels of MnO, which is different to that of Chinese Co-rich minerals that tend to contain high levels of MnO.[11] In relation to the vessel shape, the samples in Group I are mainly dishes with or without patterns.

In addition, the contents of CuO and ZnO in samples 1988–0125 and 1988–0138 are similar. Samples 1988–0141, 1988–0126 and 1988–0125 have relatively high levels of PbO.

Group II: the samples in this group have relatively high levels of Fe_2O_3 and MnO, nd most samples contain no detected CoO. Although Two samples have very high levels of MnO: the glass has decolorised by changing Fe^{3+} to Fe^{2+}. The decolorising technique employed was widely used in Roman times around the Mediterranean.

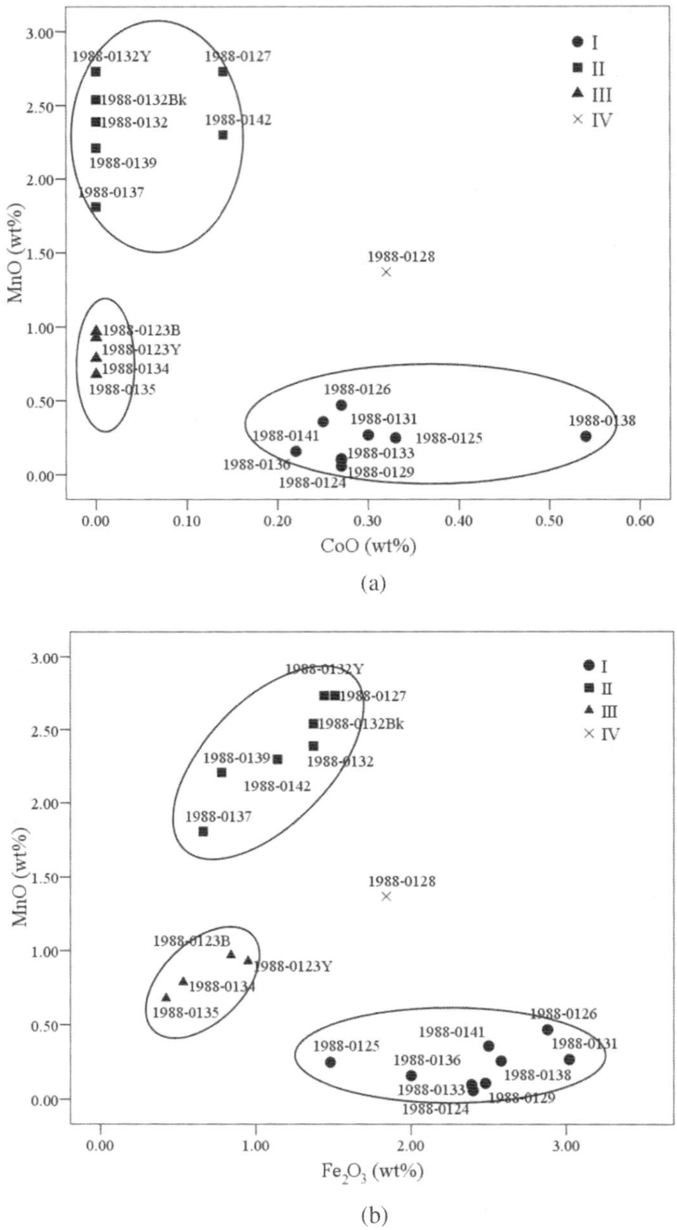

Fig. 8.4 Plots of samples found in the underground Palace of Famen Temple (a)CoO *vs* MnO (b) Fe_2O_3 *vs* MnO

Table 8.6 Glass vessels belonging to different groups

Groups	Sample number
Group I	1988–0124, 1988–0126, 1988–0131, 1988–0125, 1988–0138, 1988–0141, 1988–0133, 1988–0136, 1988–0129
Group II	1988–0127, 1988–0142, 1988–0139, 1988–0137, 1988–0132
Group III	1988–0134, 1988–0135, 1988–0123
Group IV	1988–0128

Group III: the samples in this group have medium levels of Fe_2O_3, MnO and no detected CoO. The samples in this group include a cup, a saucer and a bottle.

Group IV: Only one sample (1988–0128) belongs to this group. The sample is a blue glass dish with eight petal patterns and high levels of CoO, Fe_2O_3 and MnO. It is not certain if the raw materials containing MnO were added with those containing Co or not.

It is worth noting that the concentrations of Zn for samples 1988–0128 and 1988–0125 are 4187 μg/g and 3051 μg/g, respectively, which are higher than that of other samples. It has been noted elsewhere that many Islamic cobalt blue glasses are characterised by correlated levels of zinc and lead oxides.[12]

Combining the chemical compositions and vessel shape, it can be suggested that the samples in group I are close to that of Group IV, while the samples in Group II are close to that of Group III.

8.3.4 *A comparison with contemporaneous glass samples from Islamic world*

Some scholars believe that the glass vessels found in the underground palace of Famen Temple were transported to ancient China from the Islamic word.[1] Therefore, published chemical data for some Islamic glass samples have been selected so as to compare them with our data obtained by using the low vacuum detection unit. The glass samples chosen are contemporaneous with or close in date to that of the glass samples found in the underground palace. The first batch of glass samples chosen for comparison is from Fort, Raya and dated to 9th ~ 10th century AD[13]; the second from Nishapur, Iran and dated to 9th ~ 10th century AD.[14] And the third from Tell Zujaj, Raqqa, Syria and dates to late 8th ~ 9th century AD.[15]

As shown in Fig. 8.5, the CaO contents of the glass samples from Nishapur, Iran are higher than that of the glass samples from Fort, Raya, and have relatively constrained concentrations for the both places. The levels of Na_2O for the both are similar. For the samples from Tell Zujaj, Raqqa, the Na_2O contents have a restricted variation compared to that of glass samples from Nishapur, Iran and Fort, Raya, while the contents of CaO vary far more. The data for the samples found in the Famen Temple are denoted as black triangles in Fig. 5, and the levels of Na_2O and CaO are closest to many of the glass samples from Nishapur, Iran; the lowest levels of Na_2O in the Famen samples may partly result from the technique used though the total alkali (Na_2O+K_2O) in sample 1988–0132 of 13.13 weight % is

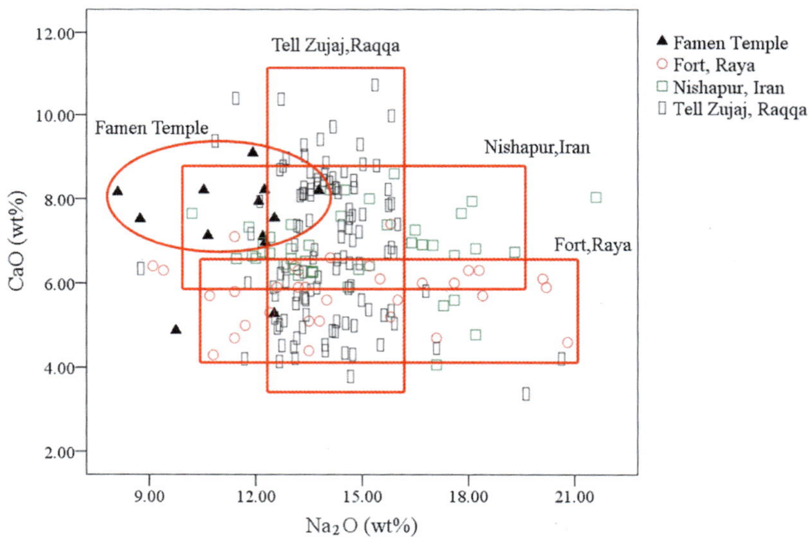

Fig. 8.5 A comparison of relative levels of Na_2O and CaO in glasses from the Famen Temple with glasses from other places in Islamic world: Na_2O vs CaO

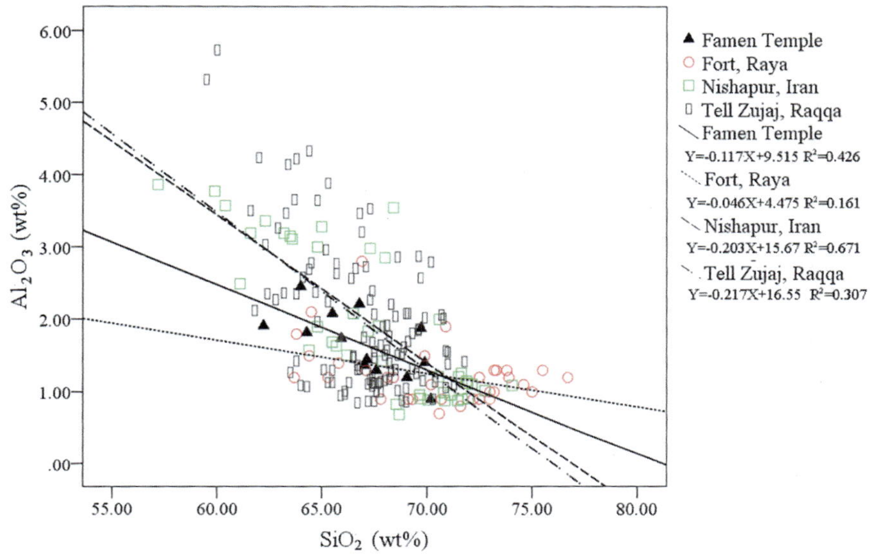

Fig. 8.6 Comparison of relative SiO_2 and Al_2O_3 levels in glasses from the Famen Temple with glasses from other places in Islamic world

close to some of the others; it contains the highest K_2O level of any of the glasses analyzed (4.39 wt%) and may possibly have in Afghan origin.[16]

Figure 8.6 is a plot of SiO_2 versus Al_2O_3 for the samples found in the Famen Temple and other published Islamic glass data. Fitted equations and coefficients (R^2) for the samples from each place are also given in Fig. 8.6. As shown in Fig. 8.6, the R^2 coefficent for glass from Raya is 0.161, which indicates that the contents of SiO_2 and Al_2O_3 have a minimal linear relationship. The values of the correlation coefficients for the samples from the other three places show that the contents of SiO_2 and Al_2O_3 have negative linear relationships. The slopes of the fitted equations for the samples from Famen Temple and Nishapur, Iran are –0.117 and –0.203, respectively, and those for Tell Zujaj, Raqqa samples is –0.217 This indicates that the relationship between SiO_2 and Al_2O_3 for Famen Temple glasses is closest to one of the groups of Nishapur glasses.

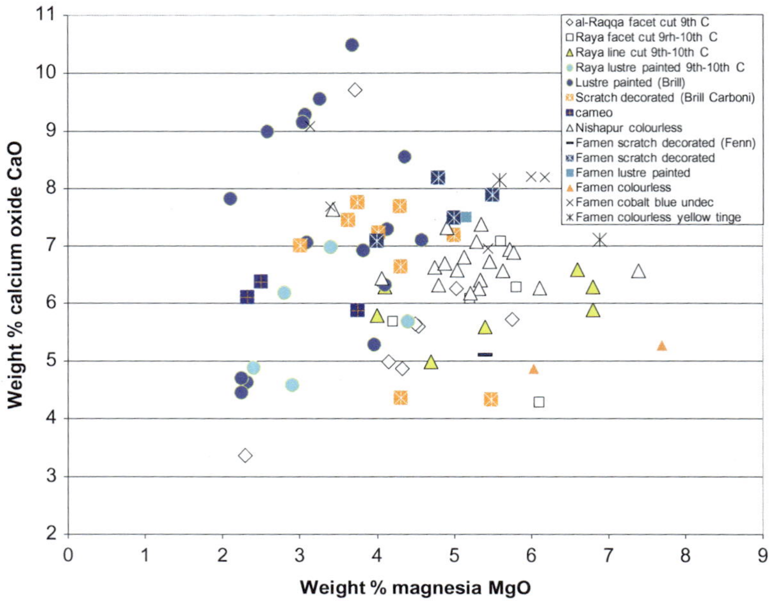

Fig. 8.7 A comparison of relative levels of MgO and CaO in glasses from the Famen Temple with glasses from other places in Islamic world: MgO vs CaO

As mentioned above, the glass samples found in the underground palace in the Famen Temple have a very close relationship to the glass samples from Nishapur, Iran.

So as to investigate the relationship between chemical composition and the type of vessel analysed we have plotted MgO vs CaO according to vessel type in Fig. 8.7. As can be seen there is a clear distinction between primarily colourless glasses with high magnesia and relatively low calcium oxide and the rest of the glasses. Such colourless glasses are commonly attributed to production zones in Iran and/or Iraq. For example there is a large concentration of them in Nishapur, northern Iran.[17] One colourless cut vessel (137) and the rim of a colurless vessel (139) found in the Famen temple fall into this area of the plot where results of other colourless, including linear cut and facet cut vessels are plotted using a range of published data.[3,5,13,15] Another well-known type of 9th century Islamic vessel is what is known as 'scratch-decorated'. We have quantitative analyses for four examples of scratch-decorated vessels and MgO and CaO levels for these have been plotted with published results for other scratch-decorated vessels in Fig. 8.6.[3,18,19] The Famen examples fall close to many of the other results for scratch-decorated vessels bolstering the claim for the existence of a specialised production site or zone for such vessels — perhaps in Syria[20] where some have been found. However another example from Nishapur was probably made there. Based on the plotted results in Fig. 8.7 lustre-decorated vessels, including one from this study, appear to have been made in more than one centre, probably in Syria or possibly Egypt but not Iran/Iraq.[21] This seems also to be the case for the undecorated cobalt blue vessels from the Famen temple. These results are interesting but need to be further investigated with trace element analysis.

8.4 Conclusions

(1) According to our results, all the samples analyzed in this paper can be classified as soda-lime silica, and all MgO and K_2O contents are higher than 1.5 wt% which indicates that they are all plant ash glasses. The comparisons of data obtained in this paper and published data show that the glass samples found in Famen Temple are similar to samples from Nishapur, Iran in terms of their relative soda and calcium oxide levels. The relative levels of MgO and CaO however provides some evidence that some scratch decorated vessels, which form a constrained compositional group, may have been made in Syria.

(2) Although the glass samples found in the underground palace in Famen Temple have a certain degree of weathering, they are mainly well preserved. The loss of flux Na_2O appears to be quite low.

(3) High lead silicate glasses typical of the Tang dynasty in China has not been found amongst the samples tested from the underground palace in the Famen Temple. It is likely that all the glass samples found are imported from the West and are not indigenous. The presence of pontil scars shows that the glass vessels were blown.

(4) The colours for the glass vessels analyzed can mainly attributed to the presence of transition metal ions including those of Co, Fe, Mn and Cu. The blue colour is mainly caused by Co ions. The transparent light greenish-yellow colors are caused by Fe. The color of brown buttons decorating sample 1988–0123 is caused by a combination of Mn and Fe ions.

References

1. Ren Xinlai, *WenBo* **1**, 68 (2011).
2. Chinese Archaeology 1999, 168: 484.
3. S. Carboni, *Glass of the Sultans*, The Metropolitan Museum of Art in association with The Corning Museum of Glass, Benaki Museum, and Yale University Press, New York, USA, 2001, p. 162.
4. W. Han, Primary studies of Islamic glass wares found in the underground palace of Famen Temple, in *Land Routes of the Silk Roads and the Cultural Exchanges Between the East and West Before the 10th Century*, eds. United Nations Educational, Scientific and Cultural Organization and Institute of Archaeology, CASS (New World Press, Beijing, 1996), pp. 393–402.
5. P. M. Fenn, R. H. Brill and M. G. Shi, Appendix to Chapter 4, in *Scientific research in early Chinese glass*, eds. R. H. Brill and J. H. Martin (Corning, The Corning Museum of Glass, New York, USA, 1991), pp. 59–64.
6. A. Sokolov, A. Loupilov and V. Gostilo, *X-Ray Spectrom* **33**, 462 (2004).
7. S. Liu, Q. H. Li, F. X. Gan *et al.*, *Journal of Archaeological Science* **39**, 2128 (2012).
8. S. Liu, Q. H. Li, Q. Fu *et al.*, *X-Ray Spectrometry* **42**, 470 (2013).
9. S. Liu, Q. H. Li and F. X. Gan, *Spectroscopy Letters* **48**, 302 (2014).
10. S. Liu, Q. H. Li, F. X. Gan *et al.*, *X-Ray Spectrom* 40, 364 (2011).
11. X. H. Huang, *China Ceramics* **50**, 87 (2014).
12. J. Henderson, *Ancient Glass, an interdisciplinary exploration*, Cambridge University Press, Cambridge, UK, 2013, p. 73, Fig. 3.5.
13. N. Kato, I. Nakai and Y. Shindo, *Journal of Archaeological Science* **37**, 1381 (2010).
14. R. H. Brill, Chemical analyses of some glass fragments from Nishapur in The Corning Museum of Glass. Appendix 3, in *NISHAPUR: Glass of the Early Islamic Period*, ed, J. Kröger (The Metropolitan Museum of Art, New York, 1995), pp. 211–233.
15. J. Henderson, S. D. McLoughlin and D. S. McPhail, *Archaeometry* **46**, 439 (2004).
16. R.H. Brill, Some Thoughts on the Chemistry and Technology of Islamic Glass, in *Glass of the Sultans*, ed, S. Carboni (The Metropolitan Museum of Art in

association with The Corning Museum of Glass, Benaki Museum, and Yale University Press, New York, USA, 2001), pp. 25–45.

17. J. Kröger, *Nishapur: Glass of the early Islamic period*, The Metropolitan Museum of Art, New York, USA, 1995.
18. R.H. Brill, *Chemical analyses of early glasses, volume 2, tables of analyses*, The Corning Museum of Glass, New York, USA, 1995.
19. J. Henderson, *Ancient Glass, an interdisciplinary exploration*, Cambridge University Press, Cambridge, UK, 2013, pp. 295–296, Fig. 10.11.
20. S. Carboni, *Glass of the Sultans*, The Metropolitan Museum of Art in association with The Corning Museum of Glass, Benaki Museum, and Yale University Press, New York, USA, 2001, p. 80.
21. S. Carboni, *Glass of the Sultans*, The Metropolitan Museum of Art in association with The Corning Museum of Glass, Benaki Museum, and Yale University Press, New York, USA, 2001, p. 52.

Chapter 9 第九章

Chemical and Lead Isotope Analysis of Glass Wares Found in Lijiaba Site, Yunyang County, Chongqing City

*Jianfeng Cui**, *Deyun Zhao*†, *Wei Huang*† *and Yuanhong He*†
*School of Archaeology and Museology,
Peking University, Beijing 100871, China
†Department of Archaeology,
Sichuan University, Chengdu 610064, China

9.1 Introduction

The site of Lijiaba is located in Yunyang County, Chongqing City. It is one of the largest sites in the Three Gorges Reservoir area (see Fig. 9.1). The site is dated from Shang Dynasty to Qing Dynasty. The archaeological excavation in 1999 has been awarded the "China Ten Most Important New Archaeological Finds in 1999". According to the archaeological finds, Lijiaba was an important inland water port on the Yangtze River for almost three thousand years. The site was also one of the most important waterway transport hubs along the Yangtze River, which linked Hunan and Hubei Provinces with Sichuan Province. Many artifacts were found during the excavations, including bronzes, potteries, porcelains, stone tools and glass wares. The finds indicated that Lijiaba was once very flourishing and had a well developed commerce.[1]

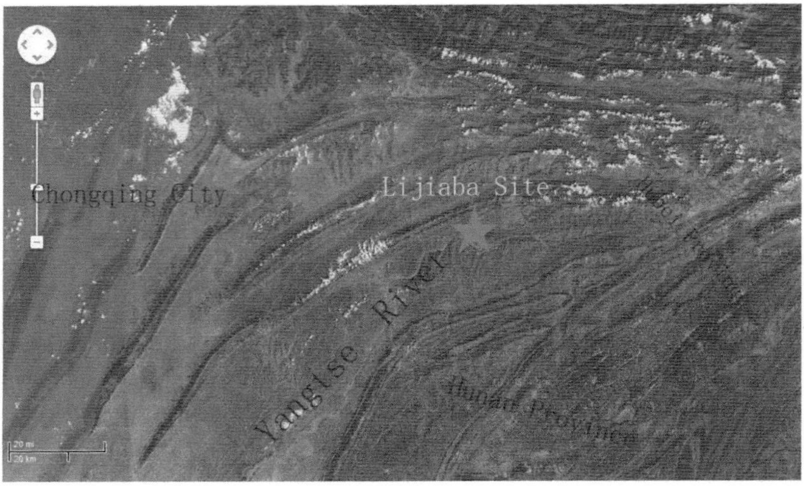

Fig. 9.1 The location of Lijiaba (from Google earth)

In this study, we carried out chemical analysis and lead isotope analysis on some glass finds found during the archaeological excavations of Lijiaba in 2003. The glass wares include nine ear pendants and four eye–beads. The tombs which these wares were found in date from the early East Han Dynasty to the Three Kingdoms period (see Table 9.1)

9.2 Methodology

9.2.1 *Chemical analysis*

The chemical compositions of Lijiaba glass wares were determined using Laser Ablation Inductively Coupled Plasma Atomic Emission Spectrometry (LA–ICP–AES).

Twenty–two elements were determined in the glass matrices, including Si, Al, Fe, Mg, Ca, Na, K, Ti, Mn, Pb, Ba, P, Cu, Sn, Sr, Co and Sb. The Corning C glass standard was used as a standard reference material. The Corning D glass standard was employed as a means of checking the quantitative accuracy: it was analysed after every seven samples. According to the analytical results for the major and minor elements, most relative standard deviations are less than 1%

Table 9.1 Catalogue of the samples

No.	Sample	Date	Analytical items	Sample photo
03YLIV M2:4	pendant	Late Eastern Han	CA, LIA	
03YLIV M3:2	pendant	Late Eastern Han	CA, LIA	
03YLIV M3:3	pendant	Late Eastern Han	CA, LIA	

(*Continued*)

Table 9.1 (*Continued*)

No.	Sample	Date	Analytical items	Sample photo
03YLIV M3:4	pendant	Late Eastern Han	CA, LIA	
03YLIV M5:16	pendant	Late Eastern Han	CA, LIA	
03YLIV M8: 53–1/2	pendant	Three Kingdoms Period	CA, LIA	
03YLIV M17:33	pendant	Three Kingdoms Period	CA, LIA	
03YLIII M21:7	pendant	Late Eastern Han	CA, LIA	
03YLIV M12: 5(A/B/ C/D)	Eye beads	Late Eastern Han	CA, LIA for Bead A	

(**CA**: chemical; **LIA**: lead isotope analysis)

for major elements, and less than 5% for minor and trace elements. Usually, the data totals for major elements were 100±5%, and for minor and trace elements 100±20% (see the details in [2]).

9.2.2 *Lead isotope analysis*

Lead isotope ratios were measured using a multi–collector inductively coupled plasma mass spectrometer (MC–ICP–MS) using a VG Elemental system in the School of Earth and Space, Peking University. The sample preparation procedure is described in detail elsewhere.[2]

Based on the repeated analysis of SRM981, it was determined that the overall analytical 2σ error for all lead isotope ratios was less than 0.04%. The standard deviation for the 20 measurements of each of five ratios carried out during the analysis of a sample was less than 0.04%.

9.3 Results

9.3.1 *Chemical analysis*

The chemical compositional data are shown in Table 9.2. Each eye bead was analysed twice, for the white and greenish–blue components. The results suggest that all samples are $PbO–BaO–SiO_2$ glasses except ear pendant M3:4, which is of $K_2O–SiO_2$ glass. Thus the results indicate that they are all traditional ancient Chinese glasses.[3]

9.3.2 *Lead isotope analysis (LIA)*

We carried out lead isotope analyses on six lead–barium ear pendants and one eye bead (M12:5). Table 9.3 shows the lead isotope ratios of all glass samples. The data are also displayed as binary–plots of $^{207}Pb/^{206}Pb$ vs. $^{208}Pb/^{206}Pb$ in Fig. 9.2.

Although small in number, the LIA results show wide variations of all the lead isotope ratios for the lead–barium glasses, including the highest and the lowest groups mentioned by Brill (1991).[4] Thus, the LIA data indicate that there might be diverse ore sources for these glasses. However, as we pointed out elsewhere,[2] since 'low ratios' and 'high ratios' always appear together and the spread of the data are

Table 9.2 Chemical results of Lijiaba glass wares (wt.%)

No.	sample	SiO_2	Al_2O_3	Fe_2O_3	MgO	Ca O	Na_2O	K_2O	P_2O_5	BaO	PbO	CuO	TiO_2	MnO	Co_2O_3	SrO	V_2O_5	ZrO	Ag_2O	SnO_2	Sb_2O_5	NiO	B_2O_3	ZnO
03YLIVM2:4	ear pendant	47.07	0.57	0.29	0.43	1.69	4.89	0.20	0.065	14.02	29.81	0.09	0.049	0.131	0.009	0.233	0.005	0.005	0.005	0.002	0.429	0.002	0.005	0.004
03YLIVM3:2	ear pendant	50.44	0.29	0.19	0.29	0.86	5.38	0.06	0.023	12.46	29.54	0.02	0.019	0.026	0.016	0.337	0.001	0.000	0.000	0.003	0.041	0.002	0.011	0.002
03YLIVM3:3	ear pendant	54.91	0.39	0.34	0.50	1.84	6.93	0.13	0.052	15.13	19.19	0.02	0.025	0.051	0.019	0.384	0.001	0.001	0.000	0.004	0.063	0.002	0.015	0.003
03YLIVM3:4	ear pendant	78.62	1.54	2.53	0.36	1.62	0.15	10.62	0.148	0.44	0.00	0.05	0.178	3.395	0.222	0.011	0.019	0.002	0.001	0.005	0.002	0.027	0.045	0.006
03YLIVM5:16	ear pendant	45.89	0.31	0.16	0.46	1.69	3.76	0.03	0.059	15.26	31.51	0.12	0.014	0.010	0.008	0.227	0.005	0.000	0.005	0.002	0.476	0.002	0.003	0.005
03YLIVM8:53–1	ear pendant	49.14	0.65	0.34	0.54	2.27	5.17	0.27	0.122	13.13	27.38	0.05	0.056	0.157	0.010	0.218	0.006	0.005	0.004	0.006	0.447	0.003	0.005	0.005
03YLIVM8:53–2	ear pendant	47.55	0.34	0.24	0.51	1.82	4.55	0.01	0.068	13.29	30.77	0.04	0.024	0.026	0.012	0.213	0.005	0.001	0.003	0.004	0.517	0.002	0.005	0.005
03YLIVM12:5A (blue)	eye bead	53.06	1.85	0.46	0.81	0.57	5.54	0.10	0.000	8.29	26.37	2.62	0.025	0.003	0.000	0.158	0.000	0.000	0.001	0.006	0.076	0.002	0.000	0.046
03YLIVM17:33	ear pendant	47.11	0.40	0.18	0.53	1.89	3.72	0.12	0.081	14.01	30.69	0.26	0.023	0.031	0.007	0.212	0.006	0.001	0.005	0.004	0.701	0.003	0.004	0.007
03YLIIIM21:7	ear pendant	46.49	0.92	0.33	1.31	2.06	4.86	0.06	0.070	8.91	34.12	0.60	0.035	0.006	0.012	0.069	0.001	0.001	0.006	0.023	0.087	0.002	0.005	0.006
03YLIVM12:5A (white)	eye bead	50.20	0.47	0.17	0.42	0.42	5.20	0.05	0.025	12.87	29.85	0.05	0.012	0.002	0.001	0.225	0.000	0.000	0.001	0.003	0.018	0.001	0.002	0.001
03YLIVM12:5A (blue)	eye bead	48.88	0.90	0.18	0.82	0.46	6.44	0.11	0.000	13.18	27.81	0.91	0.014	0.001	0.001	0.249	0.000	0.000	0.000	0.007	0.018	0.002	0.003	0.010
03YLIVM12:5B (blue)	eye bead	49.15	0.35	0.10	0.56	0.43	7.11	0.06	0.000	9.44	31.02	1.48	0.009	0.002	0.000	0.194	0.000	0.000	0.001	0.018	0.030	0.002	0.001	0.040
03YLIVM12:5C (blue)	eye bead	50.69	0.67	0.17	0.82	0.45	6.97	0.01	0.000	9.20	29.48	1.27	0.015	0.002	0.000	0.201	0.000	0.001	0.001	0.003	0.020	0.001	0.001	0.026
03YLIVM12:5C (white)	eye bead	52.31	0.30	0.14	0.51	0.39	5.66	0.22	0.014	7.98	32.19	0.06	0.012	0.001	0.001	0.151	0.000	0.000	0.002	0.022	0.023	0.001	0.001	0.002
03YLIVM12:5D (blue)	eye bead	50.06	0.89	0.15	0.83	0.38	6.20	0.09	0.000	11.40	28.89	0.84	0.014	0.001	0.000	0.211	0.000	0.000	0.000	0.005	0.018	0.002	0.002	0.010
03YLIVM12:5D (white)	eye bead	57.14	0.83	0.30	0.44	0.44	4.76	0.08	0.135	10.10	25.51	0.02	0.028	0.010	0.001	0.177	0.001	0.000	0.001	0.004	0.008	0.002	0.000	0.002

Table 9.3 Lead isotope ratios for the lead–barium glasses from Lijiaba site

Sample no.	Names	$^{207}Pb/^{206}Pb$	$^{208}Pb/^{206}Pb$	$^{206}Pb/^{204}Pb$	$^{207}Pb/^{204}Pb$	$^{208}Pb/^{204}Pb$
03YLIVM8:53–1	ear pendant	0.7615	1.9757	20.861	15.886	41.215
03YLIVM8:53–2	ear pendant	0.7691	1.9853	20.647	15.880	40.990
03YLIVM3:3	ear pendant	0.8279	2.0678	19.019	15.746	39.327
03YLIVM5:16	ear pendant	0.7579	1.9687	20.964	15.889	41.271
03YLIVM17:33	ear pendant	0.7649	1.9787	20.752	15.873	41.061
03YLIVM2:4	ear pendant	0.7657	1.9787	20.741	15.881	41.040
03YLIIIM21:7	ear pendant	0.8734	2.1655	17.775	15.525	38.492
03YLIVM12:5	eye bead	0.9101	2.2010	16.867	15.351	37.123

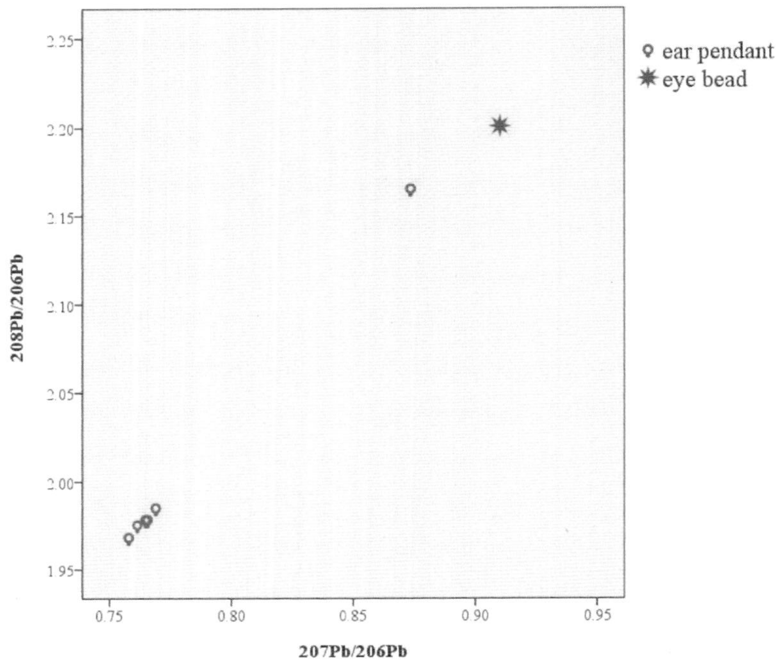

Fig. 9.2 Lead isotope ratios for the Lijiaba lead–barium glasses

similar to the two stage evolution model proposed by Stacey and Kramers,[5] we consider that they may have the same origin.

9.4 Discussion

9.4.1 *Lead–barium–silica glass*

For the lead–barium–silica glasses, the range of contents of lead oxide is between 19%–32%, and barium oxide is between 8%–16%. The silica ranges from 45% to 55%. All these lead–barium–silica glasses also all have more than 4% sodium oxide. The major differences between ear pendants and eye beads are the calcia contents. Ear pendants all have more than 1% CaO, while eye beads contain less than 1%.

As shown in Fig. 9.3, there is a negative correlation between SiO_2 and PbO, and their sum often reaches 70%. Most of the ear pendants

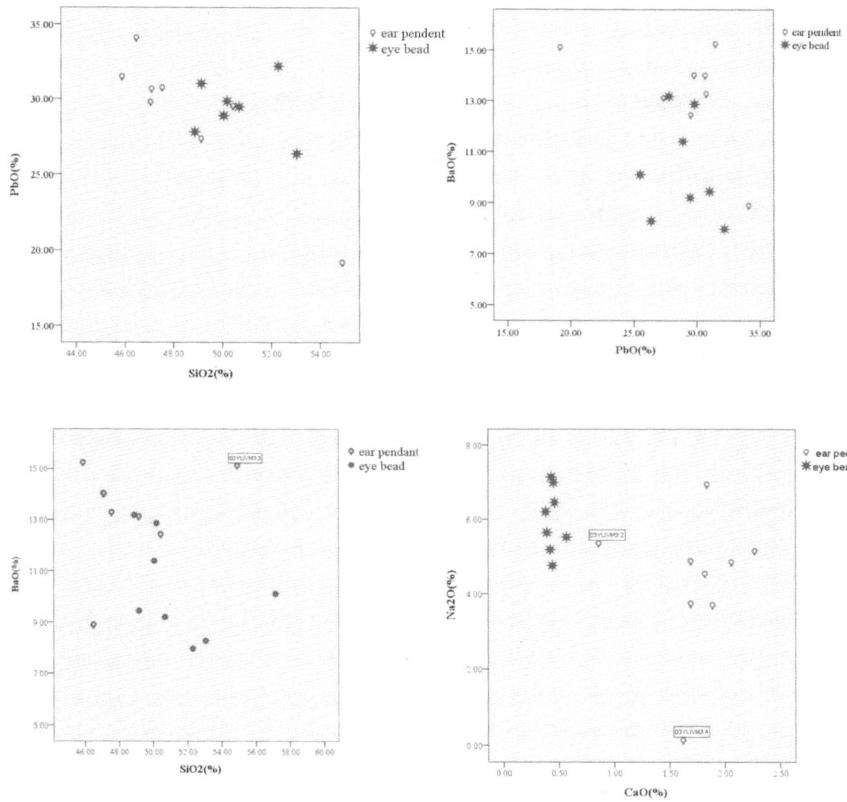

Fig. 9.3 Binary–plots of some element oxides in the glasses

contain around 30% PbO or more except for M3:3. The concentrations of lead oxide in most of the eye beads is less than 30%; the SiO_2 contents of most of the ear pendants is less than 48%, and those of the eye beads more than 50%. This suggests that these two oxides are the most important independent ingredients of the glasses, and the level of their contents has been controlled. However, the BaO contents for all except one sample (M21:7) are all about 14%. The results indicate that BaO might also be a primary ingredient and that its contents were carefully controlled. However, as we pointed out, during the Warring States period, the BaO contents varied randomly and spanned a large concentration range.[2] This means that the Eastern Han Dynasty craftsman understood clearly more functions of BaO in lead–barium–silica glasses and that they controlled the addition of BaO-containing ores more carefully than in the Warring States period. Thus one can conclude that the technology of making lead–barium–silica glass advanced from the Western Han Dynasty to the Eastern Han Dynasty.

All lead–barium ear pendants are light blue, thus copper or cobalt has probably been used as colorants. According to their chemical compositions, the contents of copper and cobalt are all very low. However, the cobalt oxide levels are 100 ppm or more. Since the bivalent cobalt ion is a very strong colourant, the blue colours of the ear pendants could be all due to these trace cobalt levels. The bluest one of all pendants is 03YLIVM3:3. It only contains about 0.02% cobalt oxide and 0.02% copper oxide. Thus the copper ions make almost no relative contribution to the glass colour, the blue colour being mainly due to the cobalt ions (although reduced iron can not be ruled out). However, some ear pendants have copper contents more than 0.1%, such as 03YLIVM5:16, 03YLIVM17:33 and 03YLIIIM21:7. They have a greenish blue colour which indicates that they are coloured by copper ions. It is worth noting that there are also trace contents of MnO in all of the pendants. This suggests that the cobalt could have been 'earth' cobalt ore, which is a kind of Chinese cobalt ore that was used in blue and white porcelain during the Ming and Qing dynasties.[6]

The main colours of all of the eye beads are white and green. According to the chemical compositions, the green parts of the eye beads all contain higher amounts of CuO showing that copper ions were used as the colourant. According to the ligand filed theory, this may be due to the ratio between SiO_2 and PbO+BaO. This ratio determines whether the copper ion (Cu^{++}) replaces silica as a network former or replaces the flux ions as a network modifier.[7] Thus, as the lead and barium contents increase, the colour will change from blue to green. When a certain critical value is exceeded, the colour will change to green. Thus the copper ions in these eye beads are mainly network modifiers.

The white parts of the beads have nothing different with the green part except their lower copper contents. Thus the low levels of colorant ions contribute to their white appearances. It is much more likely that a white crystalline opacifier causes the colour.

9.4.2 *Lead isotope analysis of lead–barium glasses*

In 1966, Brill and Wampler applied the lead isotope method to the provenance of ancient glasses. They found that the lead isotope field of ancient Chinese glasses can be considered as distinct from all the other signatures identified anywhere else. Their study also showed that the lead in some early Chinese glasses fell into two main isotopic groups, and these groups contained the highest and lowest lead isotope ratios in their database.

On the plot of $^{207}Pb/^{206}Pb$ vs. $^{208}Pb/^{206}Pb$ (Fig. 9.2), the lead isotope data for Lijiaba glasses occupy all the ranges in Brill's database, especially the lowest and highest ones.[4,8] According to Brill's database, no glass with the low ratio lead isotope occurs after the Eastern Han Dynasty. Thus the three Lijiaba glass objects (M8:53a,b and M2:4) of the Three Kingdoms period are up to now the latest examples of lead barium glasses with the low lead isotope ratio.

According to former studies, the earliest lead barium glasses with low lead isotope ratios belongs to the 4th–3rd BC. Now we can conclude that this kind of lead had been used to make lead–barium glass for almost 500 years. The low lead isotope ratios mentioned by Brill *et al.* are indicative of so–called 'high radiogenic lead'. This type of lead contains higher radiogenic isotope ratios than the commonest lead type.[9] In fact, this type of lead is very rare, meaning that there was only one source of glass that contained this lead. It is remarkable that in most circumstances where the low lead isotope ratios have been identified, the 'high lead ratios' were found; this aspect indicates that these two types of lead could originate from the same source. This suggests that there should have been a lead–barium–glass making centre dating to between the Warring States period and the Three Kingdoms period (or at least to late Eastern Han Dynasty) which used the 'high radiogenic lead' as their raw material.

From the chemical and lead isotope analysis, we considered that the early Chinese lead–barium glasses with the lowest and highest $^{207}Pb/^{206}Pb$ values may have been produced in the southern area of the ancient Chu Kingdom during the Warring States Period.[2] Now, we believe that this glass–making centre could have existed for almost 500 years and that the Lijiaba lead–barium glasses were also made in the centre. We also think that the disappearance of the early lead–barium glass making technology must have had some relationship with the shutdown of this glss-making centre in the southern Chu area.

9.4.3 *The potash silicate glass*

03YLIVM3:4 is the only K_2O–SiO_2 glass found in this study. The major components in this ear pendant include K_2O and SiO_2, with high levels of colourants such as Fe_2O_3, MnO and CoO. However, MgO, CaO and Na_2O are both very low, which indicates that the K_2O could come from a mineral such as saltpetre (KNO_3).[10] The content of Al_2O_3 is also very low. This means that the major source of SiO_2 could be a material characterised by very pure quartz, such as quartz sand or vein quartz.

It is interesting that the contents of MnO, Fe_2O_3 and CoO are all so high. This makes the colour of the pendant a deep translucent purple colour, resembling a 'black' colour. Thus it contained mixed colourants, in which the Mn ions could be the major colourant. It is difficult to tell what kind of colour the craftsman wanted to achieve: 'black', purple or blue. The chemical composition shows that the raw material of the colourant should be the 'earthy' cobalt., as mentioned above.

According to Xiong and Li [11] the potash glass classification that this pendant belongs to is m–K_2O–CaO–Al_2O_3–SiO_2 glass. It also contains about 110 ppm Sr. This is much like the low–Rb potash glass also noted by Xiong and Li. [11] On the basis of their study, the m–K_2O–CaO–Al_2O_3–SiO_2 potash glass with low Rb contents was probably made in the Guangxi area. In addition, Xiong and Li also pointed out that the cobalt–blue potash glasses of this type often have high MnO contents too, which suggests that the 'earth' cobalt was added to the glass recipes to make them blue. The phenomenon can also be seen in this Lijiaba potash glass. Thus this pendant is very likely to have been imported from Guangxi area.

9.5 Conclusion

The chemical results suggest that silica, a lead–bearing material, and a barium–bearing material were the main primary raw materials used to make Lijiaba lead–barium glasses. The carefully controlled BaO contents indicates that the technology of making lead–barium glass advanced from the Warring States Period to the Eastern Han Dynasty.

The colourants that gave the blue colours to the pendants were cobalt–bearing ores, although the cobalt contents were all very low. The bivalent copper ions are the main reason for the green appearance of the eye–bead glasses.

LIA results show that the lead–barium glasses have the same source with only early lead–barium glasses containing the high radiogenic lead. We consider that this kind of lead–barium glass may have been produced in the southern area of the ancient Chu Kingdom. We also think that the disappearance of lead–barium glass making technology in ancient China must be related to the closure of this centre.

The main ingredients of the potash–silica glass include silica, saltpeter and 'earth' cobalt. Compared with the published data of ancient potash–silica glasses, this Lijiaba pendant may originate from the Guangxi area.

Acknowledgments

This work is financially supported by the National Social Science Foundation of China (Project No: 10CKG003) and the Culture Heritage Conservation Science and Technology Foundation of China State Administration of Culture Heritage. The authors would like to thank Prof. F. X. Gan and Prof. Q. H. Li for their kind invitation to the ISAG' 2012 conference. Mrs. Huang B.L. is gratefully acknowledged for the measurements of lead isotope ratios

References

1. Department of Archaeology, Sichuan University and Yunyang Institute of Culture Relics Administration, A brief archaeological report on excavation 1999 of Lijiaba Site in Yunyang County, Chongqing, in *Southern Ethnology and Archaeology*, eds. Sichuan University Museum, Department of Archaeology, Sichuan University and Chengdu Institute of Archaeology (Science Press, Beijng, China, 2011), Vol. VII, pp. 495–526.
2. J. F. Cui, X. H. Wu and B. L. Huang, Chemical and Lead isotope analysis of some lead-barium glasses of Warring States period unearthed from Chu tombs

in Changde City, Hunan Province, China, *Journal of Archaeological Science* **38**, 1671(2011).

3. F. X. Gan, Evolution of the chemical composition and the origin of making technology of Chinese ancient glass, in *Development of Chinese Ancient Glass*, ed. F. X. Gan (Shanghai Scientific and Technical Publishers, Shanghai, China), 2005, pp. 220–240,
4. R. H. Brill, I. L. Barnes and E. C. Joel, Lead isotope studies of early Chinese glasses. in *Scientific research in early Chinese glass*, eds. R. H. Brill and J. H. Martin (Elmira Quality Printers, New York, 1991), pp. 65–83.
5. J. S. Stacey and J. D. Kramers, Approximation of terrestrial lead isotope evolution by a two-stage model. *Earth and Planetary Science Letters* **26**(2), 207–221 (1975).
6. J. G. Li, *A history of science and technology in China, Volume of Ancient Pottery and Ceramic*, Scientific Press, Beijing, China, 1998, pp. 375–386.
7. A. M. Pollard and C. Heron, Archaeological Chemistry, The Royal Society of Chemistry, Cambridge, UK, 1996, pp. 163–173.
8. R. H. Brill and H. Shirahata, The second Kazuo Yamasaki TC–17 lecture on Asian glass: Recent lead isotope analyses of some Asian glasses with remarks on strontium isotope analyses, in *Ancient glass research along the Silk Road*, eds. F. X. Gan, R. H. Brill and S.Y. Tian (World Scientific Publishing, Singapore, 2009), pp. 149–164
9. J. F. Cui, and X. H. Wu, Lead isotopic archaeology: provenance study of bronzes unearthed from Yunan Province and Vietnam, Cultural Relics Press, Beijing, China, 2008, pp. 24–39.
10. M. G. Shi, O. L. He and F. Z. Zhou, *Journal of Chinese Ceramic Society* **14**, 307 (1986).
11. Z. M. Xiong and Q. H. Li, *Archaeological, scientific and technical study on the glass artefacts of the Han Dynasty unearthed from Guangxi, China*, Cultural Relics Press, Beijing, China, 2011, pp. 71–164.

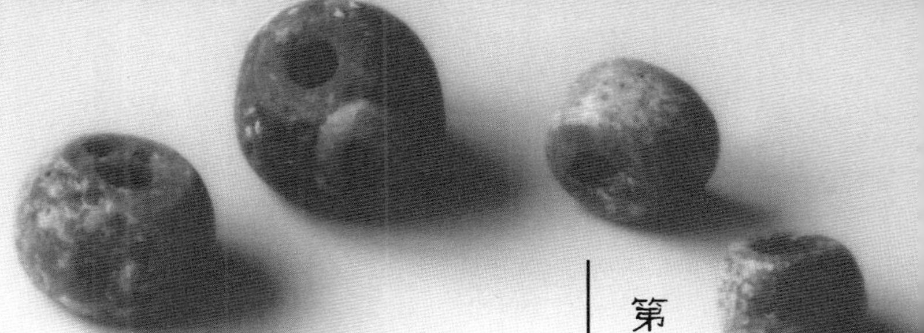

Chapter 10 第十章

Scientific Analysis of a Glass Vessel and an Eye Bead from the Zhagunluke and the Shanpula Sites from the Overland Southern Line of the Silk Road

Qian Cheng, Jinlong Guo†, Bo Wang† and Jianfeng Cui‡*

*Chinese Academy of Cultural Heritage,
Beijing 100029, China

†Xinjiang Uygur Autonomous Region Museum,
Urumqi 8300006, China

‡Perking University, Beijing 100871, China

10.1 Introduction

The Xinjiang region was a trade and cultural hub between the East and the West in ancient times. The study of excavated relics has already been an important topic that has received much attention worldwide. In parallel with pottery and bronze wares, glass is particularly unique and colourful amongst the various types of objects, and is one of the three earliest man-made materials.

Qiemo county, on the southern route of the Silk Road, was located in the southern Tarim Basin oasis in the Xinjiang region. It was the main activity zone of the ancient residents of the Qiemo Kindom. The Chinese reference is "Hanshu (汉书)"; it was recorded that the ancient Qiemo Kingdom was a small country, with an eastern connection to Ruoqiang (若羌), and westward to Jingjue (精绝).

With a population of about 610 and 120 soldiers, the small country was ruled by a Han Dynasty king.[1] However, there is very little historical documentation about the history and culture of the small country. Until the excavation of the Zhagunluke tomb site it has been difficult to gain much information about the small ancient kindom and its relationship with other kindoms on the Silk Road.

One hundred and sixty-seven tombs were excavated at Zhangunluke by archaeologists from the Xinjiang Uygur Autonomous Museum and the local heritage administration between 1985 and 1988.[2] The Zhagunluke tomb site is five kilometers to the southwest of Qiemo county (Fig. 10.1). Precious archaeological finds such as iron and wooden objects, pottery, textiles and glass were unearthed, which reflect the history, culture and economy of the ancient Qiemo

Fig. 10.1 Map of the Xinjiang Region. The Zhagunluke and the Shanpula tomb site are located near Qiemo county and Khotan city respectively, on the ancient Southern Line of the Silk Road

Kingdom and the relationship between the East and the West during the 1st and the 6th century AD.

Another county should be mentioned in the Lop county which belongs to the Hotan Oasis in the southern part of Tarim Basin of Xinjiang region. It was the main activity zone of ancient Khotan residents. The Shanpula tomb site (Fig. 10.1) is 14 kilometres to the southwest of Lop county and has been regarded as one of the most important archaeological discoveries in Xinjiang since the 1980s. A huge number of precious archaeological finds such as iron and wooden objects, pottery, textiles and glass were unearthed. The tomb period of the site was between the 1st century BC and the 4th century AD.[3]

The glass discovered were mainly in form of beads which were strung into necklaces, and also bracelets and earrings. Those glass accessories were always mixed with stone and bone beads. The colour and pattern of the glass beads were highly varied: the patterns including moving clouds, oblique stripes, concentric circles, as well as dragonfly-eye and bi-coloured beads in 15 assortments.

In this paper, the chemical composition, provenance and manunfacture of two interesting glass objects: a glass vessel and an eye bead, from the Zhagunluke site (Qiemo) and the Shanpula site (Khotan), will be disccussed.

10.2 Sample descriptions

10.2.1 *The glass cup*

The Zhagunluke tomb site consists of 167 tombs of local ancient residents. Some glass finds have been discovered, like quite brightly coloured yellow, red, green and blue monochrome beads made of opaque or translucent glass. However, from the tomb M49, a delicate glass cup was excavated (Fig. 10.2). It was the only glass vessel discovered from whole tomb complex. Tomb M49 is dated between the Han Dynasty and the Wei-Jin Dynasty (the 1st–6th century AD), according to the tomb structure and the other archaeological finds, such as wooden objects, lacquer objects and textiles.

Fig. 10.2 The glass cup excavated from tomb M49 of Zhagunluke. Photo by Xinjiang Museum.

The vessel is a cylindrical facet-cut glass cup, resembling a honeycomb. The outer-surface of the cup is covered with three rows of facets. From top to bottom, there are 13 oval cut facets in the first and second rows and seven facets in the third row. The facets are in staggered arrangement such as two facets above and two below a central facet. A large round facet forms the cup base.

The cup surface is partly weathered and present a light brown color. However, through the cross-section of the broken rim, the vessel can be seen to be colorless glass of high quality. It can be infered that this cup used to be a transparent and delicate vessel that was buried with its owner. The diameter of the glass cup is 68.45 mm; the base is 18.05 mm. The height is 66.64 mm, and the thickness is slightly uneven, between 3–3.4 mm.

10.2.2 *The eye bead*

The dragonfly eye bead (Fig. 10.3) was discovered in tomb M45, at the Shanpula tomb site near Khotan city. This eye bead was made in brick red monochrome base glass decorated with bi-color spots in the form of eyes. One brown and two white concentric circles plus a

Fig. 10.3 Dragonfly eye bead excavated from the tomb M45 of the Shanpula site

brown pupil compose one eye. The three eyes were embedded in the red bead. The outer and the hole diameters of the bead are 6.74 and 1.99 mm, with the height being 4.61 mm. The diameter of the eye is 4.59 mm. There are two types of eye-bead making. One type is the flush stratified eye; the other one is the mosaic eye.[4] This bead seems to be the second type, formed by applied sections cut from a mosaic cane with an eye design.

10.3 Scientific analysis

Two techniques were used to analyse the glass cup, a 3D video microscope (3D-Microscope) and Laser Ablation Inductively Coupled Plasma Atomoic Emission Spectrometry (LA-ICP-AES).

10.3.1 *3D video microscope*

The microscope was used to observe the details of glass vessel manufacture providing very interesting observations. The equipment was the KH-300 produced by HIOX Company, Japan.

10.3.2 *LA-ICP-AES*

With the aim of better understanding the glassmaking technology of the two glass samples, it was hoped that LA-ICP-AES analysis would

Table 10.1 Instrument and measurement parameters

Prodigy High Dispersion ICP-AES: Leeman Labs Inc. USA	
RF generator: 40.82MHz	RF power: 1.1KW
Argon flow rate	
Plasma: 1.4l/min	Auxiliary pressure: 0 psig
Nebulizer pressure: 35 psig	
Laser ablation system: UP-266 Marco, New-Wave Inc. USA	
Laser: ND-YAG	Laser mode: Q-switched
Laser wavelength: 266nm	Output energy: 10±1mJ
Facular aperture: 515μm	Helium flow rate: 800ml/min

suggest the origin and provenance. Due to the low detection limits, good precision and accuracy, and the possibility of a simultaneous quantification of major and trace elements, LA-ICP-AES is increasly used in the analysis of archaeological glass finds. The instrument and measurement parameters are listed in Table 10.1. Normally, a sample's dissolution is performed and analysis is carried out on the solution. However, it is not an especially accurate result because if weathered glass is included in the sample, the light elements like sodium and potassium will have been leached out. Laser ablation is a useful method to avoid this deviation. By ablating the weathered layer of the sample in a certain depth, a fresh layer is exposed producing reliable data. The Corning glass B standard was employed as a quantative controlling reference glass.

10.4 Results and discussion

The results obtained by LA-ICP-AES analysis are given in Table 10.2. The chemical composition of silicon oxide is 66.40%, soda is 20.50% and lime is 7.4%. The glass is therefore a typical soda-lime-silica glass, with the relatively low contents of potassium and magnesium oxides (0.42% and 0.66% respectively). Characteristics of the so-called low magnesium glass composition (LMG)[5] show that it is a glass made in the 'Roman tradition'.

Table 10.2 Composition of the analyzed glass samples (wt%)

Sample	SiO_2	Al_2O_3	Fe_2O_3	MgO	CaO	Na_2O	K_2O	MnO	P_2O	TiO_2	Sb_2O_3	CuO	PbO	CoO	BaO	SnO_2	SrO	ZnO	V_2O_5
Glass cup	66.40	2.26	0.63	0.66	7.40	20.50	0.42	1.42	n.d.	n.d.	21ppm	n.d.	n.d.	10ppm	0.07	20ppm	660ppm	n.d.	0.02
Eye-red	42.24	1.57	0.37	0.39	3.43	8.55	0.39	0.34	0.00	0.04	1.06	8.05	33.458	20ppm	0.017	410ppm	300ppm	0.002	0.004
Eye-black	70.28	2.58	1.11	0.88	8.69	12.79	0.97	1.70	0.12	0.07	0.35	0.03	0.178	20ppm	0.062	80ppm	970ppm	0.011	0.005
Eye-white	53.12	1.95	0.67	0.58	6.46	9.38	0.77	0.78	0.09	0.06	17.04	0.03	8.898	10ppm	0.040	150ppm	600ppm	0.009	0.003

n.d. means not detected.

Information on the kind of the sand used in ancient glass manufacture can further be obtained by considering strontium (Sr) contents. In fact, calcium and strontium (Sr) are found in marine shells. A relatively high Sr level should be featured in glass made using coastal sand rather than inland sand. In this glass cup a Sr content of 660 ppm has been detected. According to Inn Freestone's research,[6] Sr contents above 300 ppm should rule out the use of limestone bearing sand together with an evaporate flux to make the glass. The result indicates that coastal sand was use to make the LMG cup. Therefore, it is very possible that, this colorless glass cup was made from sand of the Syria-Palestine coast and produced during the Roman period which is about the 1st–5th century AD. Before the scientific analysis was carried out, local archaeologists had predicted that it was possibly produced in the later Sasanian period by the 7th century AD according to its appearance.[2] However, with this scientific analysis, the maunfacturing date within the Roman period is much earlier than the assumptions from archaeologists.

10.4.1 *The manufacture of colourless glass*

The use of high-purity sands and the addition of decolorizers in the production of colourless glass have implications for the organization of production. In 1963, Sayre and Smith proposed a model for the use of decolorizer in Roman colourless glasses. They found that colourless glasses 'from the Syrian coast' are characterized by the increasing use of manganese oxide (MnO) rather than antimony pentoxide (Sb_2O_5), in concentrations of about 1% MnO towards the end of the Roman period.[8] This model has since been used widely to characterize the use of decolorizers in colourless glasses throughout the Roman Empire.

The iron oxide content in the glass is only 0.63%. An iron impurity in blue-green glass is assumed to have derived from sand. Antimony has been detected at 21 ppm. However, the manganese oxide has been detected at 1.42% and this probably reflects Roman glass decolorizing technology. The Romans favoured beach sands and a natron flux, the sand typically deriving from Mediterranean coastal beaches some of which produced low-impurity colourless glass. Therefore, manganese was utilised instead of antimony with weak decolorizing effect. In Jackson's study,[9] she demonstrated that in the later Roman period, decolorizer used was gradually moved from antimony (Sb) to manganese (Mn) in the fourth century. Therefore, it is very possible that the glass cup was produced in the later Roman period due to Mn as a decoloriser.

Natron is a crystalline mineral composed predominantly of sodium compounds with few impurities. It has been suggested that decolorizing could have been achieved by calcining the iron-containing raw materials before melting, thus oxidizing the iron to its more colorless state.[10] It can be argued that colourless glasses are a good subject to demonstrate the skill of the Roman glassmaker and illustrate the influence of raw materials choice upon the final glass composition.

10.4.2 *Facet-cut decoration of the cup*

It is very interesting to observe the manufacture details of the glass vessel by the 3D video microscope. It is more suitable to explore such a glass object than an optical microscope. The microscope can construct 3D images by imaging the different components and the surface morphology. It has the same superior resolutions observed with an optical microscope and some information can be revealed more clearly, such as bubbles, impurities in the glass and the traces left by glass working.

According to the Fig. 10.4, it is obvious that by microscope, those fine parallel lines indicate the polishing traces during the facet processing. Similar tools were probably used for the gemworking.

In the Roman period, luxurious colorless facet-cut glass vessels were popular amongst the noble people in the upper classes.[11] A class of specialist artisans possessing the skills required to cut and engrave glass sometimes cooperated with glassblowers to fashion and decorate vessels. A talented glasscutter, using many of the same techniques

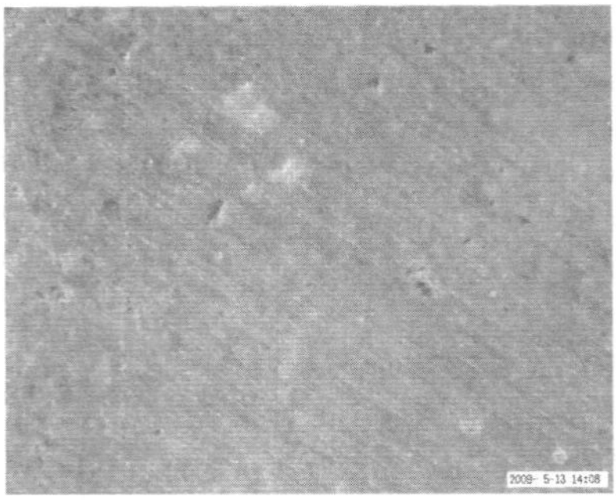

Fig. 10.4 The polishing traces on the glass vessel base by 3D video microscope, (magnification 50 times)

employed by gemcutters on precious stones, could produce a broad range of decorative motifs from simple geometric patterns.[12]

10.4.3 *Analysis of the eye bead*

The chemical compositions of the tri-coloured eye-bead determined by LA-ICP-AES are given in Table 10.2. The results show that the brick-red colour of the bead is composed of 42.24% silica, 8.55% soda and 3.43% lime. Potash and Magnesia contents are below 1% and this suggests that this delicate and slightly weathered eye bead, dating from between the 1st century BC and the 4th century AD, could have been made using an evaporite flux such as that used by Roman glass-makers. The brick-red base glass contains lead and copper oxides at extremely high levels: 8.05% and 33.46% respectively. It seems to be a typical red soda-lime-silica glass from the Roman Period. Cuprite plays an important role in glass making as a simultaneous colorant and opacifier. In order to prepare the glass it is very necessary to keep the glass in reducing conditions. Therefore, the opacifier is precipitated out as a suspension and is not dissolved in the glass. A charcoal blanket is useful to produce the required reducing conditions.[13]

In the eye zone, the thick brown circle (Fig. 10.2) is obviously of a soda-lime-silica composition with relatively low potash and magnesia at 0.88% and 0.97% respectively. It is interesting that the manganese oxide is rather high at 1.7%.[14] In this situation, Mn (2+) would have been deliberately used as a purple-brown colorant in glass, mixed with Fe (2+) at a concentration of 1.11%, as this would lead to a translucent brown color.[15]

In each eye, there are two thin circles of opaque white glass. The two white circles present similar characteristics as the base glass: soda-lime glass with low potash and magnesia. However, calcium and antimony are at extremely high levels of 6.46% and 17.04% respectively. It seems that calcium antimonite ($Ca_2Sb_2O_7$) crystals were formed as colorant and opacifier in glass. When antimony is added to soda-lime-silica glass, it will bond with the calcium in the glass producing opaque white crystals of calcium antimonite. Calcium antimonate belongs to the opacifier 'Type A' in white colour so that it is also a white colourant.[16]

Shortland[17] has proved that antimony was used in the very first Egyptian glasses. The antimonies are not only used in the earliest glasses, but also part of the most popular colours used in the 90% of decorated Roman glass. In the Roman Period, many vessels, tesserae and enamels were found to have been opacified with $Ca_2Sb_2O_7$ crystals. Henderson's study[18] shows that during the 1st–4th century AD, all the samples of blue, turquoise and green enamels were found to be opacified with $Ca_2Sb_2O_7$ crystals. All this evidence shows that the Roman-age enamels were manufactured using an antimony-rich glass. Furthermore, Henderson has discovered several opaque white antimonate of the 9th century dating from Viking Ribe in Jutland.[19] Therefore it is very possible that antimony was used from the 15th century BC to the 9th century AD.

According to the results, it can be assumed that the eye bead was probably produced in eastern Mediterranean region in the Roman period and transferred from the west to the Shanpula site.

10.5 Conclusions

According to the analytical results by LA-ICP-AES, the glass cup was probably produced on the syro-palestine coastal area during the Roman period. Manganese (MnO) was used as decolorant to purify the glass. Moreover, facet-cut decoration was a typical decorative style on Roman glass vessels. All of the evidence indicate that the glass cup was probably produced on the east coast of Mediterranean Sea during Roman Period and was transfered to Qiemo along the Southern Line of the Silk Road.

An eye bead was analyzed using ICP-AES. The results indicate that the red monochrome base and the eye zone are of a soda-lime composition with low contents of potassium and magnesium oxides. The bead decorated with a provincial Roman pattern perhaps suggests a Mediterranean origin.

It is interesting that manganese was intentionally added to the glass used to make the glass cup as a decolorant offsetting the yellow-green colour. It shows a high technological skill in controlling the furnace atmosphere so as to oxidize Fe (2+) to Fe (3+) through the chemical action of oxygen. However, the Mn (2+) has caused a totally different purple-brown colour in the circle of the eye bead. The furnace atmosphere would have been adjusted to produce reducing conditions so as to keep the Mn (2+) ion in glass. By comparing the composition of the glass cup and eye bead, it is useful to find out the glass artisans were very skillful to controlling the ionic composition, temperature and furnace atmosphere in order to obtain the specific type of glass they produced.

The glass cup and glass bead provide the archaeological evidence for a transfer of western glass to ancient eastern countries along the Silk Road. The glass reflects the existance of commerical and cultural communications between the west and the east via the southern line of the Silk Road during the 1st BC — the 5th AD. It is necessary to use trace element analysis and radiogenic isotope analysis to explore further information about potential manufacturing regions.

Acknowledgement

The financial supports from the National fund of Philosophy and Social Science and Fund of State Administration of Cultural Heritage are gratefully acknowledged. It was also grateful for Chinese Ceramic Society and Shanghai Institute of Optics and Fine Mechanics, CAS to invite my to participate the 2012 International Symposium on Ancient Glass.

References

1. G. Ban, *Hanshu Vol. 12* (ancient Chinese reference), Zhonghua Book Company, Beijing, China, 1983.
2. Xinjiang Uygur Autonomous Regoin Museum, *Xinjiang Relics* **1**, 1 (2003).
3. Xinjiang Uygur Autonomous Regoin Museum Xinjiang Institute of Archaeology, *Shanpula in Xinjiang of China: Discovery and Study of ancient Khotan Civilization*, Xinjiang People's Publishing House, Xinjiang, China, 2001.
4. M. Spare, *Ancient Glass in the Israel Museum: Beads and Other Small Objects*, The Israel Museum, Jerusalem, 2001.
5. J. Herderson, *Oxford Journal of Archaeology* **4**, 267 (1985).
6. I. C. Freestone, K. A. Leslie, M. Thirlwal *et al.*, Archaeometry **45**, 19 (2003).
7. R. H. Brill, *Chemical analysis of early glasses*, Vols. I and II, Corning Museum of Glass, New York, 1999.
8. E. V. Sayre, The intentional use of antimony and manganese in ancient glasses, in *Advances in glass technology, part 2*, eds. F. R. Matson and G. Rindone (Plenum Press, New York. 1963), pp. 263–282,
9. C. M. Jackson, *Archaeometry* **47**, 763 (2005).
10. W. B. Stern, The composition of Roman glass: problems of non-destructive analysis, in *Annales du 11e Congrès de l'Association Internationale pour l'Histoire du Verre*, ed. (AIHV, Amsterdam, 1990), pp. 37–41.
11. H. Tait (ed.), *Five Thousand Years of Glass*, British Museum Press, London, 1991.
12. R. A. Crossmann, *Ancient glass: A guide to the Yale Collections*, Yale University Art Gallery, New Haven (CT) 2002.
13. J. Henderson, *Oxford Journal of Archaeology* **4**, 267 (1985)
14. P. Mirti and M. Pace, *Archaeometry* **50**, 429 (2008).
15. C. Y. Wang and Y. Tao, *Design and Adjustment of Glass Composition*, Chemical Industry Press, Beijing, China, 2006.
16. W. E. S. Turner and H. P. Rooksby, *Further historical studies based on X-ray diffraction methods of the reagents employed in making opal and apaque glasses*, Jahrbuch des (Romisch-Germanischen Zentra museums, Mainz, 1961), Vol. 8, pp. 1–16.
17. A. Shortland, *Archaeometry* **44**, 517 (2002).
18. J. Henderson, Chemical characterisation of Roman glass vessels, enamels and tesserae, in *Materials Issuses in Art and Archaeology II: Symposium Held April 17–21, 1990, San Francisco, California, U.S.A.* (*Materials Research Society Symposium Proceedings)*, eds. P. B. Vandiver, J. Druzik, and G. S. Wheele (Materials Research Society, Pittsburgh, 1991), pp. 601–607.
19. J. Henderson, X-ray fluorescence analysis of Iron Age glass, Ph.D. thesis, University of Bradford (1982).

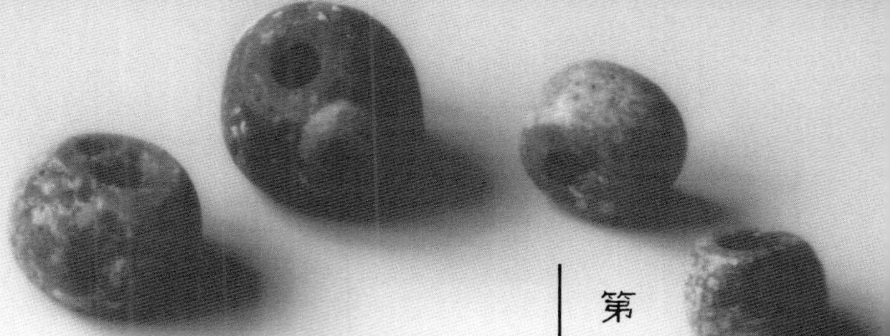

Chapter 11 第十一章

Early Byzantine Glass Supply and Consumption: The Case of Dichin, Bulgaria

*Thomas Smith, Julian Henderson and Edward W. Faber**
*University of Nottingham,
University Park, Humanities Building, Nottingham, NG7 2RD

11.1 Introduction

Throughout the fourth to sixth centuries, societies within the late Eastern Roman Empire underwent gradual social and economic changes, retaining varying elements of continuity. Early Byzantine society was defined by widespread economic reforms, the like of which had feedback effects on important commodities. One such commodity was glassware, formed from batches of ingredients into raw glass at primary production sites, before being transported further afield to be fashioned into vessels at secondary workshop sites. The application of scientific techniques to excavated finds has previously illustrated large variations in glass compositions between the fourth and sixth centuries.[1] Two broad compositional groupings have emerged; Levantine I, produced in the Levant, and high iron, manganese and titanium (HIMT), possibly produced in Egypt.[2] However, the currently available data is insufficient to answer many primary research questions. Not only do global questions about relations between glass production, distribution, dating and chemical variations remain unanswered, but regions such as the Balkans have seen little research.

The primary aim of this research is to supplement the study of early Byzantine glass compositions, so as to generate a broader compositional database for glass artefacts dating to the fifth and sixth centuries within the Balkans. To this end, the early Byzantine fortress of Dichin, Bulgaria, was chosen for the focus of study because a large assemblage of glass has been excavated from securely stratified contexts. The site was excavated by a joint British-Bulgarian team between 1996 and 2003. It occupies an elevated position on the south bank of the river Rositsa and is located *c.* 11 km west of Nicopolis ad Istrum. Interestingly the site is not connected to a roadway, the closest one being to the north of the river Rositsa, and it has a regular layout, as opposed to the sprawling distribution of buildings commonly found on other contemporary Bulgarian sites.[3] The site's initial occupation occurred *c.* 410–490 AD before it was destroyed deliberately by fire. It was re-established again from *c.* 520–575 AD, but it is unclear if it continued to function as a defensive stronghold.[4] By chemically analysing a range of glass vessel types from Dichin using electron probe microanalysis and comparing them with results from other sites it was hoped that patterns of production, trade and provenance for it would result. More broadly it was hoped that the results for glass could provide valuable insights into the economic and cultural connections that Dichin had with its local region and with the wider world as well as potentially providing evidence for trade along the early Byzantine frontier.

11.2 The glass assemblages

The glass from the British excavations at Dichin have been categorised by Cholakova, primarily being based on morphological characteristics (fragments of decorated walls, rims, bases, or handles) and vessel shapes.[5] Overall 270 fragments of glass were recovered from the British excavations, of which 119 were diagnostic to specific vessel forms. The majority of finds dated to the first phase of occupation (410–490 AD) were found in area E, with a few finds also coming from areas C, D, G, H, and K. Finds dating to the second phase of occupation (520–575 AD) were, as already noted, fewer in number, there were fewer vessel forms and restricted to only being found in areas C, D, E, and K. Notably the assemblage features a wide array of common forms used during the fifth and sixth centuries, but only in certain cases was it possible to reconstruct the whole profile of a vessel (Samples 7, 9, 28, 71 and 97); not a single vessel could be fully reconstructed. Some forms correlated with those identified by Isings,[6] including; conical bowls, conical cups, cone lamps, cylindrical phials, hemispherical bowls, rounded bowls, stemmed goblets and straight cups. Other fragments were identified as belonging to more general forms, such as cups, jars, bowls, lamps, jugs and pots. There are also several fragments that cannot be identified. Some fragments from the first occupational phase have been interpreted as being raw glass or waste from the process of glass blowing. It has been suggested that the population of the fortress collected these raw materials and also attempted to work the glass, although there is no evidence of glass furnaces on site. The glass colours include aqua blue, dark blue, green blue, and amber yellow-brown and range from very pale to a deep translucent tints. However, the majority are colourless, a characteristic that is found throughout both periods of occupation. Most vessels are free blown, with no decoration. In rare cases vessels have applied decoration such as threads or have been cold worked producing ground and polished decorations.

The typological forms present in the earliest fifth century contexts occurred throughout the first occupation of the site until *c*. 490 AD. As noted above, during the second phase of occupation (*c*. 520–575 AD), the majority of the glass forms, colours, and decorations found in the first occupation were absent. In this respect Dichin differs from the nearby sites of Novae, Iatrus and Nicopolis ad Istrum, in that these sites exhibit continuity in the vessel forms found throughout this period of time. Cholakova has suggested that the technical characteristics of the glass vessels across both periods of occupation show that they were worked on a local or regional scale, with few vessels being imported.[5] The assemblage appears to feature typical forms and features of production technology that are common throughout fourth to sixth century contexts. A notable change in the distribution of vessel types between the two periods of occupation has been suggested as an indicator by Cholakova (ibid.) of general changes in the lifestyle of the inhabitants, including the economic status of the region, the supply of glassware to Dichin, and differences in glass consumption between the inhabitants of the first and second periods of occupation.

11.3 Previous chemical analyses of Byzantine glasses

During the 5th–7th centuries Levantine I glass appears to be the dominant early Byzantine glass type made in the Levant (Freestone *et al.* 2000). The type is characterised by low soda (~15%) and high lime (~8.5%) contents, often also featuring low levels of iron oxide (~0.4%) and relatively high alumina (~3%). Further compositional variations are exhibited by manganese oxide levels. Levantine Ia variants contain low levels of manganese oxide (up to ~0.2%), whereas Levantine Ib contain relatively high levels (~1.2%). Brill concluded that levels of manganese oxide below 0.4% could be attributed to natural impurities, and glasses that feature higher levels are assumed to result from its deliberate addition in order to colour the glass.[7]

HIMT was manufactured first in the late forth century, becoming widespread by the fifth. It is characterised by high levels of iron, manganese and titanium oxides, with a strong positive correlation between iron and titanium, and iron and manganese oxides. The primary production area for HIMT glasses has been suggested as Egypt.[2] This attribution has been supported also by characteristic lead, neodymium, oxygen, and strontium isotope signatures. A sand source with mixed geological signatures, such as around the Nile delta was probably used to make it. However, the composition of HIMT glass is far more variable than that of the Levantine I type. Foster and Jackson's analysis of Romano-British HIMT glasses identified two subgroups of the HIMT, HIMT 1 and HIMT 2.[8] HIMT 1 glasses were characterised as containing double the concentration of iron, manganese and titanium oxides than found in HIMT 2 glasses (in the case of Romano-British glasses averaging 1.36%, 1.71% and 0.33% as opposed to 0.72%, 0.98% and 0.12% respectively). Both groups are characterised by correlations between iron, alumina, titanium and manganese oxides, although this is stronger in HIMT 1 glasses. Foy *et al.* also identified two groups of HIMT glasses, termed groups 1 and 2.[9–11] Group 2 appears to correspond with Foster and Jackson's HIMT 1,[8] while group 1 features lower levels of iron, alumina, titanium, and manganese oxides than is present in both HIMT 1 and HIMT 2, creating a third variant of HIMT glass (HIMT G1). The decolouration of HIMT glasses complicates the issue further, with Foster and Jackson describing four sub types.[12] HIMT colourless 1 features antimony as a decolouriser (~0.29%), HIMT colourless 2a features manganese (~1%), but low levels of alumina (~1.81%) and high magnesia (~0.67%) whereas HIMT colourless 2b also features manganese (~1%), but differs in that it contains high alumina (2.37%) and low magnesia (0.41%). Finally, colourless 3 contains ~0.35% manganese oxide and ~0.30% antimony oxide.

Rehren and Cholakova analysed ten samples of fifth century Dichin glass from the Bulgarian team's excavations; they selected especially deep green or olive-amber coloured glasses.[13] Despite a macroscopic examination which suggested that the samples belonged to two groups based on colour, production technique, and quality, chemical analysis showed there were three: two extreme variations of the HIMT composition and a high iron and titanium (HIT) composition. It was therefore anticipated that these HIMT compositions would be found amongst our analytical results.

11.4 Materials and methods

Seventy-nine vessel fragments from well stratified contexts were sampled, representing the full range of vessel forms, colouration, and decoration, for both occupational periods within each area excavated by the British team. Because the phase I occupation contexts in area E produced such a large amount of material compared with other areas, the other areas were prioritised for sampling so as to avoid bias as much as possible. As a result, virtually all stratified samples from the second phase of occupation were selected for study. Each individual sample is listed together with its site area code, vessel form and vessel colouration, for both phases of occupation in Table 11.1.

Table 11.1 Archaeological information and the EPMA results for the Byzantine glass from Dichin, excluding results for SrO, SnO_2 and BaO which were measured but all results were below the minimum detection limit. Results are presented as weight % oxide

No.	Site area	Colour	Form	SiO_2	Na_2O	CaO	MgO	K_2O	Al_2O_3	FeO	TiO_2	MnO	P_2O_5	SO_3	Cl	PbO	ZrO_2	NiO	CoO	CuO	Sb_2O_5	Ag_2O	V_2O_5	Cr_2O_3	ZnO	As_2O_5
5th Century HIMT 1 glass																										
2	E	Y	CC	64.3	17.9	5.23	1.03	0.48	2.95	2.24	0.38	2.44	0.05	0.23	1.01	<mdl	0.02	<mdl	<mdl	0.08	0.03	<mdl	0.023	0.008	<mdl	<mdl
4	E	Y	Cup	62.85	18	6.07	1.17	0.41	2.91	2.94	0.65	2.23	0.09	0.28	0.94	<mdl	0.05	<mdl	<mdl	0.02	<mdl	<mdl	<mdl	0.01	<mdl	<mdl
22	E	GB	U	66.7	20.62	5.03	1.28	0.42	3.27	1.36	0.54	0.12	0.04	0.32	1.22	<mdl	0.03	<mdl	<mdl	<mdl	<mdl	<mdl	<mdl	0.01	<mdl	<mdl
24	E	Y	U	63.17	19.1	6.09	1.19	0.39	2.96	2.98	0.63	2.24	0.09	0.28	0.94	<mdl	0.03	0.009	<mdl	0.25	<mdl	<mdl	0.02	0.01	<mdl	<mdl
38	E	DB	U	65.6	18.37	5.86	1.09	0.37	2.7	2.25	0.36	0.09	0.04	0.23	1.17	0.25	<mdl	0.01	0.24	0.47	0.03	<mdl	<mdl	0.01	<mdl	<mdl
40	E	C	Bowl	66.69	18.79	4.92	1.05	0.43	3.06	2.25	0.39	1.88	0.08	0.24	1	<mdl	0.02	0.01	0.01	0.17	<mdl	<mdl	<mdl	0.01	<mdl	<mdl
53	E	Y	U	66.96	18.67	4.93	1.04	0.44	3.1	2.25	0.41	1.86	0.08	0.22	1.01	<mdl	0.02	0.01	<mdl	0.15	<mdl	<mdl	<mdl	0.009	<mdl	<mdl
54	E	Y	U	65.61	17.81	5.6	1.06	0.42	2.91	1.37	0.48	2.39	0.04	0.24	0.86	<mdl	0.03	<mdl	<mdl	0.11	<mdl	<mdl	0.02	0.01	<mdl	<mdl
74	D	Y	SG	66.38	15.13	5.79	1.52	0.46	3.51	3.44	0.62	1.62	0.11	0.11	0.9	<mdl	0.03	0.01	<mdl	0.16	<mdl	<mdl	0.02	0.013	<mdl	<mdl
92	K	AB	CP	67.94	18.36	6.55	0.84	0.65	2.46	1.26	0.19	0.97	0.07	0.23	1.02	<mdl	<mdl	0.01	<mdl	0.05	<mdl	<mdl	<mdl	<mdl	<mdl	<mdl
101	G	Y	CC	64.23	18.17	7.16	1.34	0.68	2.95	2.44	0.38	1.72	0.13	0.29	0.9	<mdl	0.02	0.01	<mdl	0.09	<mdl	<mdl	<mdl	<mdl	<mdl	<mdl
103	G	Y	U	64.45	17.38	5.09	1.16	0.4	3.05	2.96	0.38	2.22	0.1	0.24	1.02	<mdl	<mdl	<mdl	<mdl	0.31	<mdl	<mdl	<mdl	<mdl	<mdl	<mdl
107	C	Y	U	62.66	18.43	6.81	1.04	0.52	2.62	2.42	0.34	1.8	0.11	0.38	1.01	<mdl	0.03	<mdl	<mdl	0.02	<mdl	<mdl	0.02	0.008	<mdl	<mdl
108	C	Y	U	64.16	19.03	6.03	1.18	0.49	2.92	2.4	0.48	2.16	0.08	0.32	0.91	<mdl	0.03	<mdl	<mdl	0.04	<mdl	<mdl	0.03	0.01	<mdl	<mdl
111	C	GB	Jar	66.67	19.42	6.43	0.86	0.64	2.5	1.28	0.2	0.96	0.08	0.24	1.01	<mdl	<mdl	0.01	<mdl	0.11	<mdl	<mdl	<mdl	<mdl	<mdl	<mdl
112	C	DB	U	67.86	18.97	5.68	0.86	0.52	2.75	1.64	0.23	0.25	0.04	0.24	1.06	0.14	<mdl	<mdl	0.1	0.36	<mdl	<mdl	0.02	<mdl	<mdl	<mdl
5th Century HIMT 2 glass																										
6	E	C	CB	65.59	20.94	7.69	0.84	0.46	2.09	1.02	0.09	1.05	0.05	0.39	1.17	<mdl	<mdl	<mdl	<mdl	0.05	<mdl	<mdl	<mdl	<mdl	<mdl	<mdl
7	E	C	SG	65.06	18.08	8.37	0.7	0.51	2.04	0.58	0.08	0.87	0.04	0.3	1	<mdl	<mdl	<mdl	<mdl	0.1	<mdl	<mdl	0.02	0.01	<mdl	<mdl
9	E	C	Bowl	66.36	18.17	6.63	0.7	0.66	2.37	0.67	0.09	0.76	0.05	0.26	1.09	<mdl	<mdl	<mdl	<mdl	0.13	0.03	<mdl	<mdl	<mdl	<mdl	<mdl

10	E	C	Bowl	68.44	19.41	6.32	0.47	0.46	1.78	0.49	0.04	0.91	0.03	0.37	1.13	<mdl	<mdl	<mdl	<mdl	0.25	<mdl	<mdl	0.03	<mdl	<mdl	<mdl
12	E	C	Cup	65.51	20.41	6.74	0.73	0.52	2.11	0.89	0.08	0.95	0.06	0.36	1.1	0.11	<mdl	<mdl	0.05	0.23	<mdl	<mdl	<mdl	<mdl	<mdl	<mdl
13a	E	DB	SC	66.17	19.88	5.3	0.81	0.43	2.45	1.97	0.13	0.18	0.03	0.26	1.35	0.2	<mdl	0.009	0.18	0.46	<mdl	<mdl	<mdl	<mdl	<mdl	<mdl
23	E	C	HB	63.97	19.21	7.25	0.82	0.59	2.39	0.8	0.09	0.78	0.05	0.32	1.21	<mdl	<mdl	<mdl	<mdl	0.09	0.04	<mdl	<mdl	<mdl	<mdl	<mdl
28	E	C	Lamp	65.02	21.11	6.58	0.66	0.42	1.86	0.46	0.06	0.87	0.04	0.36	1.17	<mdl	<mdl	<mdl	<mdl	0.22	<mdl	<mdl	0.03	0.01	<mdl	<mdl
31	E	C	Pot	66.36	18.72	6.47	0.61	0.48	2.05	0.66	0.08	0.62	0.03	0.27	1.16	<mdl	<mdl	0.009	<mdl	0.07	<mdl	<mdl	<mdl	<mdl	0.02	<mdl
32	E	GB	Lamp	69.02	19	6.65	0.71	0.56	2.04	0.61	0.07	0.66	0.06	0.21	1.13	<mdl	<mdl	<mdl	<mdl	0.18	0.03	<mdl	<mdl	<mdl	<mdl	<mdl
34	E	C	Lamp	65.67	20.13	6.52	0.64	0.44	1.85	0.48	0.05	0.86	0.04	0.36	1.19	<mdl	<mdl	<mdl	<mdl	0.31	<mdl	<mdl	<mdl	<mdl	0.02	<mdl
36	E	C	CP	65.19	18.15	7.39	0.75	0.56	2.26	0.65	0.12	1.2	0.07	0.3	0.85	<mdl	<mdl	<mdl	<mdl	0.18	<mdl	<mdl	<mdl	<mdl	<mdl	<mdl
44	E	AB	Waste	68.51	19.54	6.52	0.69	0.52	2	0.71	0.14	0.65	0.05	0.32	1.11	<mdl	<mdl	<mdl	<mdl	0.03	<mdl	<mdl	<mdl	<mdl	<mdl	<mdl
49	E	C	Lamp	65.48	19.72	5.6	0.66	0.46	2.23	0.5	0.07	1.22	0.03	0.36	1.04	<mdl	<mdl	<mdl	<mdl	0.06	<mdl	<mdl	<mdl	<mdl	<mdl	<mdl
50	E	C	Lamp	65.59	20.61	5.87	0.73	0.42	2.14	0.69	0.07	0.94	0.02	0.29	1.39	<mdl	<mdl	<mdl	<mdl	0.05	<mdl	<mdl	<mdl	<mdl	<mdl	<mdl
51	E	AB	U	67.9	19.41	6.74	0.77	0.55	2.17	1.05	0.16	0.9	0.07	0.31	1.02	<mdl	<mdl	<mdl	0.01	0.25	<mdl	0.021	<mdl	<mdl	<mdl	<mdl
52	E	DB	U	65.31	18.89	6.9	0.76	0.46	2.17	1.32	0.1	0.46	0.07	0.32	1.05	0.12	<mdl	0.01	0.06	0.43	<mdl	<mdl	<mdl	<mdl	<mdl	<mdl
59	E	C	U	67.25	21.33	6.88	0.79	0.42	1.97	0.84	0.08	0.64	0.05	0.37	1.21	<mdl	<mdl	<mdl	<mdl	0.22	<mdl	<mdl	0.03	0.008	<mdl	<mdl
64	D	C	HB	68.99	18.23	6.67	0.53	0.38	1.92	0.45	0.06	0.86	0.03	0.25	1.12	<mdl	<mdl	<mdl	<mdl	0.03	0.03	<mdl	<mdl	<mdl	<mdl	<mdl
65	D	C	HB	69.08	18.29	6.69	0.53	0.38	1.9	0.43	0.08	0.82	0.02	0.25	1.11	<mdl	<mdl	<mdl	<mdl	0.04	<mdl	<mdl	<mdl	<mdl	<mdl	<mdl
68	D	C	SG	66.18	20.08	6.16	0.51	0.35	1.85	0.54	0.04	1.24	0.01	0.36	1.24	<mdl	<mdl	<mdl	<mdl	0.26	<mdl	<mdl	<mdl	<mdl	<mdl	<mdl
70a	D	DB	Cup	65.01	20.23	6.86	0.7	0.43	1.92	0.51	0.06	0.86	0.04	0.31	1.14	<mdl	<mdl	<mdl	<mdl	0.13	<mdl	<mdl	<mdl	<mdl	<mdl	<mdl
70b	D	C	Cup	66.95	18.59	8.64	0.71	0.52	1.97	0.63	0.05	0.92	0.03	0.31	1.04	<mdl	<mdl	<mdl	<mdl	0.09	<mdl	<mdl	<mdl	<mdl	<mdl	<mdl
86	K	C	HB	67.67	19.38	5.88	0.37	0.52	1.81	0.4	0.07	0.77	0.03	0.33	1.08	<mdl	<mdl	0.01	<mdl	0.23	<mdl	<mdl	0.02	<mdl	<mdl	<mdl
87	K	C	SG	70.63	19.76	5.26	0.61	0.5	2.01	0.42	0.06	0.78	0.03	0.27	1.15	<mdl	<mdl	<mdl	<mdl	0.02	0.03	<mdl	<mdl	<mdl	<mdl	<mdl
88	K	C	Cup	64.21	19.06	7.64	0.85	0.66	2.32	0.8	0.08	0.77	0.04	0.34	1.19	<mdl	<mdl	<mdl	<mdl	0.08	0.03	<mdl	0.02	<mdl	<mdl	<mdl

(Continued)

Table 11.1 (*Continued*)

No.	Site area	Colour	Form	SiO_2	Na_2O	CaO	MgO	K_2O	Al_2O_3	FeO	TiO_2	MnO	P_2O_5	SO_3	Cl	PbO	ZrO_2	NiO	CoO	CuO	Sb_2O_5	Ag_2O	V_2O_5	Cr_2O_3	ZnO	As_2O_5
89	K	C	U	68.56	20.6	6.45	0.6	0.34	1.89	0.45	0.04	1.04	0.03	0.39	1.17	<mdl	<mdl	<mdl	<mdl	0.21	<mdl	<mdl	<mdl	<mdl	<mdl	<mdl
93	K	C	Raw	66.69	18.95	6.81	0.66	0.5	2.07	0.75	0.1	0.62	0.05	0.28	1.22	<mdl	<mdl	<mdl	<mdl	0.11	<mdl	<mdl	<mdl	0.008	<mdl	<mdl
105	C	GB	U	65.01	18.47	7.14	1.35	0.93	2.66	0.86	0.1	1.71	0.17	0.34	0.9	<mdl	0.02	<mdl	<mdl	0.12	0.06	<mdl	<mdl	<mdl	<mdl	<mdl
106	C	GB	RB	67.13	19.8	7.11	0.56	0.37	1.79	0.39	0.04	0.8	0.03	0.31	1.15	<mdl	0.02	<mdl	<mdl	0.03	<mdl	<mdl	0.02	<mdl	<mdl	<mdl
109	C	C	SG	66.02	19.68	7.06	0.53	0.37	1.78	0.4	0.05	0.81	0.03	0.3	1.16	<mdl	0.02	<mdl	<mdl	0.3	<mdl	<mdl	<mdl	0.009	<mdl	<mdl
115	H	DB	U	65.65	18.92	6.53	0.8	0.54	2.21	1.13	0.18	1.06	0.07	0.3	1.05	0.04	<mdl	<mdl	0.01	0.15	<mdl	<mdl	<mdl	<mdl	<mdl	<mdl
5th Century HIMT G1 glass																										
11a	E	DB	RB	68.17	18.58	6.14	0.48	0.45	1.79	0.56	0.05	0.4	0.03	0.3	1.12	0.15	<mdl	<mdl	0.07	0.39	<mdl	<mdl	<mdl	0.01	<mdl	<mdl
35	E	C	CP	68.68	19.69	6.94	0.63	0.48	1.91	0.51	0.07	0.59	0.04	0.32	1.16	<mdl	<mdl	0.01	<mdl	<mdl	<mdl	<mdl	<mdl	<mdl	<mdl	<mdl
47a	E	DB	U	69.42	19.07	6.11	0.43	0.54	1.68	0.59	0.04	0.57	0.04	0.32	1.07	<mdl	<mdl	<mdl	<mdl	0.02	<mdl	<mdl	<mdl	<mdl	<mdl	<mdl
97	K	C	Lamp	69.36	19.85	6.75	0.61	0.39	1.94	0.51	0.06	0.4	0.03	0.3	1.27	<mdl	<mdl	<mdl	<mdl	0.18	0.03	<mdl	<mdl	<mdl	<mdl	<mdl
5th Century HIT glass																										
14	E	AB	Bowl	67.32	17.8	5.2	1.25	0.42	3.25	1.43	0.63	0.05	0.04	0.22	1.07	<mdl	0.03	0.02	<mdl	<mdl	<mdl	<mdl	0.02	0.01	<mdl	<mdl
15	E	AB	CB	69.65	18.34	5.12	1.26	0.4	3.25	1.41	0.64	0.05	0.02	0.21	1.06	<mdl	0.05	0.04	<mdl	<mdl	0.03	<mdl	<mdl	0.01	<mdl	<mdl
16	E	AB	CB	69.49	18.23	5.1	1.23	0.4	3.18	1.43	0.62	0.05	0.04	0.22	1.05	<mdl	0.05	0.04	<mdl	<mdl	<mdl	<mdl	0.02	0.01	<mdl	<mdl
17	E	AB	SG	68.78	17.74	5.19	1.26	0.4	3.2	1.41	0.62	0.05	0.02	0.24	1.07	<mdl	0.04	0.04	<mdl	<mdl	<mdl	<mdl	0.02	0.01	<mdl	<mdl
19	E	AB	SG	69.54	18.51	5.15	1.29	0.4	3.25	1.45	0.63	0.05	0.03	0.24	1.05	<mdl	0.03	0.02	<mdl	<mdl	0.03	<mdl	0.02	0.01	<mdl	<mdl
20	E	GB	U	69.36	18.19	5.12	1.25	0.4	3.23	1.42	0.63	0.04	0.03	0.24	1.05	<mdl	0.03	0.03	<mdl	<mdl	<mdl	<mdl	<mdl	0.01	<mdl	<mdl
21	E	GB	Bowl	68.83	18.49	5.17	1.28	0.39	3.27	1.42	0.62	0.05	0.04	0.22	1.09	<mdl	0.04	0.035	0.01	<mdl	<mdl	<mdl	<mdl	0.01	<mdl	<mdl
29	E	AB	CL	69.21	18.67	5.13	1.26	0.43	3.2	1.42	0.62	0.05	0.03	0.24	1.04	<mdl	0.03	0.03	<mdl	<mdl	<mdl	<mdl	0.02	0.01	<mdl	<mdl
57	E	AB	U	67.36	17.83	4.96	1.23	0.4	3.22	1.38	0.62	0.04	0.02	0.22	1.06	<mdl	0.04	0.03	<mdl	<mdl	0.03	<mdl	<mdl	0.01	<mdl	<mdl
71	D	AB	Bowl	67.08	18.08	5.14	1.26	0.4	3.27	1.44	0.62	0.05	0.04	0.22	1.06	<mdl	0.04	0.03	<mdl	<mdl	<mdl	<mdl	<mdl	0.01	<mdl	0.02

5th Century Levantine Ia glass																										
11b	E	C	RB	63.54	18.21	7.55	1.21	0.87	2.85	1.36	0.11	0.96	0.15	0.3	0.81	<mdl	<mdl	<mdl	<mdl	0.32	0.06	<mdl	<mdl	<mdl	<mdl	<mdl
30	E	C	Jug	69.39	14.35	8.74	0.81	0.6	3.24	0.41	0.05	0.02	0.11	0.06	1.02	<mdl	<mdl	<mdl	<mdl	0.17	<mdl	<mdl	<mdl	0.008	<mdl	<mdl
37	E	AB	CP	68.99	14.01	8.74	0.81	0.61	3.2	0.42	0.05	0.02	0.12	0.06	1.03	<mdl	<mdl	<mdl	<mdl	0.15	<mdl	<mdl	<mdl	<mdl	<mdl	<mdl
81	D	C	RB	68.79	17.22	7.47	0.55	0.66	2.53	0.4	0.04	0.32	0.11	0.2	1.09	<mdl	<mdl	0.01	<mdl	0.22	0.26	<mdl	<mdl	<mdl	<mdl	<mdl
113	C	C	Lamp	63.9	18.37	9.96	1.17	0.68	2.72	0.99	0.14	1.01	0.1	0.41	0.82	<mdl	<mdl	<mdl	<mdl	0.02	0.04	<mdl	<mdl	<mdl	<mdl	<mdl
5th Century Levantine Ib glass																										
13b	E	C	SC	64.8	17.29	7.43	1.3	0.93	2.71	0.92	0.12	1.56	0.16	0.32	0.83	<mdl	<mdl	<mdl	<mdl	0.26	0.06	<mdl	<mdl	<mdl	<mdl	<mdl
43	E	Y	Moil	63.42	18.05	9.17	1.44	0.93	2.51	0.92	0.12	1.59	0.16	0.34	0.78	<mdl	<mdl	0.01	<mdl	0.14	0.06	<mdl	0.02	<mdl	<mdl	<mdl
47b	E	C	U	64.89	18.7	7.93	1.12	0.73	2.46	0.84	0.11	1.45	0.09	0.33	0.88	<mdl	<mdl	<mdl	<mdl	0.026	0.05	<mdl	0.02	<mdl	<mdl	<mdl
55	E	AB	RB	64.64	18.54	8.04	1.13	0.72	2.51	0.84	0.11	1.47	0.1	0.34	0.88	<mdl	<mdl	<mdl	<mdl	0.03	0.05	<mdl	<mdl	<mdl	<mdl	<mdl
67	D	C	SG	62.96	18.23	8.18	1.19	0.74	2.59	0.81	0.1	1.62	0.1	0.28	0.94	<mdl	<mdl	<mdl	<mdl	0.08	0.04	<mdl	<mdl	<mdl	<mdl	<mdl
6th Century HIMT 1 glass																										
3	E	Y	CC	64.43	17.32	5.54	1.31	0.4	3.22	2.65	0.71	2.37	0.09	0.23	0.95	<mdl	0.04	<mdl	<mdl	0.03	<mdl	<mdl	0.03	0.01	<mdl	<mdl
6th Century HIMT 2 glass																										
8	E	C	SG	65.45	17.99	7.61	0.93	0.54	2.15	0.66	0.09	1.25	0.05	0.36	0.86	<mdl	<mdl	<mdl	<mdl	0.15	0.06	<mdl	<mdl	<mdl	<mdl	<mdl
41	E	AB	CP	66.61	19.04	6.54	0.66	0.5	2.14	0.57	0.05	1.13	0.05	0.3	0.99	<mdl	<mdl	<mdl	<mdl	0.12	0.04	<mdl	0.02	<mdl	<mdl	<mdl
75	D	C	SG	68.82	20.21	5.69	0.55	0.42	2.07	0.54	0.06	1.2	0.04	0.33	1.19	<mdl	<mdl	<mdl	<mdl	0.16	<mdl	<mdl	<mdl	<mdl	<mdl	<mdl
90	K	C	SG	63.2	18.49	7.09	1.29	0.94	2.59	0.86	0.1	1.51	0.16	0.36	0.9	<mdl	0.02	0.01	<mdl	0.21	0.05	<mdl	0.02	<mdl	<mdl	<mdl
98	K	C	U	65.66	19.89	7.83	0.77	0.53	2.2	0.72	0.09	1.19	0.08	0.38	1.03	<mdl	<mdl	<mdl	<mdl	0.13	<mdl	<mdl	0.02	0.01	<mdl	<mdl
6th Century Levantine 1a glass																										
80	D	C	RB	69.08	14.11	8.97	0.59	1.15	2.99	0.35	0.03	0.08	0.11	0.13	0.85	<mdl	<mdl	<mdl	<mdl	0.02	0.06	<mdl	<mdl	<mdl	<mdl	<mdl
110	C	C	U	68.66	15.16	7.73	0.55	0.97	2.81	0.36	0.04	0.09	0.08	0.13	0.92	<mdl	<mdl	<mdl	<mdl	0.23	0.06	<mdl	0.02	<mdl	<mdl	<mdl

(Continued)

Table 11.1 (*Continued*)

No.	Site Area	Colour	Form	SiO_2	Na_2O	CaO	MgO	K_2O	Al_2O_3	FeO	TiO_2	MnO	P_2O_5	SO_3	Cl	PbO	ZrO_2	NiO	CoO	CuO	Sb_2O_5	Ag_2O	V_2O_5	Cr_2O_3	ZnO	As_2O_5
6th Century Levantine 1b glass																										
78	D	C	SG	64.62	19.64	7.95	1.26	0.63	2.69	0.87	0.13	1.77	0.06	0.39	0.87	<mdl	<mdl	<mdl	<mdl	0.08	0.06	<mdl	<mdl	<mdl	<mdl	<mdl
79	D	C	SG	64.89	19.73	7.98	1.18	0.72	2.63	0.83	0.11	1.47	0.1	0.36	0.87	<mdl	<mdl	<mdl	<mdl	0.12	0.05	<mdl	<mdl	<mdl	<mdl	<mdl
104	C	C	U	65.08	19.46	7.71	1.17	0.65	2.45	0.71	0.12	1.72	0.07	0.33	1.06	<mdl	<mdl	<mdl	<mdl	0.14	<mdl	0.02	0.031	<mdl	<mdl	<mdl

Inclusions and crystallites

No.	SiO_2	Na_2O	CaO	MgO	K_2O	Al_2O_3	FeO	TiO_2	MnO	P_2O_5	SO_3	Cl	PbO	ZrO_2	NiO	CoO	CuO	Sb_2O_5	Ag_2O	V_2O_5	Cr_2O_3	ZnO	As_2O_5
32x	2.52	1.7	0.69	0.16	0.05	0.1	86.06	<mdl	0.13	0.08	0.24	0.08	<mdl	<mdl	0.01	0.21	0.2	<mdl	<mdl	<mdl	<mdl	<mdl	0.04
32y	52.95	10.48	21.28	0.65	0.29	1.28	14.55	0.04	0.57	0.71	0.18	0.52	<mdl	<mdl	<mdl	<mdl	0.2	<mdl	<mdl	<mdl	<mdl	<mdl	<mdl
71x	67.17	18.16	5.12	1.25	0.4	3.23	1.42	0.61	0.06	0.03	0.25	1.07	<mdl	0.03	<mdl	<mdl	0.14	<mdl	<mdl	0.017	0.013	<mdl	<mdl
71y	68.05	17.26	5.18	1.26	0.46	3.29	1.41	0.63	0.05	0.03	0.26	1.07	<mdl	0.03	0.03	<mdl	<mdl	<mdl	<mdl	<mdl	0.018	<mdl	<mdl
71z	66.99	16.75	5.25	1.01	0.49	2.85	3.34	0.46	0.22	0.06	0.25	1.05	<mdl	0.04	0.04	<mdl	<mdl	<mdl	<mdl	<mdl	0.008	<mdl	<mdl
79x	52.82	18.83	24.63	0.47	0.22	0.85	0.31	<mdl	0.81	0.14	0.11	0.13	<mdl	<mdl	0.01	<mdl	<mdl	<mdl	<mdl	<mdl	0.009	<mdl	<mdl

Key to table 11.1:

Colours: AB = Aqua blue, C = Colourless, DB = Dark blue, GB = Green blue, Y= Amber yellow-brown

Forms: CB = Conical bowl, CC = Conical cup, CL = Cone lamp, CP = Cylindrical phial, HB = Hemispherical bowl, RB = Rounded bowl, SG = Stemmed goblet, SC = Straight cup, and U = Rims, bases, and body fragments of an unidentifiable vessel form.

Samples of 0.5 mm were removed from the fragments. Fragments 11, 13, 47, and 70 had threads applied to rims, resulting in an extra four samples and a total of 83 samples. The glass samples were then mounted in cold-setting epoxy resin, before being ground and polished using standard sample preparation procedures down to a 0.25 micron diamond suspension. The samples were coated with a thin film of carbon prior to analysis to prevent surface charging and distortion of the electron beam during analysis. The polished samples were analysed by EPMA-WDS using a JEOL JXA-8200 electron microprobe (Department of Archaeology, University of Nottingham). Quantitative compositional analyses were carried out using an analytical set-up of a 20 kV accelerating voltage and a 50nA beam current with a 50 μm defocused beam, and the quantitative results were produced using a PRZ correction routine. The counting times were 30 seconds on the peak and 15 seconds on the background either side of the peak. A defocused beam is used to reduce the effect of the migration of alkalis within the samples.[14] The EPMA-WDS was calibrated against a combination of certified mineral, pure metal and synthetic standards. The composition of 26 elements were presented as oxide weight percentage: Na_2O, MgO, Al_2O_3, SiO_2, P_2O_5, SO_3, Cl, K_2O, CaO, TiO_2, V_2O_3, Cr_2O_3, MnO, FeO, CoO, NiO, CuO, ZnO, As_2O_5, SrO, ZrO_2, Ag_2O, SnO_2, Sb_2O_5, BaO, PbO. Three areas of interest at ×1000 were analysed for the main glass phase of each sample and the results were averaged. The results for any crystallites or inclusions were produced from a single analysis at ×1000.

To provide a check on the accuracy and precision of the EPMA-WDS and to monitor any drift in the instrument, repeat analyses of a secondary standard, Corning B glass, were included during the analytical run (at the start and end of the analysis, and in between every other polished block).[15] The following relative analytical accuracies were obtained using the Corning B standard as unknown: 3% for Na_2O; 2.5% for SiO_2; 1.5% for K_2O; 1.5% for CaO and 2% for PbO. For minor elements the accuracy was 5% for MgO, 2.5% for Al_2O_3, 15.5% for P_2O_5, 6.5% for FeO, 5% CuO and up to 13% for Cl. For trace elements the levels of detection varied from 75 ppm for Mn and Cr to 455 ppm for Ba.

11.5 Results

Bi-plots of the results are presented in Figs. 11.1–11.4 together with published data: results for each sample are given in Table 11.1. Seventeen samples were of the HIMT 1 compositional type, constituting 21% of the results. The majority of these glasses are of an amber yellow-brown colouration and are mainly cups- or the vessel forms are unidentifiable. The highest proportion of the samples analysed in this project are of HIMT 2 composition (35 analyses; 43% of the results). The majority of these results are of colourless glasses and include a notably wide variety of vessel forms, the only raw glass sample, and melted glass fragment. The applied rims of samples 13a and 70a, and the vessel body of 70b, were also identified as HIMT 2. A further four results were of the HIMT G1 composition type, making up 5%. They are the least represented compositional type, all of which are colourless. The group is constituted of a cylindrical phial, a lamp, a rounded bowl, and a vessel of unknown form, the latter two both featuring an applied dark blue rim. Ten HIT glasses were identified:

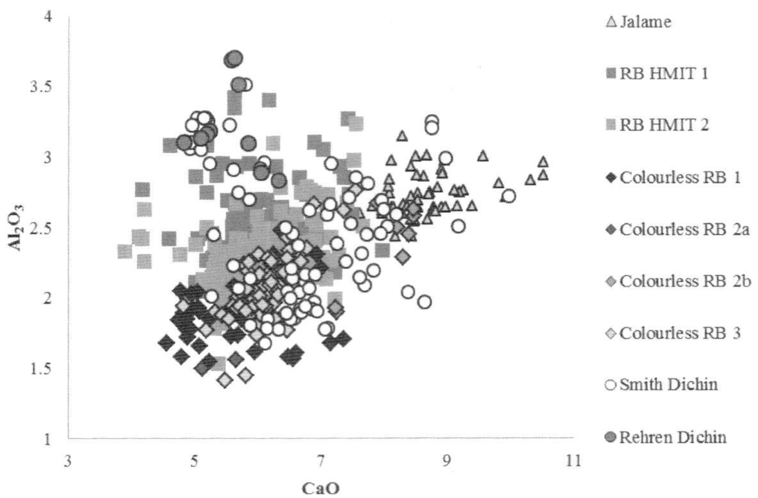

Fig. 11.1 Bi-plot of alumina and calcium oxide results from Dichin compared with Levantine glasses from Jalame (Brill, 1988), Romano-British HIMT (Foster and Jackson, 2009, 2010), and Rehren and Cholakova's (2010) preliminary Dichin analysis

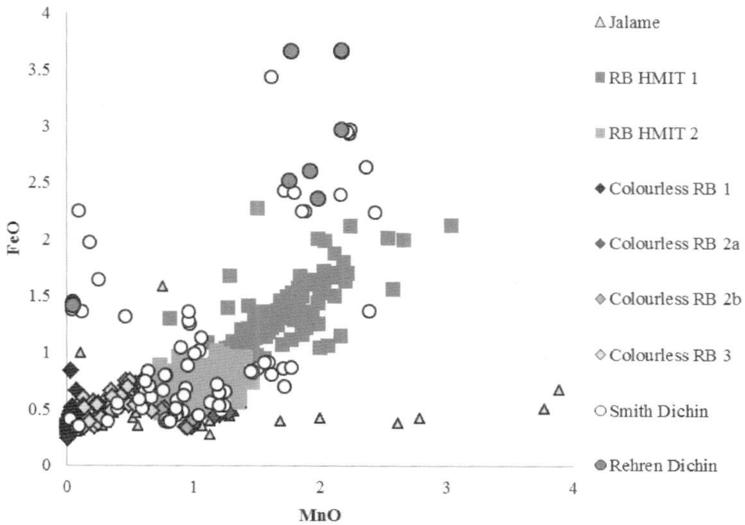

Fig. 11.2 Bi-plot of iron and manganese oxide results from Dichin compared with Levantine glasses from Jalame (Brill, 1988), Romano-British HIMT (Foster and Jackson, 2009, 2010), and Rehren and Cholakova's (2010) preliminary Dichin analysis

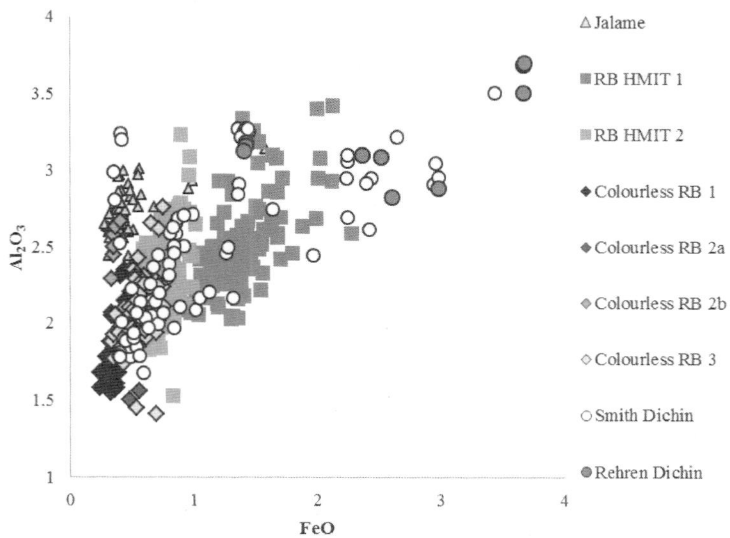

Fig. 11.3 Bi-plot of alumina and iron oxide results from Dichin compared with Levantine glasses from Jalame (Brill, 1988), Romano-British HIMT (Foster and Jackson, 2009, 2010), and Rehren and Cholakova's (2010) preliminary Dichin analysis

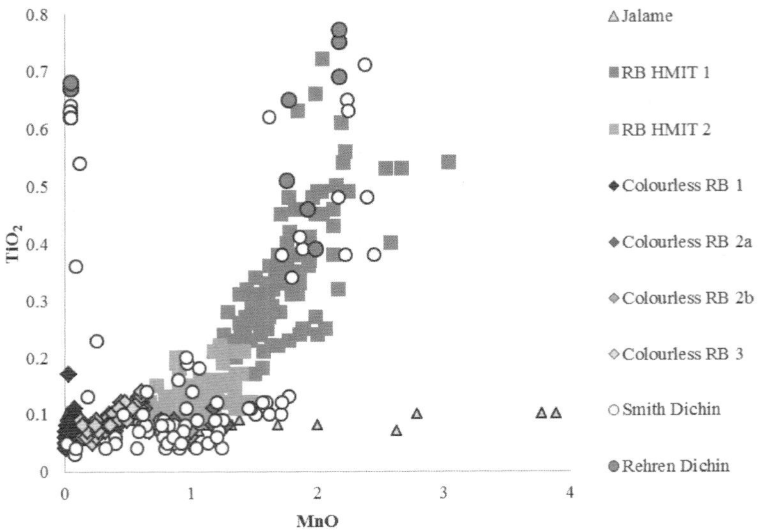

Fig. 11.4 Bi-plot of titanium and manganese oxide results from Dichin compared with Levantine glasses from Jalame (Brill, 1988), Romano-British HIMT (Foster and Jackson, 2009, 2010), and Rehren and Cholakova's (2010) preliminary Dichin analysis

eight were aqua blue and two blue-green. In total these equalled 12% of the assemblage, and were composed of bowls, goblets or unidentifiable forms. Finally, seven examples of Levantine Ia and eight examples of Levantine Ib glasses were identified, equalling 9% and 10% of the total respectively. The majority of these glasses, which include a moil fragment, are colourless and were made into a large variety of vessel forms. Sample 13 is the only example of a straight cup from the site and it has an applied string applied to the rim: the vessel body is of a Levantine Ib composition and an applied string HIMT 2.

Overall, it has been found that there was no significant difference in the compositions between fifth century and sixth century samples from Dichin. Furthermore, only a single instance of antimony decolourisation was found-in sample 81. All other samples contained <0.07 antimony; if colourless they were decolorised with manganese oxide. During the analysis numerous inclusions and compositional banding was identified within the glasses. Those that proved of interest are listed at the bottom of Table 11.1. Results 32x and 32y are for an iron inclusion associated with a fluxing component as revealed by BSE imagery. Results 71x, 71y and 71z are analyses of subtly banded HIT and HIMT 1 glasses within a glass vessel sample (See Fig. 11.5). Furthermore, on analysis an inclusion in sample 79 (79x) was found to be limestone. The significance of inclusions is discussed in more detail below.

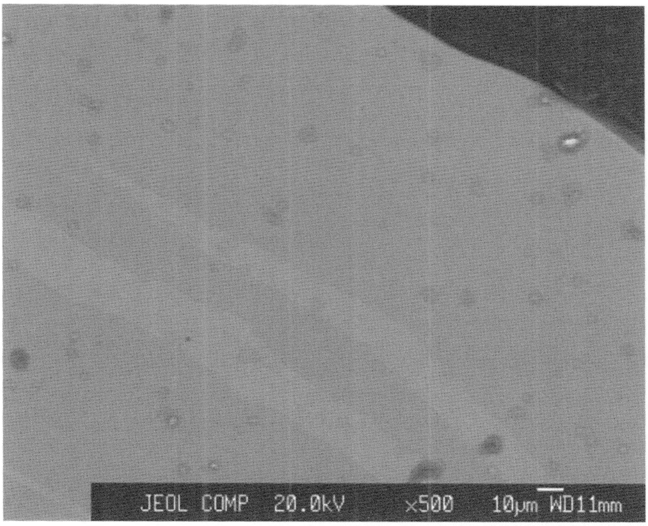

Fig. 11.5 Sample 71, x500 zoom — BSE image of HIMT 1 and HIT banding within the glass, as identified by results 71x, 71y, and 71z

11.6 Discussion

11.6.1 *Provenance*

All of the samples analysed featured low levels of potassium and magnesium oxides that are typical for natron glass (see Table 11.1). Levels of alumina and calcium were compared so as to provide an indication of possible different production zones (Fig. 11.1). On the assumption that CaO and Al_2O_3 were introduced in the sands it is possible to conclude from this comparison that the HIMT, HIT and Levantine compositional types are each produced with different sources of silica, and therefore probably in different regions. Our results for HIMT and Levantine glass compositional types correlate with published results suggesting that the glasses where produced in Egypt[2] and the Levant[7] respectively. Of note is the significant proportion of the Dichin HIMT results that share similar calcium and alumina levels with colourless Romano-British HIMT 2a, 2b and 3 variations. This suggests that they were produced within the same production centre. However, until now there has not been enough HIT data to investigate whether it was produced in a different zone from HIMT glass. Unfortunately no comment can be made on how silica affects this judgement, as a suitable dataset that includes silica results for HIMT glasses was not available for comparison with Dichin results . Despite this, by combining our analyses of the Dichin glass with Rehren and Cholakova's analyses,[13] it can be seen that HIT glasses appear to have been made from a silica source with lower variations in CaO and Al_2O_3 than HIMT and Levantine glasses.

11.6.2 *HIMT glass*

A large proportion of the HIMT 1 type glasses from Dichin have extended the typical composition of HIMT 1 because of their very high iron oxide contents. The manganese oxide content within these glasses is also relatively high in relation to other published HIMT data. This is most clearly seen for sample Nos. 32x and 32y (Table 11.1). Furthermore, levels of titanium oxide do not increase in a correlated way as is the case for many other published examples of HIMT 1 glasses. Freestone *et al.* have previously suggested that the purpose of the high manganese content within some HIMT glasses is not in order to neutralise the iron and decolourise the glass, but rather to minimise the reaction of sulphides with the iron.[2] This prevents the sulphides from darkening the glass and turning it black. It is therefore plausible that the increased iron and manganese content is derived from a colouration process intended to specifically create the richly-coloured yellow glasses using a reducing environment. There is only a single other example of an amber yellow-brown coloured glass that plots outside the HIMT 1 group.

Our HIMT 2 results compare closely to colourless Romano-British HIMT types 2a, 2b and 3, already defined and published by Foster and Jackson.[12] However, bar one or two possibilities, the presence of Romano-British HIMT type 1 colourless glass has not been identified at Dichin. This can most clearly be seen in Figs. 11.2 and 11.4. From these Figures we can show that there are differences in the percentages of manganese oxide in different compositional types and in colourless as opposed to coloured glasses. It is also possible that the four previously defined composition types could each have their own separate production zones. Furthermore, the high variation of vessel forms made from HIMT 2 glasses suggests that it was being supplied to large, and possibly multiple, secondary workshops. Schibille *et al.* have noted a similar phenomenon for mosaic tesserae, attributing it to the possibility that secondary workshops diversified from the homogenous Roman manufacturing style in the early Byzantine period, becoming increasingly specialised.[16] The high frequency and diversity of HIMT 2 glass suggests that it may have been utilised more for vessels with everyday functions, rather than fine wares that HIMT 1 is often used for. Perhaps sample 44, interpreted as a piece of waste glass from glass working, and sample 93, interpreted as a piece of raw glass, are potential indicators that HIMT 2 was being used for repairs or was reworked by the local population.

HIMT G1 is the least represented composition type. Only four samples were identified, but comfortably fit within Foy *et al.'s* definition of the composition.[9–11] Due to the limited number of samples, and reference material for the composition type, it is difficult to draw inferences about where the glass was made. The significance of the use of HIMT G1 for vessels with applied rims and their colouration, is discussed further below.

11.6.3 *HIT glass*

Similar to the HIMT 1 glasses, the majority of the HIT glasses appear to have been used to make glasses with specialised colouration and forms. Lower iron and manganese oxide contents are combined with higher levels of titanium and worked to create dominantly aqua blue colourations (see Table 11.1). The Dichin materials are not directly comparable to potential HIT materials from Butrint, Albania, as described by Conte *et al.*[17] The Butrint samples feature a higher iron content than those at Dichin, and this is attributed by Conte *et al.* to them being deliberately coloured aqua blue partly by using a cobalt-rich mineral, a somewhat surprising suggestion. Furthermore, HIT was used for a limited range of vessels at Ditchin- predominantly bowls or goblets. This suggests that HIT was being coloured and worked by increasingly specialist secondary workshops, as discussed previously.

The analytical results for samples 71x, 71y and 71z are fascinating. They show subtle banding of HIT and HIMT 1 compositional types within an aqua blue HIT bowl (Fig. 11.5). This shows that the craftsman that made the bowl had access to both HIT and HIMT 1 glasses at the same time, and that both compositions were mixed/blended in a secondary production centre. It is possible that this may be testament to a further change in practice at secondary workshop sites, where glass workers combined chunks of glass from various primary production sites. This could be because of the scarcity of raw glass of particular colours and/or because of their intrinsic values related to their 'exotic' origins. If we then take into account the published discussions about HIMT recycling, such as Foy *et al* and Aerts *et al.*,[11,18] our suggestions could help to explain why its distribution is so varied in different zones and between individual sites.

If it were not for the low levels of manganese oxide then these HIT glasses would have similar compositions to HIMT 1 glasses. Unlike other glass compositional types found at Dichin HIT and HIMT 1 both share fluctuating levels of ZrO_2. It is suggested by Conte *et al.* that the similarity between HIT and HIMT 1 places them into the same HIMT 1 sub-group.[17] This is a possibility, especially given our evidence of blending HIT and HIMT 1 (analyses 71x, 71y, and 71z) with the craftsman making the conscious decision to mix them. Furthermore, Freestone *et al.* have shown HIMT glass is the product of mixing two compositionally different glasses, one produced with a modern Sr isotope signature of marine shells and the other with an older non-marine (limestone) source of strontium.[2] Given the lack of candidates from the range of published glass compositions, HIT may be worth considering as a candidate for one of the glasses that is being mixed to create the HIMT composition. It may be for this reason that HIMT 1 shares so many characteristics with HIT glasses. Nevertheless we should question why HIT is currently not well represented within the archaeological record. Its scarcity of HIT could possibly be explained if the primary reason for making it was to use it as a component of HIMT glass, particularly if mixing HIT with natron glass to create HIMT was a well known process amongst secondary craftsmen. The use of more sensitive analytical techniques may help to address these issues.

11.6.4 *Levantine glass*

The majority of the results from the Levantine glasses correlate with previously published data. However, in Figs. 11.2 and 11.3 it can be seen that some feature a distinctly elevated level of iron oxide. These are: 5th century Levantine 1a samples (11b and 113), all of the 5th century Levantine 1b samples (13b, 43, 47b, 55, and 67), and all of the 6th century Levantine 1b samples (78, 79, and 104) from Dichin. In the case of the Levantine 1a samples this also correlates with an increase in manganese oxide.

This shows that a relatively impure source of sand was used to make some of these Levantine glasses found at Dichin. Since these glasses were made at the same time as other Levantine glasses using purer sand perhaps there is a Levantine primary production site that is yet to be identified. Even though the samples have all the other compositional characteristics of Levantine glasses this distorts the interpretation of the glasses by plotting them within the HIMT 2 region. The fact that the samples have compositional characteristics that are similar to HIMT 2, and that the majority of these glasses were colourless and were made into a wide variety of vessel forms, does not make things easier to understand. However, since they contain very high levels of calcium for an HIMT glass, and in some cases high alumina and titanium, it is safe to conclude that these glasses are correctly identified as being Levantine products. The high levels of iron are probably due to impurities in the sand used in the batch. The compositionally coherent Dichin vessels made from high iron Levantine glasses may have been worked at the same time from the same batch of Levantine glass. A 5th century Levantine 1b moil fragment (sample 43), would have resulted from the blowing of Levantine glass either at Dichin or elsewhere.

11.6.5 *Vessels with applied rims*

The vessels with applied string rims all have colourless bodies (samples 11b, 13b, 47b, and 70b); the rims are dark blue (samples 11a, 13a, 47a, and 70a). Samples 11a and 11b were from a rounded bowl, with 11a having a HIMT G1 composition and the body Levantine 1a. Samples 13a and 13b are from a straight cup, with 13a (the rim) being HIMT 2 and the body being Levantine 1b. Samples 47a and 47b are from an unidentifiable vessel form, with 47a, the rim, being HIMT G1 in composition and the body being Levantine 1b. Samples 70a and 70b are from a cup and were both of HIMT 2 composition.

The question is, why were different glass compositions use to make such vessels? The craftsmen that were making these vessels evidently had access to a wide variety of imported glass types. Although we have only four examples to discuss it is notable that three of the bodies were made from Levantine glasses and that all of the applied rims are of an HIMT composition. Furthermore, these different glasses were therefore mixed and matched as the craftsmen saw fit, and the blue glasses were from a non-Levantine source. These vessel characteristics could suggest that they were made in a limited number of secondary workshops in which the craftsmen had access to these raw coloured glasses. The absence of HIMT 1 and HIT is significant, since it may suggest, although with a small number of examples, that the workshops producing these vessels with applied string rims did not have, or did not want access to HIMT 1 and HIT glasses. These four sets of Dichin results show that the HIMT 2 composition is dominant. During the phase I occupation HIMT 1, HIMT 2 and HIMT G1 were at least being traded, if not still being produced, on a variety of scales. Since the chemical compositions ***and glass coloration*** of HIMT 1 and HIT are significantly different from glasses used to make the vessel with applied rims perhaps HIMT 1 and HIT glasses were uncommon specialist products, had specific uses with a social value attached to them, were more difficult to obtain, or were simply more expensive.

11.6.6 *Inclusions in the glasses*

The iron inclusion, combined with a fluxing component, (results 32x and 32y) is in a green-blue HIMT 2 glass lamp; perhaps it was introduced as an impurity during glass working. The limestone inclusion (result 78x) is in a colourless Levantine Ib stemmed goblet: perhaps it was introduced during low temperature glass mixing, since it would have dissociated at temperatures used to fuse glass raw materials. The latter is of interest, given that samples 43, 44, and 93 have been interpreted by Cholakova as being a raw piece of glass, waste from the direct working of glass and a moil, respectively.[5]

11.6.7 *Recycling*

Elemental levels within all the glasses do not indicate extreme levels of recycling. However, it is worth noting that levels of CuO are typically raised slightly for HIT glass samples, with the a few samples containing notably higher levels; the Levantine materials featured elevated levels of Sb_2O_5. Similar patterns have been noted by Gallo *et al.* in glass from Aquileia, Italy.[19] These slightly raised levels may therefore be attributable to a low level of recycling which is to be expected given the previous discussion on the production of HIMT.

11.7 Conclusions

The results from Dichin provide evidence for a site which was receiving varying qualities of glass vessels in a variety of forms, probably from a range of areas in early Byzantine world. Poulter states that the wide variety of materials found at Dichin is not reflected at nearby Nicopolis ad Istrum, and that, combined with the fort's unusual positioning and layout, Dichin may have been a supply base for the collection of locally produced wares as *annona* (a tax).[4] The collected materials would then be shipped eastwards along the river Rositsa, which flows into the river Iantra and ultimately into the Danube; supplies could be delivered to the frontier garrisons along its southern bank. The links along the inter-connected river system would certainly explain why such an unusual collection of glassware has been found at the fortress. However, the influence of trade with the core Byzantine market place via the Danube cannot be ignored. The variations in chemical compositions and vessel forms amongst the Dichin glasses are likely to have been the result of local glass-working of a variety of imported glass chunks including both those regarded as being of 'high quality' and/or rarer as well as more readily available composition types. Most compositional types were widely distributed throughout each context within the site of Dichin, the exception being HIT which was centred on what has been interpreted as store houses (area E). Other areas on the site typically featured larger amounts of HIMT 2 and Levantine glasses than HIMT 1 and HIMT G1 glasses. Glasses with HIT and HIMT 1 compositions appear to have been used in more specialised roles in early Byzantine society than other compositional types. They are found in lower frequencies, in fewer contexts, have been used to make specific glass colours that are less likely to be found in other compositional variations, and are more likely to be formed into specific vessel forms. Collectively this indicates that, in its first phase of occupation, Dichin was a settlement with good trade links, comparable in terms of glass chemistry to the river port of Thamusida, Morocco.[20] Analysis of glass from Thamusida revealed that it had a wide variety of glass compositions, including both Levantine and HIMT types both traded along the Mediterranean coast. Little can be inferred about the second occupation at Dichin from the glass results presented here. The very low numbers of glass finds from the period suggests that the occupants either didn't desire glass vessels or did not have access to the same trade routes that previously existed.

With respect to the larger picture for HIMT, HIT and Levantine composition types, it now seems possible to suggest that large early

Byzantine primary production workshops were connected to specialist secondary production workshops where the glass, including raw chunks, was coloured, worked, and, potentially blended on local and regional scales. The vessels found at Dichin are testament to this. Firstly, vessels with applied rims have been made using a combination of glasses from different primary workshops and come in a variety of vessel forms. This demonstrates that glass was being supplied by multiple primary workshops to a single, or multiple, specialist secondary workshops that were producing these vessels. It would seem that the secondary glass-workers made a conscious decision to collect and combine these multiple glass types. It is clear from the uniform use of colourless glass bodies and dark blue applied rims that this decision was based purely on aesthetics and not the primary the source of the glass. However, the lack of HIT glass within these combinations may be yet another indicator that it was set apart from the other glass types. Perhaps the most interesting point is that blending of multiple glasses from different primary workshops occurred at these specialist secondary workshops. This is evident within the visual and compositional banding of HIMT 1 and HIT compositions within sample 71 at Dichin, though it is difficult to know how common this practice was. This suggests that the supply of certain raw glasses was more restricted than others, and could indicate that some primary glass production sites were more specialised in the producing certain raw glass types than is currently understood. Other important factors that would have impacted on the supply of raw glass are political control and the geographical factors affecting access to the glass.

It is probable that HIT was produced at some distance from the HIMT and Levantine production centres, but the production location is currently unknown. Analysis of glass from Bubastis, northern Egypt, shows that, somewhat later, HIMT dominated due to the site's proximity to the primary production sites near Wadi Natrun.[21] This pattern is not observable amongst the Dichin glass compositions, and suggests that none of them are likely to have been fused from raw materials locally. It is more likely that Dichin had a direct link to a variety of secondary production sites, perhaps via the river Danube. One such secondary production site had access to HIT glass.

It should be stressed that the economy of this Balkan region is still not fully understood during this period of transition. The supply of such a wide range of glass compositions to Dichin reflects, in important ways, the sites economic ties, its economic status and the use of glass as part of the site's function in early Byzantine society.

Acknowledgments

We are grateful to Anastasia Cholakova of the Bulgarian Academy of Sciences for assisting us with advice about vessel forms and sampling methodology.

References

1. N. Schibille, Supply routes and the consumption of glass in the first millennium CD Butrint (Albania), *Journal of Archaeological Science* **38**, 2939–2948 (2011).
2. I. C. Freestone, S. Wolf and M. Thirlwall, The production of HIMT glass: Elemental and Isotopic evidence, *Annales du 16e Congres de l'Association Internationale pour l'Histoire du Verre*, 153–157 (2005).
3. V. Dintchev, The fortresses of Thrace and Dacia in the Early Byzantine Period, In *The Transition to Late Antiquity: on the Danube and beyond*, ed. A. G. Poulter (Oxford University Press, Oxford, 2007), pp. 479–546.
4. A. G. Poulter (ed.), *The Transition to Late Antiquity: on the Danube and beyond*, Oxford University Press, Oxford, 2007.
5. A. Cholakova, Part VI. Glass vessels from Gradishteto, In *Bulgarian-British archaeological excavations of Gradishteto near the village of Dichin, Veliko Tarnovo region, 1996–2003* (*the results of the Bulgarian team* (in Bulgarian)), eds. V. Dinchev, G. Kuzmanov, P. Vladkova, A. Cholakova and Ts. Popova (Excavation and reports 39, Sofia, 2009), pp. 257–307.
6. C. Isings (ed.), *Roman glass: From dated finds*, J. B. Wolters, Groningen, 1957.
7. R. H. Brill, Scientific Investigations of the Jalame Glass and Related Finds, In *Excavations at Jalame: Site of a glass factory in Late Roman Palestine,* ed. G. D. Weinberg (Missouri Press, Columbia 1988), pp. 257–294.
8. E. H. Foster and C. M. Jackson, The composition of 'naturally coloured' late Roman vessel glass from Britain and the implications for models of glass production and supply, *Journal of Archaeological Science* **36**, 189–204 (2009).
9. D. Foy, M. Vichy and M. Picon (eds.), *Lingots de verre en Me´ diterrane´e occidentale (IIIe sie` cle av. J.-C.–VIIe sie` cle ap. J.-C.): Annales du 14e Congre` s de l'Association Internationale pour l'Histoire du Verre, Venezia-Milano 1998*, Lochem: AIHV, 2000, pp. 51–57.

10. D. Foy, M. Picon and M. Vichy (eds.), *Les matie` res premie` res du verre et la questiondes produits semi-finis. Antiquite´ et Moyen A^ ge: Arts du Feu et productions Artisanales. XXe Rencontres Internationales d'Arche´ ologie et d'Histoire d'Antibes. E´ditions*, Antibes: APDCA, 2000, 99. 419–432.
11. D. Foy, M. Picon, M. Vichy and V. Thirion-Merle (eds.), Caracte´ risation des verres de la fin de l'Antiquite´ en Me´ diterrane´e occidentale: l'e´mergence de nouveaux courants commerciaux, In *E´changes et commerce du verre dans le monde antique. Actes du colloque de l'AIHV, Aix-en-Provence et Marseille, juin 2001*, eds. D. Fory and M. D. Nenna, Vol. 24 (Monographies Instrumentum, 2003), pp. 41–85.
12. E. H. Foster and C. M. Jackson, The composition of late Romano-British colourless vessels: Glass production and consumption, *Journal of Archaeological Science* **37**, 3068–3080 (2010).
13. T. Rehren and A. Cholakova, The Early Byzantine HIMT glass from Dichin, Northern Bulgaria, *Interdisciplinary Studies* **22/23**, 207–216 (2010).
14. J. Henderson, Electron probe microanalysis of mixed-alkali glasses, *Archaeometry* **30**, 77–91 (1988).
15. A. Meek, J. Henderson and J. Evans, Isotope analysis of English forest glass from the Weald and Staffordshire, *Journal of Analytical Atomic Spectrometry* **27**, 786–795 (2012).
16. N. Schibille, P. Degryse and M. Corremans *et al.*, Chemical characterisation of glass mosaic tesserae from sixth-century Sagalassos (south-west Turkey): chronology and production techniques, *Journal of Archaeological Science* **39**, 1480–1492 (2012).
17. S. Conte, T. Chinni, R. Arletti *et al.*, Butrint (Albania) between eastern and western Mediterranean glass production: EMPA and LA-ICP-MS of late antique and early medieval finds, *Journal of Archaeological Science* **49**, 6–20 (2014).
18. A. Aerts, B. Velde, K. Janssens *et al.*, Change in silica sources in Roman and post-Roman glass, *Spectrochimica Acta B* **58**, 659–667 (2003).
19. F. Gallo, A. Marcante, A. Silvestri *et al.*, The glass of the "Casa delle Bestie Ferite": a first systematic archaeometric study on Late Roman vessels from Aquileia, *Journal of Archaeological Science* **41**, 7–20 (2014).
20. E. Gliozzo, A. S. Barebone and F. D'acapito, Waste glass, vessels and windowpanes from Thamusida (Morocco): Grouping Natron-based blue-green and colourless Roman glasses, *Archaeometry* **55**, 609–639 (2013).
21. D. Rosenow and T. Rehren, Herding cats — Roman to Late Antique glass groups from Bubastis, northern Egypt, *Journal of Archaeological Science* **49**, 170–184 (2014).

Chapter 12 第十二章

Glass Corrosion: Issues and Approaches for Archaeological Science

Nikolaos Zacharias and Eleni Palamara
Department of History,
Archaeology and Cultural Resources Management,
University of the Peloponnese 24100 Kalamata, Greece

12.1 Introduction

Major glass-technology achievements and the spread of glass artifacts started during the 16th to 12th century BC,[1] and the advances appear to originate mostly from the Mesopotamian region and Egypt. During the Bronze Age era, the Peloponnese stands as one of the major areas of the Mycenaean world that characterize mainland Greece and Aegean civilizations. From the end of the prehistoric era to the Hellenistic period, the picture for the technological history of glass in East Mediterranean and Greece remains rather sketchy. Though what is clear was the advanced use of luxurious core-formed and cast glass vessels, a practice that was adopted by the early Roman glass industry and characterized by artefacts in a broad range of bright colors. By the mid second century AD, and with the adoption of blowing as the main form of production, glasses were mass produced and in use as tablewares, window glasses, artefacts, etc.

Although glass is considered to be a material that ages well, the majority of historical glasses show macroscopic effects of weathering and for archaeological specimens more severe weathering effects from being buried are evident. The aim of this study is to highlight the presence of corrosion phenomena and their effect on archaeological glass studies through the presentation of specimens from three case studies, each one examined and presented from different analytical perspectives.

12.2 The study of Mycenaean glass objects using microscopy

Three objects, forming part of an extensive glass collection of several hundred glass items are presented in the study; these are mainly blue beads and decorated glass plaques recovered during excavations at the Mycenaean (c. 1,500–1,100 BC) cemetery of Voudeni, Achaia (Table 12.1).

Prior to the scientific investigation, documentation and characterisation of the material is usually carried out with digital macro photography. A second step should include the microscopic examination of the state of the material which nowadays can be achieved with the use of a portable digital, low magnification optical microscope. In our study a fibre optic microscope (abbreviated hereafter as FOM) consisting of an array of led optical lenses (i-Scope-Moritex) was used for the *in situ* characterisation of the objects. Further, scanning electron microscopy (SEM), an advanced way of studying microtopography, homogeneity and corrosion patterns, was used; when the SEM is equipped with an energy dispersive spectrometer (EDS) a chemical analysis of the studied sample can result. For the study, a SEM type JEOL JSM-6510LV coupled with EDS (Oxford Systems) provided the analytical data using the INKA software.

In Table 12.1 and using macro, but more clearly FOM and SEM photos, various types of corrosion are clearly observed, like white spots (e.g. FOM at T29/31a, T26/5), black spots (FOM, T26/5), iridescence (T29/31b), signs of pitting (T26/5), micro- or larger-cavities (T26/5), hollows (T29/31b) and sol-gel texture (T29/31b), but also areas showing homogeneity in terms of colour and texture (T26/5) that are potential indicators of healthy, intact glass. It can also be noted from the macroscopic examination that both sides (front and back) of the artefacts show identical patterns.

Table 12.1 Macro, FOM and SEM photos of the Mycenaean glass objects

Sample	Macro front	Macro back	FOM (x50) front	FOM (x50) back	SEM (x55, x60, x170)
T29/31a (1.7x1.1 cm)			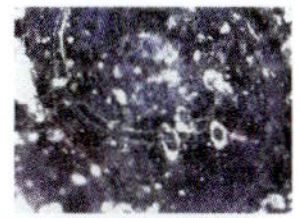	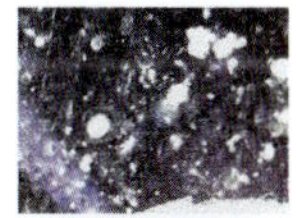	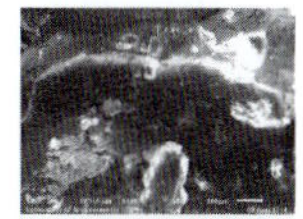
T29/31b (1.5x0.9 cm)					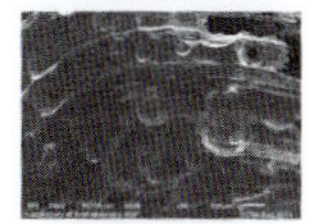
T26/5 BEB 1146 (1.6x0.2 cm)			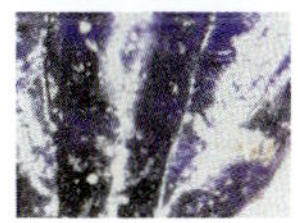	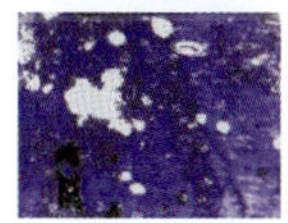	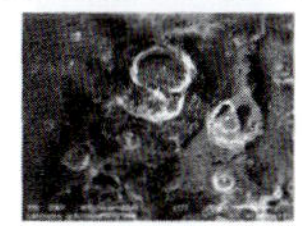

Since the corrosion state of the artefact will impose constraints on its management from the perspective of the analyst and the conservator, the robust, multi-elemental technique of prompt-gamma activation analysis (PGAA)[2] was used; an advancement on the method critical to the study of archaeological glass is that PGAA is a bulk method, therefore the analytical results are not affected by the variation in the surface composition as much as in other surface analytical techniques. The analytical data obtained (Table 12.2) point to a soda plant ash glass as the primary glass used for the production of the objects, the use of cobalt for the dark blue colour — and in sample T29/31a the additional use of copper — and the presence of iron at

Table 12.2 PGAA analytical data in weight %, normalised to 100% and their standard deviations (σ)

PGAA Analytical data	T29/31a		T29/31b		T26/5 BEB 1146	
	wt%	σ	wt%	σ	wt%	σ
H_2O	0.40	0.01	0.79	0.02	0.29	0.11
Na_2O	18.00	0.36	18.60	0.41	18.10	0.42
MgO	4.60	0.14	4.00	0.24	4.20	0.01
Al_2O_3	2.00	0.08	2.10	0.08	2.00	0.08
SiO_2	63.30	0.65	65.10	5.60	64.20	5.14
K_2O	2.14	0.04	1.84	0.03	1.84	0.04
CaO	6.93	0.21	6.09	0.18	6.93	0.17
SO_3	0.43	0.03	0.41	0.04	0.43	0.01
TiO_2	0.11	0.01	0.11	—	0.10	—
MnO	0.16	0.01	0.12	—	0.18	—
Fe_2O_3	0.70	0.03	0.56	0.03	0.58	0.03
CoO	0.05	0.01	0.07	0.01	0.08	0.01
NiO	n.d.	—	n.d.	—	0.09	—
CuO	0.46	0.02	0.10	0.01	0.23	0.01
B	0.02	—	0.02	—	0.01	—
Cl	0.65	0.01	1.02	0.01	0.70	0.01

levels corresponding to natural impurity of the sand. The values of hydrogen (H) and boron (B) given in Table 12.2, indicate that no severe hydrolysis of the glass network has occurred (hydrogen) and also that corrosion of the glass matrix due to environmental contamination (boron) is at minimal levels.

A final step in the examination of the glass objects included the study of micro-areas spotted during the first microscopic inspection in an attempt to extract additional information about the manufacturing technology. In most of the cases, inhomogeneity is caused by combined alteration (material) and contamination (environmental) processes which require careful evaluation. Of interest are the results from the study of sample T29/31b (Fig. 12.1) for which the analytical (SEM-EDS) data are given in Table 12.3. In Fig. 12.1 a large formation with clearly observed whitish round or elongated spots can be seen in the left part of the photo.

Despite acknowledging that the resulting data are with large uncertainties due to interferences from the silica matrix, the following observations can be made: the near 50 wt% of the calcite content for the large spot indicates that a calcitic mineral was used for the decoration of the objects. For the small areas scanned, the combined values of aluminum, titanium and iron point to the use of bauxite minerals. In a recent study of Mycenaean dark blue relief fragments from the Argolid,

Table 12.3 EDS analytical data in weight %, normalised to 100%, and their standard deviations (σ). For the large spot, the values are the mean ones from three measurements; the 'small whitish' values are the average from the measurement of three different whitish spots

	Na_2O	MgO	Al_2O_3	SiO_2	SO_3	Cl	K_2O	CaO	TiO_2	FeO
Large spot										
Mean	4.77	0.12	15.23	27.10	0.01	n.d.	1.83	44.52	3.72	2.85
σ	0.33	0.01	0.75	1.00	—	—	0.04	1.85	0.20	0.05
Small whitish										
Mean	5.78	2.57	26.40	35.51	n.d.	0.01	3.47	14.52	4.45	7.30
σ	0.45	0.01	1.50	1.75	—	—	0.08	0.80	0.30	0.12

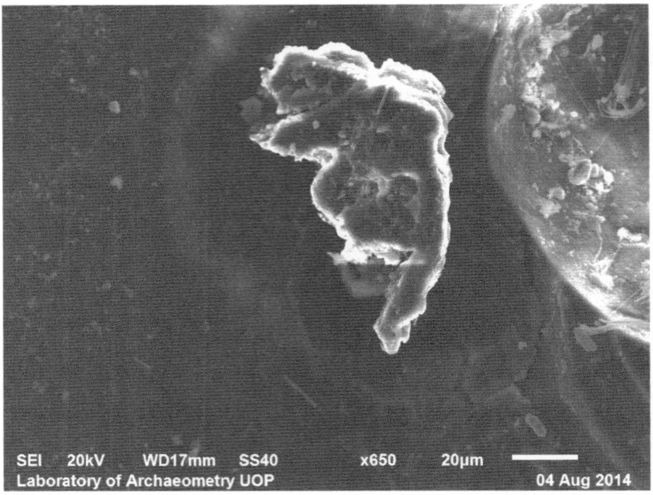

Fig. 12.1 SEM photo (x650) showing a characteristic spot found in many areas of sample T29/31b

Peloponnese,[3] the application of RAMAN and IR spectroscopy led to the assumption that a borate component related to the gilding process of the artefacts could explain the presence of the elevated calcite content found also in our study. Nevertheless, the application of an extra substance to the surface of the object could possibly explain the intense iridescence that this sample exhibits and is a subject for further research.

12.3 The study of Classical glass vessels using spectroscopic analysis

A glass assemblage consisting of 12 vessel fragments, that was recovered during the excavation of the Manisalis field in the city of Thebes, is presented here.[4] The vessels are dated to the Classical period (5th c. BC) and were manufactured using the core-forming technique. They have an intense blue colour and are decorated with white and/or yellow zig-zag trails (Fig. 12.2).

The chemical composition of the samples was determined using SEM-EDS on polished surfaces.[4,5] The concentrations of the major elements of the body of the glass vessels is presented in Table 12.4, along with the relative standard deviations (σ).

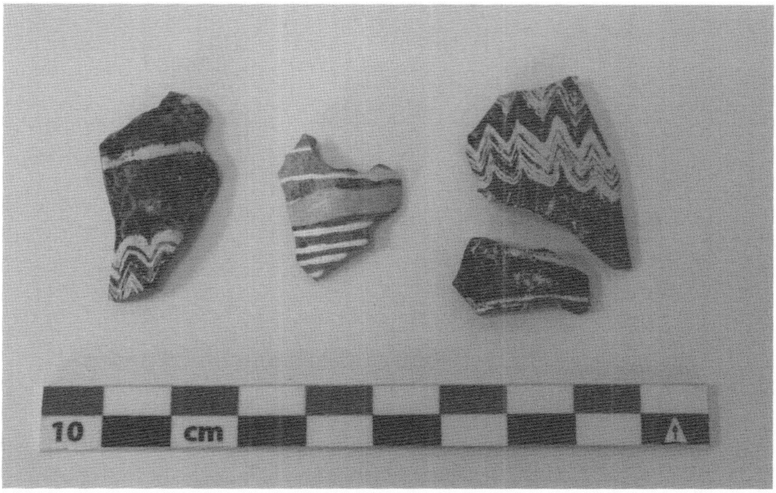

Fig. 12.2 Characteristic examples of decorated vessel fragments from the Thebes collection

Table 12.4 EDS compositional analysis (weight %, normalised to 100%) of the body of vessel fragments

	Na_2O	MgO	Al_2O_3	SiO_2	SO_3	Cl	K_2O	CaO	TiO_2	MnO	FeO	CoO	CuO
Mean	20.45	0.70	2.63	67.21	0.68	1.08	0.40	5.53	n.d.	n.d.	0.68	0.32	0.31
σ	1.33	0.14	0.12	1.43	0.08	0.27	0.06	0.50	—	—	0.13	0.12	0.12

All samples demonstrate visible signs of weathering, in the form of thin whitish or iridescent coatings and micro-pitting. The techniques of Raman and Infrared (IR) spectroscopy were applied to further examine the corrosion effects visible on the samples. Raman spectra were obtained by a Renishaw high confocal micro-Raman spectrometer, using the 488 and 514 nm lines of an argon ion laser and the 633 nm line of a He-laser. IR spectra were obtained by a Hyperion infrared microscope.

The SiO_4 tetrahedra are present in the glass matrix with different tetrahedral arrangements, each with characteristic Raman and IR signatures. Isolated SiO_4 tetrahedra are referred to as Q^0 and tetrahedra with 'n' common bridging oxygen atoms as Q^n.

Spectroscopic analysis initially focused on polished surfaces of pristine glass detected during the microscopical observation of the

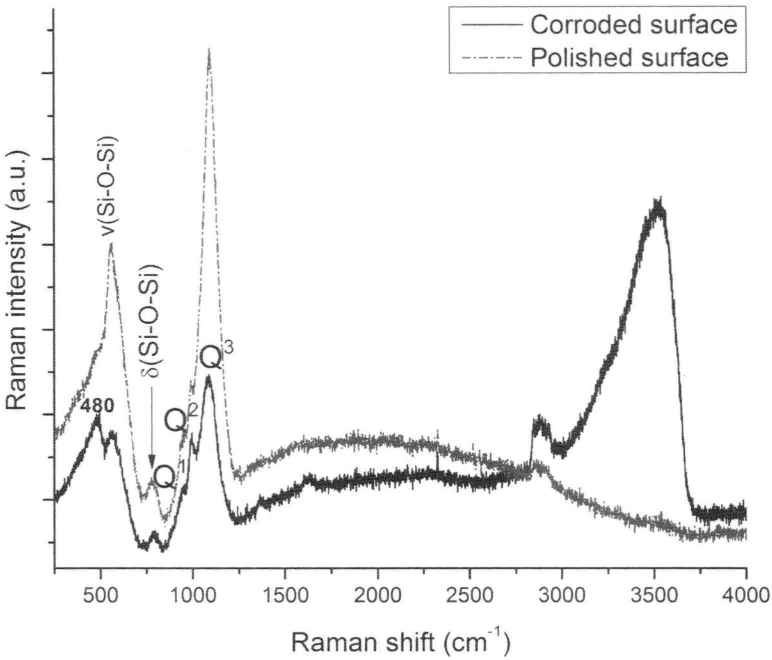

Fig. 12.3 Raman spectra from a blue vessel fragment (sample MAN01) taken from a polished (solid line) and a non-polished (dashed line) surface

samples. The basic glass bands of both Raman and IR spectra of the polished surfaces demonstrate significant similarities with the spectra of modern glass samples of similar composition.

Analysis was then repeated focusing on intact surfaces that had not been previously polished or treated by any method. Figure 12.3 shows a characteristic Raman spectrum from a corroded surface of a vessel fragment, sample MAN01 (solid line). For comparative reasons, a spectrum from a polished surface of the same sample is also shown (dashed line). In the corroded surface, the Q^3 band is shifted towards the left, while Q^1 and Q^2 bands have an increased intensity relative to the Q^3 band. Also, a new band at 480 cm^{-1} appears, which is related to corrosion processes. All the above mentioned alterations are indicative of increased break up of the glass network.

The IR spectra of the non-polished surfaces demonstrate a different behaviour. Figure 12.4 shows a characteristic example of an IR

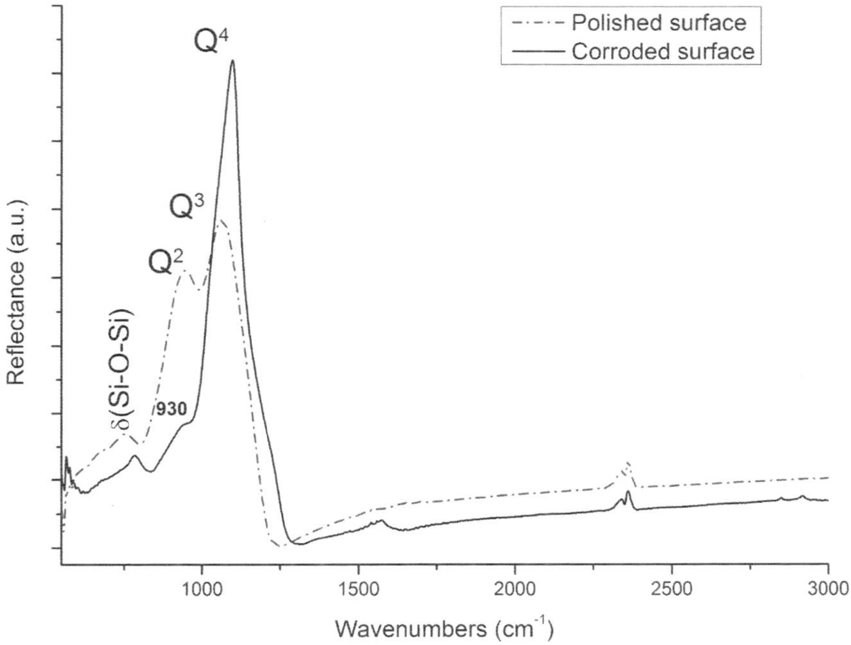

Fig. 12.4 IR spectra from a blue vessel fragment (sample MAN01) taken from a polished (solid line) and a non-polished (dashed line) surface.

spectrum from a non-polished surface (solid line) of sample MAN01. Again, a spectrum from a polished surface is depicted for comparative reasons (dashed line). The (Si-O-Si) band has shifted towards the right, while Q^2 band is altogether absent. Taking into consideration both the shift towards the right and the appearance of the shoulder at ~1200 cm^{-1}, the strong band at ~1093 cm^{-1} should be attributed to the stretching vibrations of Si-O-Si bridges in Q^4 silicate units instead of Q^3 units. Also, a new band appears at ~930 cm^{-1} which could be attributed to the Si-O stretching vibration of the bonds of free (Si-OH) groups. These alterations suggest an increase in polymerization of the glass network and a likely increase in the presence of water.

Leaching experiments on alkali-silicate glasses have shown that the initial break up of the glass network can be followed with longer leaching times by the formation of a gel-layer which transforms into a silica type network.[6,7] Taking into consideration that the Raman beam probes deeper into the samples than IR reflectance, the above mentioned effects can be explained by the formation of two successive corrosion layers: (a) a thin outer layer, showing an extreme degree of 'polymerization', evident in the IR spectra and (b) a second layer, showing intense 'depolymerization', evident in the Raman spectra. At greater depth, the glass network is not altered by corrosion. Similar behaviour has previously been noted by Möncke *et al.* in an assemblage of Mycenaean glass samples.[3]

12.4 Corrosion due to the content in Roman colourless glass vessels

During a systematic excavation of a Roman funerary complex (2nd c. AD) at the city of Patras, Achaia, an assemblage of luxurious glass vessels was recovered, found placed together in a wooden chest. Some of the items are considered to be unique amongst Roman products found in the Aegean. The vessels have a wide variety of colours, from transparent colourless to translucent blue, green and purple, but here we will focus only on the colourless samples of the assemblage.

The chemical composition of the samples was determined using SEM-EDS (cf. part 12.2) given in Table 12.5. All colourless samples belong to the general type of soda-lime-silica glass and have a typical concentration for colourless Roman glass of the 2nd–3rd c. AD.[8] The alkali source used is natron and a combination of both antimonate and manganese is introduced as decolourisers.

A residue of an intense pink colour was found attached to the inner surface of a number of the colourless samples (Fig. 12.5). The chemical analysis of the residue showed that it is an aluminosilicate substance, with a relatively high sulphur content (Table 12.6).

Table 12.5 EDS compositional analysis of the body of colourless vessel fragments (weight %, normalised to 100%)

	Na_2O	MgO	Al_2O_3	SiO_2	SO_3	Cl	K_2O	CaO	TiO_2	MnO	FeO	Sb_2O_3
Mean	19.54	0.42	2.42	68.94	0.25	1.00	0.34	6.15	0.19	0.50	0.24	0.79
σ	0.50	0.19	0.25	0.79	0.19	0.11	0.10	0.25	0.14	0.18	0.14	0.49

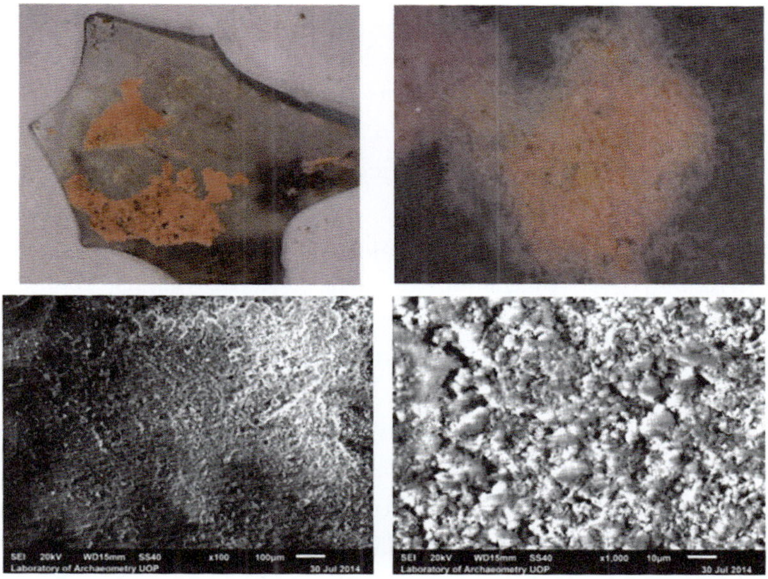

Fig. 12.5 FOM microscope (Left: x10; Right: x50) and SEM images (Left: x100; Right: x1000) of the pink residue

Table 12.6 EDS compositional analysis of the pink residue (weight %, normalised to 100%)

	MgO	Al_2O_3	SiO_2	SO_3	K_2O	CaO	TiO_2	FeO
Mean	2.23	45.34	41.74	2.66	1.34	2.64	0.38	3.68
σ	0.08	0.92	0.31	0.22	0.20	0.06	0.20	0.30

Both the colour of the residue and its chemical composition resemble madder lake, a red lake pigment which was used during the Roman period.[9] The pigment is extracted from the common madder plant Rubia tinctorum and can contain two organic red dyes: alizarin and purpurin. Since madder lake is in the form of a very fine powder and has very little bulk it must be precipitated on an inert pigment or lake base, in order to make it suitable for brushing. Clay or alum was often used for this purpose.

In order to securely identify the residue, the spectrometric technique of Laser Desorption–Ionization-Time of Flight-Mass Spectrometry (LDI-ToF-MS) was applied. The analysis securely identified the presence of alizarin as the main component of the organic extract and therefore verified the characterization of the pink residue as a madder lake.

The presence of the aluminosilicate residue has resulted in the extensive corrosion of the inside surface of the vessels. In fact, the inner surface seems significantly more corroded than the outer one, which was not in direct contact with the madder lake as long as the vessels remained intact.

Figure 12.6 shows a concentration profile of the major elements conducted on a fresh section of a colourless sample, focusing specifically on two zones of 40 μm on the inside and outside edge of the vessel's wall. In both cases, a layer is formed which shows a high loss of alkali and a consequent increase in the concentration of silica and alumina. On the inner corrosion zone, an increase in calcium is also noted. However, the external corrosion layer seems to be very thin, approximately 2 μm in depth, while the inner corrosion layer has a depth of over 15 μm. We suggest, therefore, that the original content played a critical role in the corrosion effects of the glass samples, even in cases where the vessels broke at some unknown point in the past.

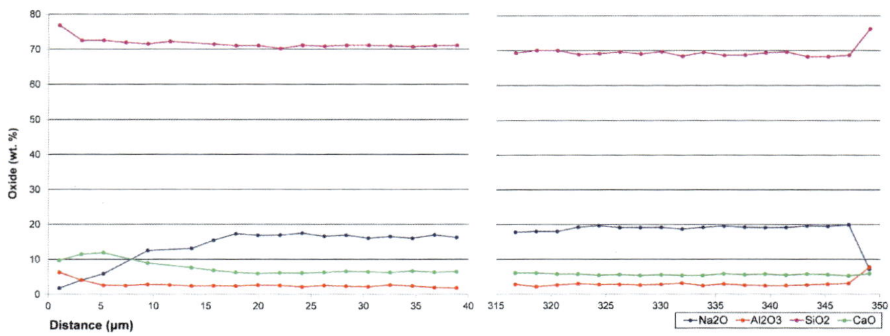

Fig. 12.6 Concentration profile by SEM/EDS (0 μm: Inner surface; 350 μm: Outer surface)

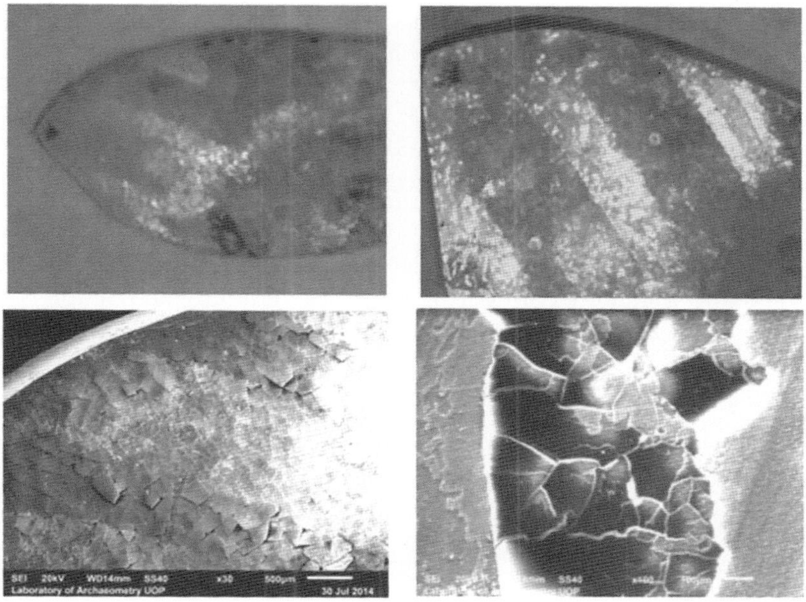

Fig. 12.7 LED microscope (Left: x10; Right: x50) and SEM images (Left: x30; Right: x100) of iridescent layers on the inner surface of colourless fragments

Another corrosion alteration which is likely to be associated with the madder lake residue is the formation of areas that show high iridescence and have a flaky appearance under the SEM (Fig. 12.7). The formation of iridescent areas is common in weathered glass.[10] However, certain aspects of the formations under examination are of particular interest. The iridescent layers are only present in the inner surfaces of the colourless samples and did not flake away during the process of cleaning. Moreover, they appear in well-formed stripes with areas of clear glass between them. The chemical composition of both the clear and the iridescent areas is characterized by an almost complete loss of alkalis and, to a smaller degree, of calcium, and a consequent increase in the concentration of silica and alumina (Table 12.7). However, the iridescent areas have a twice as high concentration of aluminium, suggesting precipitation of alumina from the surrounding. The exact way the iridescent stripes were

Table 12.7 EDS compositional analysis of clear and iridescent surfaces on the colourless vessel fragments (weight %, normalised to 100%)

	Na_2O	MgO	Al_2O_3	SiO_2	SO_3	Cl	K_2O	CaO	TiO_2	MnO	FeO	Sb_2O_3
Clear	0.25	0.57	4.86	86.25	0.28	0.52	0.22	5.77	0.05	0.56	0.28	0.39
Iridescent	0.59	0.58	9.56	81.55	0.52	0.42	0.28	4.99	0.08	0.43	0.35	0.64

formed and the likely connection with the madder lake residue is still under investigation.

12.5 Conclusion

To summarize all the issues considered in this study, we highlight the use of various analytical approaches that can be used for the non-destructive characterization of archaeological glasses where microscopy used at both low (Optical Microscopy) and high magnification (SEM) offers direct and great potential. The study of corrosion patterns provides vital information on the history, technology and use of the artefact while the analytical information is unique for its appropriate conservation and preservation.

Acknowledgments

The authors acknowledge granted permits to study the material jointly given by the Ministry of Culture and the Archaeological Ephorates of Achaia and Boeotia. NZ acknowledges a CHARISMA FP7 Grant Nr.320 to visit the Budapest Neutron Center of the Hungarian Academy of Science for the implementation of the PGAA measurements.

References

1. J. Henderson (ed.), *Ancient glass: an interdisciplinary exploration*, 1st edn. (Cambridge University Press, Cambridge, New York, 2013).
2. L. Szentmiklósi, T. Belgya, Z. Révay and Z. Kis, Upgrade of the prompt gamma activation analysis and the neutron-induced prompt gamma spectroscopy

facilities at the Budapest Research Reactor, *Journal of Radioanalytical and Nuclear Chemistry* **286**, 501–505 (2010).

3. D. Möncke, D. Palles, N. Zacharias, M. Kaparou, E. I. Kamitsos and L. Wondraczek, Formation of an outer borosilicate glass layer on late Bronze Age Mycenaean blue vitreous relief fragments, *Physics and Chemistry of Glasses: European Journal of Glass Science and Technology* **54**(1), 52–59 (2013).
4. Ar. Oikonomou, K. Beltsios and N. Zacharias, Analytical and technological study of an ancient glass collection from Thebes, Greece: An overall assessment, *Proceedings of the 18th Symposium of the Association Internationale pour l'Histoire du Verre, AIHV*, eds. D. Ignatiadou and A. Antonaras (Thessaloniki, 2012), pp. 75–80.
5. N. Zacharias, K. Beltsios, Ar. Oikonomou *et al.*, Thermally and optically stimulated luminescence properties of an archaeological glass collection from Thebes, Greece, *Journal of Non Crystalline Solids* **354**, 761–767 (2008).
6. S. A. MacDonald, C. R. Schardt, D. J. Masiello and J. H. Simmons, Dispersion analysis of FTIR reflection measurements in silicate glasses, *Journal of Non-Crystalline Solids* **275**, 72–82 (2000).
7. M. E. Lynch, D. C. Folz and D. E. Clark, Use of FTIR reflectance spectroscopy to monitor corrosion mechanisms on glass surfaces, *Journal of Non-Crystalline Solids* **353**, 2667–2674 (2007).
8. I. C. Freestone, M. Ponting and M. J. Hughes, The origins of Byzantine Glass from Maroni Petrera, Cyprus, *Archaeometry* **44**(2), 257–272 (2002).
9. N. Eastaugh, V. Walsh, T. Chaplin and R. Siddall (eds.), *The pigment Compendium, A dictionary of historical pigments*, Elsevier, Oxford (2004).
10. S. Davison (ed.), *Conservation and restoration of glass*, Butterworth-Heinemann, Oxford (2003).

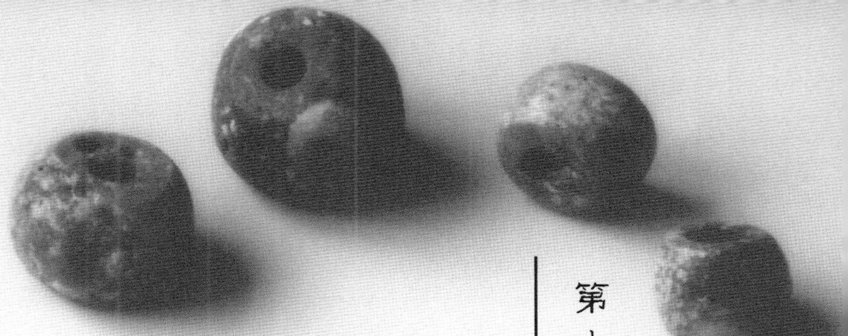

Chapter 13

第十三章

Raman Identification of Glassy Matrices in Ancient Glasses from China

*Hongxia Zhao and Qinghui Li**

* Shanghai Institute of Optics and Fine Mechanics,
Chinese Academy of Sciences, Shanghai 201800, China

13.1 Introduction

The micro/nanostructure of ancient glasses contains information on their provenance, production process, and conservation conditions. From the analysis of these micro/nanostructures, many different facets of science, technology and culture can be traced. Thus, salient features covering the history of the glass industries can be extracted from amorphous silicate networks and pigments from different products. As for precious historical relics, greater objectivity is required for identification and conservation purposes. Therefore, a non-destructive micro-probe method for examining these kinds of samples is needed. Among the available methods, Raman spectroscopy is one of the most versatile techniques. In the past 20 years, Raman spectroscopy gradually has become an established technique for the non-destructive identification and study of cultural heritage materials, especially those made of glass.[1] Since the 1950s, a large number of articles have been published on natural silicate glasses, whose structures were interpreted in terms of a 3D network of SiO_4 units. Therefore, Raman spectroscopy can be used for the characterization of crystalline or glassy structures consisting of covalent vibrational units like silicates.[2–4] The first valid quantitative work by Raman dealt with a series of high-temperature porcelain glazes. It was published in the year 2005.[5] Philippes Colomban, who has a wide range of experience in analyzing various artifacts (Phoenician, Roman, Islamic, medieval and modern glass, Asian, European and Islamic faience, fritware, stoneware, and porcelain), proposed a tentative guide to recognize and classify the different types of glassy silicates as a function of their composition.[6] The polymerisation index [7] is now being used by different groups of researchers, but information is still lacking on the use of Raman spectroscopy to study ancient glasses from China; moreover, few papers have focused on ancient Chinese ceramics.[8]

In this paper, we present the main parameters extracted from Raman spectra for nineteen ancient glasses from Xinjiang and Guangxi Provinces. The relationship between the composition of glassy silicates and the parameters of their Raman signatures (the polymerisation index and gravity centre of the Si-O stretching mode) was verified. The processing techniques used in these ancient glass beads and relative issues were discussed.

13.2 Samples and experiments

13.2.1 *Samples*

A total of 19 ancient glass beads were selected for analysis. These beads were excavated from different sites in China. Among 10 glasses from Xinjiang, 7 samples were excavated from different tombs in the Kizil reservoir cemetery (the Spring and Autumn Period). Another three samples were unearthed from tomb M1 in the Wanquan cemetery (dating from the Western Zhou Dynasty to the Spring and Autumn Period), which is located in E'min county in the Tacheng region. For comparison, nine glasses (dating from the Han to Jin Dyansties) excavated from Hepu, Guangxi were also analyzed. Most of the samples are opaque or translucent and were uneven monochrome beads. The bead colours were mainly purple, blue, green, bluish green, yellow or light yellow. The shapes of these samples were irregular, and the outer diameters were approximately 0.5 cm to 1.5 cm. Their description, inlcluding chemical compositions and Raman parameters of the samples are presented in Table 13.1.

Table 13.1 Description of ancient Chinese glasses of different chemical type and their Raman data

No.	Location	Date	$Ip = \frac{A_{500}}{A_{1000}}$	ν_{max} Si-O stretching (cm^{-1})	R_2O*	CaO	Chemical type of glass
XJ-1A	Kizil reservoir cemetery, Baicheng, Xinjiang	the Western Zhou Dynasty to the Spring and Autumn Period	0.522	1087	20.25	6.8	Na_2O-CaO-SiO_2
XJ-29C	Kizil reservoir cemetery, Baicheng,Xinjiang	the Western Zhou to the Spring and Autumn Period	0.536	1087	15.28	7.12	Na_2O-CaO-SiO_2
XJ-44Q	Wanquan cemetery M1,E'min, Xinjiang	the Spring and Autumn Period	0.536	1087	18.86	7.29	Na_2O-CaO-SiO_2
XJ-44M	Wanquan cemetery M1,E'min,Xinjiang	the Spring and Autumn Period	0.544	1087	8.4	9.82	Na_2O-CaO-SiO_2
XJ-3A	Kizil reservoir cemetery, Baicheng, Xinjiang	the Western Zhou to the Spring and Autumn Period	0.58	1087	12.75	6.6	Na_2O-CaO-SiO_2
XJ-44R	Wanquan cemetery M1, E'min, Xinjiang	the Spring and Autumn Period	0.582	1087	14.16	9.55	Na_2O-CaO-SiO_2
XJ-3B	Kizil reservoir cemetery, Baicheng, Xinjiang	the Western Zhou to the Spring and Autumn Period	0.606	1088	17.93	8.41	Na_2O-CaO-SiO_2

(*Continued*)

Table 13.1 (*Continued*)

No.	Location	Date	$Ip = \frac{A_{500}}{A_{1000}}$	ν_{max} Si-O stretching (cm^{-1})	R_2O*	CaO	Chemical type of glass
XJ-4M	Kizil reservoir cemetery, Baicheng, Xinjiang	the Western Zhou to the Spring and Autumn Period	0.61	1082	6.64	7.19	Na_2O-CaO-SiO_2
XJ-33M	Kizil reservoir cemetery, Baicheng, Xinjiang	the Western Zhou to the Spring and Autumn Period	0.657	1087	8.91	5.9	Na_2O-CaO-SiO_2
XJ-1D	Kizil reservoir cemetery, Baicheng, Xinjiang	the Western Zhou to the Spring and Autumn Period	0.672	1091	11.09	11.25	Na_2O-CaO-SiO_2
10HFPM4-20	Firework factory, Hepu, Guangxi	the later of Western Han Dynasty	1.101	1093	13.89	1.44	K_2O-SiO_2
11HFPM8-40	Firework factory, Hepu, Guangxi	the later of Eastern Han Dynasty	1.111	1101	11.81	4.73	K_2O-SiO_2
11HFPM5-25	Firework factory, Hepu, Guangxi	the early of Eastern Han Dynasty	1.168	1098	12.27	1.33	K_2O-SiO_2
09HYGM15-9	Civil servant estate, Hepu, Guangxi	the Jin Dynasty	1.195	1097	12.35	2.60	K_2O-SiO_2

(*Continued*)

Table 13.1 (*Continued*)

No.	Location	Date	$Ip = \frac{A_{500}}{A_{1000}}$	v_{max} Si-O stretching (cm^{-1})	R_2O*	CaO	Chemical type of glass
10HTQM66-2	Auto gears factory in Tuguiling, Hepu, Guangxi	the later of Western Han Dynasty	1.234	1098	12.69	1.05	K_2O-SiO_2
11HFPM10-5	Firework factory, Hepu, Guangxi	the later of Eastern Han Dynasty	1.275	1096	14.81	0.87	K_2O-SiO_2
12HLWM6-1-2	Luowu village, Hepu, Guangxi	the Jin Dynasty	1.283	1096	10.55	1.09	K_2O-SiO_2
10HTQM2-12	Auto gears factory in Tuguiling, Hepu, Guangxi	the later of Western Han Dynasty	1.315	1102	14.75	1.62	K_2O-SiO_2
09HYGM18-5-A	Civil servant estate, Hepu, Guangxi	the later of Eastern Han Dynasty	1.557	1094	12.24	2.35	K_2O-SiO_2
Quartz crystal	—	modern sample	—	—	—	—	quartz

*R_2O is Na_2O in Na_2O-CaO-SiO_2 glass, and is K_2O in K_2O-SiO_2 glass. They are both obtained on fresh surfaces.

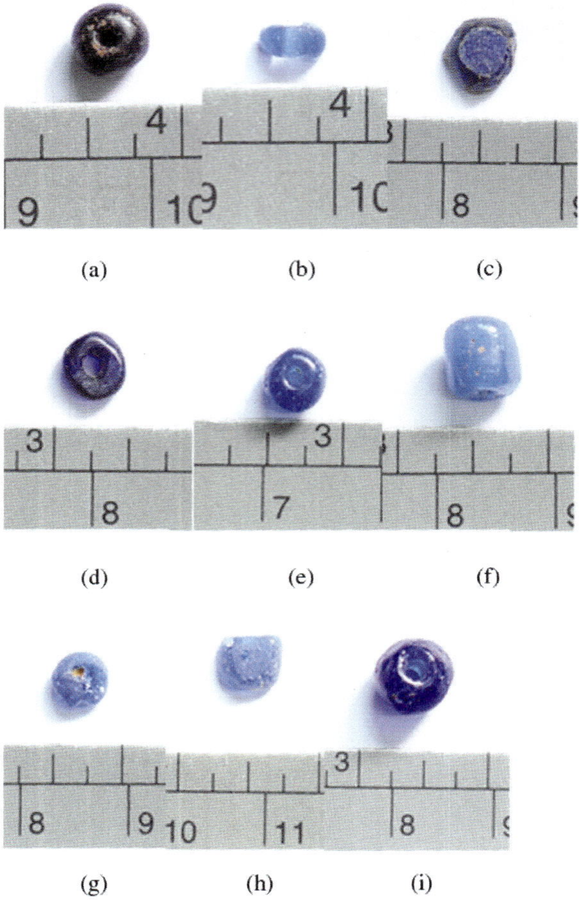

Fig. 13.1 Photographs of ancient Chinese glasses. (a) 10HFPM4-20, (b) 11HFPM8-40, (c) 11HFPM5-25, (d) 09HYGM15-9, (e) 10HTQM66-2, (f) 11HFPM10-5, (g) 12HLWM6-1-2, (h) 10HTQM2-12, (i) 09HYGM18-5-A

Photographs of the samples are given in Fig. 13.1 One modern quartz sample was also studied.

The chemical compositions of the 20 samples were determined using a portable energy dispersive X-ray spectrometer (pXRF). The beads from Xinjiang are soda-lime-silica glasses, whereas those from Guangxi are potash-silica glass (Table 13.1). Technical details of the pXRF set-up are described elsewhere.[9]

We selected glasses in this study for which Raman signatures from pigments and precipitated crystalline phases is small or nil.

13.2.2 *Micro-Raman spectroscopy*

The vibrational bands of ancient glasses were examined nondestructively with a LabRam XploRA confocal Raman microspectrometer equipped with a laser source of 532 nm, 1800 gr/mm diffraction grating and a 50× long working distance objective. The laser was projected onto the focal spot (with a beam diameter of ~1 μm) on the artifact and maintained at a power of less than 15 mW. The spectral resolution was less than 2 cm^{-1} over the spectral range of 100 cm^{-1} to 4000 cm^{-1}. The detection system consisted of a multi-channel air-cooled charge-coupled detector. The system was calibrated with a 520.7 cm^{-1} Si line from a single crystal Si wafer at the beginning of each measurement.

13.3 Results and discussion

13.3.1 *Quartz crystal and silicate glass*

Glasses are silicate networks consisting of SiO_4 tetrahedral joined together by oxygen atoms located at the vertices. These SiO_4 tetrahedral connections are modified by the incorporation of aluminum, magnesium, iron, and alkali/alkaline earth metallic ions which change the properties (color and viscosity) of glasses.[7] As the building block of all silicates (amorphous or crystalline), the SiO_4 tetrahedron is a covalent entity with a well-defined vibrational signature, particularly in Raman Scattering.[6] Thus, these modifications of the SiO_4 tetrahedral vibrational units should be revealed by the Raman spectra of the glasses. As ancient glasses are all silicate glasses, a modern sample (quartz crystal), a soda-lime-silica glass bead (XJ-44R), and a potash-silica glass bead (11HFPM8-40) were selected for comparison by Raman spectroscopy. Fig. 13.2 shows that the spectra of ancient glasses (baseline subtracted) and the quartz crystal varied. In the well-connected SiO_4 tetrahedral framework (Fig. 13.2c), the Raman line at

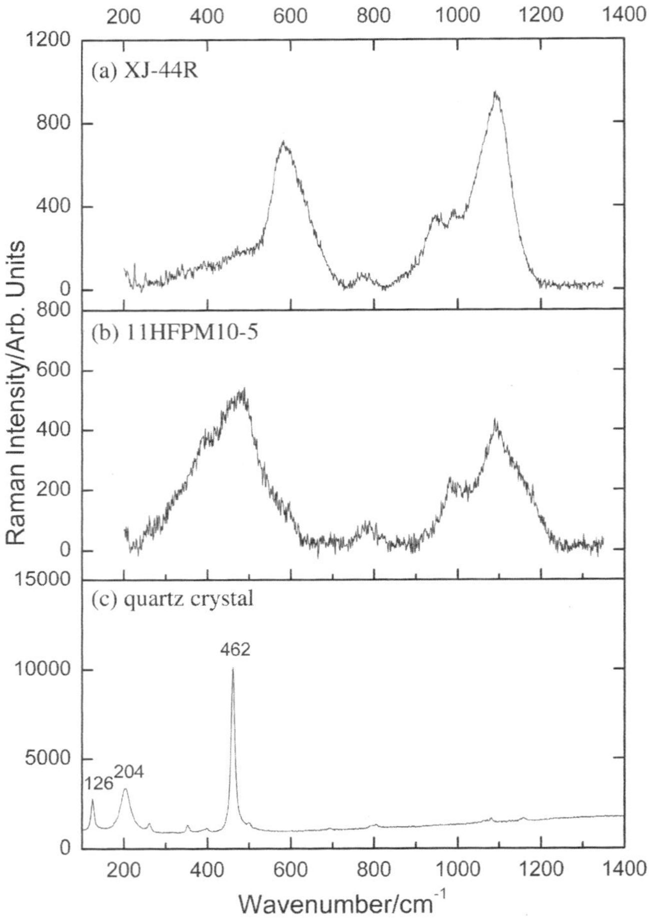

Fig. 13.2 Representative Raman spectra of ancient glasses: (a) XJ-44R, soda-lime-silica glass from Xinjiang (after baseline subtraction), (b) 11HFPM10-5, potash-silica glass from Guangxi (after baseline subtraction), and (c) quartz crystal, modern sample

462 cm^{-1} was sharp and strong and originated from the bending vibration in the isolated SiO_4 tetrahedron (υ_2 mode).[7] In soda-lime-silica (XJ-44R, Fig. 13.2a) and potash-silica (11HFPM10-5, Fig. 13.2b) glass beads this mode remained strong but was much broader and had various components. We can imagine the motion of the bridging oxygen atom in a polymerized TO_4 network as

a combination of the bending vibrations in the two adjacent TO_4 tetrahedron.[4] Thus, we can discriminate amorphous and crystalline silicate samples using Raman spectroscopy. The identification of ancient glasses having different chemical compositions is discussed in the next Section.

13.3.2 *The influence of weathering on Raman spectra of ancient Chinese glasses*

All ancient glasses were weathered to different degrees, which leads to the loss of flux. Surface weathering can directly affect the identification of the chemical type of ancient glass, but few papers have discussed the impact of surface weathering on the qualitative analysis by Raman spectra. To understand this phenomenon, eight samples from Xinjiang and Guangxi were selected to compare the differences between weathering and fresh surfaces (Table 13.2). In Table 13.2, both the polymerisation index I_p and the Si-O stretching peak maximum are given as average values for a few testing points. Using a diamond-coated fine grinding file, small areas of the sample surface were abraded within an area between 25 and 55 mm^2. The outer 100 microns of weathered glass were removed.

Analysis showed that glass weathering induced crack formations on the glass surface resulted in the formation of a corroded silica-rich layer on the surface exposed to pollution.[10,11] Such weathering can occur in museum reserves.[12] From Table 13.2, we can see that the polymerisation index I_p values of the fresh surfaces of these eight samples were higher than those of weathering surfaces. For the soda-lime-silicate glasses from Xinjiang, the error was approximately 40% and reached as high as 90% (for XJ-1D). However, besides 09HYGM18-5 (error = 53%), the errors for the potash-silica glass from Guangxi ranged from 2% to 20%, which was lower than those for the samples from Xinjiang. Thus, the impact of weathering on soda-lime-silica glasses was stronger than that on potash-silica glasses. The difference in Si-O stretching peak between fresh and weathering surfaces in XJ-1A and XJ-3A from Xinjiang (2 cm^{-1} to 3 cm^{-1}) was low, but the values in XJ-29C and XJ-1D were 13 and 30 cm^{-1}, respectively. For

Table 13.2 Comparison between weathering and fresh surfaces of ancient Chinese glass from Xinjiang and Guangxi

No.	Chemical type of glass	surface	Ip = A_{500}/A_{1000}	ν_{max}Si-O stretching	Error*
XJ-1A	Na_2O-CaO-SiO_2	fresh	0.522	1087	0.408
		weathering	0.309	1089	
XJ-29C	Na_2O-CaO-SiO_2	fresh	0.536	1087	0.701
		weathering	0.160	1074	
XJ-3A	Na_2O-CaO-SiO_2	fresh	0.580	1087	0.641
		weathering	0.208	1090	
XJ-1D	Na_2O-CaO-SiO_2	fresh	0.672	1091	0.876
		weathering	0.083	1061	
11HFPM8-40	K_2O-SiO_2	fresh	1.111	1101	0.177
		weathering	0.914	1095	
09HYGM15-9	K_2O-SiO_2	fresh	1.195	1097	0.199
		weathering	0.957	1087	
10HTQM2-12	K_2O-SiO_2	fresh	1.315	1102	0.017
		weathering	1.292	1085	
09HYGM18-5	K_2O-SiO_2	fresh	1.557	1094	0.527
		weathering	0.737	1095	

*error = (I_p of fresh surface-I_p of weathering surface)/I_p of fresh surface.

potash-silica glasses from Guangxi, the wavenumbers of Si-O stretching peaks of fresh and weathering surfaces in 09HYGM18-5 were very close, but an obvious offset (6 cm^{-1} to 17 cm^{-1}) was observed in the other three samples. Thus, the loss of flux caused by the weathering of ancient glasses can affect the Raman spectra of different chemical compositions to varying degrees. Moreover, such loss even affected the related identification of families and the deduction of firing temperatures. Therefore, the following discussion was based on results obtained using fresh surfaces to achieve Raman spectra that describe the true characteristics of ancient glasses.

13.3.3 *Classification of different chemical types in silicate glasses*

To ensure the accuracy of the analysis, we selected Raman signatures for fresh polished surfaces of ancient glasses in which the contribution of pigments and precipitated crystalline phases was small or nil. Representative examples are presented in Figs. 13.3 and 13.4. Within a family, the Raman signatures were very similar, but among the different families, the Raman signatures can be associated with the main fluxing agent or such a similar cation. This result agrees with the hypothesis that the Raman signature is determined by the polymerisation degree of the silicate network.[6]

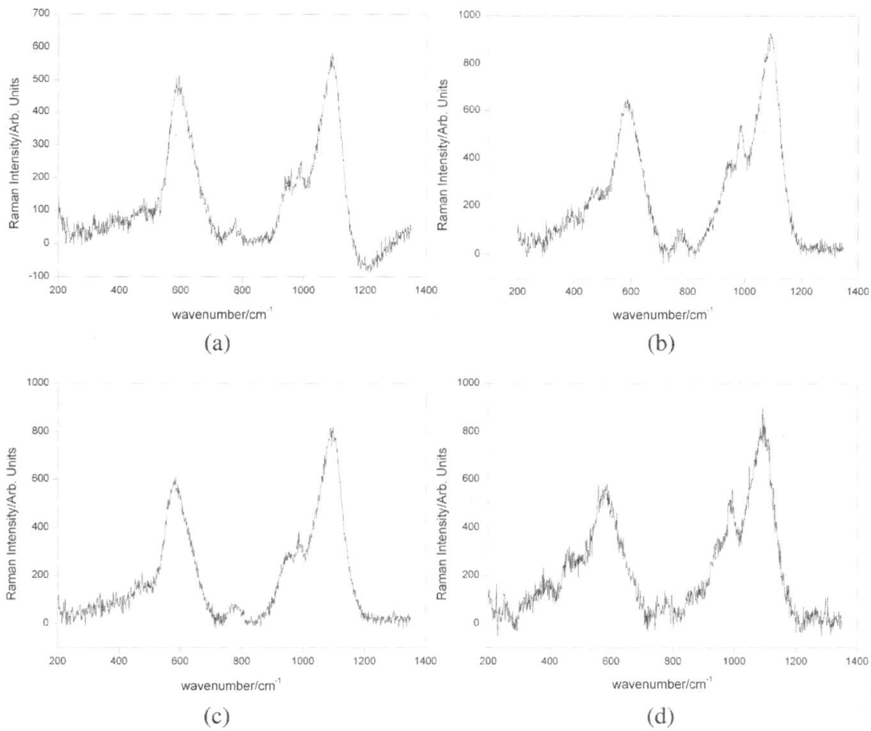

Fig. 13.3 Representative Raman spectra of ancient glasses from Xinjiang, after baseline subtraction: (a) XJ-1A, (b) XJ-29C, (c) XJ-44Q, and (d) XJ-44M

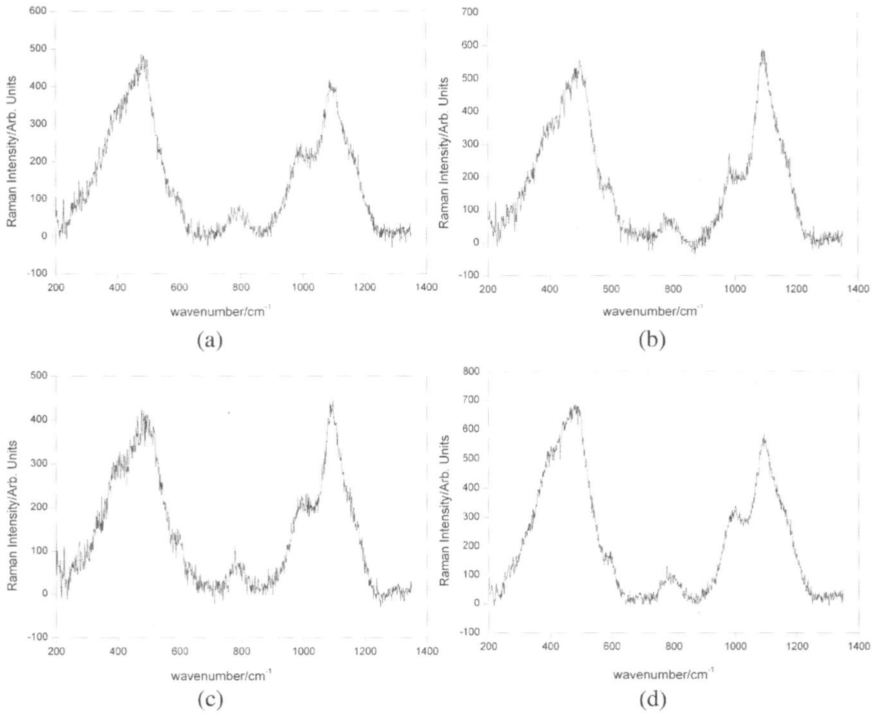

Fig. 13.4 Representative Raman spectra of ancient glasses from Guangxi, after baseline subtraction: (a) 10HFPM4-20, (b) 11HFPM5-25, (c) 09HYGM15-9, and (d) 10HTQM66-2

The intensity ratio of the two Raman envelopes, at ca. 1000 and 500 cm^{-1}, depended on the components and structures existing in ancient glasses. It was possible to identify the various glasses for the connectivity of the (SiO_4) polymeric units, which can be analyzed through the relative intensities of Si-O stretching and bending modes. In Xinjiang soda- lime-silica glass spectra (Fig. 13.3, Table 13.2), the highest intensity band was the stretching mode envelope (1087 cm^{-1} to 1091 cm^{-1}), and that for XJ-4M was around 1082 cm^{-1}, thereby describing the characteristics of structures made of “isolated” and poorly connected tetrahedras. Moreover, two small envelopes appeared to the left of the strongest peak. The mean I_p value ranged from 0.522 to 0.672. Fig. 13.4 shows that the intensity observed at ca.

500 cm^{-1} significantly increased compared with stretching mode. Moreover, the width broadened. Only one small envelope was found to the left of the stretching mode. Table 13.2 shows that the corresponding values of the index of polymerisation covers a large range (from 1.101 to 1.557). This range of values was higher than that obtained for the samples from Xinjiang, which showed a shift in the Si-O stretching mode to a high wave number (1093 cm^{-1} to 1102 cm^{-1}) simultaneously.

The empirical relationship between I_p, the glass composition, and the processing temperature is well-documented in previous papers.[5,9] In Fig. 13.5, the I_p of 19 ancient glass samples increased with the time and the evolution of chemical composition. It also means that the processing temperature gradually increased. According to the seven families recognized by Colomban, the two parameters (the Si–O stretching peak maximum and the polymerisation index) were

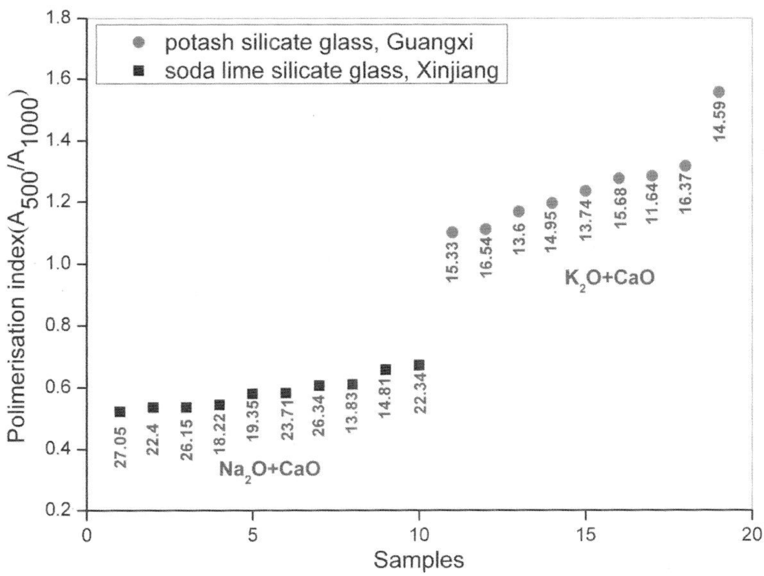

Fig. 13.5 The polymerisation index measured for a series of ancient glasses from China with different chemical types

sufficient to classify the glassy silicates as a function of compositional types. The I_p of soda-lime-silica glass was in the range 0.522 to 0.672, which mostly corresponded to the second family ($0.5 < I_p < 0.8$) processed at medium temperature (800 °C).[6] Given the firing temperature of early faience from China (~900 °C) and the good glassy structure observed in these samples, we deduced that the 10 ancient glasses dating from the Western Zhou Dynasty to the Spring and Autumn Period should be fired at ~1000 °C. The I_p of potash-silica glass, by contrast, was in the range 1.101 to 1.557, and we deduced that this glass may be fired at around 1000 °C according to results of Colomban.[6] Based on reasonable melting temperatures for potash-lime silica glass made in the Warring State Period (1100 °C to 1200 °C),[13] the nine samples from Guangxi should be fired at 1100 °C or even higher. In Fig. 13.5, we have labeled the main flux and the corresponding wt% contents of Na_2O, K_2O, and CaO. The values of flux content decreased from left (soda- lime-silica-glass: 18.22% to 27.05%; XJ-4M: 13.83%; and XJ-33M: 14.81%) to right (potash-silica glass: 11.64% to 16.54%), i.e., from low-temperature melting to high-temperature melting. The Raman index of polymerisation is related closely to the nature and content of the fluxing cation. Substituting Si^{4+} atoms by cations of a different nature (M^+, M^{2+}), such as alkali or alkaline-earth metals and some other elements, caused the depolymerization of the network breaking the Si-O connectivity and creating an ionic bond between the oxygen atoms and the cations.[1] Consequently, with a decreasing number of Si-O bridges and connectivity of the network, the melting temperature decreased.

To check the validity of the classification, confirmation was provided by plotting the polymerisation index as a function of the position of the main Si-O stretching component wave number in Fig. 13.6. From left to right, the two compositional families could be viewed as a function of the main flux: soda-lime-silica glasses to the left and potash- silica glasses to the right. Then, from top to bottom, the two families could be seen as a function of the melting/processing temperature. Therefore, the classification was obvious and was in

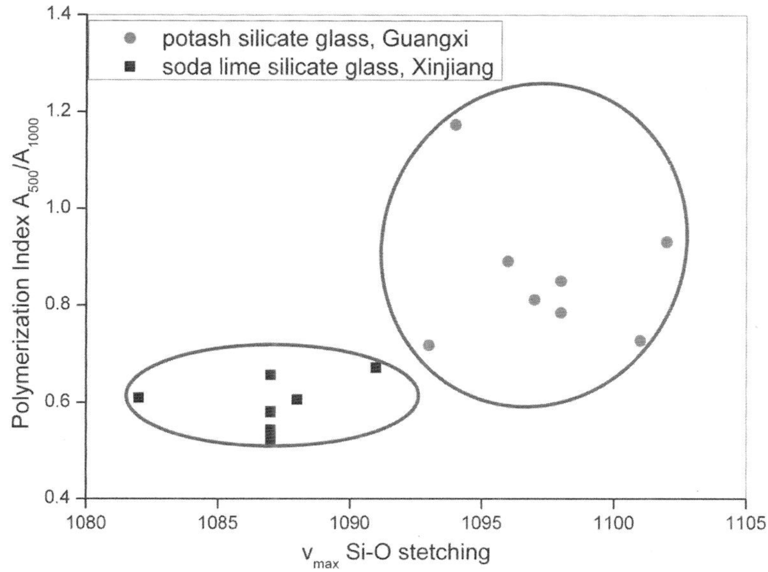

Fig. 13.6 Plots of the polymerisation index of different chemical types of ancient glasses from China as a function of the main Si-O stretching component (wavenumber)

agreement with the results mentioned above. As a preliminary conclusion, the type of composition of any glassy silicate could be associated with a specific Raman signature, but the pertinent parameters for classification may change according to the set of samples being considered.

13.4 Conclusion

Nineteen ancient glasses from Xinjiang and Guangxi, dating from the Western Zhou to the Jin Dynasties, have been analyzed in detail by Raman spectroscopy. Besides discrimination of amorphous and crystalline silicate samples, the classification of different chemical compositions in these samples were identified through two Raman parameters (the Si–O stretching peak maximum and the polymerization index) based on the investigation about weathering and fresh surfaces. With increasing I_p, the melting temperature gradually increased and the contents of the main flux decreased. System errors were present, but the overall trends were consistent with the results of previous investigations on glassy silicates from the West. According to the experimental non-destructive results, with no sample preparation, and combined with micro-probe chemical characteristics it is confirmed that micro-Raman spectroscopy is a unique tool that can be used to determine glass structure. Further studies are required to discriminate between sets of artifacts with different origins and dates of production.

Acknowledgments

This research was supported by the Compass Plan Project from the State Administration of Cultural Heritage, China (Grant No. 20110303), the National Key Technology Support Program (Grant No. 2013BAK08B08) of the Chinese Ministry of Science and Technology and the Shanghai R&D Public Service Platform Construction Project (Grant No. 13DZ2295800). The authors would like to thank the relative provincial institutes of cultural relics and archaeology in Xinjiang Uygur Autonomous Region and Guangxi Zhuang Autonomous Region for allowing the researchers to examine precious artifacts.

References

1. Ph. Colomban, *J. Cul. Herit.* **9**, e55 (2008).
2. B. O. Mysen, D. Virgo and C. M. Scarfe, *Am. Mineral.* **65**, 690 (1980).
3. D. Michel, Ph. Colomban, S. Abdhassani *et al.*, *J. Eur. Ceram. Soc.* **16**, 161 (1996).
4. N. Q. Liem, N. T. Thanh and Ph. Colomban, *J. Raman Spectrosc.* **33**, 287 (2002).
5. Ph. Colomban, O. Paulsen, *J. Am. Ceram. Soc.* **88**, 390 (2005).
6. Ph. Colomban, A. Tournie and L. Bellot-Gurlet, *J. Raman Spectrosc.* **37**, 841 (2006).
7. Ph. Colomban, *J. Non-Cryst Solids* **323**, 180 (2003).
8. L. Zhao, X.Ch. Zhao, H. Zheng *et al.*, *Palace Museum Journal* **151**, 153 (2010).

9. S. Liu, Q.H. Li, F.X. Gan *et al.*, *J. Archaeol. Sci.* **39**, 2128 (2012).
10. Ph Colomban, M.-P. Etcheverry, M. Asquier *et al.*, *J. Raman Spectrosc.* **37**, 614 (2006).
11. A. Tournie´, P. Ricciardi, Ph. Colomban, *Solid State Ionics* **179**, 2142 (2008).
12. L. Robinet, C. Coupry, K. Eremin *et al.*, *J. Raman Spectrosc.* **37**, 1278 (2006).
13. S. Liu, Q.H. Li, F.X. Gan, *Spectrosc Lett* **48**, 302 (2015).

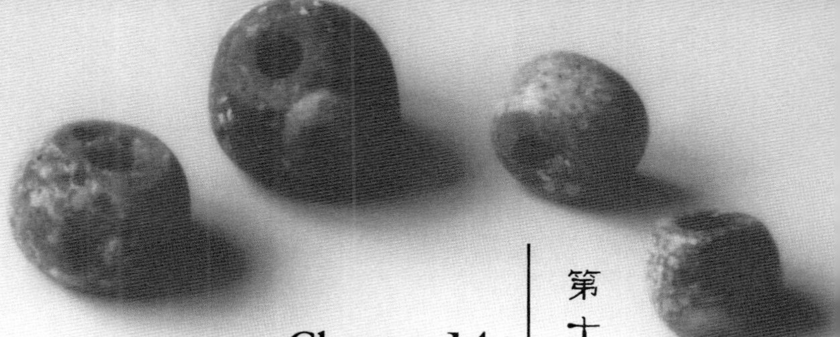

Chapter 14 第十四章

Scientific Analysis of Ancient Glass: Answering the Questions and Questioning the Answers

J. W. Lankton, B. Gratuze,† K. Tantrakarn,‡ Q. H. Li§ and S. Liu§*

*UCL Qatar, Doha +974, Qatar

†Institut de recherche sur les archéomatériaux, CNRS, Orleans 45000, France

‡Techno-X Co, Osaka 540-8570, Japan

§Shanghai Institute of Optics and Fine Mechanics, Shanghai 201800, China

14.1 Introduction

The study of ancient glass requires an integrated approach, including knowledge of archaeological context, textual clues, typology and classification, interpretation of technological traces in order to determine how, and possibly where and by whom, an object was made, and, not least, the scientific analysis of the glass itself. None of these are entirely straightforward, and it is rare that all are well known. Most importantly, the significance of each aspect of glass study depends largely on the archaeological question being asked.

One useful theoretical framework for glass study includes the three broad headings of Production, Exchange and Consumption. Our questions about early glasses often fit into one or more of these general categories. In this regard, two of the most common questions from archaeologists are 'Where did this glass come from?' and 'How old is it?' For the first question, roughly subsumed under the topic of provenance (or provenience, the most commonly used term in the United States), we must investigate both production and exchange, with less emphasis on consumption, or how the object was used. For the second question, that of age, we would be interested not only in production, because certain glass types were made for a relatively circumscribed period, and possible exchange, but also the consumption of the object in society. For example, if the object were found in a grave, it is possible that it was already several hundred years old by the time it was deposited. So let's take a closer look at our three categories of knowledge about ancient glass- production, exchange and consumption, with an emphasis on how scientific analysis can help to answer questions in each area.

14.2 Production, exchange and consumption

14.2.1 *Primary production*

The separation of production into two stages- 'primary' and 'secondary,' has been one of the most important achievements of scholars studying glass from Roman and Late Antique/Byzantine sites in western Europe and the Mediterranean world.[1] Primary production is the melting or 'smelting' of glass from its raw materials, most commonly silica, whether from sand or quartz pebbles, combined with a fluxing agent to lower the melting temperature of the silica. In the early western world, soda (Na_2O) was the most common flux. By studying the chemical compositions of glasses of known date, Sayre and Smith were able to demonstrate that the earliest glass from the Late Bronze Age in Mesopotamia and Egypt was fluxed with soda from a plant-ash source.[2] Then, over the course of the 1st millennium BCE this plant-ash soda was replaced, east of the Euphrates River, by soda from a mineral source, most commonly trona or natron, an efflorescent salt mainly composed of a mixture of sodium carbonate and sodium bicarbonate, found at seasonal lakes near Wadi Natrun in Egypt. Glass compositions changed again in the 9th c. CE, following the Arab conquest, when plant-ash soda again became the norm in the glassmaking centers of western Asia and Egypt.

Primary glass production has been the topic of intensive study in Europe and the Middle East, with an emerging picture of glass smelting, often on a large scale, in specialized centers. During the Late Bronze Age (ca 1550–1200 BCE) in the Near East, these centers are thought to have been concentrated in Mesopotamia, the Levant and Egypt. Over time, these same areas remained the heartland of glass production, so that by the Late Roman/Early Antique period of the 4th to 9th CE, large semi-subterranean tank furnaces along the Levantine coast supplied much of the raw glass that was used at multiple, smaller, units where craftworkers fashioned this raw glass into useable objects.[1,3]

14.2.2 *Secondary production*

This secondary production of individual objects from already-prepared raw glass was a highly skilled craft, often requiring many years of apprenticeship. The workers and the *chaînes opératoires* were different from those required for the more 'industrial' primary production of glass. In many cases the locations were also different, with numerous secondary workshops supplied by a limited number of primary production sites. Among the best evidence for this separation has been the discovery of several Mediterranean shipwrecks, where raw glass, often broken into chunks, formed a prominent part of the cargo.[4] The evidence for the geographical separation of primary and secondary glass production in Asia has been less clear, although the recent recovery of glass ingots in a shipwreck off the southern coast of Sri Lanka, dated to late BCE/early CE, provides excellent evidence that such separation did at least sometimes occur.[5]

In order to answer archaeological questions about glass production, it is essential to remember this separation between primary and secondary production, since different categories of evidence are required for each. Study of the area of primary production must be based on the glass itself, and not the form of the finished object. We would emphasize the evidence from archaeological investigation, particularly of possible primary production sites and any texts referring to glass produced in a certain area, although here we must consider that ancient authors may not have made the same primary/secondary distinction that we make today. Object typology and technology would be less helpful, since any object could, in theory at least, be made from any type of primary glass. Scientific analysis would be the cornerstone for any answer regarding primary production, based on the theory that many of the batch materials, particularly heavier ingredients like sand or quartz pebbles, would have come from the area around the primary production site, thus lowering transport costs. By combining knowledge of local geochemistry with chemical analysis of the glass, we might get a good idea of whether or not the glass could have come from a certain area. More commonly, we compare the chemical composition of an object with the chemical compositions of glasses thought to have been made in a particular geographical area. In addition to chemical composition, we may compare the ratios of isotopes of selected chemical components of the glass.[6] In recent years strontium (Sr) isotope ratios have been used to reflect the lime (CaO) source, generally linked to the flux, while neodymium (Nd) ratios will reflect neodymium-containing heavier minerals contained in the silica source. Lead (Pb) isotopes may also be useful to help to source lead-related materials, whether in the primary batch or associated with added colorants.

Answers for questions related to secondary production will most often come from the study of the typology of the object and the technology used for its production. Just as some chemical compositions tend to be regional, technologies may also have a regional distribution. If we consider for a moment beads made in Asia, we know that the production of small, monochrome, drawn beads was centered in South India, with likely spread to Southeast Asia by the late centuries BCE.[7] While there were some drawn beads made at North Indian sites, beadmaking by winding hot glass around an iron mandrel, or even by lapidary techniques, appears to have been much more common. Wound beads were also the norm in China, Europe and the Middle East. Thus, by studying the technology of how the object was produced, we may have some idea of where it might have been made, regardless of the source of the raw glass.

14.2.3 *Exchange*

While glass was certainly not unique in early long-distance trade, it has become one of the most important materials for studying the archaeological remains of such trade, since glass, as opposed to other luxury goods such as textiles, drugs and spices, and even people, is well preserved at many archaeological sites. In addition, because glass can be studied by different but complementary methods, as mentioned above, the glass remains are often the most important evidence for trade.

Chemical analysis or isotope determinations may give us a good idea of the ultimate origin of the raw glass, while the technology of the object helps us to know where, or at least in which technological tradition, the object was made. However, neither helps to answer the question of how the object got from one place to another. In order to answer questions of exchange, it is important to consider the forms and technologies of any objects found between the source area and the final area of consumption. Ideally, we would need to study the types of glass objects, including their chemical compositions, at multiple stops along possible trade routes. To the extent that such information is available, we might be able to at least exclude certain routes, if not to identify the actual routes themselves. For this reason, it is very difficult to study the glass from a single area in isolation. Glass was so commonly an object of exchange in the ancient world that a broad knowledge of glass from multiple areas is certainly advantageous, if not required.

14.2.4 *Consumption*

Glass objects, including vessels and ornaments, were among the most treasured of ancient objects, and, as such, can tell us many things about the societies in which they were used. For example, glass found within settlement areas may help to identify the uses of particular buildings or even who might have lived or worked there, while the distribution of glass grave goods can help to identify high and low prestige burials. In Asia, while glass vessels are quite rare, glass beads are found at most sites, and detailed study of these seemingly simple artefacts may provide insight into many aspects of ancient life. While questions of glass consumption may be of profound importance for understanding early societies, the answers may depend more on careful archaeological investigation and comparison with other sites, as well as detailed typologies of the excavated objects. However, chemical compositions may also be very helpful, even though ancient people most likely were not aware of the chemical compositions of various objects. An example might be bead distributions from a particular cemetery, where certain types of beads, identified by their typologies, were associated with higher-prestige graves. In this case, chemical analysis might reveal that these higher-prestige beads were all imported from a particular area, raising the possibility that not only was the form of the objects important, but also their source may have held a special meaning for the society.

14.3 Chemical analysis: A consumer's guide

We can see in the above discussion that chemical analysis of glass may be helpful to answer many archaeological questions. The problem then becomes deciding which technique would be best suited to answer the particular question. We can divide the various techniques into two broad categories: those using an electron, proton or X-ray beam in order to excite the atomic structure of the specimen, after which the energy given off as the glass returns to its equilibrium state is measured, and those based on the complete ionization of a small sample of glass in a plasma torch, with subsequent measurement of atomic composition by some type of spectrometer, usually a mass spectrometer (MS), but sometimes an atomic emission spectrometer (AES). These two main types of instruments, along with their advantages and disadvantages, will be considered below, but first, it is important to understand two issues that complicate the analysis of ancient glass. The first is that ancient glasses, as opposed to modern glass or glass standards, are seldom completely homogeneous, so that the composition of one part of a sample, however 'correct' the analysis, may be different from the composition of another part of the samples. Sometimes this heterogeneity is nothing more than the uneven dispersal of coloring agents. However, more important differences in base composition may exist as well, either because of incomplete mixing of the glassmaking ingredients, or, less commonly, because two quite different types of glass were mixed together in a secondary workshop; along this line, we have analysed beads where most of the bead was one composition, but small areas actually had an entirely different base glass. Heterogeneity may also be introduced by weathering, the second major issue for ancient glass analysis. All ancient glass samples, regardless of their external appearance, will show chemical signs of weathering, changes resulting from long-term exposure to the matrix in which the glass has been stored. If the atmosphere is very dry, or if the soil is very acid, the changes of weathering will be reduced. Repeated cycles of wet and dry, particularly when combined with alkaline soil, may alter the sample so much that very little or no healthy glass remains. Light or mild weathering will leave the glass surface shiny and strong, but will always cause some change in the chemical composition near the exposed surface.[8] The most common changes are in the light elements, with a loss of alkaline oxides- soda (Na_2O) and potash (K_2O- and thus a relative increase in less soluble oxides such as alumina (Al_2O_3), lime (CaO) and silica (SiO_2). With heavier weathering, the surface will become dull, powdery or even iridescent, and the heavier elements will be affected as well, often in an unpredictable way.[9] Analytical laboratories, no matter how good their results with modern glass standards, will produce erroneous results for ancient glass unless these two issues, particularly weathering, are taken into account, as we shall see in the discussion below.

14.3.1 *Excitation-based instruments*

Perhaps the most commonly used excitation-based instrument for the chemical study of glass is a scanning electron microscope linked to an energy- or wavelength-dispersive spectrometer, namely SEM-EDS and SEM-WDS, the latter also known as an electron microprobe, abbreviated EPMA for electron probe microanalysis. Either instrument may produce quantitative values, with a limit of quantification ca 0.1 wt% for SEM-EDS and ca 0.05 wt% for EPMA. These limits are averages and both instruments will be better for some elements than for others. EPMA, in particular, is capable of high precision when used under appropriate circumstances. Both SEM-EDS and EPMA measure only the surface glass, with depth of analysis ranging from a few microns for light elements like sodium, to a few hundred microns for heavier elements. This surface analysis, coupled with the excellent imaging provided by the scanning electron microscope, has the advantage of WYSIWYG (what you see is what you get), and means that very small surface areas, such as pigment or mineral inclusions, may be measured with high spatial resolution. In addition, if the surface is well prepared, it is usually possible to distinguish between weathered and healthy glass with the electron image, so that only healthy glass may be analysed. Although the measurement itself is entirely nondestructive, the surface layer, usually ca 100 microns, must be removed in order to reach this 'healthy' glass, since measurement of the unprepared surface of an ancient sample may give very misleading results. In most laboratories seeking quantitative results, the glass samples are embedded in epoxy resin, then polished to achieve a flat, shiny, surface. The embedding is usually irreversible, and considerable amounts of sample may need to be polished away to achieve proper results. Thus, although the technique itself is nondestructive, the procedures needed to achieve quantitative results are certainly invasive, even when minimally destructive.

Another major technology using an excitation beam forms the basis of X-ray fluorescence, or XRF. In this case an X-ray beam is pointed at the sample, resulting in excitation of the atomic structure. As the atoms 'relax' to their original energy state, electromagnetic particles or waves, with energies characteristic for the atomic structure, are released. The energy released as X-rays will be measured by the XRF spectrometer, and the number of 'counts' at particular X-ray energies will be proportional to the concentration of the element for which this energy is characteristic. Quantitative laboratory-based XRF, based on measuring glass beads prepared by mixing the ground-up sample with a glass-forming agent such as lithium borate, may achieve high accuracy with quite low detection limits. Less accurate bench-top instruments may measure the glass samples directly, with best results when the surface of the sample has been prepared by polishing to both flatten the surface and to remove any weathering. In this case, measurement under vacuum may approach the results achievable with SEM-EDS, although without the surface resolution since incident X-ray beam sizes are generally a few mm. Like SEM-EDS, XRF is a surface technique, with the depth of analysis depending on the energy of the emitted X-rays. For light elements such as sodium, the depth of analysis will be less than 20 microns, while for heavier elements it may be as high as a few mm. One consequence of this depth variability is that quantitative results from very thin samples are particularly elusive.

While laboratory XRF continues to be useful for glass chemical analysis, the most commonly used XRF instruments are small enough to be carried to the archaeological site, thus avoiding the difficulty of removing glass samples from their area of origin, a major issue in glass study in many parts of the world. These portable XRF (pXRF) instruments may be used for many materials and are easily transported, and, in many cases, hand-held during the measurement process. There is a confusing array of hand-held pXRF instruments available on the market, and they are a tempting purchase for archaeological teams wanting to characterize excavated materials. However, all of these instruments share the same limitations as laboratory XRF in that they require a fresh glass surface in achieve best results, particularly for light elements. In addition, unless either a vacuum or helium flow is used during the measurement, the light elements that are important to characterize the base glass composition, such as sodium and magnesium, will not be detected; even with vacuum conditions these elements are difficult to quantify. Although pXRF has proved very useful in obsidian studies, where only a few, heavier, elements are required to match the sample to a source area, pXRF for glass study has been less useful, primarily because of issues of sample preparation, requirement for vacuum, and difficulties in calibration of portable instruments so that X-ray counts may be translated into weight percent concentrations. One notable exception has been the pXRF instrument developed by Professor Izumi Nakai at the Tokyo University of Science, in collaboration with Ourstex.[10] For these instruments that might be described as portable, but certainly not hand-held, with use limited to Tokyo and the Shanghai Institute of Optics and Fine Mechanics, excellent results have been possible using measurement under vacuum conditions and careful calibration with glass standards of known composition.

14.3.2 *Ionization-based instruments*

Glass samples prepared as liquids by dissolving in hydrofluoric acid may be ionized in a plasma flame at 8,000 °C, with the resulting ions separated by mass and charge for measurement in a mass spectrometer. In the typical liquid inductively coupled plasma-mass spectrometer (ICPMS) instrument, the number of counts for mass/charge isotopes of many elements may be measured within a short period of time, with these counts then translated to actual concentrations based on the isotopic abundance of particular species and the ease or difficulty of ionization for particular elements. The technology breakthrough making these instruments more popular for the analysis of archaeological materials came in the 1990s, when laser sampling devices were coupled with ICPMS resulting in laser ablation-ICPMS (LA-ICPMS).[11] Laser sampling allows the study of actual archaeological samples, removing the requirement for invasive mounting or surface preparation, since it is possible to measure only the sub-surface component of the laser-liberated sample. With laser beam diameters ranging from 30 to 100 microns, considerable spatial resolution is also possible. The great advantage of LA-ICPMS is that many elements, often up to 60, may be measured almost simultaneously, with limits of quantification below one part per million (1 ppm or 0.0001 wt%) for most elements. In addition, the multiple signal interferences that complicate XRF and SEM-EDS are much less important. However, these limits of quantification do vary by element, with limits in the 0.05 to 0.1 wt% range for some light elements, including sodium, potassium and magnesium, present in small amounts, where the background counts may be high.

While LA-ICPMS has a number of advantages, as outlined above, there are of course disadvantages as well. One of these disadvantages results from the nature of the sampling. In an effort to analyse only healthy glass, most operators will start to measure the MS signal only after a lag time of approximately 20 seconds, so that the laser can dig approximately 100 microns into the sample thus avoiding the outer layer of weathered glass. As a result, even with excellent microscopic imaging of the sample surface, it is not possible to see what is actually being analyzed, with some risk of analyzing non-homogeneous areas within the glass. Because of this, it is very important to carefully observe the raw signal coming from the mass spectrometer, in order to be sure that the concentrations of various elements do not change greatly during the analytical period, usually about 60 seconds. It is possible to get excellent results using LA-ICPMS, even comparable to those from EPMA for the elements that EPMA can measure (see below), but it is also possible to get results not indicative of the overall sample. As for all quantitative analytical methods, careful attention to details, whether in terms of sample preparation or in operation of the instrument itself, is required to meet the challenges of analyzing ancient samples.

The second drawback of techniques like LA-ICPMS that are laboratory-based and unlikely to ever become portable, is that the sample must go to the lab. It is becoming increasingly difficult to export archaeological materials, even small glass fragments, and many of these techniques are not available in large parts of the world. There may be some promise in remote sampling methods, even with a portable laser to take micro-samples. In addition, in some cases such techniques as LA-ICPMS may already be available in university or industrial geology laboratories, and might be adapted for the analysis of ancient glass. Here, the conditions for measurement and the calibration methods used for geological specimens may differ from those required for ancient glass, so that initial results must be compared with those from facilities with extensive experience in studying archaeological materials.

14.3.3 *Trueness, precision and uncertainty*

All analytical techniques have their advantages and disadvantages, as mentioned above, but no matter what the technique we must consider the three parameters of trueness, precision and uncertainty. These are gradually supplanting the more familiar terms accuracy and precision. Trueness refers to how close the average of a large number of observations, from a particular technique, comes to the accepted value. The observations need not have been made all at the same time, but could even be the result of several months of analysis. Trueness is generally calculated as the difference between the observed average value and the accepted value, divided by the accepted value and expressed as a percentage. Thus, a trueness of 10% (actually a deviation from trueness of 10%), means that the observed value will be within plus or minus 10% of the accepted value. Precision is perhaps the easiest to understand, and has the same meaning in the older and newer terminologies. The precision of a measurement tells us how close are the results of a series of measurements, regardless of how different they might be from the accepted value. Precision is also expressed as a percentage, in this case the standard deviation of the measurements divided by the mean of the measurements. Higher or 'better' precision will have a lower percentage value. Very precise instruments may have a precision (again, the number calculated is actually deviation from precision) in the range of 1%, while for many measurement techniques, 10 to 20% may be acceptable. While 'accuracy' is a term traditionally used to reflect deviation from trueness, the reasoning for the new terminology is that a technique cannot be called accurate unless both trueness and precision are taken into consideration. In the new terminology, trueness and precision combine for estimates of total uncertainty, more or less synonymous with the lay person's notion of accuracy. There is no one formula that applies to all types of uncertainty, but in general the total estimated uncertainty may be taken as the sum of the deviation from trueness plus some multiple of the deviation from precision. The trueness value tells us on average how far from 'true' our observed value might be, and if we add the multiple of 1.65 times the calculated precision, we can expect that 90% of our observed values will be no greater than this distance from the actual value. In a recent large study comparing the two techniques of EPMA as practiced at the UCL Institute of Archaeology in London and LA-ICPMS at the Institut de recherche sur les archéomatériaux in Orleans, France,[12] we found that both techniques have estimated uncertainties (90% confidence limits) in the range of 5 to 10% for major elements in excess of 1 wt%, with EPMA uncertainty about 5% lower than for LA-ICPMS because of better precision. For elements in the 0.1 to 1 wt% range, the estimated uncertainties were approximately 20%, while for elements below 0.1 wt% the EPMA uncertainties were considerably higher than those for LA-ICPMS. Surprisingly, this is the only large study that we are aware of that compares two techniques not only for the analysis of glass standards, but also for the analysis of actual archaeological samples. Clearly, many more such studies are needed before we can be confident that the results from one technique and even one laboratory will be comparable to those from another.

14.4 Representative research cases

The remainder of this chapter will present two cases where scientific analysis, primarily chemical analysis, may be used to answer particular archaeological questions. The first example is the study of Late Bronze Age glass in Mesopotamia and Egypt, and the second example concerns the study of early glass in Southeast Asia and China.

14.4.1 *Example 1. Late Bronze Age glass and the mystery of the Uluburun glass ingots: Answering the questions*

LBA glass in Mesopotamia and Egypt

One of the most fertile areas for the study of ancient glass has been that of the very early glass objects produced in Late Bronze Age Mesopotamia, Egypt and Greece. As early as 1964, Sayre had noted similarities between Egyptian and Aegean glasses,[13] and by 1999

Brill was able to suggest that there were compositional, based on major elements alone, and lead-isotope indicators for separate productions at least in Mesopotamia and Egypt, and that the very earliest glasses (pre-1391 BCE) in Egypt were similar to Mesopotamian examples.[14] The 1984–1994 excavation of a Late Bronze Age shipwreck dated to the late 14th c. BCE, with a cargo including at least 175 bowl-shaped glass ingots, at Uluburun off the southern coast of Turkey, raised obvious questions as to where the ingots found on the ship were made and for which markets they were intended.[15] However, in spite of chemical analyses for major (>1 wt%) and minor (0.1 to 1 wt%) elements, it was very difficult to link individual ingots with either a Mesopotamian or Egyptian source. There was too much variability within the sample groups for secure identification, so that in order to answer the question of the origin of the Uluburun ingots, even high quality chemical analysis equivalent to EPMA for what were then the standard elements, was not sufficient. A breakthrough came with the publication by Shortland *et al.* presenting trace element (present at levels below 0.1 wt%) data for selected glasses thought to have been made in either Mesopotamia or Egypt. [16] Shortland *et al* concluded that there was too much variation in major elements such as sodium, magnesium, potassium and calcium to be able to distinguish the groups on this basis. Trace elements associated with either the copper or cobalt colorants were likewise not useful to characterize a source area for the glasses. In contrast, they suggested that such elements as titanium (Ti), zirconium (Zr), lanthanum (La), chromium (Cr), scandium (Sc), vanadium (V), cerium (Ce), thorium (Th) and iron (Fe), not generally associated with colorants, could be used to form distinct compositional groups. An important point here is that it is preferable to compare the ratios of one element to another, rather than the actual concentrations of the elements. Many of these elements correlate with one another, indicating that they entered the glass raw material batch as contaminants of a single ingredient, most likely the silica source. Thus, the variation in individual elements may simply reflect differential proportions of the same ingredients, not a different source area. On the other hand, if the correlation curves between two elements differed by region, this might be a clue to geographic characterization of the samples. Shortland *et al*'s Figure 4 demonstrates the different correlations between La and Cr for Mesopotamian and Egyptian samples, while their Figure 6 provides a convincing separation of their study samples into Mesopotamian and Egyptian groups based on a plot of Cr/La vs Zr/Ti.

Solving the mystery of the Uluburun glass ingots

After Shortland *et al* provided a reasonable basis for the separation of Mesopotamian and Egyptian Late Bronze Age glasses, Jackson *et al* published their analyses of samples from three of the Uluburun ingots, using the techniques of SEM-EDS for major elements and LA-ICPMS for traces.[17] When these results were combined with Shortland's samples on the same Cr/La by Zr/Ti graph, all three Uluburun samples fell within the 'Egypt' range, thus indicating that at least these three ingots were produced in Egypt (re-plotted here as Fig. 14.1). Thus, for the example of Late Bronze Age glass in Mesopotamia and Egypt, plus the identification of the source of at least some of the Uluburun glass ingots, the most basic questions regarding production and exchange could only be answered using an analytical technique capable of measuring trace elements in the low

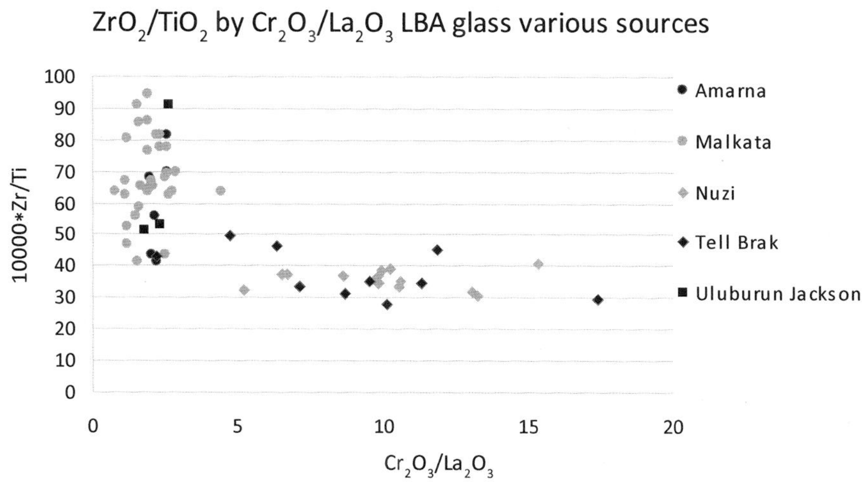

Fig. 14.1 Late Bronze Age glass from Egypt[16] (Amarna, Malkata), Mesopotamia[16] (Nuzi, Tell Brak) and the Uluburun shipwreck[17] plotted as in Shortland *et al.*[16] All three of the Uluburun ingot fragments are similar to the Egyptian samples

parts per million range, in spite of many years of detailed knowledge of major and minor glass components.

14.4.2 *Example 2. Soda and potash glass in Southeast Asia and southern China: Questioning the answers*

Based on pioneering work by Dussubieux and others, Lankton and Dussubieux were able to propose a basic scheme for the compositions, based on major and minor elements, of early glasses found in the maritime trade from India to southern China. [18] However, several important problems resisted solution when only major elements were available. Although soda glasses seemed to be characteristic for Indian production, it was not clear where in India a particular soda glass might have been made, thus making it very difficult to link chronology with geography. For potash glasses, although these were ubiquitous at early Southeast Asian sites, there were no clear source areas. Although it was possible to suggest at least three types of potash glass, based on levels of lime and alumina, there was considerable overlap, particularly in the low-lime glasses, and without secure identification it was difficult to link groups with possible production areas. More recently, the availability of high-quality trace element analyses using LA-ICPMS has led to important refinements for the classification of soda glasses, with Dussubieux *et al* identifying at least five groups of Asian soda glasses based on levels on uranium and cesium. [19] Using these trace elements, it was now possible to distinguish North Indian production (high uranium, high cesium) from South Indian production (moderate uranium, low cesium). For potash glass, the situation is less clear, although the ratio of rubidium to strontium does appear to help separate groups in the low-lime range.

One of the most exciting developments in the study of Asian glasses has been the use of high quality portable XRF (pXRF) instruments, pioneered by Izumi Nakai and his students (see above for more details regarding pXRF).[10] The Ourstex 100FA-IIL instrument has made it possible to study many samples that might not have been available for other techniques perceived as more invasive or requiring export permits. Under vacuum conditions, reasonable values for major, minor and a few trace elements may be obtained for glass standards of known composition.[20] Although Tantrakarn *et al* did include one sample studied by both pXRF and LA-ICPMS, showing excellent correspondence between the two techniques, [20] we thought it would be useful to examine a larger number of samples, including both soda and potash glasses, in order to evaluate not only the quality of the agreement, but also whether the glasses would be assigned to the same compositional groups using the two techniques of pXRF and LA-ICPMS. We focused on three sample groups and the two techniques of pXRF as practiced in Tokyo and Shanghai, and LA-ICPMS as practiced in Orleans.

Sample groups

Group 1. 9 samples, mostly bead fragments, dating to the 1st c. BCE to 2nd c. CE, from Guangxi Province, China. These samples were analysed by Li and Liu (pXRF) and Lankton and Gratuze (LA-ICPMS).

Group 2. 10 samples including possible cullet, melted beads, bangle fragments and vessel fragments, dating from the 2nd to the 7th c., from Khuan Lukpad (KLP), southwest Thailand. These samples were analysed by Tantrakarn and Nakai (pXRF) and by Lankton and Gratuze (LA-ICPMS).

Group 3. 26 samples, mostly cullet and vessel fragments, dating from the 1st c. BCE to the 4th c. CE, all but one found at Tha Chana (TCN), southeast Thailand. These samples were analysed by Liu and Li (pXRF) and Lankton and Gratuze (LA-ICPMS).

Methods

See above for introductions to the two techniques. More detailed explanations available in Liu *et al*,[21] Tantrakarn *et al*[20] and Gratuze.[22]

Results

The pXRF and LA-ICPMS results are given in Tables 14.1–14.3, along with the relative difference (Δ) between the two techniques, calculated using the formula Δ=(pXRF-LA-ICPMS)/LA-ICPMS.

Table 14.1 Sample group 1 chemical compositions by LA-ICPMS and pXRF, along with relative differences (Δ) calculated as (pXRF-LA-ICPMS)/LA-ICPMS. Glass types estimated by algorithm in Lankton and Dussubieux,[18] along with trace elements Rb_2O and SrO, with different results for JZLM5 17. Major elements in weight percent (%), minor and trace elements in parts per million (ppm). See text for glass type definitions. nd not detected

Number	Source	Technique	Glass type	SiO_2	Na_2O	K_2O	MgO	CaO	Al_2O_3	P_2O_5	TiO_2	MnO	Fe_2O_3	CoO	NiO	CuO	ZnO	Rb_2O	SrO	SnO_2	BaO	PbO
FMLM 26 2	Guangxi	LA-ICPMS	mKCA	806209	4879	135500	2377	14470	17193	6360	1578	556	6899	14	19	58	69	78	115	31	200	6
GX 4	Guangxi	LA-ICPMS	mKCA	756273	1110	151635	3410	14484	36980	2216	1744	7661	14468	321	50	1396	45	350	50	444	808	2488
JZLM5 10	Guangxi	LA-ICPMS	mKA	744254	1693	156466	1242	5221	49863	1000	2621	14053	16255	257	30	2319	25	492	15	29	1407	10
JZLM5 17	Guangxi	LA-ICPMS	mKCA	727422	1668	154578	2779	11587	32066	2028	1329	45161	14623	71	19	667	118	410	81	223	1353	1092
JZLM5 5 AES	Guangxi	LA-ICPMS	mKA	770147	1685	149333	1530	6816	37762	1357	1695	14138	9355	262	37	749	40	485	27	153	1805	472
XZHM 06 01	Guangxi	LA-ICPMS	mKCA	770776	6590	169499	2831	16569	17198	4589	1498	544	7343	12	19	51	50	80	125	18	199	5
XZHM 06 05	Guangxi	LA-ICPMS	mKA	745179	4266	126922	1310	6773	65054	928	2648	22359	17168	653	82	268	40	735	22	27	2573	15
XZHM 06 06	Guangxi	LA-ICPMS	mKA	776687	1686	141276	1593	7522	36852	1488	1651	15122	10010	288	41	763	38	500	26	152	1750	421
XZHM 06 07	Guangxi	LA-ICPMS	mKCA	876539	277	32752	2898	19666	17814	2805	1760	17942	21767	1001	183	300	50	84	90	15	1347	5
FMLM 26 2	Guangxi	pXRF	mKCA	815100	8560	162470	4780	15530	23030	5545	1190	570	5790	n.d.	11	40	36	70	107	62	1570	42
GX 4	Guangxi	pXRF	mKCA	629780	5080	134450	6810	12220	29790	1442	950	5810	9840	346	94	890	17	260	22	242	510	2402
JZLM5 10	Guangxi	pXRF	mKA	748450	4960	159520	4360	7420	50680	3004	1680	13980	15280	471	50	1805	8	460	n.d.	63	1010	35

JZLM5 17	Guangxi	pXRF	mKA	579260	3420	114740	4390	8100	24440	1084	860	35200	14180	338	140	330	25	280	36	n.d.	1360	722
JZLM5 5 AES	Guangxi	pXRF	mKA ?	792270	8990	171130	4830	9100	40710	4012	1680	14110	9450	396	69	517	48	500	12	63	3490	491
XZHM 06 01	Guangxi	pXRF	mKCA	807701	7840	169570	4690	16560	28330	5751	1240	590	5740	n.d.	17	63	61	70	113	23	540	36
XZHM 06 05	Guangxi	pXRF	mKA ?	710760	8990	224150	2950	12490	61670	2040	2130	19070	14420	575	121	180	39	660	6	n.d.	2540	39
XZHM 06 06	Guangxi	pXRF	mKA ?	789890	5710	171530	4360	10030	42450	3375	1620	14190	9300	295	49	547	46	500	8	n.d.	2840	475
XZHM 06 07	Guangxi	pXRF	mKCA	761210	4170	174630	4540	18650	22130	3240	1320	13140	15530	672	118	189	20	80	42	182	860	31
FMLM 26 2	Guangxi	Δ		1%	75%	20%	101%	7%	34%	–13%	–25%	3%	–16%	n.d.	–43%	–31%	–48%	–11%	–7%	100%	685%	636%
GX 4	Guangxi	Δ		–17%	358%	–11%	100%	–16%	–19%	–35%	–46%	–24%	–32%	8%	88%	–36%	–62%	–26%	–56%	–46%	–37%	–3%
JZLM5 10	Guangxi	Δ		1%	193%	2%	251%	42%	2%	200%	–36%	–1%	–6%	83%	67%	–22%	–68%	–6%	n.d.	118%	–28%	240%
JZLM5 17	Guangxi	Δ		–20%	105%	–26%	58%	–30%	–24%	–47%	–35%	–22%	–3%	376%	654%	–51%	–79%	–32%	–56%	n.d.	0%	–34%
JZLM5 5 AES	Guangxi	Δ		3%	434%	15%	216%	34%	8%	196%	–1%	0%	1%	51%	86%	–31%	19%	3%	–55%	–59%	93%	4%
XZHM 06 01	Guangxi	Δ		5%	19%	0%	66%	0%	65%	25%	–17%	8%	–22%	n.d.	–11%	24%	22%	–13%	–10%	31%	171%	612%
XZHM 06 05	Guangxi	Δ		–5%	111%	77%	125%	84%	–5%	120%	–20%	–15%	–16%	–12%	48%	–33%	–2%	–10%	–73%	n.d.	–1%	165%
XZHM 06 06	Guangxi	Δ		2%	239%	21%	174%	33%	15%	127%	–2%	–6%	–7%	3%	19%	–28%	20%	0%	–69%	n.d.	62%	13%
XZHM 06 07	Guangxi	Δ		–13%	1408%	433%	57%	–5%	24%	15%	–25%	–27%	–29%	–33%	–36%	–37%	–60%	–4%	–53%	1114%	–36%	501%

Table 14.2 Sample group 2 chemical compositions by LA-ICPMS and pXRF, along with relative differences (Δ) calculated as (pXRF-LA-ICPMS)/LA-ICPMS. Glass types estimated by algorithm in Lankton and Dussubieux.[18] Major elements in weight percent (%), minor and trace elements in parts per million (ppm). See text for glass type definitions. nd not detected, tr trace

Number	Source	Technique	Glass type	SiO_2	Na_2O	K_2O	MgO	CaO	Al_2O_3	TiO_2	MnO	Fe_2O_3	CoO	NiO	CuO	ZnO	Rb_2O	SrO	ZrO_2	SnO_2	PbO
KL003	Khuan Lukpat	LA-ICPMS	mNA 1	658174	165203	33422	797	12228	120652	3036	227	6945	1	6	16	23	54	469	169	5	61
KL007	Khuan Lukpat	LA-ICPMS	vNC	675655	143412	33708	34242	79405	12080	628	365	7096	6	35	9434	115	8	549	27	726	1334
KL013	Khuan Lukpat	LA-ICPMS	mNCA	750086	143779	9900	3965	43360	29916	1233	7758	7944	497	104	162	46	18	464	148	49	314
KL023	Khuan Lukpat	LA-ICPMS	mNA 1	610368	160657	33626	3595	20547	101269	4878	384	14788	10	10	166	32	59	438	786	9614	42678
KL024	Khuan Lukpat	LA-ICPMS	vNC	688655	148999	33759	30640	78652	10397	563	344	6331	3	21	18	30	9	473	20	3	4
KL029	Khuan Lukpat	LA-ICPMS	mNCA	728856	127183	12948	4921	57486	58328	2649	344	7844	3	4	20	31	25	465	472	6	11
KL030	Khuan Lukpat	LA-ICPMS	mKCA	773020	10244	144398	5269	23090	30355	1161	642	10856	18	11	39	29	136	159	53	6	11
KL042	Khuan Lukpat	LA-ICPMS	mNA 1	650773	141433	32679	20320	29543	68305	4309	2451	10476	28	97	29598	183	94	336	933	3670	5937
KL049	Khuan Lukpat	LA-ICPMS	mNA 1	633329	186684	16033	2939	19267	102778	4062	14337	16405	7	30	5169	25	24	545	741	866	348
KL050	Khuan Lukpat	LA-ICPMS	mNCA	743301	96802	14717	6506	55526	71717	2546	545	9024	6	6	14	22	24	684	405	2	15
KL003	Khuan Lukpat	pXRF	mNA	560148	215605	28910	nd	15904	91975	2894	832	4681	nd	nd	98	tr	83	424	431	nd	1178
KL007	Khuan Lukpat	pXRF	vNC	687460	147193	32428	38533	72822	15203	839	855	5775	nd	nd	9231	tr	tr	454	183	976	1803
KL013	Khuan Lukpat	pXRF	mNCA	672725	170752	10557	tr	40452	18323	1623	8872	7364	501	nd	247	tr	19	421	244	nd	1301

KL023	Khuan Lukpat	pXRF	mNA	581432	144934	22314	tr	19547	67280	nd	955	7914	nd	nd	245	tr	tr	254	360	3660	29391
KL024	Khuan Lukpat	pXRF	vNC	753817	144475	35493	31624	61880	14378	805	859	3817	nd	nd	99	tr	12	460	169	nd	1160
KL029	Khuan Lukpat	pXRF	mNCA	691527	133688	15111	nd	51204	45132	2265	868	5849	nd	nd	88	tr	44	437	404	nd	1161
KL030	Khuan Lukpat	pXRF	mKCA	783217	3000	141286	tr	23242	21255	1195	1109	7055	nd	nd	110	tr	203	118	160	nd	1148
KL042	Khuan Lukpat	pXRF	mNA	584221	147917	31504	11416	29161	51147	3880	2524	8927	nd	nd	25832	tr	72	205	452	1677	3982
KL049	Khuan Lukpat	pXRF	mNA	566004	228965	17289	nd	22492	75079	3786	16459	17900	nd	tr	5599	nd	31	441	526	1028	1315
KL050	Khuan Lukpat	pXRF	mNCA w.	670220	99073	13938	tr	45733	50858	2097	1038	6733	nd	nd	105	tr	24	569	380	nd	1165
KL003	Khuan Lukpat	Δ		−15%	31%	−13%	nd	30%	−24%	−5%	267%	−33%	nd	nd	500%	tr	53%	−9%	155%	nd	1841%
KL007	Khuan Lukpat	Δ		2%	3%	−4%	13%	−8%	26%	34%	135%	−19%	nd	nd	−2%	tr	tr	−17%	585%	34%	35%
KL013	Khuan Lukpat	Δ		−10%	19%	7%	tr	−7%	−39%	32%	14%	−7%	1%	nd	52%	tr	7%	−9%	65%	nd	315%
KL023	Khuan Lukpat	Δ		−5%	−10%	−34%	tr	−5%	−34%	nd	149%	−46%	nd	nd	48%	tr	tr	−42%	−54%	−62%	−31%
KL024	Khuan Lukpat	Δ		9%	−3%	5%	3%	−21%	38%	43%	149%	−40%	nd	nd	449%	tr	38%	−3%	733%	nd	27214%
KL029	Khuan Lukpat	Δ		−5%	5%	17%	nd	−11%	−23%	−15%	153%	−25%	nd	nd	334%	tr	76%	−6%	−14%	nd	10941%
KL030	Khuan Lukpat	Δ		1%	−71%	−2%	tr	1%	−30%	3%	73%	−35%	nd	nd	185%	tr	49%	−26%	201%	nd	10340%
KL042	Khuan Lukpat	Δ		−10%	5%	−4%	−44%	−1%	−25%	−10%	3%	−15%	nd	nd	−13%	tr	−23%	−39%	−52%	−54%	−33%
KL049	Khuan Lukpat	Δ		−11%	23%	8%	nd	17%	−27%	−7%	15%	9%	nd	tr	8%	nd	25%	−19%	−29%	19%	278%
KL050	Khuan Lukpat	Δ		−10%	2%	−5%	tr	−18%	−29%	−18%	91%	−25%	nd	nd	655%	tr	−1%	−17%	−6%	nd	7498%

Table 14.3 Sample group 3 chemical compositions by LA-ICPMS and pXRF, along with relative differences (Δ) calculated as (pXRF-LA-ICPMS)/LA-ICPMS. Glass types estimated by algorithm in Lankton and Dussubieux,[18] along with trace elements Rb_2O and SrO, with different results for TCN D 2. Major elements in weight percent (%), minor and trace elements in parts per million (ppm). See text for glass type definitions. w. weathered sample, nd not detected, tr trace

Number	Source	Technique	Glass type	SiO_2	Na_2O	K_2O	MgO	CaO	Al_2O_3	P_2O_5	TiO_2	Cr_2O_3	MnO	Fe_2O_3	CoO	Ni O	CuO	ZnO	Rb_2O	SrO	SnO_2	BaO	PbO
BKN 128	Bang Kluay	LA-ICPMS	mKA	759235	2173	145649	1775	6257	28098	1523	1564	30	23634	22659	585	76	348	59	233	39	102	3310	296
TCN 100 18	Tha Chana	LA-ICPMS	mKCA	788487	7294	154625	1872	14486	22255	1276	738	9	355	7186	19	11	14	15	153	119	26	144	23
TCN 100 22	Tha Chana	LA-ICPMS	mKCA	789058	4191	130922	1826	12045	14407	957	2232	28	20368	18030	968	94	246	48	179	52	37	2678	nd LA-ICPMS
TCN 100 5	Tha Chana	LA-ICPMS	mKA	764692	7737	142253	1397	5925	46043	1023	2712	34	646	7534	40	24	17590	39	506	14	1535	87	1717
TCN 100 7	Tha Chana	LA-ICPMS	mKA	606984	3159	168991	2701	10643	26695	1891	1108	21	1603	17489	24	58	14879	85	379	39	25793	127	132455
TCN 100 8	Tha Chana	LA-ICPMS	mKCA	692971	1527	204908	3690	24445	22877	2605	2089	56	16738	22551	1044	191	414	45	81	124	15	1630	nd LA-ICPMS
TCN 100.3	Tha Chana	LA-ICPMS	mKA	756776	4827	149150	1647	4438	43401	1653	1732	32	890	14383	24	23	22068	87	498	20	1185	150	404
TCN 16 1	Tha Chana	LA-ICPMS	mKA	677148	2797	198290	1705	5891	48631	2576	2240	38	347	51789	6	18	1761	22	673	15	5701	73	289
TCN 16 3	Tha Chana	LA-ICPMS	mKA	762155	6146	128683	1764	3887	52525	1948	3241	45	177	33382	21	15	4243	34	451	20	45	80	188
TCN 16 4	Tha Chana	LA-ICPMS	mKA	749200	11783	131825	1606	4115	55271	1259	2210	45	220	35989	24	16	4772	37	434	18	74	71	124
TCN 17 2	Tha Chana	LA-ICPMS	mKCA	705115	1243	198009	3854	21918	20543	2311	1925	56	13120	26223	1012	133	537	39	102	109	13	1723	nd LA-ICPMS
TCN 20 2	Tha Chana	LA-ICPMS	mKCA	709001	4685	185717	4432	12187	51375	1928	2638	34	9810	12572	660	91	942	59	476	53	95	682	161

(*Continued*)

Table 14.3 *(Continued)*

TCN 313 2	Tha Chana	LA-ICPMS	mKCA ?	521515	973	132975	2317	9718	22923	1822	1005	19	591	8010	19	36	22027	36	269	43	51413	93	255485
TCN 313 3	Tha Chana	LA-ICPMS	mKA	730092	1525	161748	1744	4041	71573	1119	1997	34	151	6574	5	16	17478	20	432	18	1293	78	2615
TCN 54 1	Tha Chana	LA-ICPMS	mKA	757640	5685	140932	2277	8579	44776	1300	2526	34	13050	16270	369	53	2390	45	484	42	387	1738	849
TCN 54.2	Tha Chana	LA-ICPMS	mKA	733124	4486	157459	1852	5275	53584	1739	1649	28	1123	16827	29	23	20934	33	447	22	1525	163	1273
TCN D 2	Tha Chana	LA-ICPMS	mKA	466750	4684	96314	916	2692	30350	876	1310	21	103	9429	5	24	2021	43	354	14	67048	39	356509
TCNX 01 1	Tha Chana	LA-ICPMS	mKA	747417	866	138577	1287	3060	60519	1003	1986	28	161	23495	4	18	18194	36	326	13	2182	68	2605
TCNX 01 3	Tha Chana	LA-ICPMS	mKA	766400	6926	135817	1540	3997	54917	1065	2402	30	643	6133	34	18	17168	19	484	15	1496	105	1938
TCNX 01 4	Tha Chana	LA-ICPMS	mKA	770844	999	143376	1922	5825	45919	1097	1952	28	11138	11952	306	52	776	36	347	27	153	1557	324
TCNX 04 1	Tha Chana	LA-ICPMS	mKCA	755048	5762	159433	2199	17182	15191	1372	1350	23	21317	15344	1038	104	307	34	164	85	21	2869	nd LA-ICPMS
TCNX 08 01	Tha Chana	LA-ICPMS	mKA	767687	1049	161534	1655	4341	47406	1393	2265	32	720	6753	10	8	1340	17	410	18	134	163	305
TCNX 08 05	Tha Chana	LA-ICPMS	mKA	767218	6375	143171	1059	2238	52748	1560	2544	38	9360	9473	283	47	177	28	623	20	21	1557	nd LA-ICPMS
TCNX 08 10	Tha Chana	LA-ICPMS	mKCA	797282	5292	142254	2894	13702	21962	4967	1661	15	469	6035	10	18	64	54	78	161	19	291	13
TCNX 09 5	Tha Chana	LA-ICPMS	mKA	745757	2890	166093	2298	6158	36052	1074	2019	32	15066	17705	915	113	293	40	743	41	21	1764	nd LA-ICPMS
TCNX 09 7	Tha Chana	LA-ICPMS	mKA	756907	7694	152070	1196	2877	49413	1166	2431	36	11374	10367	264	40	769	26	566	20	34	1689	nd LA-ICPMS

(Continued)

Table 14.3 (*Continued*)

Number	Source	Technique	Glass type	SiO_2	Na_2O	K_2O	MgO	CaO	Al_2O_3	P_2O_5	TiO_2	Cr_2O_3	MnO	Fe_2O_3	CoO	Ni O	CuO	ZnO	Rb_2O	SrO	SnO_2	BaO	PbO
BKN 128	Bang Kluay	pXRF	mKA	723278	2466	154847	7130	10098	39901	3721	1972	44	25796	22704	1379	186	477	134	283	27	50	6836	349
TCN 100 18	Tha Chana	pXRF	mKCA	782473	5043	143178	6450	15926	34258	3576	935	14	536	6461	280	10	21	55	226	168	14	622	63
TCN 100 22	Tha Chana	pXRF	mKCA	775602	5675	136091	5372	10188	23779	2815	1864	nd	18492	15183	989	211	285	62	188	34	27	4174	43
TCN 100 5	Tha Chana	pXRF	mKA	736844	15253	148564	8981	15116	46117	2434	970	82	474	3932	nd	28	31328	33	1084	44	415	nd	60
TCN 100 7	Tha Chana	pXRF	mKA ?	745209	10679	153711	5906	6468	51892	2451	1620	nd	810	6370	nd	2	18078	26	620	nd	1719	nd	2025
TCN 100 8	Tha Chana	pXRF	mKCA	706842	2385	150779	3713	24407	26158	nd pXRF	564	66	1539	9941	581	6	9717	37	141	18	7140	nd	72512
TCN 100.3	Tha Chana	pXRF	mKCA	730351	1584	164981	8084	21693	28358	2315	1635	nd	17018	19843	1340	226	386	26	61	120	nd	3070	22
TCN 16 1	Tha Chana	pXRF	mKA ?	673156	3236	199921	6772	12625	51189	3049	1548	nd	350	41181	nd	46	1641	61	729	nd	4560	2431	283
TCN 16 3	Tha Chana	pXRF	mKA	740395	5192	147317	5629	5539	50021	2213	1559	nd	292	35803	nd	89	4758	19	577	nd	47	2617	95
TCN 16 4	Tha Chana	pXRF	mKA	735589	9524	139437	5702	5502	58114	2244	1711	36	190	36825	nd	65	4931	49	622	nd	27	1508	46
TCN 17 2	Tha Chana	pXRF	mKCA	696850	2997	191591	9013	22774	29303	3012	1750	nd	13729	24409	1272	209	506	76	100	80	nd	3452	58
TCN 20 2	Tha Chana	pXRF	mKCA	697262	3577	178335	10264	17836	62909	3989	1740	nd	10109	10945	924	84	821	74	511	55	111	1259	203
TCN 313 2	Tha Chana	pXRF	mKCA ?	701660	8468	117293	3113	8696	16222	nd pXRF	600	nd	1014	7254	nd	39	16164	84	17	18	10798	383	137176

(*Continued*)

Table 14.3 (*Continued*)

TCN 313 3	Tha Chana	pXRF	mKA	745474	4638	143653	8643	5202	66164	1914	1416	57	322	6422	nd	8	16240	7	482	nd	1803	1439	3413
TCN 54 1	Tha Chana	pXRF	mKA ?	725334	3805	151773	6826	12759	56253	4040	2020	nd	14308	15789	926	160	2616	38	633	28	356	3329	1131
TCN 54.2	Tha Chana	pXRF	mKA	748205	4488	158769	6911	7303	34553	2129	1154	nd	1834	14409	nd	14	22267	12	455	3	2074	2937	1996
TCN D 2	Tha Chana	pXRF	mKA ?	716095	7590	70321	2222	2204	15874	nd pXRF	284	217	nd	6488	nd	27	1565	22	nd pXRF	15	12080	nd	197592
TCNX 01 1	Tha Chana	pXRF	mKA	722431	2495	161310	6546	9516	57188	2601	1414	37	263	21952	nd	56	16601	83	387	nd	1606	135	2487
TCNX 01 3	Tha Chana	pXRF	mKA	736027	5659	159856	6549	9589	56301	3005	1630	20	770	6199	nd	nd	16588	34	610	nd	1760	558	1968
TCNX 01 4	Tha Chana	pXRF	mKA	734229	5476	163331	5607	9195	50740	2822	1719	nd	11185	11821	588	93	856	75	447	20	102	2368	334
TCNX 04 1	Tha Chana	pXRF	mKCA	741995	5526	160202	6705	15420	24631	2232	1377	n.d.	20956	15579	1209	203	335	109	166	64	35	4456	27
TCNX 08 01	Tha Chana	pXRF	mKA	775132	3702	139380	6472	5226	53510	3081	1443	5	826	7101	nd	59	1469	nd	599	nd	112	2560	333
TCNX 08 05	Tha Chana	pXRF	mKA	753883	6092	148189	4089	4327	52634	2500	2003	nd	10346	10025	614	155	258	53	807	nd	38	5028	43
TCNX 08 10	Tha Chana	pXRF	mKCA	796552	5463	140163	8738	10899	25423	4136	1178	nd	570	5618	nd	56	25	42	77	147	nd	1098	34
TCNX 09 5	Tha Chana	pXRF	mKA ?	732715	2349	160897	6191	12003	42427	3599	1862	nd	15293	17187	1437	274	342	33	880	17	8	3806	22
TCNX 09 7	Tha Chana	pXRF	mKA	749430	8545	148503	6786	4457	51196	2541	1842	nd	11670	10006	542	78	786	26	782	nd	29	3821	33
BKN 128	Δ			–5%	13%	6%	302%	61%	42%	144%	26%	49%	9%	0%	136%	144%	37%	127%	21%	–32%	–50%	107%	18%

(*Continued*)

Table 14.3 (*Continued*)

Number	Source	Technique	Glass type	SiO_2	Na_2O	K_2O	MgO	CaO	Al_2O_3	P_2O_5	TiO_2	Cr_2O_3	MnO	Fe_2O_3	CoO	Ni O	CuO	ZnO	Rb_2O	SrO	SnO_2	BaO	PbO
TCN 100 18	Δ			−1%	−31%	−7%	245%	10%	54%	180%	27%	66%	51%	−10%	1344%	−10%	52%	258%	48%	41%	−45%	331%	172%
TCN 100 22	Δ			−2%	35%	4%	194%	−15%	65%	194%	−16%	nd pXRF	−9%	−16%	2%	125%	16%	30%	5%	−34%	−28%	56%	nd LA-ICPMS
TCN 100 5	Δ			−4%	97%	4%	543%	155%	0%	138%	−64%	140%	−27%	−48%	nd pXRF	14%	78%	−16%	114%	216%	−73%	nd pXRF	−97%
TCN 100 7	Δ			23%	238%	−9%	119%	−39%	94%	30%	46%	nd pXRF	−49%	−64%	nd pXRF	−97%	22%	−69%	63%	nd pXRF	−93%	nd pXRF	−98%
TCN 100 8	Δ			2%	56%	−26%	1%	0%	14%	nd pXRF	−73%	20%	−91%	−56%	−44%	−97%	2249%	−17%	73%	−85%	49118%	nd pXRF	nd LA-ICPMS
TCN 100.3	Δ			−3%	−67%	11%	391%	389%	−35%	40%	−6%	nd pXRF	1812%	38%	5425%	896%	−98%	−70%	−88%	515%	nd pXRF	1949%	−95%
TCN 16 1	Δ			−1%	16%	1%	297%	114%	5%	18%	−31%	nd pXRF	1%	−20%	nd pXRF	156%	−7%	181%	8%	nd pXRF	−20%	3230%	−2%
TCN 16 3	Δ			−3%	−16%	14%	219%	43%	−5%	14%	−52%	nd pXRF	64%	7%	nd pXRF	510%	12%	−44%	28%	nd pXRF	4%	3171%	−49%
TCN 16 4	Δ			−2%	−19%	6%	255%	34%	5%	78%	−23%	−20%	−13%	2%	nd pXRF	301%	3%	31%	43%	nd pXRF	−63%	2029%	−63%
TCN 17 2	Δ			−1%	141%	−3%	134%	4%	43%	30%	−9%	nd pXRF	5%	−7%	26%	57%	−6%	97%	−1%	−26%	nd pXRF	100%	nd LA-ICPMS
TCN 20 2	Δ			−2%	−24%	−4%	132%	46%	22%	107%	−34%	nd pXRF	3%	−13%	40%	−8%	−13%	26%	7%	4%	16%	85%	26%
TCN 313 2	Δ			35%	770%	−12%	34%	−11%	−29%	nd pXRF	−40%	nd pXRF	72%	−9%	nd pXRF	9%	−27%	134%	−94%	−59%	−79%	310%	−46%
TCN 313 3	Δ			2%	204%	−11%	396%	29%	−8%	71%	−29%	65%	113%	−2%	nd pXRF	−51%	−7%	−67%	12%	nd pXRF	39%	1743%	30%

(*Continued*)

Table 14.3 (*Continued*)

TCN 54 1	Δ			–4%	–33%	8%	200%	49%	26%	211%	–20%	nd pXRF	10%	–3%	151%	200%	9%	–16%	31%	–34%	–8%	92%	33%
TCN 54.2	Δ			2%	0%	1%	273%	38%	–36%	22%	–30%	nd pXRF	63%	–14%	nd pXRF	–37%	6%	–62%	2%	–86%	36%	1705%	57%
TCN D 2	Δ			53%	62%	–27%	143%	–18%	–48%	nd pXRF	–78%	918%	nd pXRF	–31%	nd pXRF	12%	–23%	–49%	nd pXRF	4%	–82%	nd pXRF	–45%
TCNX 01 1	Δ			–3%	188%	16%	409%	211%	–6%	159%	–29%	33%	63%	–7%	nd pXRF	212%	–9%	133%	18%	nd pXRF	–26%	100%	–5%
TCNX 01 3	Δ			–4%	–18%	18%	325%	140%	3%	182%	–32%	–33%	20%	1%	nd pXRF	nd pXRF	–3%	82%	26%	nd pXRF	18%	433%	2%
TCNX 01 4	Δ			–5%	448%	14%	192%	58%	10%	157%	–12%	nd pXRF	0%	–1%	92%	80%	10%	111%	29%	–27%	–34%	52%	3%
TCNX 04 1	Δ			–2%	–4%	0%	205%	–10%	62%	63%	2%	nd pXRF	–2%	2%	16%	96%	9%	220%	1%	–25%	68%	55%	nd LA-ICPMS
TCNX 08 01	Δ			1%	253%	–14%	291%	20%	13%	121%	–36%	–85%	15%	5%	nd pXRF	631%	10%	nd pXRF	46%	nd pXRF	–16%	1470%	9%
TCNX 08 05	Δ			–2%	–4%	4%	286%	93%	0%	60%	–21%	nd pXRF	11%	6%	117%	229%	46%	89%	29%	nd pXRF	81%	223%	nd LA-ICPMS
TCNX 08 10	Δ			0%	3%	–1%	202%	–20%	16%	–17%	–29%	nd pXRF	22%	–7%	nd pXRF	217%	–62%	–22%	0%	–9%	nd pXRF	278%	158%
TCNX 09 5	Δ			–2%	–19%	–3%	169%	95%	18%	235%	–8%	nd pXRF	2%	–3%	57%	142%	17%	–17%	19%	–57%	–62%	116%	nd LA-ICPMS
TCNX 09 7	Δ			–1%	11%	–2%	467%	55%	4%	118%	–24%	nd pXRF	3%	–3%	106%	92%	2%	–2%	38%	nd pXRF	–14%	126%	nd LA-ICPMS

These relative differences may be either positive, in which case the pXRF value was greater than the LA-ICPMS value, or negative, where the pXRF value was less. A positive difference of 50% means that the pXRF value was 1.5 times the LA-ICPMS value, while a positive difference of 100% indicates that pXRF is twice LA-ICPMS. On the negative side, a relative difference of minus 50% means that the pXRF value was half of the LA-ICPMS value, while for a relative difference of minus 70% the pXRF value was 30% of the LA-ICPMS value. The most negative possible relative difference is minus 100%, where the pXRF value is zero, while there is no upper limit to the positive relative differences.

For major elements, agreement between the two techniques for lime (CaO) and alumina (Al_2O_3) tends to be good, often within 20%, for LA-ICPMS values above 1%. However for lower CaO values the pXRF determinations are usually high, in one case up to four times the LA-ICPMS value. Potash (K_2O) values are also generally close, even with K_2O as low as 1 wt%. Among the major elements, soda (Na_2O) and magnesia (MgO) appear to have the least agreement; this picture is likely the result of very low Na_2O and MgO levels in the potash glasses, the majority of the samples, reported here. For Na_2O values above a few percent the agreement was excellent, as was the agreement for MgO above 1 wt%. For minor oxides present between 0.1 and 1 wt% (1000 to 10000 ppm), there is good agreement, generally within 50%, for titanium (TiO_2) and iron (Fe_2O_3). For the colorants manganese (MnO), cobalt (CoO) and copper (CuO) at levels of 0.05 wt% (500 ppm), pXRF appears to be a useful indicator, with many of the relative differences below 50%. Rubidium (Rb_2O) and strontium (SrO) are both very useful for glass classification (see below), and pXRF agreement is quite good for both oxides at levels at least as low as 100 ppm (0.01 wt%). For the heavier oxides tin (SnO_2), barium (BaO) and lead (PbO), the samples we compared do not provide a very good test. Most of the values are quite low, so a lack of agreement is not surprising.

Two methods for a graphical comparison of the results are plots of pXRF vs LA-ICPMS values for the same oxides, and plots of the

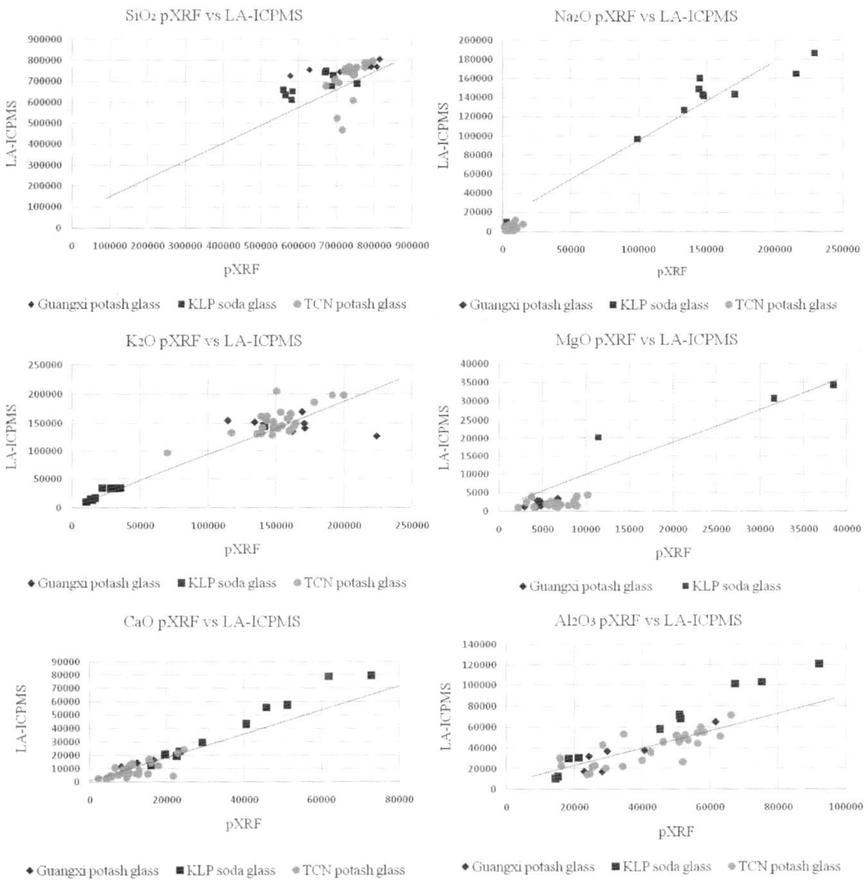

Fig. 14.2 pXRF and LA-ICPMS values for major oxides for glasses from Guangxi, Khuan Lukpad (KLP) and Tha Chana (TCN). Perfect correlation would fall along the lines provided. All values in parts per million (ppm), where 10000 ppm equals 1 weight percent (wt%)

relative differences vs either pXRF or LA-ICPMS. Fig. 14.2 shows pXRF *vs* LA-ICPMS for major oxides, while Fig. 14.3 presents relative difference *vs* pXRF for selected major, minor and trace oxides. Fig. 14.2 includes reference lines that would indicate perfect agreement.

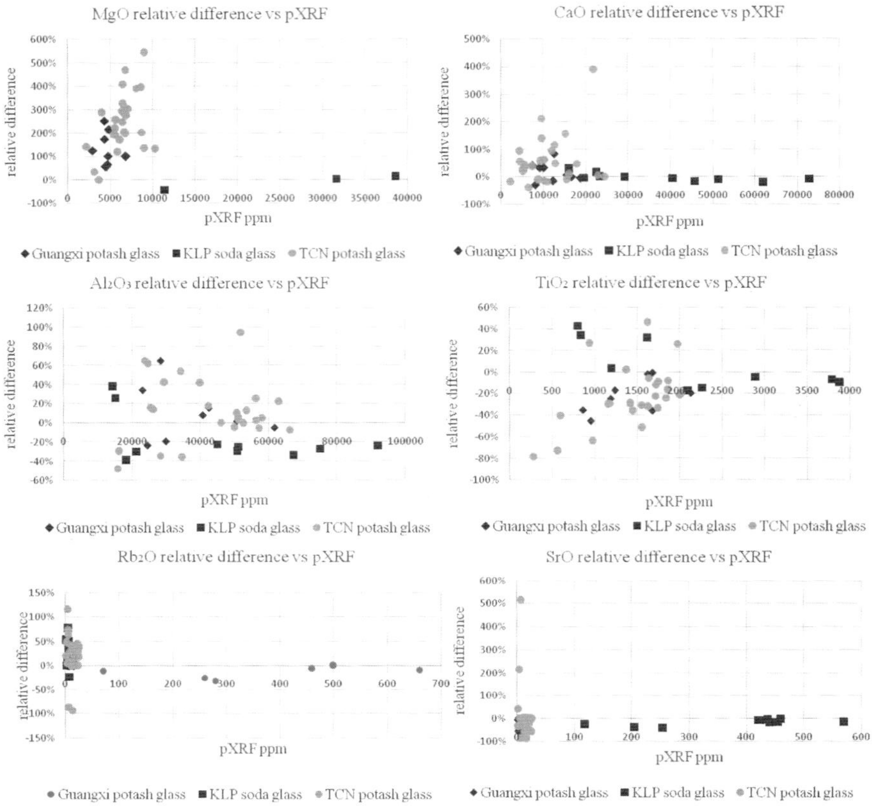

Fig. 14.3 pXRF values and associated relative differences calculated as (pXRF-LA-ICPMS)/LA-ICPMS for selected major (>1 wt%), minor (0.1-1 wt%) and trace elements (<0.1 wt%). pXRF values in parts per million (ppm) where 10000 ppm equals 1 wt% and 100 ppm equals 0.01 wt%). See Figure 14.2 for abbreviations and text for further explanation.

14.5 Discussion and interpretation

While comparison studies such as ours can be very useful, the study itself will not tell us which of the two values is more likely to be correct. For this, we would need results from the two techniques for the oxides of interest for the same glass standards of known composition, ideally over the full range of values observed in the archaeological samples. This would allow us to estimate both trueness and precision for the two techniques. This data is not yet available, so that for the interpretation of the comparisons for actual archaeological samples, it may be necessary to assume that one technique or the other is more likely to be accurate, in terms of greater trueness and/or greater precision. We do have good data for LA-ICPMS, over a wide range of values, indicating trueness in the range of 5–10% or less, and seldom exceeding 20%, for most oxides. Precision for LA-ICPMS will depend on how many spots are measured and the actual procedures used for the measurements, but with our system in Orleans, focused more on rapid throughput than high precision, the precision values are usually in the range of 10–20%, depending on the oxide measured. Based on published studies, pXRF measurements of trueness and precision on standard glasses appear to be in the same range for most oxides present in levels greater than 1–2 wt%.[20] Detection levels will also vary by element, with some elements easier to measure by pXRF than others. In our study, we noted significant issues with pXRF values for MgO when the LA-ICPMS measurement was less than 1 wt%, and for CaO at LA-ICPMS values below 1–1.5 wt%, as indicated in Fig. 14.3. On the other hand, pXRF values for TiO_2, Rb_2O and SrO agreed well with LA-ICPMS down to levels of 2000 ppm, 50 ppm and 100 ppm, respectively. In the absence of direct comparison studies on the same standard glasses, it seems reasonable to conclude that, based on the relative difference plots in Fig. 14.3, at oxide levels above 2 wt% the difference between pXRF and LA-ICPMS for these samples may be within the estimated total uncertainties of the two techniques. For oxides present at actual levels below 2 wt%, and particularly below 1 wt%, it is very likely that the LA-ICPMS values more accurately reflect the actual value. This conclusion is in line with those of Tantrakarn *et al.*,[20] who suggested pXRF detection limits for Na_2O of 1.2 wt% and for MgO 0.3 wt%. The graphs in Fig. 14.3 are organized from the point of view of the pXRF operator. From the plots of relative difference *vs* pXRF measurement, we can observe, for example, that if the observed pXRF value for CaO is below about 1.5 wt%, there is a good chance that this value is between 50% and 150% higher than the value measured by LA-ICPMS. In contrast, for Rb_2O and SrO, the pXRF values above 100 ppm tend to be a bit low, compared to LA-ICPMS. However, even very low (less than 25–50 ppm) measurements for these oxides may be useful; for RbO these low values are usually overestimates of the LA-ICPMS values, while for SrO the pXRF values are generally much lower than LA-ICPMS.

14.6 Questioning the answers

Most important for the archaeologist is not how close are the pXRF and LA-ICPMS measurements, but rather do both datasets lead to the same answers to the questions of interest. A fundamental question necessary for answering any others is 'What are the general glass types of these samples?' In answer to this question, in Tables 14.1–14.3 we have estimated the glass type for each samples, based on the analytical data provided by the particular technique, and following the general guidelines suggested in Lankton and Dussubieux.[18] The Group 1 and Group 3 samples are all potash glasses, and we have identified these as either mineral (m) potash (K_2O) glass moderate in lime (CaO) and alumina (Al_2O_3), mKCA, or mineral potash glass low in lime (usually below 1 wt%) and relatively high in alumina (usually close to 4 wt% or higher), mKA. There were no examples of the third major group of potash glasses, very low in alumina and relatively high in lime. While major oxides are very useful in making these distinctions, when CaO levels are between 0.8 and 1.2 wt% the two groups may overlap based on major elements. Here, trace elements make an important contribution, since mKA glasses are usually very high in rubidium (Rb_2O) and low in strontium (SrO), while mKCA glasses are moderate in both. In our comparison study, the two techniques led to the same glass type identification for all but one of the potash glasses, even though there was a bit more question with some of the pXRF examples, usually because of relative overestimate of low CaO values by pXRF. In particular, the pXRF RbO and SrO measurements were very helpful. Our conclusion from this study would be that in terms of identifying the basic type of potash glass, pXRF does an excellent job. However, from our data here, it is not possible to say whether answers to other questions, such as whether or not it is possible to identify subgroups within these major groups, would be the same for the two techniques.

For the nine soda glasses in Group 2, three important compositional groups would be mineral soda (Na_2O) glass with relatively low lime and high alumina, mNA, mineral soda glass with moderate lime and alumina, mNCA, and plant-ash (v) soda glass with high lime and low alumina, vNC. Based only on major elements, these identifications are similar for the two techniques. However, we now know that there are several sub-groups of high-alumina soda glass, and that these sub-groups indicate very different production zones.[19] For the mineral soda glasses with moderate lime and alumina, a further division into possibly Roman glass and Asian glass is equally important. Unfortunately, the trace elements necessary to make these distinctions, uranium and cesium, are present in levels not detectable by pXRF. Even for zirconium, also useful for sub-groups and measured by pXRF for the Group 2 samples, the pXRF results do not seem to be very reliable. This inability of pXRF to measure particularly uranium makes the pXRF results for mineral soda glasses much less successful. Whether or not there are substitutes for uranium that might be measured by pXRF remains to be studied. On the other hand, the pXRF measurements do provide a means to distinguish mineral soda from plant-ash soda glasses, and this is important for many archaeological questions. During the 9th c. in the Mediterranean area the soda glass compositions changed from mineral soda (natron) glass to plant-ash soda glass, and pXRF measurements have been very useful to mark these changes. [23]

14.7 Summary and conclusions

This chapter is a hybrid between theoretical considerations, reviewing past studies and providing data from a new study. The most important point that we want to make is that the answers we need, particularly from the point of view of scientific analysis, will depend heavily on the archaeological questions being asked. Analytical techniques that are useful for one question, for example what type of potash glass, may not be useful for another, such as source areas for soda glasses found in Asia. At the same time, one technique may supplement or make another more useful. A future example here might be that now that we can separate sub-groups of high-alumina soda glass with some confidence, based on trace elements available only from LA-ICPMS, can we then find a way to use pXRF data to get some approximation of the same groupings? Only by understanding how measurements by LA-ICPMS and pXRF are likely to compare, in the way studied here, will we make progress in this important work.

In her dying words, Gertrude Stein, a celebrated American writer living in France during the first half of the 20th century, asked her companions 'What is the answer?' When no one replied, she said 'In that case, what is the question?'[24] As field archaeologists or laboratory archaeological scientists, we would all do well to remember this.

References

1. I.C. Freestone, Glass production in Late Antiquity and the Early Islamic period: a geochemical perspective, in *Geomaterials in Cultural Heritage*, eds. Maggetti, M. and Messiga, B. (Geological Society, London, 2006, Special Publications 257), pp. 201-216.
2. E.V. Sayre, and R.W. Smith, *Science* **133**, 1824 (1961).
3. I.C. Freestone, Y. Gorin-Rosen, M.J. Hughes, *La route du verre, Travaux de la Maison de l'Orient méditerranéen* **33**, 65 (2000).
4. Ch. Avronidaki, The Glassware, in *The Shipwreck off Antikythera Catalogue of the Exhibition*, ed. N. Kaltsas (National Archaeological Museum, Athens, 2012).
5. R. Muthucumarana, A.S. Gaur, W.M. Chandraratne *et al.*, *Journal of Maritime Archaeology* **9**, 41 (2014).
6. P. Degryse, J. Henderson, G.Hodgins, *Isotopes in Vitreous Materials*, Leuven University Press, Leuven, 2009.
7. P. Jr. Francis, *Asia's Maritime Bead Trade. 300 B.C. to the Present*, University of Hawai'I Press, Honolulu, 2002.
8. C.M. Jackson, D. Greenfield, L.A. Howie, *Archaeometry* **54**, 489 (2012).
9. L. Dussubieux, P. Robertshaw, M.D. Glascock, *International Journal of Mass Spectrometry* **284**, 152 (2009).
10. I. Nakai, S. Yamada, Y. Terada *et al.*, *X-ray Spectrometry* **34**, 46 (2005).
11. B. Gratuze, A. Giovagnoli, J.-N. Barrandon *et al.*, *Revue d'Archéométrie* **17**, 89 (1993).
12. J.W. Lankton, B. Gratuze, A. Cholakova *et al.*, Glass chemical analysis: Assessing the new heterarchy, oral presentation of preliminary results at the International Symposium on Archaeometry, Los Angeles, May 23, Manuscript in preparation. (2014).
13. E.V. Sayre, *Brookhaven National Laboratory, Atomic Energy Commission USA Reports*, Brookhaven National Laboratory, Upton, NY, 1964, pp. 1–25.

14. R.H. Brill, *Chemical Analyses of Early Glasses, Vol. 1*, The Corning Museum of Glass, Corning, NY, 1999, pp. 280.
15. G.F. Bass, D.A. Frey, C. Pulak, *International Journal of Nautical Archaeology and Underwater Exploration* **13**, 271 (1984).
16. A.J. Shortland, N. Rogers, K. Eremin, *Journal of Archaeological Science* **34**, 781 (2007).
17. C.M. Jackson, and P.T. Nicholson, *Journal of Archaeological Science* **37**, 295 (2010).
18. J.W. Lankton, and L. Dussubieux, *Journal of Glass Studies* **48**, 121 (2006).
19. L. Dussubieux, B. Gratuze, M. Blet-Lemarquand, *Journal of Archaeological Science* **37**, 1648 (2010).
20. K. Tantrakarn, N. Kato, I. Nakai, *Archaeometry* **54,** 508 (2012).
21. S. Liu, Q.H. Li, F. Gan *et al.*, *Journal of Archaeological Science* **39**, 2128 (2012).
22. B. Gratuze, Application de la spectrométrie de masse à plasma avec prélèvement par ablation laser (LA-ICP-MS) à l'étude des recettes de fabrication et de la circulation des verres anciens, in *Circulation des matériaux et des objets dans les sociétés anciennes*, eds. Dillmann, Ph. and Bellot-Gurlet, L. (Collection Sciences Archéologiques, Éditons Archives Contemporaines, Paris, 2014), pp. 259–291.
23. N. Kato, I. Nakai, Y. Shindo, *Journal of Archaeological Science* **36**, 1698 (2009).
24. A.B. Toklas, *What is Remembered*, Holt, Rinehart & Winston, New York, 1963, pp. 173.

Chapter 15

第十五章

Scientific Research on the Earliest Chinese Glazed Pottery

Junqing Dong, Fuxi Gan,* Xiaowei Xia,† Qinghui Li,* Song Liu* and DonghongGu**

*Shanghai Institute of Optics and Fine Mechanics, Chinese Academy of Sciences, Shanghai, 201800, China

†Hongshan Museum, Wuxi, Jiangsu Province, 214145, China

15.1 Introduction

Glazed pottery, more popular in the ancient world be it East or West, is a kind of pottery covered with glaze for decoration and protection. It can be divided into two types: the low-temperature glazed pottery and the high-temperature glazed pottery, both of them used in ancient Western countries. However, in ancient China, the low-temperature glazed pottery was the common type. Such as the burial pottery of the Han Dynasty, tri-colored glazed pottery of Tang Dynasty and architectural glazed pottery of the Ming and Qing Dynasties.

Since the 1950s, the low-temperature glazed pottery has become one of the important focuses of research in China and foreign countries. The chemical composition analysis was commonly used in studying Chinese low-temperature glazed pottery from the 1980s to 1990s. However, the analyzed objects were mostly the tri-colored glazed pottery of the Tang Dynasty, and a few being related to the early glazed pottery. After the 1990s, Zhang Fukang and Li Jiazhi summed up the flux systems of glazed potteries of successive dynasties in China,[1,2] and suggested a view that lead glaze is an independent innovation started from the Han Dynasty.

The low-temperature glazed pottery started from the Western Han Dynasty belongs to a type of high-lead-glazed pottery. It developed and prospered in the Eastern Han Dynasty. After a period of much development in the Six Dynasties Period, the tri-color glazed pottery appeared — one of the most famous of Chinese low-temperature glazed potteries — and reached its peak in the Tang Dynasty. In fact, during the late Spring and Autumn and Warring States Periods in ancient China, the handicraft industry reached an unprecedented high. In particular, the ceramic industry, the glass and metallurgy industry entered a new phase. The low-temperature glazed pottery industry also emerged around then. To date only a few glazed pottery vessels of the Warring States Period (Table 15.1) were reported, such as the pottery jars and pottery eye beads inlaid with "dragonfly eye" conserved in museums outside of China: the British Museum in England, the Nelson-Atkins Museum of Art, the Boston's Museum of Fine Arts in the United States, and the Tokyo National Museum (TNM) in Japan.

In view of its precious and fragile nature, destructive sampling analysis is not applicable to the early glazed pottery. So, only a non-destructive analysis *in situ* can be used. As a result, most of the ancient glazed potteries were not tested, except for a few archaeological artifacts. In order to study the origin and chemical composition system of the low-temperature glazed pottery, the glazed potteries of the Warring States Period unearthed from Wuxi, Jiangsu Province, China have been non-destructively analyzed *in situ* by us using portable energy dispersive X-ray fluorescence spectrometer (pXRF).

15.2 Samples

Among some glazed potteries excavated from the noble burial tomb of Yue State at Qiuchengdun site in Hongshan, Wuxi, Jiangsu, China, there are four color glazed potteries called "color glazed ceramic openwork snake balls" (hereinafter the "color glazed balls") by the

Table 15.1 Profile of several earlier Chinese low-temperature glazed potteries of the Warring States Period

Name	Number	Museum or collector	Time	Unearthed site	References
Transparent greenish-brown glazed pottery vessel	1	Nelson-Atkins Museum of Art in Kansas, American	Late Warring States Period	Jincun near to Luoyang in Henan	[3]
"Glass paste" jar	1	Nelson-Atkins Museum of Art in Kansas, American	Late Warring States Period		[3]
Glazed pottery jar	1	Boston's Museum of Fine Arts in American	Warring States Period		[3]
Glazed pottery jar	1	British Museum in England	Late Warring States Period	Xunxian in Henan	[3]
"Glass paste" ornamented pottery jar	1	Tokyo National Museum in Japan			[3]
"Polychrome glazes faience bead"	1	Nigel Wood, England	Warring States Period		[4]
"Gaience eye beads"	4	Japan	Warring States Period		[4]
Glazed pottery eye bead	1	Henan Province Antique Shop	Warring States Period		
Glazed pottery huang	1	Hubei Museum	Early Warring States Period	Jiudian, Jiangling, Hubei	[5]
Glazed eye bead with red body and white glaze	1	Hubei Museum	Middle and Late Warring States Period	Jiudian, Jiangling, Hubei	[5]
Glazed eye bead with grey body and blue glaze	1	Hubei Museum		Jiudian, Jiangling, Hubei	[5]
Multi-coloured glass eye beads with pottery core	41	Simon Kwan, Hong Kong	Warring States Period		[6]
Multi-coloured glass eye beads with pottery cores	5	Simon Kwan, Hong Kong	Warring States Period		[6]

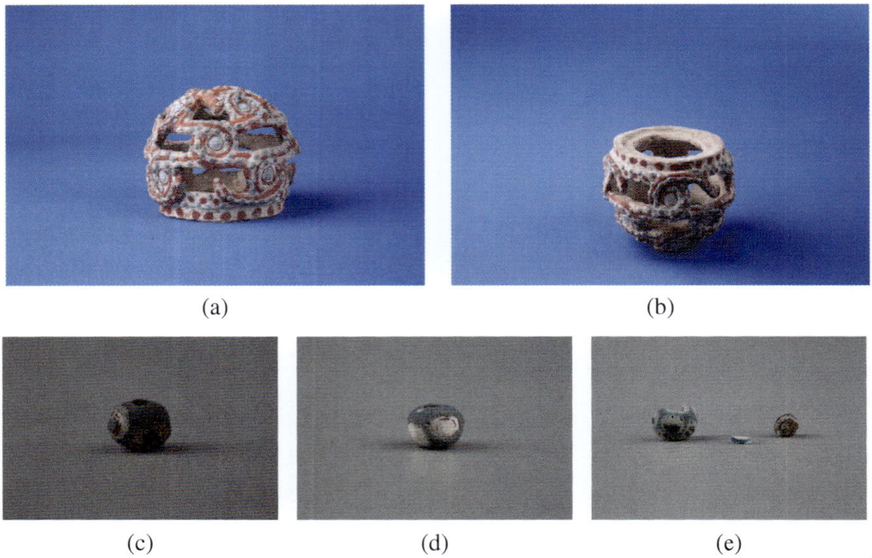

Fig. 15.1 Photographs of the glazed potteries of the Warring States Period unearthed from Qiuchengdun, Hongshan, Wuxi, Jingsu Province (a, b: the front view and side view of color glazed pottery ball of sample HSB-1; c:glazed pottery eye bead of sample HSB 147-1; d: glazed pottery eye bead of sample HSB 147-2; e: glazed pottery eye bead of sample HSB 147-5)

excavation report.[7] As the rarest ritual vessels of Yue State, they are regarded as artifacts symbolizing monarchy or theocracy. The color glazed ball is of hollow spherical shape with grayish clay body and low ring foot. It is composed of eight-snake patterns coiled into a circle to form a ball and one snake biting another. The head and body of snake are colored with blue glaze spots, spaced with red color.[7] In this study, one of the color glazed balls and three glazed pottery eye beads in Hongshan Museum, Wuxi (Fig. 15.1), are non-destructively *in situ* analyzed. The analyzed results are listed in Table 15.2.

15.3 Method

The pXRF spectrometer used in this analysis is an OURSTEX 100FA portable energy dispersive XRF spectrometer made by OURSTEX

Table 15.2 Descriptions of glazed potteries of Warring States Period unearthed from Qiuchengdun, Hongshan, Wuxi, Jiangsu Province

Sample No.	Name	Description
HSB-1	Color glazed pottery	grayish clay body, the surface decorated blue punctiform glaze alternate with red color
HSB 147-1	Glazed pottery eye bead	The inner is white sandwich layer, and the surface is glazed green glass glaze. The eye colored brown, white and green and lack of eyeballs.
HSB 147-2	Glazed pottery eye bead	The inner is white sandwich layer, and the surface is glazed green glass glaze. The eyeballs are lost.
HSB 147-5	Glazed pottery eye bead	The sample is incomplete and badly weathered. The colors of the eye are white and green.

Co., Ltd. Japan, and modified by the Scientific Archeology Center, Shanghai Institute of Optics and Fine Mechanics (SIOM), Chinese Academy of Sciences (CAS). The pXRF spectrometer uses a palladium (Pd) excitation tube with voltage of 40 kV, and current of 0.5 or 1.0 mA. The maximum power is 50 W. The x-ray beam spot focused on the sample is about 2.5 mm in diameter. The instrument is composed of four separate parts: an X-ray spectrometer, a compact digital processing unit controlled by a microcomputer, a compact power unit including high-voltage transformer and power supply and a vacuum sample chamber. The X-ray spectrometer mainly provides the high voltage for x-ray source. The compact digital processing unit consists of a control software and qualitative and quantitative analysis software. A silicon drift detector (SDD) is specially designed for light element detection. The spectral resolution is 145 eV. The detector in air accurately and quantitatively analyzes the elements after Al, but it is not good for determining light elements (Na and Mg). However, its advantage is being applicable to any sizes of samples, which gives us more freedom to select the measured area. In order to lower down the effect of atmospheric absorption on the characteristic spectrum lines of light elements, a vacuum pump is equipped in this spectrometer. The lowest air pressure is 400~600 Pa.[8] Based on the size of sample non-destructively *in situ* analysis is performed in air or in vacuum.

15.4 Results

The color glazed ball is one of the first antique rarity samples analyzed non-destructively *in situ* by pXRF in air. The results show in Table 15.3. The major component SiO_2 content varies from 46.91%~63.58%, while the minor element oxide Al_2O_3 content is from 5.52% to 11.56%. In the blue-green areas (test point a and point b), the fluxes are lead oxide and barium oxide, the PbO content is 6.78% and 6.30%, and the BaO content is 10.37% and 3.84% respectively. There is also a certain amount of the K_2O (6.21% and 5.43%). In addition, the total content of Fe_2O_3 and CuO (Fe_2O_3+CuO) is about 5%. The blue-green coloration of the color glazed ball was associated with the ions of Fe and Cu, but may be also due to the Chinese blue ($BaCuSi_2O_6$), which requires further research. The white bottom (test point c) contains high PbO and BaO, with a content of 24.59% and 5.86% respectively. The major components of the red-white area on the head of a snake (test point d) are PbO, BaO and Fe_2O_3, their contents being 26.49%, 5.56% and 8.20% respectively. Comparing the results of K_2O content with that in blue-green areas, it can be noticed that the K_2O content in the white bottom and the red-white area on the head of snake is very low, less than 0.5%, whereas in the blue-green areas is much higher, above 5%.

Three small samples of glazed pottery eye beads are measured in low vacuum. The results are shown in Table 15.4. High contents of PbO and BaO are detected ranging from 2.80% to 16.63% and 11.21% to 26.60%, respectively. The content of PbO in the white core

Table 15.3 Major element of color painting pottery unearthed from Qiuchengdun site, Hongshan, Wuxi determined by pXRF technique in air (wt.%)

Sample	Test poiont	Al_2O_3	SiO_2	Cl	K_2O	CaO	Cr_2O_3	Fe_2O_3	CuO	ZnO	BaO	PbO
HSB-1a	Body of snake: blue-green	7.35	63.58	0.17	6.21	4.57	0.09	0.64	4.26	n.d.	6.30	6.78
HSB-1b	Body of snake: blue-green	11.56	61.79	0.12	5.43	1.84	0.03	3.29	1.70	n.d.	3.84	10.37
HSB-1c	White bottom	5.52	52.34	n.d.	0.44	8.79	0.21	2.03	n.d.	0.05	5.86	24.59
HSB-1d	Head of snake: red-white	7.73	46.91	0.18	0.22	4.63	n.d.	8.20	0.07	0.03	5.56	26.49

Table 15.4 Major element of glazed pottery eye beads unearthed from Qiuchengdun site, Hongshan, Wuxi determined by pXRF technique in low vacuum (wt.%)

Sample	Test point	Na_2O	MgO	Al_2O_3	SiO_2	P_2O_5	K_2O	CaO	TiO_2	Cr_2O_3	MnO	Fe_2O_3	NiO_2	CuO	ZnO	Rb_2O	SrO	SnO_2	BaO	PbO
HSB 147-1a	Green matrix	2.32	1.02	6.05	65.33	0.00	0.77	1.02	n.d.	0.06	n.d.	0.92	0.03	2.17	n.d.	n.d.	0.20	n.d.	13.71	6.40
HSB 147-1b	Green eyeball	2.33	1.06	7.47	60.65	0.00	1.26	0.93	n.d.	0.05	n.d.	3.46	0.03	1.51	n.d.	n.d.	0.17	n.d.	18.27	2.80
HSB 147-1c	White	1.23	0.79	10.89	54.34	0.46	2.91	0.39	n.d.	0.06	n.d.	2.29	0.02	0.02	n.d.	n.d.	0.13	n.d.	12.38	14.10
HSB 147-2a	White 1	1.36	0.47	8.77	57.42	1.22	2.99	0.16	n.d.	0.05	n.d.	1.10	0.03	0.00	n.d.	n.d.	0.13	n.d.	11.71	14.61
HSB 147-2b	White 2	1.40	0.53	9.54	54.33	1.73	3.08	0.27	n.d.	0.06	0.08	0.83	0.02	0.16	n.d.	n.d.	0.13	n.d.	11.21	16.63
HSB 147-2c	Green	3.02	1.10	4.64	65.50	0.00	0.55	1.28	n.d.	0.08	n.d.	0.87	0.02	2.42	n.d.	n.d.	0.21	n.d.	13.03	7.29
HSB 147-5a	Inner body	1.09	1.78	14.12	71.92	0.63	2.20	0.97	1.21	n.d.	0.09	5.76	0.04	0.00	0.01	0.02	0.01	n.d.	n.d.	0.14
HSB 147-5b	Blue	2.62	0.65	8.07	53.32	0.00	0.94	0.83	n.d.	0.12	n.d.	0.52	0.02	2.12	n.d.	n.d.	0.22	0.39	19.19	10.98
HSB 147-5c	Green	2.41	0.72	8.69	52.76	0.00	0.83	0.49	n.d.	0.15	n.d.	0.45	0.03	1.28	n.d.	n.d.	0.27	0.34	26.60	4.99
HSB 147-5d	White	2.09	0.72	7.74	50.47	0.00	0.87	0.32	n.d.	0.11	n.d.	0.57	0.03	0.40	n.d.	n.d.	0.21	0.92	20.00	15.56

is similar with that in the glaze layer surface, while the content of BaO in the white core is higher than that in the glaze layer surface, which may be due to the fact that the barium oxide was reduced in the weathered surface of the glaze layer. The presence of copper ion and iron ion contributes to the blue, blue-green and green colors of glazed pottery eye beads, and the different relative proportions of the two kinds of ions make the colors of glazed eye beads different.

Furthermore, for the sake of comparison, the Scientific Archaeology Center, cooperating with Xi'an Institute of Archaeology and Relics Preservation, performed a non-destructively *in situ* test to determine the chemical composition of some glazed potteries of the Western Han Dynasty unearthed from Xi'an by pXRF in air. The analysis results of partial samples are shown in Table 15.5. The results suggest that they belong to low-temperature lead-silicate system glazed pottery. The major compositions in the glazed layer are PbO and SiO_2. The PbO contents are 61.17%, 63.91% and 49.66% respectively, while the SiO_2contentsare 36.31%, 34.37% and 46.46% respectively. There are nearly free of alkalis (K_2O and Na_2O), have very low calcium oxide (less than 1.1%) and contain a small quantity of coloring agents like CuO and Fe_2O_3. The contents of CuO and Fe_2O_3 vary in different color glazes. For example, the content CuO is higher in green glaze than red glaze and brown glaze, the former containing 1.25% CuO, while the red and brown glaze are free of CuO. The Fe_2O_3 contents are 0.76%, 1.18% and 1.69% in these three color glazes, respectively.

15.5 Discussion

15.5.1 *The characteristic of low-temperature glazed pottery of the Warring States Period*

The color glazed ball and glazed pottery eye beads with lead-barium glaze found in Wuxi, Jiangsu are the extremely precious ancient glazed potteries in China. The Scientific Archeology Center, SIOM, CAS has discovered a pottery *huang* coated with light blue lead-barium glaze dating to the early Warring States Period excavated from Wangshan, Jiangling, Hubei and two lead-barium-glazed pottery eye

Table 15.5 Major element of glazed potteries unearthed from Xi'an, Shaanxi determined by pXRF technique in air (wt. %)

Sample	Name	Tomb	Test point	Al_2O_3	SiO_2	P_2O_5	K_2O	CaO	TiO_2	MnO	Fe_2O_3	CuO	BaO	PbO
Xi'an-41	Barn	Zhujiang II M3:2	Abdomen: green glaze	n.d.	36.31	n.d.	n.d.	0.39	n.d.	0.12	0.76	1.25	n.d.	61.17
Xi'an-42	Pot	Shenguangwaibao M62: 13	Abdomen: red glaze	n.d.	34.37	n.d.	n.d.	0.22	0.20	0.12	1.18	n.d.	n.d.	63.91
Xi'an-46	Jar	Zhujiang New Town II M1: 3	Shoulder: brown glaze	0.73	46.46	n.d	0.06	1.08	0.33	n.d.	1.69	n.d.	n.d.	49.66

HBWKI-2-A, glazed pottery huang HBWKI-28, red-white glazed eye bead HBWKI-35, blue glazed eye bead

Fig. 15.2 Glazed potteries of the Warring States Period unearthed from Jiangling district, Hubei

beads of the middle and late Warring States Period from Jiudian, Jiangling.[5] Their photographs show in Fig. 15.2.

The chemical composition of the glazed potteries of the Warring States Period unearthed from Jiangling district, Hubei is given in Table 15.6.[3] It can be seen from the table that the major chemical composition of the deep yellow matrix in the pottery body of ***huang*** (HBWKI-2-A) are SiO_2 (70.06%) and Al_2O_3 (16.01%), while the minor chemical components are Na_2O (1.87%), K_2O (2.72%), Fe_2O_3 (3.51%) and PbO (1.11%). The major chemical component of the blue glaze are SiO_2 (49.59%), BaO (13.48%), PbO (12.25%), while the minor chemical composition are Al_2O_3 (9.63%), CuO (7.72%), Na_2O (below 0.5%) and K_2O (below 0.5%). In the strict sense, this glaze is only a layer of lead-barium silicate.

The contents of PbO and BaO in the glazed pottery eye beads uncovered from tombs of the Warring States Period in Jiudian, Jiangling are relatively low (less than 10%). The major chemical compositions of the white in sample HBWKI-28 (dating to the middle and late Warring States Period) are SiO_2 (74.31%), BaO (6.19%) and PbO (10.61%), which belongs to lead-barium silicate layer. In the uncovered yellow inner matrix, the major chemical components are SiO_2 (64.52%), Al_2O_3 (15.78%) and Fe_2O_3 (10.68%), while the contents of Na_2O and K_2O both are less than 2%, which belongs to pottery body. The major chemical components of the yellow section of sample HBWKI-35 (dating to the late Warring States Period) are

Table 15.6 Major element of glazed potteries unearthed from Jiangling district, Hubei determined by PIXE technique [3] (wt.%)

Sample	Name	Time	Tomb	Test point	Na_2O	MgO	Al_2O_3	SiO_2	P_2O_5	K_2O	CaO	TiO_2	Cr_2O_3	MnO	Fe_2O_3	CuO	ZnO	BaO	PbO
HBWKI-2-A	Glazed pottery Huang slice	Early Warring States Period	Wangshan M2	Edge: green	0.48	0.86	9.63	49.59	3.93	0.44	0.73	0.00	0.00	0.02	0.77	7.72	0.09	13.48	12.25
				Middle: pit	1.87	1.58	16.01	70.06	0.88	2.72	0.66	1.19	0.04	0.00	3.51	0.04	0.00	0.00	1.11
HBWKI-28	Glazed pottery eye bead	Middle and late Warring States Period	Brick tomb 869:7, Jiudian	White glaze layer	0.16	0.63	4.98	74.31	1.11	0.38	0.34	0.00	0.00	0.00	0.94	0.35	0.00	6.19	10.61
				Yellow matrix	1.69	2.62	15.78	64.52	0.59	1.64	0.28	1.35	0.02	0.05	10.68	0.00	0.00	0.00	0.33
HBWI-35	Glazed pottery eye bead	Middle and late Warring States Period	Brick tomb 903:4, Jiudian	Green	3.41	2.95	12.02	57.65	1.29	1.69	1.44	0.00	0.12	0.00	1.33	4.45	0.00	9.92	3.73

SiO_2 (47.54%), Al_2O_3 (17.38%), BaO (4.03%), PbO (8.37%), and the other chemical components are MgO (3.27%), CaO (4.60%), Na_2O (1.48%), K_2O (2.64%) and Fe_2O_3 (3.44%). The above mentioned two parts of glaze layer belong to lead-barium silicate system.

The body surface of the color glazed ball and glazed pottery eye beads unearthed from Wuxi, Jiangsu is covered with the lead-barium silicate glaze, which is similar with the glazed pottery *huang* and glazed pottery eye beads excavated from Jiangling, Hubei. The glazed potteries unearthed from Jiangling district, Hubei share the common feature: low magnesium, sodium and calcium oxides and high PbO (1.11%~16.63%) and BaO (6.11%~26.60%). The purpose of applying lead oxide is to lower the melting temperature and make the glass not prone to devitrify. This is an innovation of Chinese forbears, based on their intimate understanding of the relationship between the chemical composition and property of glass. Application of barium oxide could also drop down the melting temperature effectively, but as a result, devitrification could easily occur in glass, and make the glass become ivory colored. However, it could be used for imitating jade. [9] Compared to the early lead-barium silicate glass dated from the Warring States Period to the Han Dynasty, the eight glazed potteries from Wuxi, Jiangsu and Jiangling, Hubei have much lower PbO and BaO and higher Al_2O_3. The content of PbO ranges from 20% to 50% and the BaO varies from 10% to 20%. The preliminary conclusion is that such a difference may be caused by the different manufacturing technology and raw material at that time.

It seems that the initial glazed pottery were just painted raw glazes, pots or layer on surfaces. The technique of making glazed pottery eye bead should have been influenced by Western sodium-calcium silicate glass eye bead and developed from imitating that shape. The "dragonfly eye bead" was one of the most typical beads of the Spring and Autumn and Warring States Periods in China.

As we know, the early lead-barium silicate glass has been a feature of ancient Chinese glass. [10] The low-temperature glazed pottery earlier than the Warring States Period (476 B.C.) has not been found abroad. Furthermore, the low-temperature lead-barium glazed artifacts found overseas are all of Chinese origin.[11] We can therefore conclude that the low-temperature lead-barium glazed potteries of the Warring States Period unearthed from Wuxi were homemade.

15.5.2 *The chemical composition system and possible sources*

It was generally acknowledged that the origin of Chinese low-temperature glazed pottery was from the Eastern Han Dynasty. Lead glaze, in which the PbO content ranges from 45% to 60%, without BaO and nearly free of K_2O, Na_2O and other fluxing agents, is a notable feature of low-temperature colored glazes of successive dynasties after the Han Dynasty in ancient China.[1] The results of this study indicate that Chinese low-temperature glazed pottery can be traceable to the early Warring States Period, and the glaze layers of them belong to the lead-barium silicate system. The high-lead-glazed pottery just appeared in the Western Han Dynasty. The great parallels between the color ball and eye beads unearthed from Wuxi, Jiangsu and the glazed pottery *huang* and eye beads excavated from Jiangling district, Hubei show that the former was closely related to the *Chu* Cultural area in terms of technique and source. The handicraft industries of ceramic, bronze, iron and so on in the State of *Chu* (Hubei) and the State of *Yue* (Jiangsu) were well developed during the Warring States Period, which provided the foundation of making glazed pottery. The exchange of technique, culture and economy between *Chu* and *Yue* was very frequent during that time. Consequently, it is thought that the color glazed ball, as a ritual object, was probably made locally, in view of its remarkable regional characteristic. However, not excluded is the possibility that the glazed pottery beads, as the ornaments, were imported from *Chu* State.

15.6 Conclusion

The color glazed ball and glazed pottery eye beads, dated to the early Warring States Period, unearthed from Wuxi and covered with lead-barium glaze, were the earliest batch of lead-barium silicate system low-temperature glazed potteries ever found up to now. Taking the analysis results of the glazed potteries from Jiangling district, Hubei into account together, we may conclude that the origin of low-temperature glazed pottery can be traced back to the early Warring States Period, earlier than the Han Dynasty as generally considered.

The green surfaces of the three glazed pottery eye beads of the early Warring States Period uncovered from the burials of the *Yue* State at Hongshan, Wuxi are sintered body with the minerals containing lead and barium. Based on the time, transparence, consistency of the proportioning of major chemical composition and state of preservation, the manufacturing technique of these glazed pottery eye beads were regarded to have reached a relatively high level. The dragonfly eye beads with pottery body decorated with lead-barium type glaze were also found in Jiuding, Jiangling, Hubei dating to the middle and late Warring States Period. We consider that the application of the lead-barium glaze is closely related with the lead-barium glass of the Warring States Period. The glazed pottery eye beads should be derived from Western sodium-calcium silicate glass eye bead and developed from imitating that shape.[12,13]

As an extremely important glaze system in China, the lead-barium silicate glaze provides a significant clue for studying the origin and relationship of the low-temperature glaze, glass and porcelain glaze. The lead-barium silicate glazed pottery of the Warring States laid a foundation for the origin and development of the low-temperature glazed pottery and glass in ancient China. The successive high-lead low temperature glazed pottery of the Han Dynasty opened up the second important glaze system in the history of science and technology of Chinese ceramic, which contributed to the continuous creation and innovation of series of a low-temperature colored glaze and colored material.

Acknowledgements

We are grateful to Ms Chai Yi of Xi'an Institute of Archaeology and Relics Preservation for her kind help and cooperation in the experimentation. This research was supported by the Compass Plan Project from the State Administration of Cultural Heritage, China (Grant No. 20110303) and the Major State Basic Research Development

Program of China (973 Program, Grant No. 2012CB720901, 2012CB720906).

References

1. F. K. Zhang and Z. G. Zhang, *Jouranl of the Chinese Ceramic Society* **8**, 9 (1980).
2. J. Z. Li, *History of Chinese ceramics*, Cultural Relics Press, Beijng, China, 1982, pp. 464–465.
3. W. Nigel and F. A. Ian, Preliminary Examination of a Warring States Pottery Jar with So-called "Glass Paste" Decoration, in *Science and Technology of Ancient Ceramics 3 Proceedings of the International Symposium [ISAC'95]*, ed. J. K. Guo (Shanghai Scientific and Technology Literature Publishing House, Shanghai, China, 1995), pp. 12–17.
4. W. Nigel, C. F. Ian and P. S. Colleen, Early Polychrome Glazes on a Chinese Ceramic Bead of the Warring States Period, in *Science and Technology of Ancient Ceramics 4 Proceedings of the International Symposium [ISAC'99]*, ed. J. K. Guo (Shanghai Scientific and Technology Literature Publishing House, Shanghai, China, 1999), pp. 29–37.
5. Q. H. Li, J. Q. Dong and F. X. Gan, *Journal of Guangxi University for Nationalities (Natural Science Edition)* **15**, 31 (2009).
6. S. Kwan, *Early Chinese glass*, Art Museum, the Chinese University of Hong Kong, Hong Kong, China, 2001, pp. 160–171, 174–175.
7. Nanjing Museum and Jiangsu Provincial Institute of Archaeology, *Cultural Relics Management Committee of Xishan District, Wuxi. Excavation report on the burials of the Yue State at Hongshan*, Cultural Relics Press, Beijng, China, 2007, pp. 316–317.
8. S. Liu, Q. H. Li, F. X. Gan *et al*, *X-Ray Spectrometry* **40**, 364 (2011).
9. F. X. Gan, Evolution of the chemical composition and the origin of making technology of Chinese ancient glass, in *Development of Chinese Ancient Glass Technology*, ed. F. X. Gan (Shanghai Scientific and Technical Publishers, Shanghai, China, 2005), pp. 226.
10. F. X. Gan, Evolution of the chemical composition and the origin of making technology of Chinese ancient glass, in *Development of Chinese Ancient Glass Technology*, ed. F. X. Gan (Shanghai Scientific and Technical Publishers, Shanghai, China, 2005), pp. 228–228.
11. W. Nigel, *Chinese Glazes: Their Origins, Chemistry and Recreation*, A & C Black, University of Pennsylvania Press, Landon, UK, 1999, pp. 189–191.
12. F. X. Gan, H. S. Cheng, Y. Q. Hu *et al*, *Science in China Series E: Technological Sciences* **52**, 922 (2009).
13. F. X. Gan, H. X. Zhao, Q. H. Li *et al*, *Jianghankaogu* **2**, 108 (2009).

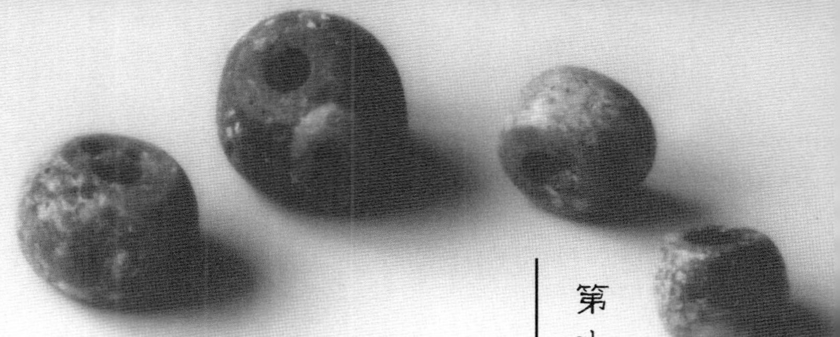

Chapter 16 | 第十六章

A Study of the Diverse Origins and Primary Development of Chinese High Temperature Porcelain Glazes

Juan Wu, Maolin Zhang,* Junming Wu,* Qijiang Li,* Jiazhi Li*[†] *and Zequn Deng*[†]

*Institute of Ancient Ceramic, Jingdezhen Ceramics Institute, Jingdezhen 333001, China;

[†]Shanghai Institute of Ceramics, Chinese Academy of Sciences, Shanghai 200050, China

16.1 Introduction

China is the only country with more than ten-thousand-years of continuous developmental history in ceramics, abundantly supported with scientific and historical evidence. The invention of glaze was an important technical breakthrough in the process of transition from pottery to porcelain. The water-tight glazed vessels endow porcelain with functions that unglazed pottery does not have. The glaze also enhances mechanical strength, aesthetic properties and durability of ceramic wares. With the emergence of celadon ware, ceramics gradually took the role of bronzes in daily life because of the impermeability of glaze. In fact, glaze is a glassy layer that adheres to the ceramic body, and there is no essential difference in material properties (micro-structure, physical or chemical) from glass. So the invention of glaze can also be considered as the invention of glass. Generally, the preparation process of glaze includes: (1) raw materials are first made into a glaze slip; (2) the glaze slip is deposited on the pottery or porcelain body; (3) the glaze slip is finally fired and melted into glaze.

The invention of porcelain glaze is a notable research topic because of its close relationship with some significant problems in the developmental history of Chinese ceramics, such as the origin of proto-porcelain, the invention of porcelain, and so on. In fact, there are different viewpoints on the origins of glaze: (1) Low refractory raw materials used in the kiln wall were melted while firing; (2) the surface of pottery bodies made from low melting point raw materials were melted at a high firing temperature; (3) the evolution of a coating used for the decoration of pottery[1]; (4) inspiration that was derived from the forming of fuel ash slag (a glassy layer on the kiln wall) or a thin glass layer on the ware in the firing process.[2] Among these viewpoints, the third and fourth are widely accepted. But which of these is the real origin of Chinese ceramic glazes? That is to say, the problem of the origin of Chinese ceramic glazes has not been completely resolved because insufficient specimens have been analyzed and because a restricted range of analytical methods have been used. Recently, archaeological excavations have consistently increased along with the rapid development of the study of the economy, and new related archaeological materials and specimens have been discovered. Meanwhile, more modern testing techniques have been widely and deeply applied in the scientific study of ancient ceramics. All these have provided powerful support to further the study of the origins of Chinese ceramic glazes. Therefore, on the basis of former studies of ancient pottery, proto-porcelain and early porcelain by the Jingdezhen Ceramics Institute and Shanghai Institute of Ceramics, Chinese Academy of Sciences,[3–6] the origin and early development of Chinese ceramic glazes are further developed and discussed in this paper.

16.2 The origins of Chinese ceramic glazes

As mentioned above, the third and fourth viewpoints on the origin of Chinese ceramic glazes are widely accepted, therefore, in this paper the relationships of the raw material recipe to manufacturing technology, pottery slips, fuel ash slags and proto-porcelain glazes are discussed.

16.2.1 *Chemical compositional characteristics of pottery slip and fuel ash slag*

The relevant archaeological excavations (such as the excavations of Nanzhuangtou in Xushui, Hebei province, and of Xianrendong in Wannian, Jiangxi province) indicate that pottery began to be manufactured about ten thousand years ago.[7] Ancient potters had accumulated a wealth of experience in the treatment of raw materials and the processes of forming and firing, over a long time. As we know, primitive pottery with rough surfaces had high water absorption, and could easily become dirty. To improve this situation a pottery slip was invented as a kind of decorative layer on the surface of pottery body (Fig. 16.1). Pottery slip is generally red or black. Sometimes ancient pottery was even directly painted with a pottery slip to produce a better decorative effect. Relevant research showed that the raw materials used for the preparation of pottery slip can be divided into two kinds: the first kind is obviously different from the raw materials used for the preparation of pottery body, the other kind is similar in the raw materials used for the preparation of pottery body, but was processed further (smaller particle sizes and higher plasticity). Generally, compared with the pottery body, the chemical compositions of the pottery slip have a higher concentration of Fe_2O_3 and lower concentrations of K_2O, Na_2O, CaO and MgO due to the selection or treatment of raw materials used (Table 16.1). Some of them are very similar to that

Fig. 16.1 Chinese ancient pottery slip[9] and fuel ash slag[10]

Table 16.1 Chemical compositions of pottery slip and fuel ash slag samples from Hubei, Jiangxi, Zhejiang province, etc.

	Number		Na_2O	MgO	Al_2O_3	SiO_2	K_2O	CaO	TiO_2	Fe_2O_3	MnO	P_2O_5
pottery slip	14	Mean/ wt%	0.48	0.95	17.99	69.84	1.82	0.92	0.72	7.08	0.10	0.19
		S.D.	0.33	0.42	1.13	2.34	0.77	0.47	0.57	2.73	0.12	0.07
slip glaze	9	Mean/ wt%	0.87	1.44	21.47	59.67	4.36	1.73	1.32	6.66	0.09	0.08
		S.D.	0.28	0.83	4.37	7.75	0.29	1.49	0.23	3.08	0.06	0.16
fuel ash slag	5	Mean/ wt%	0.28	3.05	13.13	65.88	5.35	6.24	0.75	3.35	1.32	0.91
		S.D.	0.10	1.45	2.43	9.31	1.40	4.95	0.33	0.96	1.12	0.63

of proto-porcelain glaze. Both glaze and pottery slips were used to improve the appearance and properties of the artifact as a thin layer covering the body. But it must be emphasized that pottery slip can not be melted into glassy state at low firing temperatures (~800°C, or even lower[7,8]).

It is easy to show that the pottery slip, a thin decorative layer that covers the body, is very similar to a ceramic glaze in the preparation method used. They both played a common role in improving the appearance and function.

Fuel ash slag is produced in the firing process of ceramics. When ash from burning the wood used as a fuel float down and settle on a ceramic body or kiln wall, it is possible for it to react and melt into a thin glassy layer if the temperature is high enough. It is known as fuel ash slags (Fig. 16.1). Moreover if there are high levels of fluxing elements, the associated contents of MnO and P_2O_5 are also relatively high in some fuel ash slags (Table 16.1).[2] Undoubtedly, the initial formation of fuel ash slag first benefited from the development of the firing technique used. Relevant research[7] showed that most pottery artifacts were directly fired on the ground or in primitive kiln before the Shang dynasty in China. The primitive kiln was a horizontal through-draught one, which gradually developed into the vertical through-draught kiln. But the highest firing temperature in primitive kilns was still lower than 1000°C.

Therefore, it was almost impossible for fuel ash slag to be produced at that time. Up to the Shang dynasty, more effective up-draught kilns began to be built and used, and the kiln chamber was also enlarged. The highest firing temperature could then reach up to 1200°C, which made the emergence of fuel ash slag possible.

Most of the pottery slip and fuel ash slag samples mentioned above had the possibility of melting into glaze at a certain temperature (> 1200°C) due to the high contents of fluxing elements. Also the pottery slip was similar to ceramic glaze in terms of the main preparation method (a thin layer covering a body using refined raw materials). Meanwhile, fuel ash slag also provided a direct comparison with the glassy material. In this situation, the combination of the preparation method of the pottery slip and the glassy state of fuel ash slag finally provided the technological processes which laid down a solid foundation for the invention of ceramics glazes in China.

16.2.2 *Chemical composition and technological conditions of proto-porcelain glaze in China*

Archaeological excavations have indicated that proto-porcelain was invented in the Shang Dynasty (16th–11th century B.C.). On the one hand, the modeling and firing techniques of proto-porcelain basically inherited processes from impressed-stoneware produced on a large scale during the Shang Dynasty. Sometimes proto-porcelain and impressed-stoneware were even fired in the same kiln, and the sintering temperature was up to 1200°C.[11] On the other hand, compared with impressed-stoneware, one of the most important features of proto-porcelain was a thin layer of transparent glaze on the surface of the body. The glazing technique was quite primitive at that time; it mainly consisted of two methods: brushing or dipping. And most of the glaze layers of proto-porcelain were rather thin and could easily be stripped from body[12] (Fig. 16.2). The means and standard deviations of the chemical compositions of early proto-porcelain glazes are listed in Table 16.2 for further comparison.

Generally, the glaze of the early proto-porcelain wares excavated in China belonged to the high firing temperature glaze types (~1200°C).

Fig. 16.2 Specimens of Shang Dynasty proto-porcelain and Eastern Han Dynasty Yue celadon

Table 16.2 Means and standard deviations of major and minor elements of proto-porcelain samples excavated from different sites

Site	Number		Na_2O	MgO	Al_2O_3	SiO_2	K_2O	CaO	TiO_2	Fe_2O_3	MnO	P_2O_5
Deqing,	43	Mean/ wt%	0.78	2.83	11.97	64.68	2.11	14.56	0.41	1.67	0.25	0.47
Zhejiang		S.D.	0.44	0.74	1.06	3.27	0.67	4.67	0.08	0.48	0.11	0.16
Xiaoshan,	17	Mean/ wt%	1.24	1.72	14.11	70.76	4.19	4.66	0.40	1.93	0.09	0.19
Zhejiang		S.D.	0.37	0.40	1.10	2.31	2.03	2.70	0.05	0.45	0.05	0.08
Yingtan,	8	Mean/ wt%	0.48	1.76	17.32	61.68	5.91	4.87	1.38	6.12	0.12	0.26
Jiangxi		S.D.	0.15	0.09	0.47	0.13	1.41	3.41	0.49	3.57	0.08	0.06
Qingjiang,	9	Mean/ wt%	0.68	2.01	14.70	64.05	5.12	6.50	1.16	5.21	0.34	0.29
Jiangxi		S.D.	0.36	0.90	4.17	6.24	2.25	5.11	0.55	2.67	0.31	0.24

And the proto-porcelain glaze unearthed from Jiangxi and Zhejiang provinces can be divided into two kinds (Fig. 16.3). The division is related to the main kinds of the fluxing elements used in glaze. The first kind (most of the samples) had a relatively high content of CaO (>10%) and low contents of Fe_2O_3 and K_2O (<2%). There were higher Fe_2O_3 and K_2O (>5%) contents in the second glaze (<5%). Therefore, the obvious differences between chemical compositions indicated that different raw materials were selected to make them.

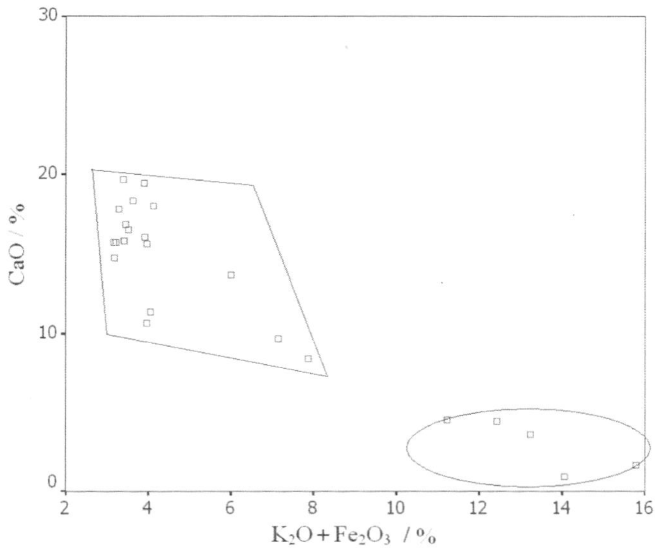

Fig. 16.3 Binary-plot of the concentrations of CaO versus $K_2O+Fe_2O_3$ in proto-porcelain glazes from Zhejiang and Jiangxi provinces

Why did early proto-porcelain glazes have two different types of raw material recipes in China? It is an inevitable question for those studying the origins of porcelain glaze in China. The specific origin of glaze should be closely related to the local natural resources (raw materials, fuel). According to the chemical compositions of some proto-porcelain glaze samples (Table 16.2), the contents of MnO and P_2O_5 in the first type of proto-porcelain glaze which used CaO as the main flux were relatively 'high' (generally > 0.4%). Although limestone with a high content of CaO is one kind of the raw material usually used for the preparation of porcelain glaze, the use of limestone would not increase the contents of MnO and P_2O_5 (generally < 0.1%) in glaze. However, as we know, firewood was the main fuel used for ceramic production in southern China, and plant ash which also can have high contents of CaO and would be easy to obtain. It was possible for ancient potters to derive inspiration from the forming of fuel ash slag. Thus plant ash which also has relatively high contents of MnO and P_2O_5 was taken as the main raw material for the preparation of the first type of proto-porcelain glaze.[2] In spite of having a certain concentration of CaO, the second type of proto-porcelain glaze also had relatively high contents of Fe_2O_3 and K_2O. Meanwhile, the contents of MnO and P_2O_5 were much lower than those of the first type of proto-porcelain glaze. So it was nearly impossible for the high concentrations of CaO in the second type of proto-porcelain glaze to have derived from plant ash. It was not difficult to find out that the chemical composition of the second type of proto-porcelain glaze was very similar to that of some pottery slip (Fig. 16.4). Such a high degree of similarity suggested that the invention of the second type of proto-porcelain glaze was closely related to the preparation of pottery slips.

Furthermore, the decorative effects of the two types of proto-porcelain glazes were quite different. The high calcium glaze (the first one) was relatively more transparent, while the glaze with a high iron content (the second one) had a darker color and covered the surface more effectively. Actually, the co-existence of different preparation methods of glaze simply indicates that the invention of glaze was closely related to the use of local resources (raw materials and fuel).

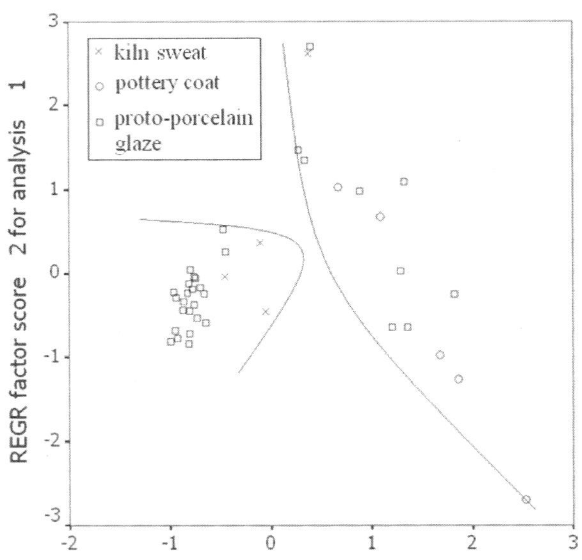

Fig. 16.4 Scatter diagram using factor analysis of the chemical compositions of proto-porcelain glaze, pottery slip and fuel ash slag

The invention of proto-porcelain glaze benefited from both the improvement of the firing technique and the raw materials used to make the glaze. Actually, the gradual improvement of the firing technique was considered as one of the three major technological breakthroughs in the development of Chinese ceramic history. The first took place in the Shang Dynasty. The highest firing temperature of impressed-stoneware reached as high as 1200°C, which was nearly 200°C higher than that of earlier pottery production.[11] It made the invention of glaze possible. For example, the so-called "slipware" excavated from the tombs of the Shang Dynasty in Jiangshan, Zhejiang province had a layer of black, rough, lustrous and permeable pottery slip on the body, which had a high content of Fe_2O_3 (Table 16.1). The sintering temperature of the samples was proven to be about 1100°C. When these pottery slip samples were further heated up to 1200°C, the pottery slip began to melt into a layer of smooth, black or brown glassy glaze.[13] Obviously, plant ash or a pottery slip could be melted into a glassy state only with a high enough firing temperature. After long-term observation and contact, ancient potters gradually realized that the glassy surface of the ceramic not only had a decorative effect but also could enhance the abrasion resistance/hardness. Above all, it had gradually taken the role of bronze-ware in daily life because of the impermeability of glaze. Undoubtedly, the breakthrough of the firing technique was a crucial factor in the invention of glaze.

16.3 The emergence and primary development of a mature porcelain glaze in China

The technology of raw material preparation for proto-porcelain took a great leap forward during the Eastern Han Dynasty (25 — 220AD), and mature porcelain — Yue celadon — was finally invented in the Yue kiln site in Zhejiang province China. Compared with proto-porcelain, one of the essential characteristics of mature porcelain was that the quality of glaze and the matching of body and glaze (glaze fit) were greatly improved.

Comparative studies have indicated that Yue celadon glaze is similar to the first type of proto-porcelain glaze in chemical composition including the use of CaO as the main flux used. The content of CaO

Table 16.3 Means and standard deviations of major and minor chemical components of Yue celadon glaze samples[14]

Site	Number		Na_2O	MgO	Al_2O_3	SiO_2	K_2O	CaO	TiO_2	Fe_2O_3	MnO	P_2O_5
Shaoxing	8	Mean/ wt%	0.90	2.34	13.69	62.06	1.87	14.56	0.76	2.30	0.32	0.77
		S.D.	0.07	0.63	0.70	3.18	0.11	2.49	0.02	0.23	0.11	0.14
Shangyu	9	Mean/ wt%	0.54	2.19	13.33	60.03	1.92	17.65	0.47	2.05	0.32	0.59
		S.D.	0.13	0.57	0.66	2.37	0.11	2.41	0.43	0.59	0.28	0.51
Shanglinhu	5	Mean/ wt%	0.82	3.86	13.33	59.08	1.58	14.84	0.71	3.14	0.58	1.46
		S.D.	0.03	2.05	0.38	2.48	0.27	0.68	0.06	1.55	0.20	0.66

in the celadon glaze is about 14 ~ 16%; some even contain up to 20% (Table 16.3). Obviously, celadon glaze still belonged to the high temperature glaze of the $CaO(MgO)\text{-}K_2O(Na_2O)\text{-}Al_2O_3\text{-}SiO_2$ series. But the highest sintering temperature of Yue celadon already reached up to 1300°C.[14] A small proportion of residual quartz or other crystals existed in the transparent and uniform celadon glaze. The thickness of the celadon glaze was generally between 0.1mm and 0.3mm. In addition, as mentioned above, proto-porcelain glaze could easily be stripped from the body, but the body and glaze of Yue celadon matched well — there was a good fit (Fig. 16.5). Most of the celadon glazes were daffodil yellow, bluish gray or cyan colours (Fig. 16.6) due to 2% content of Fe_2O_3, a certain content of TiO_2 (generally less than 1%) and different firing atmospheres. From then on, the development of Chinese porcelain entered a new stage.

More than one thousand years after the invention of Yue Celadon, especially after the Song Dynasty, Chinese high fired glaze continued to witness a constant improvement. Though it still had a high fired glaze of the $CaO(MgO)\text{-}K_2O(Na_2O)\text{-}Al_2O_3\text{-}SiO_2$ series, its content of CaO was lower and correspondingly, the K_2O content increased gradually. Thus it belongs to a high fired calcium-alkaline glaze or alkaline-calcium glaze.[15] In addition to transparent glaze, crystallized glaze and phase separation crystallized glaze have been invented through complicated physical and chemical changes in the firing process. A great variety of coloured glazes were also introduced: red glaze coloured by Cu,

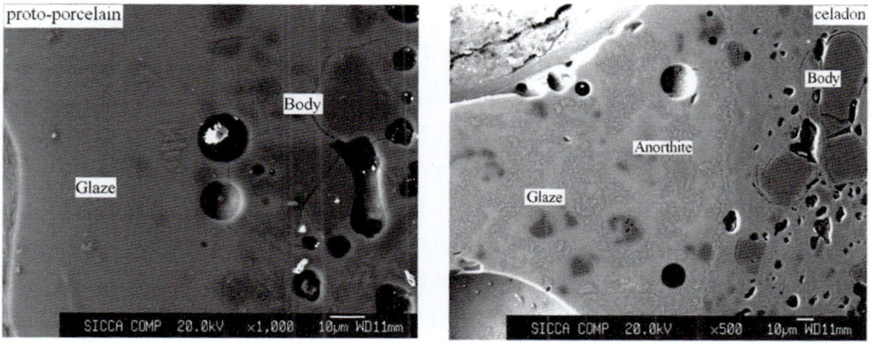

Fig. 16.5 Microstructure of typical proto-porcelain and Yue kiln celadon samples in China[3]

Fig. 16.6 Early Yue celadon samples

blue glaze coloured with Co as well as celadon glaze coloured with Fe which was still in the process of improvement. All of these have won for China the fame of being the "country of porcelain."

For example, the celadon glaze of Southern Song Guan Ware is a typical crystallized glaze. As the layer of glaze is relatively thick, when the glaze is fired, a certain amount of anorthite will be crystallized. The crystallized anorthite mixed with the residual quartz and glaze pores make the glaze semi-opaque. In addition, as it is fired in a strong reducting atmosphere, the glaze turns into a bright pinkish-blue colour which is more artistically appealing than blue jade.[16]

The Jun glaze from Henan province is a typical phase separation glaze. The ratio of RO_2/R_2O_3 contents in Jun glaze is about 10.5 which falls into the category of phase separation glaze in the CaO-Al_2O_3-SiO_2 three dimensional phase diagram. During the firing process, a liquid phase separation takes place. The diameter of one of the liquid phases ranges from 55 to 116 nm, which meets the requirement for Rayleigh Scattering. That is why Jun glaze is a blue-opalescent colour.[17]

The Hare's Fur Black Glaze of Jian Ware is a typical phase separation crystallized glaze or crystallized phase separation glaze (①liquid phase separation in the black glaze led to hematite or magnetite crystallization; ②calcium feldspar crystallization in the black glaze led to liquid phase separation and hematite or magnetite crystallization). For example, the glaze of Jianzhan which contains relatively more Fe_2O_3, when being fired can produce acicular micro-crystals and, at the same time, its Fe_2O_3 is discharged into the liquid phase, greatly increasing the amount of Fe_2O_3 in the crystallized liquid phase. This in turn increases the immiscibility of the liquid phase and during the process of cooling, the separation of liquid phase takes place. As the temperature continues to drop, the Fe_2O_3 in the liquid phase becomes saturated and Fe_2O_3 or Fe_3O_4 crystals are produced, giving hare's fur glaze different colors.[18]

16.4 Conclusions

(1) It was possible for most of the pottery slip and fuel ash slag samples to be melted into a glaze at a certain high temperature (>1200°C) due to the high content of flux. The pottery slip was similar to a ceramic glaze in terms of the main preparation methods used (such as a thin layer covering the body and the use of fine raw material processing techniques). Meanwhile, fuel ash slag also provided a direct reference to a glassy material. In this situation, the combination of the preparation method for a pottery slip and the glassy state of fuel ash slag provided the whole technological process and laid down a solid foundation for the invention of ceramic glazes in China.

(2) The proto-porcelain glazes can be divided into two kinds. The division is related to the main kinds of the flux elements used (CaO, Fe_2O_3, K_2O) in glaze. Undoubtedly, the obvious differences between

chemical compositions indicate that different raw materials were selected. The decorative effects of the two types of proto-porcelain glazes were quite different. The high calcium content glaze (the first one) was relatively more transparent, while glaze with high iron content (the second one) had a darker colour and covered the pot more effectively. Actually, the co-existence of different preparation methods for glaze just indicates that the invention of glaze was closely related to local resources (such as raw materials and fuel).

(3) The invention of proto-porcelain glaze benefited from the improvement of firing techniques and of the raw materials used to make the glazes. The breakthrough in firing technology was a crucial condition for the invention of glaze.

(4) After the Song Dynasty, Chinese high fired glazes witnessed a constant improvement. The calcium alkaline glazes or alkaline calcium glazes (which derived from calcium glazes) gave rise to crystallized glaze, phase separation glaze and phase separation crystallized glaze through complicated physical and chemical changes in the firing process. These, when mixed with different colouring elements, form a rich and colourful family of coloured glaze, winning for China the fame of being the "country of porcelain."

Acknowledgments

The present research is supported by the National Natural Science Foundation of China (Grant No. 50762006, 50962008), the Jiangxi Provincical Department of Education (Grant No. LS0908) and the Natural Science Foundation of Jiangxi Province (2009GQW0014).

References

1. J. Z. Li, *Journal of the Chinese Ceramic Society* **11**, 1 (1983).
2. F. K. Zhang, The Origin of Chinese Traditional High-Temperature Glazes, *in The Research on China Ancient Ceramics ed.* Shanghai Institute of Ceramics, Chinese Academy of Sciences (Science Press, Beijing, 1987), pp. 41–46.
3. J. Wu, J. Z. Li, J. M. Wu *et al.*, *Science in China Series E: Technological Sciences* 53, 369 (2010).

4. J. Wu, H. J. Luo, Q. J. Li *et al.*, *Journal of the Chinese Ceramic Society* **37**, 1408 (2009).
5. J. Wu, J. M. Wu, Q. J. Li *et al.*, *Science in China Series E: Technological Sciences* **39,** 074 (2009).
6. J. Wu, H. S. Wang, J. Z. Li *et al.*, *Science in China Series E: Technological Sciences* **37**, 1140 (2007).
7. J. Z. Li, *The History of China Science and Technology (Ceramics)*. Science Press, Beijing, 1998, pp. 10–20
8. C. S. Wang, *Primary study on early pottery*, *China Cultural Relic News*, 2005-11-11(7).
9. Institute of Archaeology, China Academy of Social Sciences, *Sichuan Culture Relics* **2**, 3 (2006).
10. M. Feng, G. N. Li, X. Ling *et al.*, Sciences of conservation and archaeology **16**, 29 (2004).
11. J. Z. Li, *The History of China Science and Technology (Ceramics)*, Science Press, Beijing, 1998, pp. 104–106.
12. H. J. Luo, J. Z. Li, L. M Gao *et al.*, Chemical composition and microstructure study on proto-porcelain glaze (in Chinese), in *Proc. 2nd International Symposium on Ancient Ceramics*, Shanghai, China, 1992, eds. J. Z. Li, and X. Q Chen (Shanghai Science & Technology Press, Shanghai, 1992), pp. 72–77.
13. J. Z. Li, Z. Q. Deng, Z. G. Zhang *et al.*, Study on the black slipware and proto-porcelain from Jiangshan, Zhejiang province, *in The Research on China Ancient Ceramics ed.* Shanghai Institute of Ceramics, Chinese Academy of Sciences (Science Press, Beijing, 1987), pp. 56–63.
14. J. Z. Li, *The History of China Science and Technology (Ceramics)* Beijing: Science Press, 1998, pp. 127–129.
15. H. J. Luo, J. Z. Li, L. M. Gao, *Bulletin of the Chinese Ceramic Society* **2**, 50 (1995).
16. W. D. Li Wei, J. Z. Li, Z. Q. Deng *et al.*, *Journal of Building Materials* **3**, 245 (2004).
17. J. Z. Li, *The History of China Science and Technology (Ceramics)*, Science Press, Beijing: 1998, pp. 425–432.
18. J. Z. Li, *The History of China Science and Technology (Ceramics)*, Science Press, Beijing: 1998, pp. 219–222.

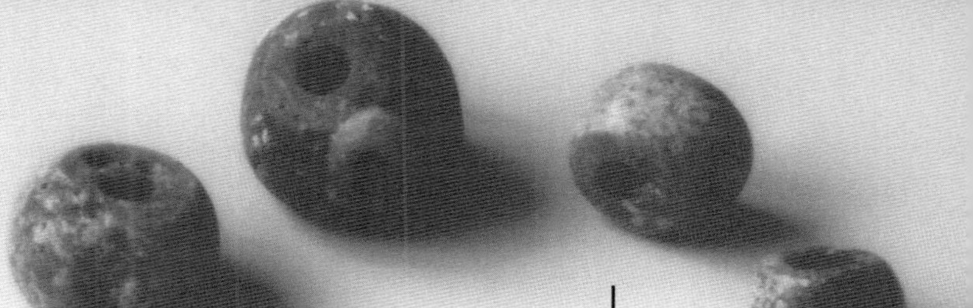

Chapter 17 第十七章

Tracing the Origin and Evolution of White Porcelain in Ancient China

Weidong Li,[*,†] *Xiaoke Lu,*[*,†] *Hongjie Luo,*[*,†]
Zhiwen Zhao[‡] *and Xinmin Sun*[‡]

*Shanghai Institute of Ceramics, Chinese Academy of Sciences, Shanghai 200050, China

†Key Scientific Research Base of Ancient Ceramics, State Administration for Cultural Heritage, China

‡Henan Provincial Institute of Cultural Relics and Archaeology, Henan 450000, China

17.1 Introduction

It has been accepted that early white porcelain is represented by the white wares excavated from the Fan Cui Tomb of the Northern Qi Dynasty (550–577 A.D.) in the city of Anyang, Henan Province. Before 2005, the Xiangzhou kiln site in Anyang was the earliest one found in Henan. Most of the excavated celadons or white porcelains were made in the Sui Dynasty (589–618 A.D.) or the Northern Qi Dynasty (550–577 A.D.), so some people thought that the production of white porcelain in ancient China started at the Xiangzhou kiln site. But the archaeological excavation of the Baihe kiln site in Gongyi county of Henan province from April 2005 to December 2007 eclipsed these ideas.[1] Many celadon shards and a few white porcelain shards of the Northern Wei dynasty (386–534 A.D.) were discovered in the fire box of kiln No. 2, brought forward the time white porcelain was invented by about one hundred years. The white porcelain shards and celadon shards were found to be fired in the same kiln, which has implications for the evolution process from celadon to white porcelain in the Northern Wei Dynasty of China.

The Baihe kiln site is located 5km to the east of Gongyi city in Henan province in China. It was named as a National Key Cultural Relics Conservation Unit. The Baihe kiln started firing porcelain in the North Dynasty (386–581 A.D.), became prosperous in the Tang Dynasty (618–907 A.D.), and declined in the Song (960–1127 A.D.) and Jin Dynasties (1115–1234 A.D.).

An archaeological excavation was conducted at the Baihe kiln site from 2005 to 2007.[1] The excavation area was 2400 square meters. The cultural deposit was 3m–6m thick. A lot of important historical remains were discovered, such as celadons, white porcelains, big tricolour potts, and several blue-and-white shards. Six kilns, including one reserved-flame dome kiln used for firing both celadons and white porcelains in the Northern Wei Dynasty were found. A kiln for firing pottery in the Han Dynasty (206 B.C.–220 A.D.) was also found in good condition. This indicates that ceramic-making in the Baihe area started no later than the Han Dynasty. The ceramic making industry here was developed in one continuous line from the Han Dynasty to the Tang Dynasty. The long-term development and inheritance of ceramic making technologies is closely related to the rise and evolution of Northern Wei white porcelains and celadons, as well as the prosperity of Tang blue-and-white porcelain and tricolor pottery.

The celadon and white porcelain from the Baihe kiln have the same vessel shapes and porcelain-making techniques as celadons and white porcelains excavated from Han-Wei ancient city of Luoyang, therefore confirming the production area of the celadons and white porcelains used by the Northern Wei imperial court. It can even be deduced that the Baihe kiln is very likely the "Luojing kiln" described in historical documents.

Baihe is the earliest kiln for white porcelain firing found so far.[1] The invention of white porcelain during the Northern Wei Dynasty was a significant breakthrough and an advancement of porcelain-making technology in ancient China, fundamentally changing the situation for celadon being the only type of porcelain since the Eastern Han Dynasty (25–220 A.D.). The era of "green porcelain in the south and white porcelain in the north" in Chinese ceramic history is supposed to have started in this period.

17.2 White porcelain and celadon excavated from the Baihe kiln site

17.2.1 *White porcelain and celadon of the Northern Wei Dynasty (386–534 A.D.)*

The Northern Wei porcelain vessels are mostly celadons; there are fewer examples of white porcelain. Celadon vessels of the Northern Wei Dynasty are mostly bowls, with a deep belly and pie-shaped foot, and had carved bow-string patterns on the outer side of the mouth. The body is relatively thick, fully glazed on the inside and half glazed on the outside; glaze accumulation on the centre on the inside and glaze dripping on the outside is quite common. Other items of celadon include plates, *dous*, pots and basins, usually with thick heavy bodies and dripped glazes.

The Northern Wei white porcelain vessels are mostly cups and bowls (Fig. 17.1), with a deep belly and pie-shaped foot. The body is thinner, whiter and finer than celadon ones. The inner side of the body is fully glazed, the outer side is half glazed over the belly with no dripped glaze. The glaze colour is usually a pale greenish white, a feature of early white porcelain in China. Both celadon and white porcelain have their foot cut obliquely around the pie-shaped foot edge. No sign of using a glaze engobe is observed. In addition, a lot of kiln furniture was unearthed in the firebox of the Northern Wei kiln and the contemporaneous rubbish pits, such as tripod supports and pillar-shaped supports. The tripod support was made by kneading the clay. It should be noted that stacked firing of white porcelain bowls and celadon bowls was discovered for the first time (Fig. 17.2), inferring that white porcelains were fired in the same kiln together with celadons. White porcelain bowl was always intentionally put on

(a) (b)

Fig. 17.1 White porcelain (a) bowls and (b) cups of the Northern Wei Dynasty from the Baihe kiln site[1]

Fig. 17.2 A white porcelain bowl was put on top of stacked celadon bowls and fired together in the same kiln

top of the stacked celadon bowls to avoid the formation of the scar from a tripod support on the inner bottom of the white porcelain bowl. This implies that white porcelain vessels were regarded as more valuable than celadon vessels in ancient times.

17.2.2 *White porcelain of the Tang Dynasty (618–907 A.D.)*

Most of the Tang vessels are white porcelains. White porcelain vessels of the Tang Dynasty are mainly bowls, pots and plates with thick bodies and jade *Bi*-shaped feet. Bottles, boxes, plates, ewers and saucers were also found. Several kinds of kiln furniture were found, including saggars in various shapes such as filler, bowl, basin, box and barrel shapes and supports in pie, ring and tripod shapes.

17.3 Experimental

In this study, the origin and development of Chinese white porcelain is to be investigated. The excavated porcelain shards of the Northern Wei and Tang dynasties from the Baihe kiln site were selected to carry out research on. The body and glaze compositions were examined by using energy-dispersive X-ray fluorescence (EDAX Eagle-IIIμProbe, USA). The firing temperatures were determined by using thermal dilatometers (NETZSCH DIL 402C, German). Microstructures were studied using field emission electron microscopy (JSM-6700F, Japan) equipped with EDS. Hunter whiteness of the glaze was determined by spectrocolorimeter (MINOLTA CMD700, Japan). Multivariate Statistics method was applied to analyze the chemical composition data.

17.4 Results and Discussion

From the experimental results, the origin and evolution of white porcelain from the Baihe kiln can be summarized in three steps: white porcelain was derived from celadon, improved gradually, and eventually became valuable.

17.4.1 *The first step: early white porcelain was derived from celadon*

From the appearance of Northern Wei wares, the early white porcelain has much in common with celadon, such as the same way of shaping, trimming and firing, showing an inherited relationship development (Fig. 17.3).

As shown in Fig. 17.4, Fe_2O_3 and CaO contents in most white porcelain bodies are lower than that of the celadon bodies, while a few

Fig. 17.3 Pictures of a celadon bowl and a white porcelain bowl of Northern Wei Dynasty from Baihe kiln site

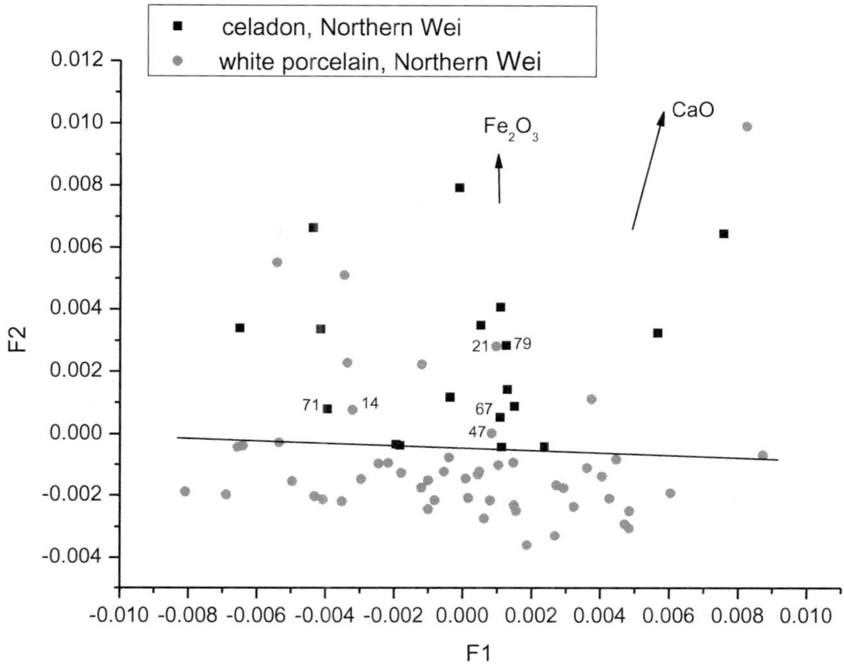

Fig. 17.4 Two dimensional correspondence analysis diagram of the main chemical compositions of the bodies of the celadons and white porcelains of the Northern Wei Dynasty from the Baihe kiln site

are similar. It is possible that a transitional period for the improvement of the body by material selection or refinement occurred, implying that the body of early white porcelain was derived from celadon.

As shown in Fig. 17.5, Northern Wei white glazes have a different recipe from celadon glazes. The glaze of white porcelain has lower Fe_2O_3 and higher MgO contents than celadon. The result of multivariate statistical analysis can almost distinguish them completely.

17.4.2 *The second step: white porcelain was improved gradually*

There are three ways of improving the whiteness of porcelain wares: the application of a low-iron glaze, the use of of a low-iron body, and

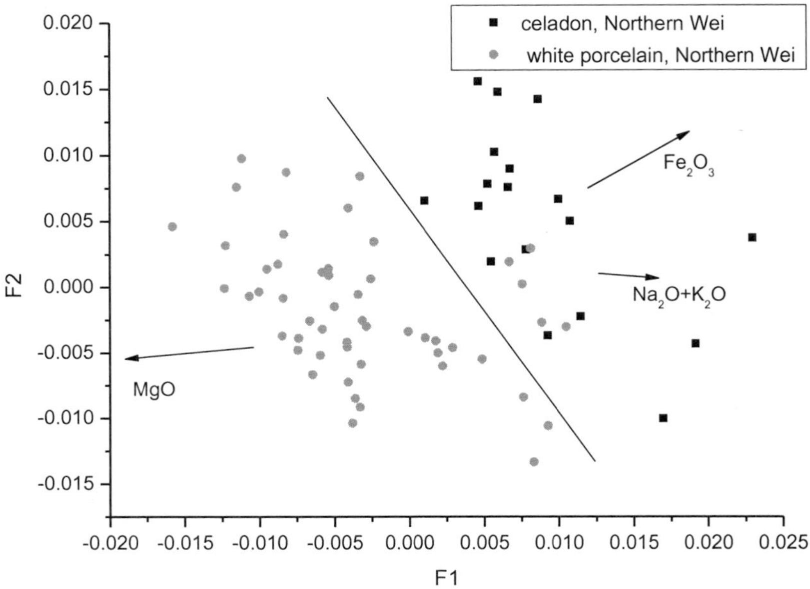

Fig. 17.5 Two dimensional correspondence analysis diagram of the main chemical compositions of celadon and white porcelain glazes of Northern Wei Dynasty from the Baihe kiln site

the application of low-iron engobe. The reduction of the Fe_2O_3 content is the key to the formation of the white porcelain.

During the Northern Wei Dynasty, the average iron content of the white porcelain glazes and bodies is obviously lower than found in celadons (Figs. 17.6 and 17.7).

As shown in Fig. 17.8 the average iron content in bodies increased over time (Fig. 17.8). The fluctuation of Fe_2O_3 in bodies is much higher than in the Northern Wei, indicating that the control over the body material used became less strict. Since Tang potters worked out another way to improve whiteness, the whiteness of the body became less important. The ancient potters of the Tang dynasty successfully used a layer of white engobe to cover the dark body so as to avoid the influence of the high-iron body on the whiteness of porcelain.

The iron content of the engobe is much lower than that of the body (Fig. 17.9). The application of engobe provided an alternative

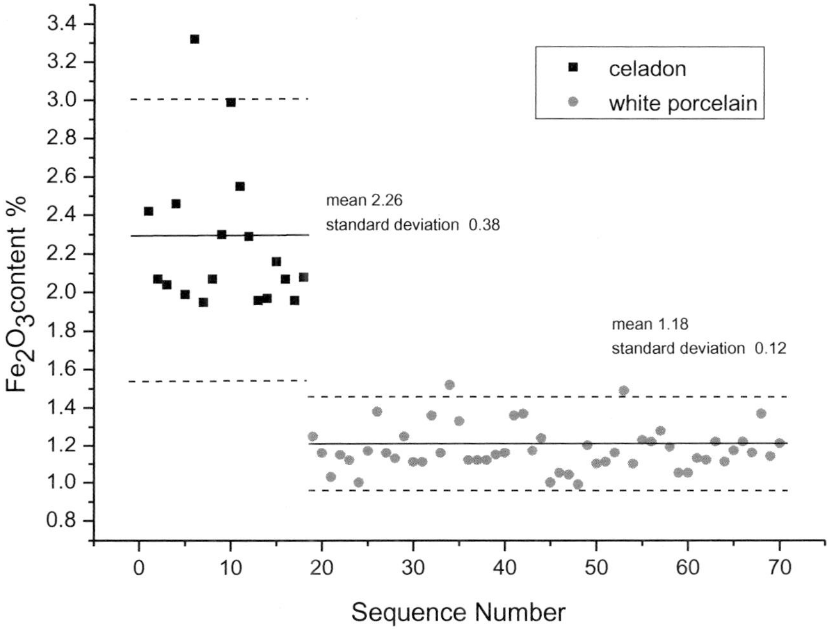

Fig. 17.6 Fe_2O_3 content analysis diagram of the glazes of celadons and white porcelains of Northern Wei Dynasty from Baihe Kiln site

solution for improving the whiteness of the whole vessel in white porcelain making.

17.4.3 *The third step: eventually becoming valuable*

By implementing the above measures, the appearance and quality of white porcelain wares from the Baihe kiln was gradually improved. The average Hunter whiteness of the white glaze increases over time (Fig. 17.10). In the late Tang dynasty the Hunter whiteness of all of the glazes exceeded 70 degrees and achieved quite a high level.

17.4.4 *New attempts to make white porcelain*

Two new products were invented on the basis of white porcelain technology: one is a new variety of fine white porcelain, the other is

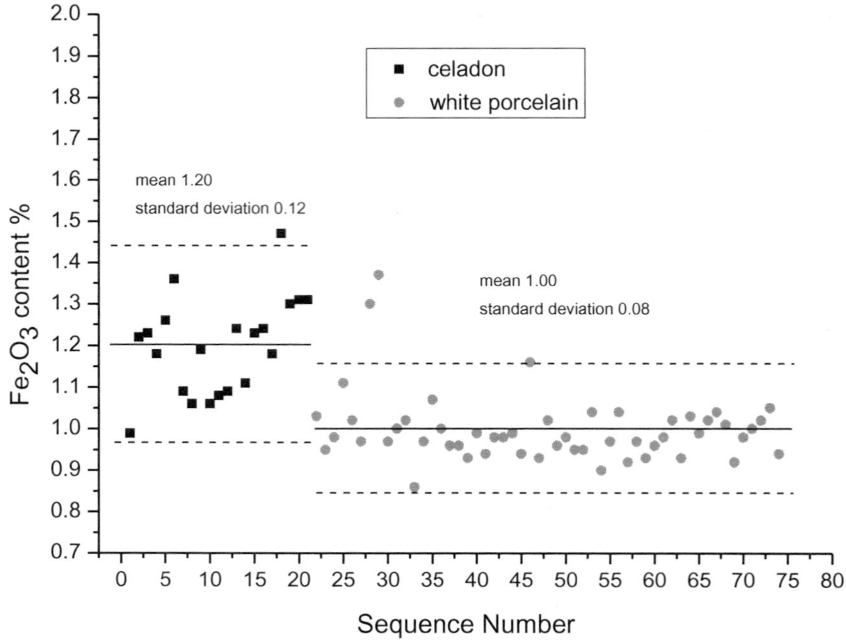

Fig. 17.7 Fe_2O_3 content analysis diagram of the bodies of white porcelains of Northern Wei and Tang dynasties from the Baihe Kiln site

blue-and-white porcelain. A study of the invention of blue-and-white porcelain will be reported in another paper.

The unique fine white porcelain (Fig. 17.11) has thin, well-sintered, translucent white body with gloss at cross-section, and with no engobe.

The body composition of this kind of unique fine white porcelain is completely different from the common white porcelain from the Baihe kiln (Fig. 17.12). Alumina and alkali contents are obviously much higher, Al_2O_3=33.21-34.64wt%, Na_2O+K_2O=6.09-7.30wt% (Fig. 17.13), the highest level among ancient Chinese ceramics. More kaolinite and feldspar were probably added to the body. The high R_2O content of body results in a greater development of a glassy phase in the body, thus increases its gloss and translucency.

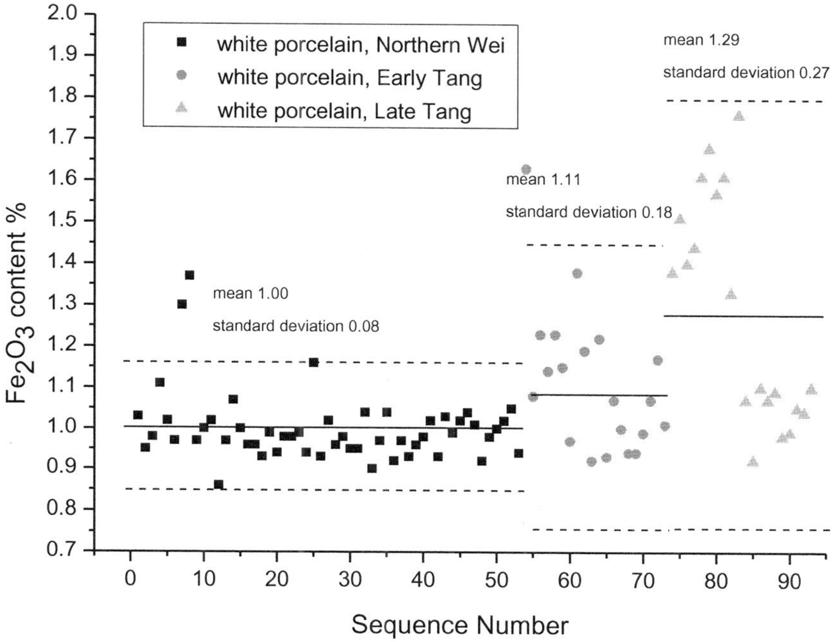

Fig. 17.8 Fe_2O_3 content analysis diagram of the bodies of white porcelains of Northern Wei and Tang dynasties from the Baihe Kiln site

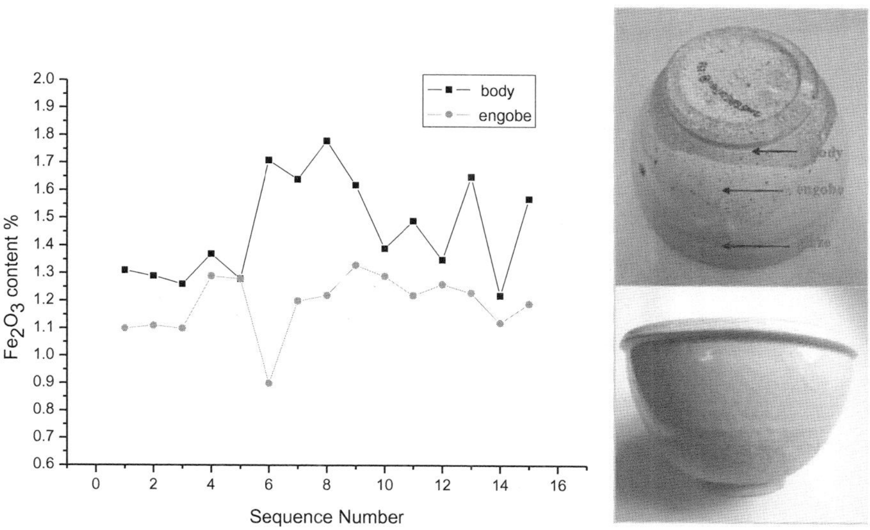

(a) (b)

Fig. 17.9 (a) Fe_2O_3 content analysis diagram and (b) a picture of the body and engobe of white porcelains of Tang Dynasty from the Baihe Kiln site

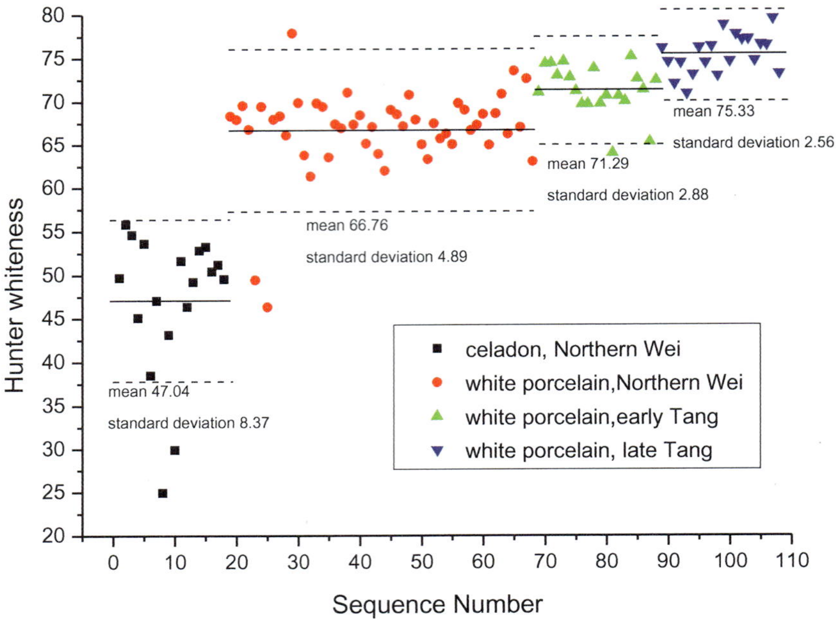

Fig. 17.10 Hunter whiteness of the porcelain samples

Fig. 17.11 Picture of a unique fine white porcelain shard discovered at theBaihe kiln site

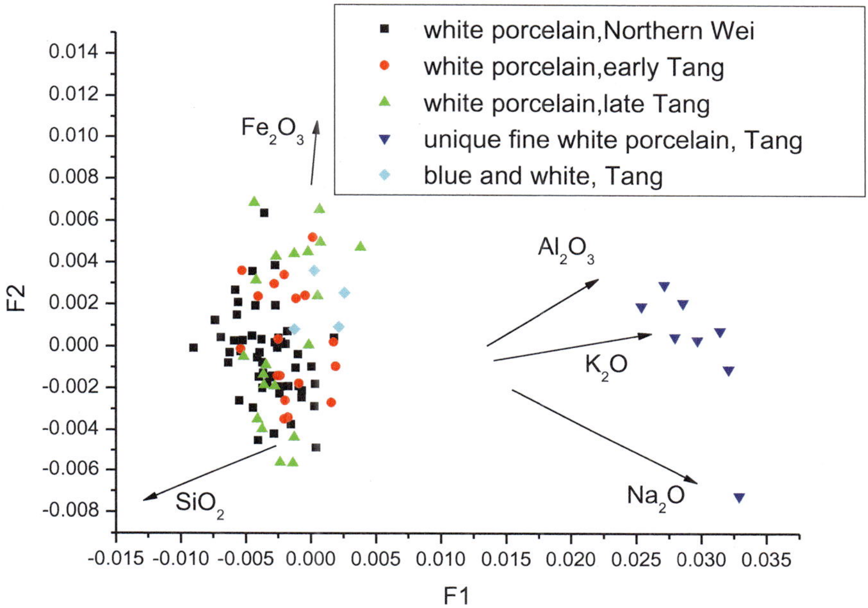

Fig. 17.12 Two dimensional correspondence analysis diagram of the main chemical compositions of the porcelain bodies from the Baihe kiln site

Similar to the body, the glaze composition also has a much higher alkali content, with an average Na_2O+K_2O of 6.50wt% (Fig. 17.14), compared with common white porcelain. Therefore feldspar was probably added to the glaze.

This is a unique and high-quality variety of white porcelain from Baihe kiln site, but no intact example has yet been found. The quality can be compared with the finest white porcelain from the Xing kiln in Hebei province. Was it used for religious purposes? Was it used as tribute for the imperial court? Its use needs to be investigated further.

17.4.5 *Evolution of porcelain firing from Northern Wei to Tang Dynasties*

The average firing temperature was on the decrease over the period under consideration: Northern Wei 1290 °C →early Tang 1258 °C →late

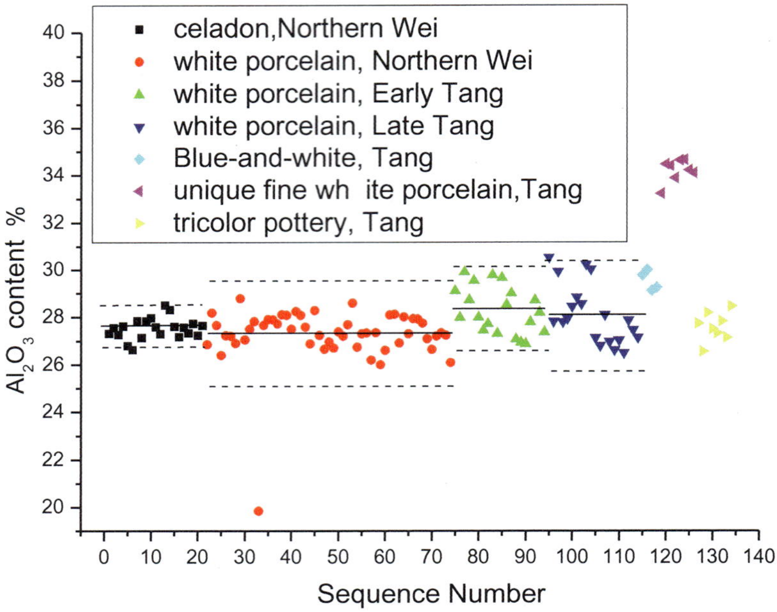

(a)

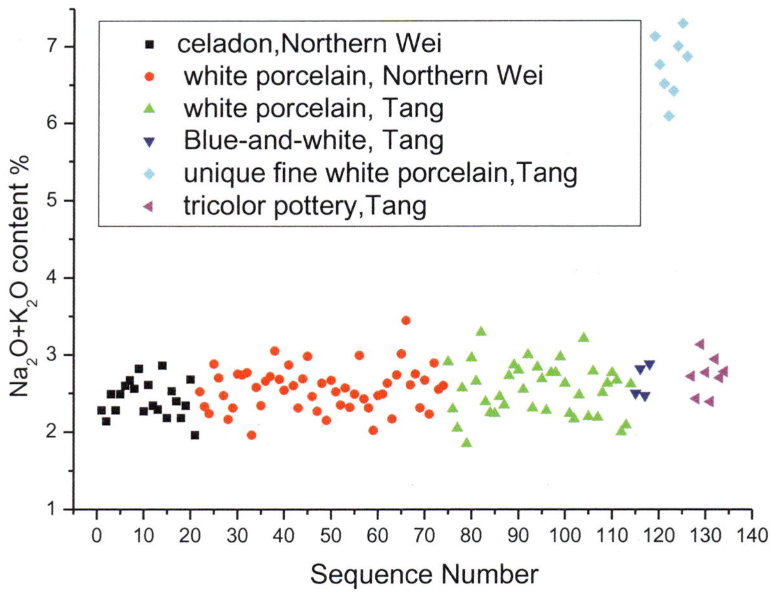

(a)

Fig. 17.13 (a) Al_2O_3 content and (b) Na_2O+K_2O content analysis diagrams of the bodies of the special fine white Tang Dynasty porcelains from the Baihe Kiln site

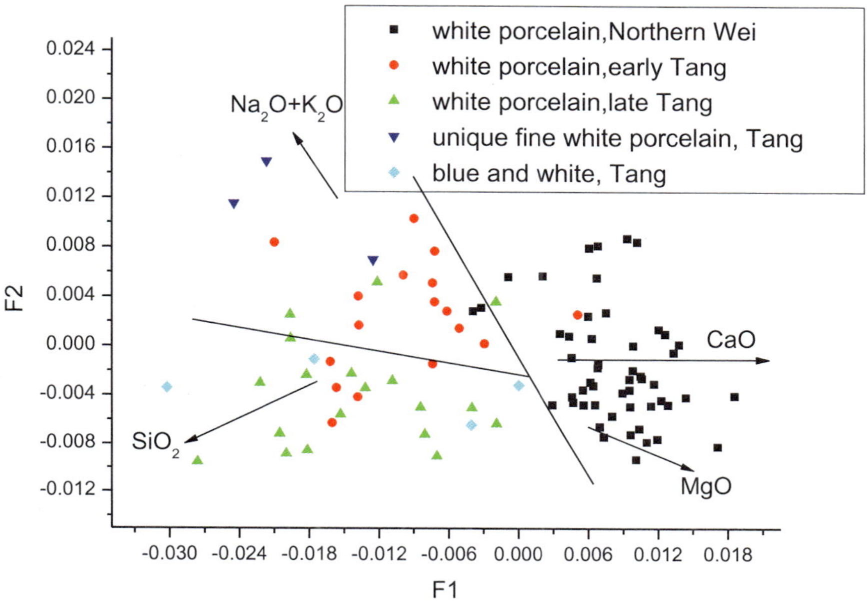

Fig. 17.14 Two dimensional correspondence analysis diagram of the main chemical compositions of the porcelain glazess from the Baihe kiln site

Tang 1251 °C (Fig. 17.15). The average flux (R_xO_y) content was also declining over this time: celadon of Northern Wei dynasty 24.94wt% →white porcelain of Northern Wei dynasty 23.74wt% →white porcelain of early Tang dynasty 18.04wt% →white porcelain of late Tang dynasty 15.66wt%, as shown in Fig. 17.16.

Glaze dripping is commonly found on Northern Wei celadon wares, because they were fired at a relatively high temperature and had high flux contents, especially CaO. Glaze dripping can also be observed on Northern Wei white glazed wares, but is less evident than on celadons. Flux contents in glazes are lower than in celadons and there is more MgO and less CaO in the glazes than in celadons. This leads to an increase in glaze viscosity. No glaze drips have been observed on Tang white porcelain because Tang wares were fired at lower firing temperatures and their flux contents are lower than in Northern Wei white porcelain and celadon.

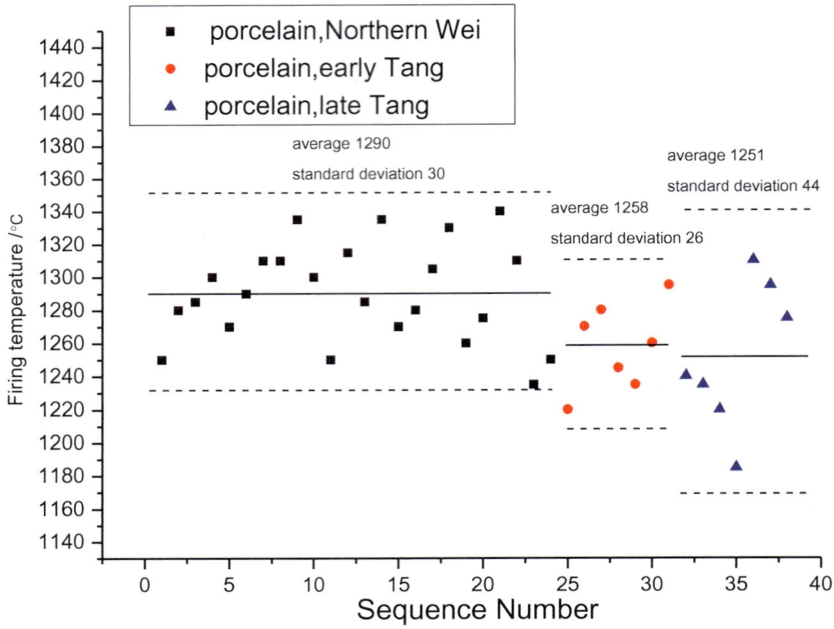

Fig. 17.15 Firing temperature analysis diagram for the bodies from the Baihe kiln

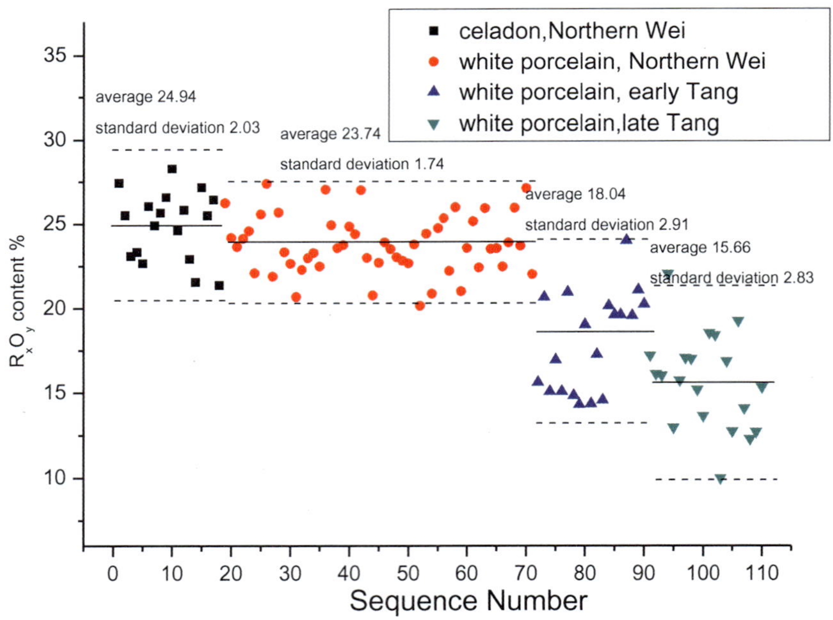

Fig. 17.16 Flux contents analysis diagram of the glazes from the Baihe kiln

17.4.6 *Microstructural characteristics of white porcelains from the Baihe kiln site*

A common characteristic of the microstructure of Baihe white porcelain is a dense interaction layer formed at the body-glaze boundary. As shown in Fig. 17.17, the dense interaction layer is composed of column-like crystals which cover the dark colour and defects of the body and improve the whiteness of the ware. The interaction layer is the reaction result between body and glaze at a high temperature. There is a 35μm thick interaction layer formed at the body-glaze boundary. The crystals in the interaction layer are column-like, well

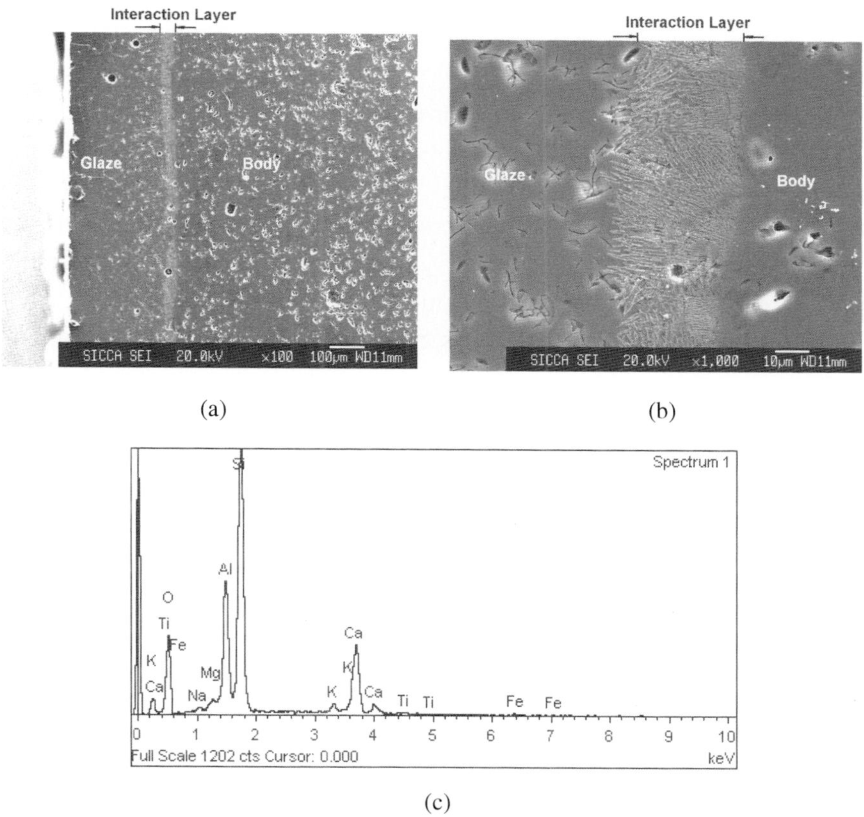

Fig. 17.17 Microstructure of the cross-section of a white porcelain bowl of the Northern Wei Dynasty from the Baihe kiln site (a) body-glaze interaction layer, (b) body-glaze interaction layer, (c) EDS spectrum of the crystals in the interaction layer

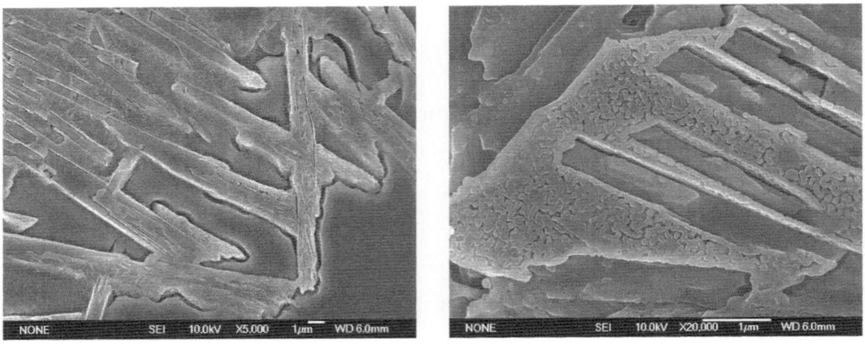

(a) (b)

Fig. 17.18 Anorthite-phase separation accompanying structure at the body-glaze boundary of a white porcelain bowl of the Northern Wei Dynasty from the Baihe kiln site

developed, and arrayed in a regular way. The crystals in the interaction layer begin to grow from the body-glaze boundary and extend to the glaze layer. The interaction layer is composed of anorthite crystals.

The second common characteristic of the microstructure of Baihe white porcelain is an anorthite-phase separation accompanying structure formed at the body-glaze boundary. In all areas of anorthite crystallization, especially in the spaces between crystal clusters and in the peripheral regions around crystals, they are always accompanied by the occurrence of the rough texture of a nano-sized liquid-liquid phase separation (Fig. 17.18). The white porcelain from Gongyi kiln has been found to be the earliest white porcelain so far which has the anorthite-phase separation accompanying structure at the body-glaze boundary.

In general, the origin and evolution of white porcelain from the Baihe kiln can be summarized as follows:

(1) Application of low Fe_2O_3 glaze materials
(2) Application of low Fe_2O_3 body material, or application of a low Fe_2O_3 engobe layer to cover the dark color of body
(3) Formation of an interaction layer at the body-glaze boundary
(4) Invention of a unique variety of fine white porcelain, and the invention of blue-and-white porcelain (a derived variety)

17.5 Conclusion

(1) The white porcelains made in the Baihe kiln of the Northern Wei Dynasty are the earliest white porcelains ever found which were discovered in company with celadons, reflecting the evolutionary pathway from celadon to white porcelain.

(2) The continuous and sustainable ceramic-making industry in the Gongyi area laid the basis for the prosperity of the Baihe kiln. Based on celadon production, white porcelain eventually came into being with the aid of intentional selection and disposal of raw materials, improvement of body and glaze recipes, and repeated practices.

(3) White porcelain went through a developmental process from rough to fine, and primary to mature. This is a great breakthrough in porcelain making technology in ancient China. The invention of white porcelain fundamentally changed the situation of celadon being the exclusive type since the East Han Dynasty.

Acknowledgments

This work was supported by key project of Natural Science Foundation of China (No. 51232008) and National Basic Research Program of China (973 Program) (No. 2012CB720901)

Reference

1. Henan Provincial Institute of Cultural Relics and Archaeology, Cultural Relics Institute of China, Nara National Institute for Cultural Properties, in *The New Archaeological Discovery at Baihe Kiln Site in Gongyi* (Elephant Press, Zhengzhou, 2009)

Chapter 18

第十八章

A New Method of Identifying the Flux in Ancient Chinese High Fired Glaze — Sr Isotopic Composition Analysis

Hongjiao Ma, Julian Henderson† and Jane Evans‡*

*Institute for Cultural Heritage and History of Science & Technology, University of Science and Technology Beijing, Beijing 100083, China

†Department of Archaeology, University of Nottingham, University Park, Nottingham NG7 2RD, UK

‡NERC Isotope Geosciences Laboratory, British Geological Survey, Keyworth, Nottingham NG12 5GG, UK

18.1 Introduction

One of the most important aims of ancient ceramic studies is to identify the raw materials used for making them. Glaze making is an intrinsic part of ceramic technology, and the choice of flux was often the major difference between the glaze making techniques carried out at different ancient ceramic production sites. In the context of Southern Chinese high fired glaze, two major fluxes, burnt limestone and plant ash, are most commonly suggested. However except for a few types of glaze, such as Jingdezhen limestone glaze for which we have multiple historical sources and which has been passed down through the continuous ceramic craft, the raw materials used in most ancient glaze types are suggested only by reference to chemical compositions. But it has been found that the glaze's chemical composition is not always a good indicator of the flux used, because the oxides that are often selected to provide a distinction between plant ash and limestone, such as P_2O_5 and MgO, are not always consistently diagnostic. A very knowledgeable scholar in the field of plant ash glaze — Robert Tichane, the author of 'ash glazes' has made the following comment[1]

> *'It would be convenient if there were some kind of internal clue that would tell us whether a glaze contained wood ash, but this does not seem likely … One could hope that some concentrated element (such as phosphorus) present in the ash, would indicate that ash was a glaze component. Unfortunately phosphorus from bone ash would also be commonly available and the presence of phosphorus in a glaze could not be identified as to source. No element in either ash or glaze would seem to be a convenient marker.'*

Such a convenient internal clue seems to have been found. In recent years in a very closely related area — ancient glass studies, Sr isotopic analysis has been developed into a very useful tool to identify the flux raw material component in the glass.

Based on a large number of Sr concentration data from previous studies and six Sr isotopic compositions analysed in the study, Wedepohl and Baumann[2] concluded that Roman imperial period glass was made by using marine molluscan shells as a calcium carbonate raw material, because limestone could not provide such a high concentration of strontium and the $^{87}Sr/^{86}Sr$ signature of the analysed samples suggested a modern seawater origin.

Freestone *et al.*[3] demonstrated how the glass making raw materials contribute to the Sr isotopic compositions of glass by studying glass from three origins — glass made from natron and coastal sand from *Bet She'an* and *Bet Eli'ezer* in Israel; glass made from natron and limestone-bearing sand from Ashmunein, Egypt; and glass made from plant ash and a lime-poor silica source found in Banias, Israel. The results show the Sr isotopic compositions of the three types of glass are distinguishable, and they are dominated by the Sr contribution from the calcium bearing raw material. Strontium from marine shell, such as found in *Bet She'an* and *Bet Eli'ezer* glass, gives a $^{87}Sr/^{86}Sr$ signature of modern seawater and high Sr concentration; Sr from limestone, such as found in Ashmunein glass, gives a lower $^{87}Sr/^{86}Sr$ signature and low Sr concentration; Sr from plant ash, as found in Banias glass, gives an $^{87}Sr/^{86}Sr$ signature inherited from the local geology and high Sr concentrations.

Henderson *et al.*[4] progressed the inquiry of how glass making raw materials contribute to the Sr isotopic compositions of glass by studying the Sr isotopic compositions of four natron glass samples, seven plant ash glass samples, two quartz pebble samples and two plant ash samples from al-Raqqa, Syria. The Sr compositions of the four natron glass showed that feldspar could be another end member in the natron glass Sr balance. The Sr isotopic composition of al-Raqqa plant ash glass and its raw material — sand and plants harvested near the site verified the suggestion that quartz pebbles contributed in a negligible way to the Sr isotopic composition in glass and that plant ash glass inherited its Sr isotopic composition from the local ashed plant.

High fired ceramic glaze is basically a thin layer of glass. So the same method of using Sr isotopic composition to identify the flux should apply to high fired ceramic glaze as well. This study is a demonstration of the initial knowledge that has been revealed in Sr isotopic analysis applied to ancient high fired Chinese ceramic glazes. The main part of this article is divided into two sections. The pilot study that focuses on two glazes with known recipes: Jingdezhen limestone glaze and Yue plant ash glaze is introduced in section one to demonstrate the basic Sr isotopic composition pattern of limestone fluxed glaze and plant ash fluxed glaze. Based on the knowledge learnt from the pilot study, a case study of the Sr isotopic compositions of Ming dynasty Dehua 'Blanc de Chine glaze', Qing Dehua blue-and-white glaze and their flux components is presented in the section two.

18.2 Methodology

Chemical analyses were carried out on our glaze samples to get the basic information. Fragments of each ceramic shard were mounted in cross-section in epoxy resin blocks and polished to a 0.25 mm diamond paste finish so as to reveal a flat analytical area. The blocks were carbon coated so as to prevent surface charging and the distortion of the electron beam during analysis.

A JEOL JXA-8200 electron microprobe in the Microanalysis Research Facility in the Archaeology department of Nottingham University was used for the chemical analysis of the samples. The system is equipped with four wavelength-dispersive X-ray spectrometers with LIF, TAP, PETJ and LIFH crystals, a single energy-dispersive X-ray spectrometer and both secondary and backscattered detectors. A defocused electron beam with a diameter of 50 μm was used so as to prevent volatilisation of light elements such as sodium. The probe was run at an accelerating voltage of 20 kV and a beam current of 5 nA. The system was calibrated with a mixture of mineral and metal standards. A commercial ZAF program was used to correct and quantify the results. The Corning B glass standard was routinely used as a secondary standard to check for accuracy and precision and to monitor any drift in the instrument.

For Sr isotope analysis, the glaze samples first needed to be separated from the ceramic bodies. Because the glaze layer on the ceramic body is very thin (200–400 μm), the samples taken are even thinner. This separation was done by 'peeling' the glaze off the porcelain using a dental drill. First the top layer of ceramic shard was cut off with a flexible diamond edged rotary dental bit, then inner surfaces were mechanically cleaned with a tungsten carbide bur to remove the adhering ceramic body. The resulting samples were then transferred to a clean (class 100, laminar flow) working area for further preparation. In a clean laboratory, the samples were first cleaned ultrasonically in high purity water to remove dust, they were rinsed twice, dried down in high purity acetone and then weighed into pre-cleaned Teflon beakers. The samples were spiked with ^{84}Sr tracer solution and dissolved in Teflon distilled 16M HNO3. Strontium was collected using Dowex resin columns.

Strontium was loaded onto a single Re Filament with TaF following the method of Birck.[5] The ^{87}Sr/^{86}Sr and strontium elemental concentrations were determined by Thermal Ionisation Mass spectroscopy (TIMS) using a Thermo Triton multi-collector mass spectrometer at the NERC Isotope Geosciences Laboratory. The international standard for $^{87}Sr/^{86}Sr$, NBS987, gave a value of 0.710250 ± 0.000006 (no 8, 2σ) during the analysis of these samples. Blank values were in the region of 100 pg.

18.3 Section one: The study of Sr isotopic compositions of Jingdezhen porcelain glaze, Yue plant ash glaze and the raw materials used to make them

18.3.1 *Research objective*

This is a pilot study in the application of Sr isotopic analysis to ancient Chinese high fired glazes. Since there is no previous knowledge about its application in this particular context (or, as far as we know, any other ancient glaze), the study needed to start with well-understood ceramic materials which could be used as references to guide us in how to interpret the results. Jingdezhen limestone glaze and Yue plant ash glaze were chosen as the examples because they are known to have been fluxed mainly from burnt limestone and from plant ashes respectively. In this study the basic Sr isotopic compositions of the two types of glazes have been revealed and they are interpreted in terms of the raw materials used in their production.

18.3.2 *Samples and relevant background*

In terms of the limestone glaze, no other site is more representative than Jingdezhen, because arguably Jingdezhen is the place where limestone glaze was first developed in South China and where it was applied for the longest time. Blue-and-white ware was the commonest Jingdezhen porcelain type which was covered with limestone glaze. Twelve blue-and-white shards from Jingdezhen were selected for this study, ten of them are Ming and Qing dynasty official kiln products from different reigns: XD1, XD3 (Xuande period: 1425–1435 AD), CH2, CH5 (Chenghua period: 1464–1487 AD), ZD5, ZD11 (Zhengde period: 1505–1521 AD), JJ1, JJ10 (Jiajing period: 1521–1567 AD), QL3, QL7 (Qianlong period: 1735–1795 AD), and the other two are high quality 'folk kiln' export porcelain wares from the Ming dynasty Wanli period (1572–1620 AD): EP1, EP2.

Plant ash glaze was the dominant type of glaze used before limestone glaze was introduced. The Yue kilns produced one of the highest quality ash glazed ceramics of all time. At the peak of its production the Yue kiln high quality products were used as court tribute and the colour of these wares was praised in poems as being mysterious of eight samples from the Silongkou Yue kiln site near the Shanglin lake kiln complex, Cixi, Zhejiang Province, four dated to the Five Dynasties and Ten Kingdoms period (907–960 AD): Ya-1, Ya-2, Ya-3, Ya-5, and four dated to early Southern Song dynasty (1127–c.1150 AD): Yb-1, Yb-2, Yb-3, Yb-5, are included in our study.

Ancient Chinese glaze recipes are made up of two components — the flux and the siliceous portion. The siliceous materials for southern Chinese high fired ceramics were commonly weathered acid igneous rock, and in many publications they are referred to as 'porcelain stone'.[6] Porcelain stone forms the major part of the glaze slip[1] as well as the body paste. Two porcelain stone samples from Jingdezhen were included in our study: they are raw porcelain stone rocks sampled from Sanbaopeng pit (SBP Rock) and Xiaoli pit (XL Rock).

The flux for Jingdezhen limestone glaze is glaze ash: a mixture of a major component of calcium carbonate powder and a minor component of fern ash. Traditional Jingdezhen glaze ash is made by first burning limestone to quicklime (CaO) in a lime kiln. The quicklime is then left for a period of time and water is sprayed on its surface; in the process some of the quicklime absorbs water and turns into slaked lime ($Ca(OH)_2$). The resulting substance is piled up with fern in alternating layers and the whole is burnt again. The last step is repeated several times until a proper powder is obtained. Finally the glaze ash is washed to remove the unburnt fern residues and the unwanted alkali salts before being used.[7] Since there are still potters in Jingdezhen today who make hand-made porcelain wares. Either as high quality artefacts or fake antiquities, glaze ash made according to

[1] Glaze slip is a suspension which is applied to the surface of ceramic biscuit. This layer of substance turns into the glaze after being fired.

the traditional way is still available to buy in Jingdezhen. Two glaze ash samples (GA1, GA2) bought from Jingdezhen pottery craft shops have therefore also been included in this study.

18.3.3 *Results*

18.3.3.1 *Chemical compositions of glaze samples*

The chemical compositions of our glaze samples are given in Table 18.1. The chemical compositions of the same types of glaze produced in the same sites have been reported in Hu *et al.*[8] and Wu *et al.*,[9] and our results agree with these published results. Jingdezhen limestone glaze samples show high contents of alkaline oxides, comparatively low content of CaO and low contents of MgO and P_2O_5. These chemical features are regarded as typical for the glazes made by mixing porcelain stone and glaze ash, and discussed in detail in Hu *et al.*[8] The Yue kiln ash glaze samples have high contents of CaO, low contents of alkali oxides and high contents of MgO and P_2O_5. These chemical features are also regarded as typical for glazes made by mixing porcelain stone and plant ash, and discussed in detail in Wu *et al.*[9] These results confirm that our samples are typical of the suggested recipe types for Jingdezhen and the Yue kiln.

18.3.3.2 *Sr isotopic compositions of glaze samples and raw materials*

The Sr isotopic compositions of our glaze samples are given in Table 18.2 and plotted as Sr ppm against $^{87}Sr/^{86}Sr$ in Fig. 18.1. The plot shows that the Jingdezhen limestone glaze and the Yue kiln ash glaze have completely different Sr isotopic compositional patterns. The $^{87}Sr/^{86}Sr$ ratios of Jingdezhen limestone glaze samples cover a wide range while their Sr concentrations are restricted, and a mixing line is clearly identifiable in the plot. On the other hand, the Yue kiln ash glaze samples have a very consistent $^{87}Sr/^{86}Sr$ signature of 0.7120–0.7128, their Sr concentrations are high and have a large variation. These two patterns are clearly a reflection of different fluxes being used to make the two different glazes. The Sr isotopic

Table 18.1 The chemical compositions of the samples analyzed

Place	Label	SiO_2	Al_2O_3	Na_2O	K_2O	CaO	FeO	MnO	MgO	P_2O_5	Total
Jingdezhen samples	XD1	70.82	16.93	3.42	3.79	4.46	0.52	0.09	0.39	0.05	100.49
	XD3	68.91	15.63	2.93	3.37	8.70	0.64	0.10	0.93	0.07	101.29
	CH2	68.46	17.44	1.84	5.67	5.35	0.77	0.33	0.35	0.07	100.29
	CH5	69.31	16.90	1.73	5.93	3.80	0.92	0.10	0.26	0.05	99.00
	ZD5	68.89	15.75	2.80	4.65	6.23	0.76	0.12	0.24	0.10	99.53
	ZD11	67.65	16.47	1.76	5.49	6.83	0.82	0.11	0.33	0.08	99.54
	JJ1	64.85	15.63	2.21	4.75	9.61	1.18	0.48	0.43	0.10	99.26
	JJ10	66.56	17.11	2.43	4.35	8.38	0.72	0.10	0.37	0.06	100.09
	EP1	63.72	16.04	1.80	3.22	13.32	0.71	0.41	0.27	0.08	99.57
	EP2	67.50	15.77	1.88	4.21	9.83	0.71	0.09	0.22	0.04	100.25
	QL3	70.55	16.47	1.81	3.87	4.92	0.86	0.11	0.25	0.03	98.87
	QL7	69.21	16.87	1.50	3.96	5.46	0.93	0.89	0.22	0.04	99.08
Yue kiln samples	Ya-1	57.58	12.28	0.87	1.58	17.16	1.44	0.36	2.74	1.07	95.08
	Ya-2	59.09	12.98	0.95	1.81	15.78	1.97	0.26	2.31	0.77	95.93
	Ya-3	57.23	11.81	0.83	1.68	17.92	1.42	0.37	2.75	0.95	94.96
	Ya-5	59.62	12.36	0.93	1.75	16.40	2.04	0.28	2.56	1.08	97.01
	Yb-1	53.96	11.72	0.77	1.29	20.34	1.86	0.57	3.99	1.43	95.94
	Yb-2	54.22	11.58	0.96	1.85	19.13	1.60	0.47	3.26	1.23	94.31
	Yb-3	54.65	12.38	0.82	1.60	18.20	2.15	0.51	3.60	1.33	95.25
	Yb-5	53.68	11.90	0.84	1.51	20.60	1.77	0.54	3.58	1.44	95.84
Dehua samples	M1	67.84	16.95	0.16	7.42	6.85	0.25	0.27	0.60	0.18	100.53
	M2	65.50	15.63	0.15	7.73	7.44	0.26	0.28	0.58	0.16	97.73
	M3	66.99	15.96	0.14	7.70	7.39	0.27	0.30	0.64	0.18	99.57
	M4	64.02	16.43	0.36	6.79	8.36	0.21	0.37	0.54	0.17	97.25
	M5	68.14	15.64	0.14	6.44	8.73	0.25	0.30	0.75	0.15	100.54
	Q1	69.97	13.89	0.38	5.26	6.33	0.73	0.06	0.24	0.05	96.90
	Q2	67.44	11.77	0.27	4.12	11.24	0.67	0.08	0.36	0.08	96.02
	Q3	70.36	13.31	0.22	4.35	8.83	0.85	0.11	0.40	0.06	98.49
	Q5	70.37	12.92	0.24	4.56	7.34	0.75	0.08	0.36	0.03	96.64

Table 18.2 The Sr isotopic compositions of the samples analyzed

Place	Label	Sr concentration (ppm)	$^{87}Sr/^{86}Sr$
Jingdezhen limestone glaze sample	XD1	71	0.737772
	XD3	98	0.729991
	CH2	64	0.748799
	CH9	40	0.755702
	ZD5	102	0.743484
	ZD11	98	0.743235
	JJ1	41	0.754651
	JJ10	65	0.749738
	EP1	68	0.743327
	EP2	46	0.748450
	QL3	38	0.757685
	QL7	50	0.759138
Jingdezhen raw material samples	SBP Rock	57	0.749616
	XL Rock	38	0.759763
	GA1	207	0.708901
	GA2	281	0.708554
Yue kiln plant ash glaze samples	Ya-1	252	0.712852
	Ya-2	549	0.712464
	Ya-3	486	0.712394
	Ya-5	279	0.712813
	Yb-1	294	0.712469
	Yb-2	401	0.712838
	Yb-3	390	0.712650
	Yb-5	375	0.712097
Dehua samples	M1	342	0.717069
	M2	268	0.716679
	M3	125	0.718470
	M4	116	0.718291
	M5	176	0.717405
	Q1	75	0.715651
	Q2	83	0.714582
	Q3	101	0.714222
	Q5	59	0.716932
Dehua raw material samples	Pine Needle Ash	317	0.708726
	Pine Needle	33	0.708553
	Rice Husk Ash	14	0.712028

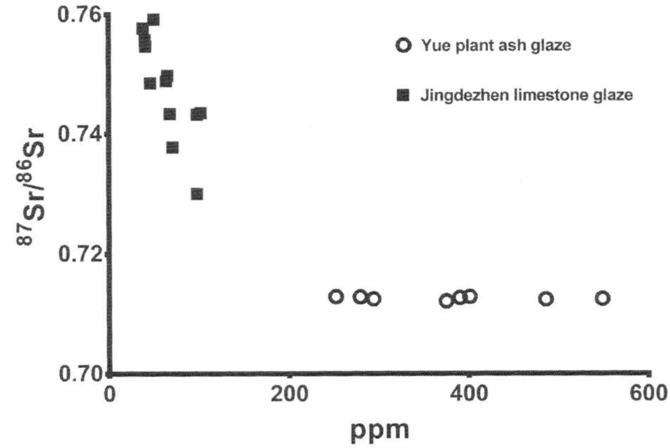

Fig. 18.1 A bi-plot of $^{87}Sr/^{86}Sr$ versus Sr concentrations (ppm) of Jingdezhen limestone glaze and Yue plant ash glaze samples in this study

compositions of glaze ash samples and porcelain stones samples are also given in Table 18.2.

18.3.4 *Discussion*

18.3.4.1 *The isotopic composition pattern of Yue kiln ash glaze*

The consistency of the $^{87}Sr/^{86}Sr$ signatures of Yue kiln samples suggests that one raw material component with a high Sr concentration and consistent $^{87}Sr/^{86}Sr$ dominates the Sr contribution to the glaze. According to Wood,[10] the Yue kiln glaze recipe is composed of 30 wt.%–50 wt.% of plant ash with the remainder being porcelain stone. The average Sr concentration retrieved from our Yue kiln samples is 378 ppm, with a maximum value of 549 ppm in sample Ya-2. The question to be asked here is which raw material component is the one that dominates the Sr contribution: porcelain stone or botanic ash?

Yue kiln porcelain stone is a weathered acid igneous rock.[11] According to the 1:1500000 geological map of Zhejiang Province, acid igneous rock dated to mid-late Jurassic is the main rock type in the area where the Yue kiln complex is located, and all kiln sites are

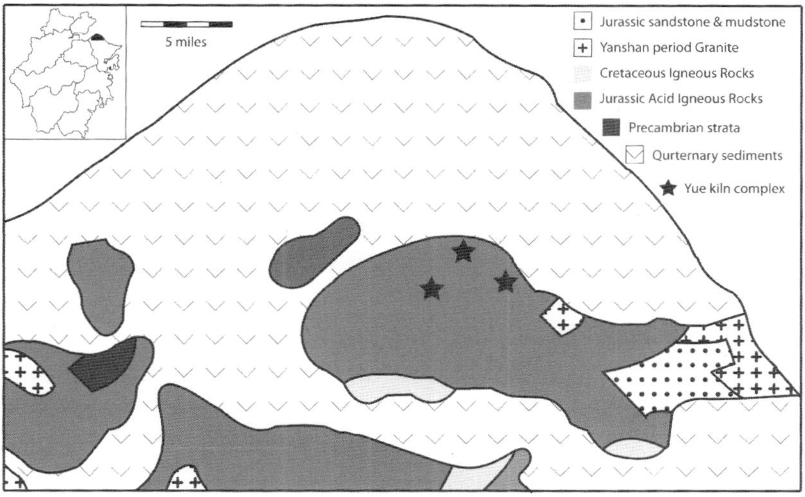

Fig. 18.2 The geology around Yue kiln complex

located on the strata of this type of rock (Fig. 18.2). So it is highly likely that the weathered portion of this rock was used as the porcelain stone for Yue kiln. Although no porcelain stone from the Yue kiln is included in this study, according to the geochemical study of Mesozoic igneous rocks in Zhejiang Province,[12] which reports the Sr contents of 11 acid Mesozoic igneous rocks (including the same rock type as found near Yue kiln site), the Sr isotopic concentrations of this type of rock are quite low, ranging from 21.8 to 196 ppm and averaging 84.5 ppm. So it is reasonable to assume that 100 ppm Sr was introduced with the porcelain stone at the Yue kiln; therefore 50 wt.%–70 wt.% of porcelain stone provides only 50–70 ppm of Sr in the final glaze. The remaining 300 ppm or more of Sr, which is the major portion, must have come from the other raw material — plant ash. So the consistent $^{87}Sr/^{86}Sr$ signature in Yue kiln glaze is a reflection of the $^{87}Sr/^{86}Sr$ signature of the ashed vegetation. This result is consistent with the results obtained from previous plant ash glass studies: plant ash is capable of providing a high concentration of Sr in the glass; the $^{87}Sr/^{86}Sr$ signature of plant ash glass is inherited from the plant used for burning the ash.[3,4]

18.3.4.2 *The Sr isotopic composition pattern of Jingdezhen limestone glaze*

The mixing line shown in the Sr isotopic compositions of Jingdezhen limestone glaze samples suggests that the Sr isotopic compositions of different raw material components provides the mixed Sr isotopic compositions of the glazes. Porcelain stone and glaze ash are the raw material components for making Jingdezhen limestone glaze, so their Sr isotopic compositions should form the ends of the mixing line. Porcelain stones and glaze ashes sampled from Jingdezhen were also analysed in this study, and their Sr isotopic compositions were plotted along with that of Jingdezhen limestone glaze as $^{87}Sr/^{86}Sr$ against 1/Sr ppm in Fig. 18.3. According to the Sr isotopic geochemistry theory, mixtures of different proportions of two mixing components with different Sr isotopic compositions form a straight line in plot of 1/ppm against $^{87}Sr/^{86}Sr$, and the Sr isotopic compositions of the mixing components are the ends of the mixing line.[13] As shown in the figure, the Sr isotopic compositions of Jingdezhen limestone glaze basically distribute along the two lines defined by the Sr isotopic compositions of glaze ashes and porcelain stones. So the mixing line shown

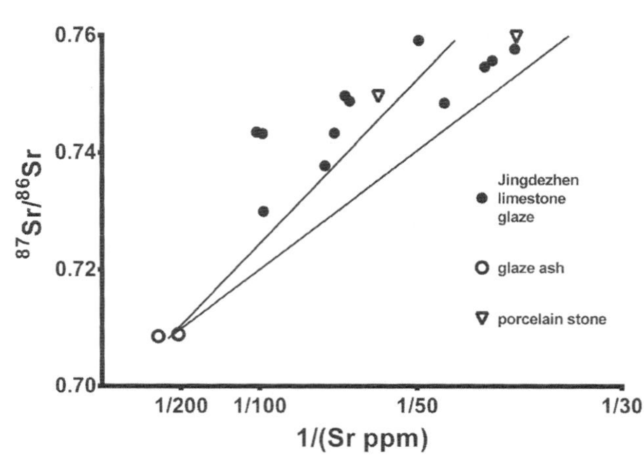

Fig. 18.3 A bi-plot of $^{87}Sr/^{86}Sr$ versus 1/Sr ppm in Jingdezhen limestone glaze, glaze ash and porcelain stone

for the Sr isotopic compositions of glazes can be used to suggest the raw materials involved in making the glazes.

18.4 Section two: The study of Sr isotopic compositions of Ming Dehua 'Blanc de Chine' porcelain glaze, Qing Dehua blue-and-white glaze and their flux components

18.4.1 *Research objective*

Two types of glazes are studied in this part of the work. They are Ming dynasty 'Blanc de Chine' glazes from the Jiabeishan kiln at Dehua and Qing dynasty blue-and-white glazes from the Meiling kiln at Dehua. The reason for choosing these two glazes is that there is a historical description of the raw materials used to make each glaze. However, there has never been any consideration of or comment on the legitimacy of these records based on scientific study. So it was hoped that by determining the Sr isotopic signatures of the glazes we could shed some new light on this issue.

18.4.2 *Samples and relevant background*

Dehua County is located in the mountainous area in middle Fujian province, and it has been a ceramic production centre in South China since the Song dynasty. Its total ceramic yield in history should be amongst the largest in China and its reputation as a ceramic production centre is inferior only to Jingdezhen.

Ming 'Blanc de Chine' is internationally very famous as a type of Chinese porcelain. During the most prosperous period of Chinese ceramic exportation, from the middle Ming to the early Qing dynasties (early 16th century — mid 17th century AD), 'Blanc de Chine' was regarded as the highest quality Chinese porcelain and sold to Europe in large quantities which is when it was given a French name. In China 'Blanc de Chine' is known as ivory white. The subtle warm tinge in its white colour and the high translucency are the signatures of this type of porcelain. Our five 'Blanc de Chine' samples, M 1–5, were excavated in 2001 from the Jiabeishan kiln, which is located in Baomei village, Xunzhong town.

'Blanc de Chine' began to decline in the early Qing dynasty, and blue-and-white became the major product in Dehua roughly from the same time. Early Qing Dehua blue-and-white was of a very high quality and was regarded in the contemporary export market as the only equal to the high quality Jingdezhen blue-and-white. Both were sold in large quantities to Europe. Our Dehua blue-and-white porcelain samples, Q 1–3 and Q 5, are from the Meiling kiln and dated to the early Qing dynasty. Meiling kiln is located in Sibin village, Sanban town. It has not been officially excavated, but multiple archaeological surveys regard it as an export blue-and-white production kiln site dating from the late Ming dynasty to the middle Qing dynasty. Our four samples were selected from the shards collected during surveys.

18.4.3 *Results*

18.4.3.1 *Major and minor chemical compositions of glazes*

The major and minor chemical compositions of the two types of glazes are listed in Table 18.1, and they basically agree with the published results of the same types of glazes from Li *et al.*[14] The chemistries of the two types of glazes are not quite the same. Firstly, Ming 'Blanc de Chine' glaze contains more alkaline oxides than Qing blue-and-white glaze. The alkaline oxides in glaze are K_2O+N_2O; along with calcium oxide they are the major fluxing oxides in ancient Chinese high fired glazes. The weight ratio of calcium oxides to alkaline oxide separates the glazes into different categories in ancient Chinese high fired glazes. The ratio in Ming 'Blanc de Chine' is around 1, so it should belong to the alkaline-calcium category of glaze; the ratio of Qing blue-and-white glaze averages 1.8, so it should belong to the calcium-alkaline category of glaze. Secondly, the Qing blue-and-white glaze is more siliceous. The SiO_2 to Al_2O_3 weight ratio ranges from 3.89 wt % to 4.36 wt % and averages 4.13 wt % in Ming 'Blanc de Chine' glaze, and ranges from 5.04 wt % to 5.73 wt % and averages 5.38 wt % in Qing blue-and-white glaze.

18.4.3.2 *Glaze Sr isotopic compositions*

The isotopic compositions of Ming 'Blanc de Chine' glaze and Qing blue-and-white glaze are listed in Table 18.2 and plotted as $^{87}Sr/^{86}Sr$ against 1/Sr ppm in Fig. 18.4. As shown in the figure the two types of glaze clearly separate into individual groups and a mixing line is suggested for each group.

By performing a linear fit to each group of isotopic compositions in the plot, the equation retrieved for Qing blue-and-white glaze mixing line is:

$$\frac{^{87}\text{Sr}}{^{86}\text{Sr}} = \frac{0.4036}{Sr\ ppm} + 0.7101.$$

And the equation retrieved for Ming 'Blanc De Chine' glaze mixing line is:

$$\frac{^{87}\text{Sr}}{^{86}\text{Sr}} = \frac{0.2885}{Sr\ ppm} + 0.7159.$$

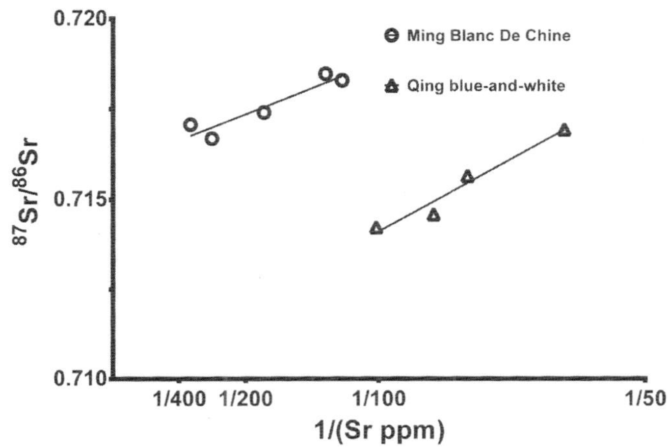

Fig. 18.4 A bi-plot of $^{87}Sr/^{86}Sr$ versus Sr concentrations (ppm) in Dehua porcelain samples

18.4.4 *Discussion*

18.4.4.1 *The flux used in Qing blue-and-white porcelain glaze*

In the ceramic industry report of Dehua written in the late Qing dynasty by Japanese observer Kitamura Yaichiro,[15] the contemporary common glaze recipe in Dehua is 'porcelain stone + a flux made of burnt limestone and rice husk ash'. The report does not specify which type of ceramic the glaze is for, but since blue-and-white porcelain was the major product in Dehua in that period, it is highly likely that this recorded recipe was the recipe for blue-and-white porcelain glaze.

This recorded Dehua glaze recipe is very similar to the Jingdezhen limestone glaze recipe; the only real difference is that fern ash was replaced by rice husk ash. Fern ash and rice husk ash actually have similar chemical compositions. They both have a high SiO_2 concentration and belong to the so-called high silica ash category used by potters. Chemical compositions of fern ash and rice husk ash quoted from Ash glazes[16] are listed in Table 18.3 as a reference. High silica ash is normally burnt from fast growing plants, such as cereal or grass, and are used to introduce extra silica in a glaze. So in this sense the recorded Dehua glaze recipe makes sense, because the techniques of blue-and-white porcelain making originate from Jingdezhen, and the glaze recipe is an intrinsic part of the whole set of techniques. The very similar glaze recipe in Dehua could have been transferred from Jingdezhen, but the plant species used for burning the ashes were slightly changed according to which species was locally available to provide the same extra silica addition.

Table 18.3 Quoted chemical compositions for three plant ashes[16]

	SiO_2	Al_2O_3	Fe_2O_3	TiO_2	CaO	MgO	K_2O	Na_2O	MnO	P_2O_5
Rice Husk	94.4	1.8	0.6		1.0		1.4			
Fern	55.0	19.3	1.7	0.3	8.6	7.4	4.8	0.6	1.4	0.9
Pine needle	3.4		7.0		23.9	6.9	30.0	0.8	7.4	16.0

However, chemical analysis is not able to identify if the glaze covering Qing blue-and-white is indeed made from such a recipe, because glazes made from different raw materials can potentially have quite similar chemical compositions. Porcelain stone was the common raw material for Southern high fired glaze, so local porcelain stone must have been involved in making the glaze. This is a non-debatable fact, so the issue is only down to whether the flux component can be identified. The Sr isotopic compositions of Dehua Qing blue-and-white glaze seem to shed some new light on this issue.

The Sr isotopic composition pattern of Dehua Qing blue-and-white glaze is very similar to that of Jingdezhen limestone glaze: a mixing line in the low Sr concentration range. The ends of a mixing line are the Sr isotopic compositions of the mixed raw materials.[13] Since the flux is the high calcium raw material in the two, and Sr is always associated with calcium, the high Sr concentration end of the mixing line is the Sr isotopic composition of the flux. Assuming the flux in Dehua blue-and-white glaze has a similar Sr concentration as Jingdezhen glaze ash which is around 250 ppm, and extrapolating the mixing line (1), the deduced $^{87}Sr/^{86}Sr$ signature of the flux is 0.7117.

The assumption that the flux of Dehua Qing blue-and-white glaze has a similar Sr concentration as Jingdezhen ash glaze is based on the fact that Dehua Qing blue-and-white glaze has a similar recorded flux to Jingdezhen glaze ash and the Sr isotopic composition patterns of the two glazes are also quite similar.

This deduced $^{87}Sr/^{86}Sr$ signature suggests that our assumption is legitimate and that the flux used for making the glaze could indeed have been a mixture of burnt limestone + rice husk ash. The typical Sr isotopic composition of a limestone is c. 460 ppm and $^{87}Sr/^{86}Sr$ = 0.7068-0.7091; the Sr isotopic composition of a rice husk ash sampled in Dehua is 14 ppm and $^{87}Sr/^{86}Sr$ = 0.7120. Our deduced Sr isotopic composition of the flux in Dehua Qing blue-and-white porcelain glaze (250 ppm and $^{87}Sr/^{86}Sr$ = 0.7117) is right in the middle between the Sr isotopic compositions of limestone and Dehua rice husk ash. This therefore suggests that the flux could have been a mixture of the two.

18.4.4.2 *The flux used in Ming 'Blanc de Chine' glaze*

The CaO concentration in Dehua Ming 'Blanc de Chine' glaze averages 7.75%, and the Sr concentration averages 205 ppm. These numbers suggest that limestone was not the major flux in Dehua Ming 'Blanc de Chine' glaze, because limestone would not have provided such a high concentration of Sr in the glaze. The CaO concentration in pure limestone ($CaCO_3$) is 56%, and its Sr concentration is around 460 ppm.[2] If limestone had been the main flux introducing the majority of the CaO in the glaze, to provide a CaO level of 7.75%, it would only have introduced 64 ppm Sr which is too low for the 205 ppm Sr concentration average found in our glazes even when considering the Sr contribution from porcelain stone (which is less than 100 ppm).[17]

The $^{87}Sr/^{86}Sr$ signature of the flux suggested by the mixing line also indicates that it has a non-marine origin. If limestone had been the flux component in the glaze, extrapolating the mixing line to 460 ppm, it would give a $^{87}Sr/^{86}Sr$ signature of 0.7165. This value is too high for a limestone. As we know limestone was formed from high calcium marine sediments, and it has a $^{87}Sr/^{86}Sr$ signature of seawater at the time of its formation. The $^{87}Sr/^{86}Sr$ signature of seawater fluctuates through geological time between 0.7068 and 0.7091.[3] So basically the $^{87}Sr/^{86}Sr$ signature of limestone should also be in this range. If diagenesis affected a limestone, its $^{87}Sr/^{86}Sr$ signature could fall outside this value range, but the value of 0.7165 is still too high for this to be the case. Therefore it indicates that limestone was not the major flux used to make the glaze.

Apart from limestone, the other high calcium flux generally used in ancient Chinese ceramic glazes was high calcium plant ash. The only historical reference regarding the glaze recipe of Ming Dehua porcelain is the entry in the famous craft encyclopaedia 'Tian Gong Kai Wu' 天工开物. The encyclopaedia states that 'pine needle ash is blended in for making glaze slip for porcelain Buda in Dehua.' Because Buddhist statues were a major theme for Ming 'Blanc de Chine', this entry probably describes the flux component in 'Blanc de Chine' glaze. No reflection or comment has been made on the accuracy of this description so far, but our newly developed Sr isotopic analysis has allowed us to derive some insight into this issue.

Pine needle ashes collected from Dehua was analyzed in this study. Because the pine needles were not collected near the Jiachunling kiln site where these 'Blanc de Chine' porcelain samples were manufactured and the $^{87}Sr/^{86}Sr$ signature of plants is highly related to the geochemistry of the immediate soil and hydrology system, so the $^{87}Sr/^{86}Sr$ signature of the pine needle ashes is not relevant to our study. But the information picked up from the Sr concentration of pine needle ashes have nevertheless provided us with some interesting results.

The Sr isotopic compositions of pine needles and pine needle ashes are shown in Table 18.1. The two $^{87}Sr/^{86}Sr$ signatures are basically the same, but the Sr concentration of pine needle ashes is significantly higher than that of the pine needles. This result has confirmed for the first time the long accepted but yet unproven hypothesis in this area that the burning process increases the Sr concentration from the plant to plant ash significantly.

Chemical compositions of pine needle ashes quoted from the meticulous study of plant ashes by Robert Tichane[18] have been used here to make an approximate estimation (Table 18.3). The quoted CaO concentration of pine needle ash is 23.9%, but this is the CaO concentration in plant ash which excludes the CO_2 content. Basically, a third of a plant ash compositional content is CO_2. To transform the quoted CaO concentration back to its real concentration, the normal practice recommended by Robert Tichane is to multiply the quoted value by 2/3. By doing this, the resulting CaO content in pine needle ash is reduced from 23.9% to 16%. Assuming pine needle ash was the flux used in making Ming 'Blanc de Chine' glaze and extrapolating the average Sr concentration in the glaze of 205 ppm proportionally by a ratio of 16%: 7.75% (the ratio of CaO content in pine needle ash to the CaO average in the glaze), we get the estimated Sr concentration in the pine needle ash as 423 ppm. The Sr concentration of pine needle ash retrieved from our analysis is 318 ppm. Considering the variable nature of plant ash and the possible practice of ash washing, which may increase the Sr concentration in the washed ash, 423 ppm can basically be considered as being at the same level as 318 ppm.

By bringing 423 ppm into the equation (2); the $^{87}Sr/^{86}Sr$ signature retrieved is 0.7166. As been pointed out earlier this signature does not suggest a marine origin for the flux, but it could definitely be passed down from a terrestrial origin, such as plant ash. So the Sr isotopic composition of the Ming 'Blanc de Chine' glaze suggests that a high calcium plant ash could have been the flux used for making this glaze. Although this approach cannot identify which plant species was ashed, it is still a positive addition to the historical record in 'Tian Gong Kai Wu' 天工开物 about pine needle ashes being blended in when making the glaze slip.

18.5 Overall conclusions

In the context of ancient Chinese high fired glaze, the Sr isotopic composition, Sr concentration and $^{87}Sr/^{86}Sr$ signature combined, of glaze is able to suggest the flux component involved in making the glaze. If a mixing line is produced by the Sr isotopic compositions of glazes, by making relevant assumptions, the Sr isotopic compositions of raw materials can be estimated and their identities suggested.

References

1. R. Tichane, in *International Conference on Ancient Chinese Pottery and Porcelain.* (Science Press, Shanghai, 1982), pp. 142–145.
2. K. H. Wedepohl and A. Baumann, The use of marine molluskan shells for Roman glass and local raw glass production in the Eifel area (Western Germany). *Naturwissenschaften* **87**, 129–132 (2000).
3. I. C. Freestone, K. A. Leslie, M. Thirlwall and Y. Gorin-Rosen, Strontium isotopes in the investigation of early glass production: Byzantine and early Islamic glass from the Near East. *Archaeometry* **45**, 19–32 (2003).
4. J. Henderson, J. A. Evans, H. J. Sloane, M. J. Leng and C. Doherty, The use of oxygen, strontium and lead isotopes to provenance ancient glasses in the Middle East. *Journal of Archaeological Science* **32**, 665–673 (2005).
5. J. L. Birck, Precision K-RB-SR isotopic analysis-application to RB-SR chronology. *Chemical Geology* **56**, 73–83 (1986).

6. Y. Guo, Raw materials for making porcelain and the characteristics of porcelain wares in north and south China in ancient times. *Archaeometry* **29**, 3–19 (1987).
7. Z. Liu and C. Xu, The making method of traditional glaze ash and the science behind. 传统釉灰的制法及其工艺原理. *Journal of Ceramics* 陶瓷学报, 35–46 (1986).
8. D. Hu, H. Zhang and S. Liu, *Researches of porcelain wares unearthed from Ming dynasty Jingdezhen imperial factory remain*. 景德镇明代御窑遗址出土瓷器研究. (Science Press, Beijing, 2011), p. 123.
9. J. Wu, H. Luo, Q. Li, W. Li and J. Wu, Composition model and microstructure characteristic of Yue ware, Longquan ware and Southern-Song Guan ware. *Journal of the Chinese ceramic society* **38**, 1408–1414 (2009).
10. N. Wood, in *From Mine to Microscope: Advances in the Study of Ancient Technology,* A. J. Shortland, I. Freestone and T. Rehren, (eds.) (Oxbow Books, Oxford, 2009), pp. 190–200.
11. N. Wood, C. Doherty and S. Rastelli, in *Proceedings of International Symposium on Ancient Ceramics 2005,* J. Guo (ed.) (Shanghai Scientific and Technological Literature Publishing House, Shanghai, 2005), pp. 143–153.
12. H. Lapierre, B. Jahn, J. Charvet and Y. Yu, Mesozoic felsic arc magmatism and continental olivine tholeiites in Zhejiang Province and their relationship with the tectonic activity in southeastern China. *Tectonophysics* **274**, 321–338 (1997).
13. G. Faure, *Principles of isotope geology.* (Wiley, New York, 1986), p. 158.
14. W. Li, H. Luo, J. Li, X. Lu and J. Guo, The white porcelains from Dehua kiln site of China: Part I. Chemical compositions and the evolution regularity. *Ceramics International* **37**, 355–361 (2011).
15. J. Chen, L. Chen, *The ceramic history of Dehua* 中国德化瓷史. (Shanghai Jiao Tong University Press, Shanghai, 2011), pp. 109–121.
16. P. Rogers, *Ash Glazes.* (A & C Black, London, 2003), pp. 231.
17. H. Ma, J. Henderson and J. Evans, The exploration of Sr isotopic analysis applied to Chinese glazes: part one. *Journal of Archaeological Science* **50** (10): 551–558 (2014).
18. R. Tichane, *Ash glazes.* (Krause Publications, Iola, 1998), p. 198.

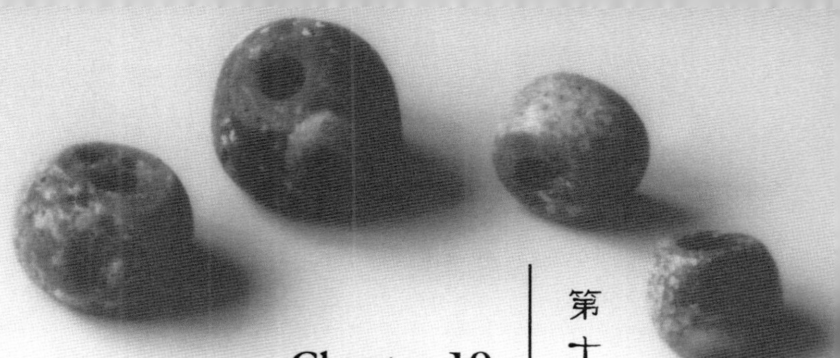

Chapter 19 第十九章

Research on Glaze-Coloring Mechanism of the Light Sky-blue Series of Ancient Jun Guan Porcelains using Modern Analysis Techniques

Guoxia Li, H. W. Sun,* R. W. Li,† X. L. Zhang,**
W. J. Zhao, W. J. Zhao,‡ Y. Yang,§ W. Li,* M. M. Cai**

*Faculty of Physical Engineering, Zhengzhou University, Zhengzhou, 450052, China;
†Department of Physics, Beijing Normal University, Beijing, 100875, China;
‡Institute of Cultural Relic and Archaeology of Henan Province, 450000, Zhengzhou;
§University of Anshun, Anshun, 561000, Guizhou Province

The Jun kiln is one of the five most famous kilns in ancient China. So far about 100 remains of Jun kilns have been found around Shenhou town, Yuzhou City in Henan Province. Amongst these, the one found at Jutai within the city of Yuzhou is called Juntai Kiln, which is believed by archaeologists to be the Jun Guan Kiln because its products were of the best quality. Jun porcelain was normally glazed twice, so the glaze was thicker and had colourful opalescence and patterns of earthworm trails. Jun porcelain presents a variety of glaze colours including sky green, sky blue, pale blue, rosy violet, begonia red, grape purple and so on. The glaze colours of Jun porcelain could not be predicted since they might change beyond imagination while being fired, even though they were of the same colour when placed in the kiln before firing. Jun porcelain has long been known for its great value; as the Chinese sayings goes, 'Gold can be priced, but the Jun porcelain is priceless', 'A great fortune could not match a piece of Jun porcelain'.[1] How the glaze-colour changed within Jun Porcelain kilns is a complex process with a variety of factors influencing one another. The study of the glaze-color changes within kilns has long been as an important research topic for ancient ceramics scholars. Kingery discovered liquid-liquid phase-separation structures[2] in glazes. After a meticulous study of these phase-separation structures, Mr. Chen found that the light scattering of these phase-separation droplets was the main cause of Jun porcelain opalescence.[3] In order to further study the characteristics of glaze-colour changes within the kiln producing the light sky blue series Jun porcelains, we measured the chemical compositions, the microscopic structures and the reflectance spectra of light sky blue, sky green and pale blue glazes Jun Guan porcelain samples. Our study produced some valuable information.

19.1 Analysis of the chemical composition of light sky blue series Jun Guan porcelain glaze samples

In order to investigate the relationship between the glaze color and the chemical composition of light sky blue series of Jun Guan porcelain glaze samples, we have selected 27 samples excavated from the site of T2H7 at the Yuzhou Juntai kiln. Among these samples, ten are light sky blue, nine sky green and eight pale blue. Energy Dispersive X-ray Fluorescence Spectroscopy (EDXRF) was used to determine the chemical compositions of each of these glaze samples. Fuzzy Clustering Analysis was applied to the data[4] to obtain the following dynamic fuzzy clustering analysis (Fig. 19.1).

We can see from the chart that these glaze samples can be roughly divided into three categories: the first category ranges from J249g to J245g with similar chemical compositions (17 samples in all with ten of them in light sky blue, five in sky green and two in pale blue); the second category covers eight samples, ranging from J243g to J222g (four of them in sky green and four others in pale blue), whose

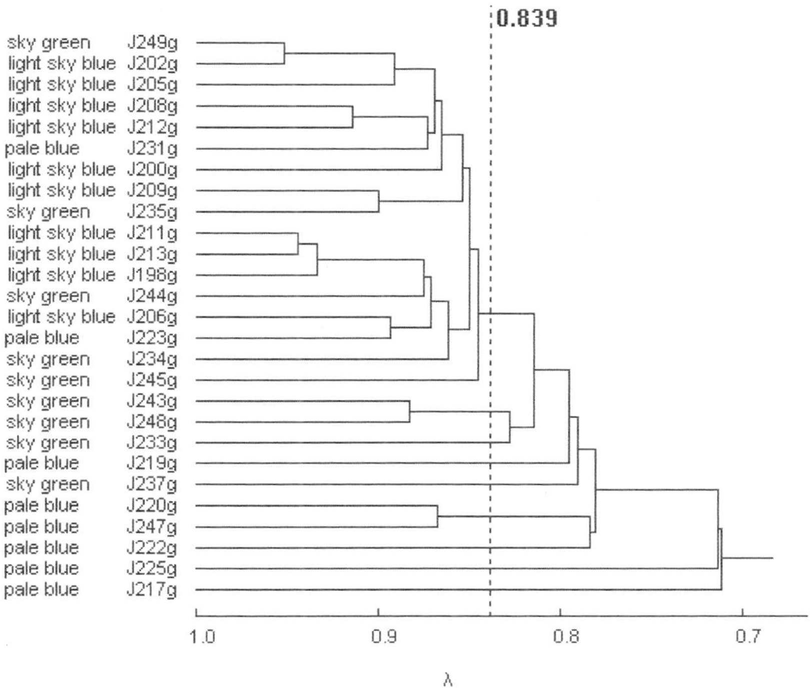

Fig. 19.1 The dynamic fuzzy clustering analysis of sky blue, sky green and pale blue Jun Guan glazes

chemical compositions are close to those of the first category; the third category contains only two pale blue samples (J225g and J217g), and their chemical compositions are different from those of the first two categories.

19.2 Microscopic Structure Analysis of light sky blue series of Jun Guan porcelain glaze samples

In order to understand the impact of micro structure on the colouring of light sky blue Jun Guan porcelain, Cold Field Scanning Electron Microscope JSM-6700F was used to observe and analyze the three color types of Jun Guan porcelain glaze samples. Three distinct micro structures were observed moving from the surface down to the bottom of the glaze. As shown in Figs. 19.2, 19.3, they

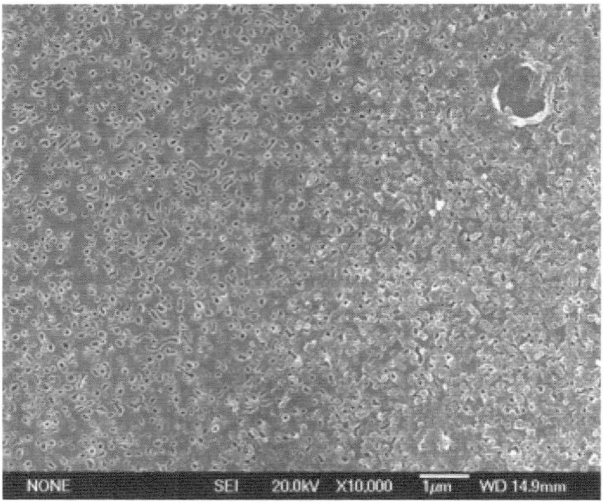

Fig. 19.2 Transition zone between the surface micro-bead-shaped phase-separation zone and the intermediate layer's three-dimensional network-shaped phase-separation zone of Jun Guan glaze

Fig. 19.3 Transition zone between the intermediate layer's three-dimensional network-shaped phase-separation zone and bottom layer's isolated micro-bead-shaped phase-separation zone of Jun Guan glaze

were respectively called the sparse and isolated micro-bead-shaped phase-separation zone, the three-dimensional network-shaped phase-separation zone, and the dense but isolated micro-bead-shaped phase-separation zone.

In this paper, we focus on how the phase-separation structure in the glaze surface impacts on glaze-colouring. Our Transmission Electron Microscope (TEM) imaging of Jun Guan Glaze surface in three glaze colour types is shown in Figs. 19.4 to 19.6

In these glazes, the optical uniformity was broken by phase-separation; therefore light scattering happens when incident light falls on them. The colouring of these phase-separation glazes abides by Rayleigh scattering and Mie scattering rules. When the diameter of the phase-separation micro particle is less than 100nm, Rayleigh scattering dominates and the glaze color shows blue; when the diameter of the micro particle is larger than the wavelength of incident light, Mie scattering dominates and the glaze color shows ivory-white. Since the light scattering of the micro phase-separation droplets in Jun Guan porcelain glaze is anisotropic, the opalescent glaze colour looks different when observed from different directions.

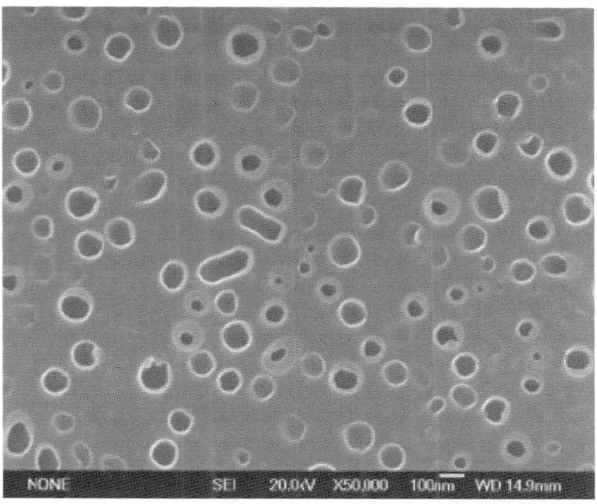

Fig. 19.4 The phase-separation structure of light sky blue glaze surface layer

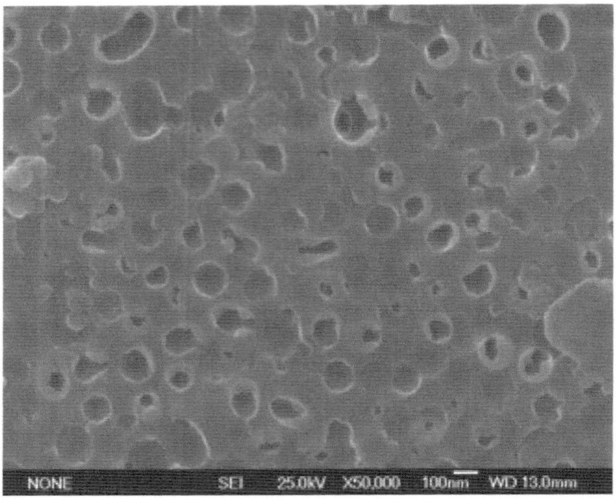

Fig. 19.5 The phase-separation structure of sky green glaze surface layer

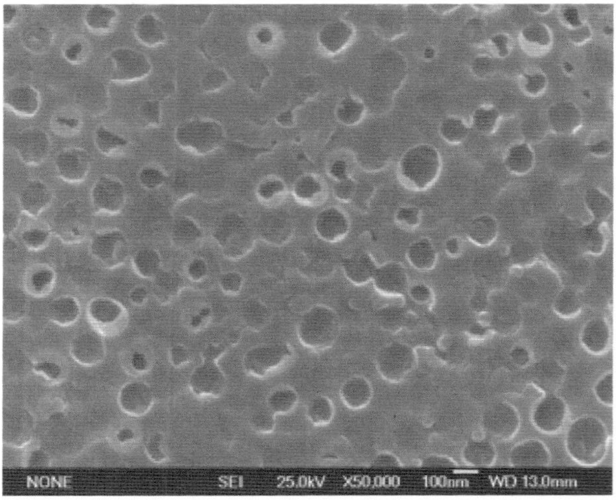

Fig. 19.6 The phase-separation structure of pale blue glaze surface layer

19.3 Analysis of the glaze colour dominant wavelength and firing temperature in the light sky blue series of Jun Guan porcelain

In order to show the color differences in light sky blue, sky green and pale blue Jun Guan porcelain glaze samples, we used the double beam

Table 19.1 The range of dominant wavelength and firing temperature of three color-types of Jun Guan porcelain glaze

Glaze color	Dominant wavelength range	Firing temperature
light sky blue	494–510nm	1260±20°C
sky green	510–520nm	1263±20°C
pale blue	520–541nm	1276±20°C

UV-visible spectrophotometer UV-3150 made by Shimadzu Corporation to measure the reflection curve of these samples and analyzed the range of their dominant wavelength. The thermo-dilatometer DIL-402C was also used to obtain the re-firing expansion curve so as to explore the influence exerted at a range of firing temperature on Jun porcelain glaze colours. The measured results of Jun Guan Porcelain samples in three color-types are shown in Table 19.1.

19.4 Analysis of the colouring mechanism in light sky blue Jun Guan porcelain

The colouring of Jun porcelain is not only related to light absorption of its colouring ion, but also to the light scattering of particles. The large heterogeneous structure in glazes reflects all wavelength of incident light, affecting only the colour saturation.

The three layers of phase-separation structures in Jun porcelain glaze (as in Fig. 19.2 and Fig. 19.3) all play some role in colouring. The bluish opalescence of Jun porcelain glaze is caused by short wavelength scattering in the visible spectrum from phase-separation due to dispersed particles in the glaze layer. In addition, we have also conducted research on the influence of different firing temperature systems and atmospheres on Jun porcelain colouring particle valence states and the coloring.[6]

The 17 samples in the first category of light sky blue, sky green and pale blue Jun Guan porcelain glaze samples in Fig. 19.1 have similar chemical compositions, indicating that their glaze source materials and formula are similar. The reason why their glaze colors differ is that their phase-separation and the valence states of their colouring particles are different, causing the difference in light scattering and absorption. The sky green and pale blue glaze samples of the second category in Fig. 19.1 have similar chemical compositions. Their glaze colors are different, which might be caused by their different phase-separation and different valence states of their colouring particles, causing different light scattering and absorption.[5] The two pale blue glaze samples of the third category have different chemical compositions compared to the light blue samples in this study. This indicates that the pale blue glaze formula is different from that of other samples. The reasons why they look pale blue might be that the light absorption of the coloring particles in them are different.

The diameter of the second phase-separation particles of light sky blue glaze samples in Fig. 19.4 to Fig. 19.6 is about 120nm or larger. Accordingly, the scattering of short wavelength light decreases and the overall saturation of the output spectrum also declines. Thus the blue colour on the exterior is reduced and it looks light blue. As for the sky green glaze, its phase-separation structures are irregularly formed and larger in diameter. Therefore, the selective light scattering further decreases and its blue colour also becomes fainter and becomes nearly white. The phase-separation structures in pale blue glazes are even larger, more irregularly shaped, and are denser. Therefore, the intensity of light scattering is no longer relevant to the wavelength and the scattering light looks white when observed from all directions. The whole glaze has an opalescent appearance and only very few remaining second phase-separation particles, which have diameters that are comparatively small, are able to selectively scatter light. This explains the slight bluish hint in the white exterior, seen as pale blue. Table 19.1 shows that the firing temperatures of light sky blue, sky green and pale blue glaze samples are different and the range of their main wavelengths are also different.

The Jun porcelain glaze has phase-separation and the firing temperatures exert great influence on the formation of phase-separation particles. Since the temperature and atmosphere inside the kiln are uneven, the size and distribution of the glaze phase-separation particles, as well as the valence states of the colouring particles, are different from one another. Therefore, the glaze's light absorption and scattering are not the same. As a result, Jun porcelain develop a rich variety of colours from being fired.

"Glaze-color changes within kiln" of Jun porcelain is a complex process. More efforts are needed to conduct experiments, tests and analysis so as to sort out the profound mystery of Jun porcelain's "Glaze-color changes within kiln".

19.5 Conclusion

Based on the analyses above we can draw the following conclusions:

[1] Most of the light sky blue, sky green and pale blue Jun Guan porcelain glaze samples have the same or similar chemical compositions while a few samples have different chemical compositions. Both the dominant wavelength and firing temperature of three colour-types of glaze samples are different and so are their microscopic phase-separation and shapes.

[2] There are three layers of different phase-separation structures in the Jun Guan porcelain glaze samples, namely, the surface layer, the intermediate layer and the bottom layer. Both the surface and bottom layers are of isolated calcium-rich phase distributed in the silicon-rich marix; the intermediate layer is of a three-dimensional network structure made of a calcium-rich phase and a silicon-rich phase.

[3] The colouring of light sky blue, sky green and pale blue Jun Guan porcelain glazes is not only related to their chemical compositions, but also to the sizes and shapes of the phase-separation particles, and valence state changes of colouring particles.

[4] The light sky blue glazes Jun Guan porcelain is has glazes containing phase-separation and the firing temperature has a great influence on the glaze phase-separations. Different temperatures and atmospheres inside the kiln will cause differences in the size and micro-shape of phase-separation particles, the changes in the valence states of the colouring particles, resulting in differences in light absorption and light scattering, and therefore the colour of glazes vary greatly from one another.

[5] Our experiments indicate that glaze materials of the same formula could generate different phase-separation structures at different firing temperatures, and develop different glaze colours accordingly.

References

1. Henan Provincial Institute of Cultural Relics and Archaeology, in *Yuzhou Juntai Kilns* (Elephant Publishing House, 2008), pp. ??-??.
2. W. D. Kingery and H. K. Bowen, *Am. Ceram. Soc. Bull* **62**, 1269 (1983).
3. X. Q. Chen, R. F. Huang, Sh. P. Chen *et al.*, *Structure and Artistic Appearance of Immiscible* ???
4. Y. M. Luo, W. Pan, S. Q. Li *et al.*, *Ceramics* **6**, 23 (2000).
5. Z. Y. Gao, J. Wang and S. H. Chen, *Nucl. Technology* **20**, 561 (1997).
6. X. L. Zhang, Zhengzhou University ?, ???, (2007).
7. Y. Yang, Zhengzhou University ?, ???, (2008).

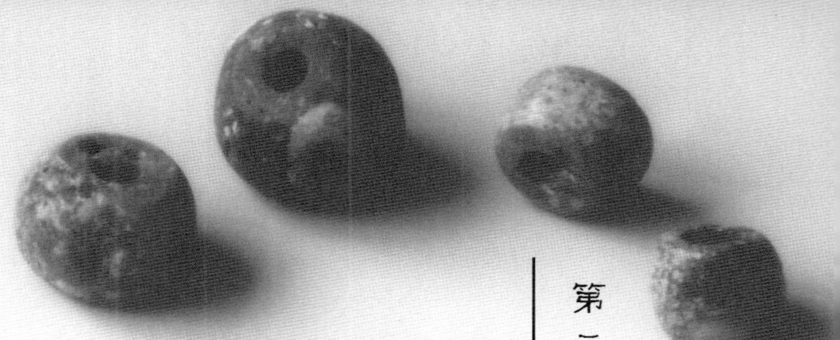

Chapter 20

第二十章

Preliminary Studies Based on the Chemical Analysis of Coulorful Glazed Pottery Unearthed from Tombs of Yue State at Hongshan Site

Xiaowei Xia, Qing Wang,* Qinghui Li,† Song Liu† and Donghong Gu†*

*Hongshan Site Museum, Wuxi 214000, China

†Shanghai Institute of Optics and Fine Mechanics, CAS, Shanghai 201800, China

20.1 Introduction

The Tombs of Yue State are located in Hongshan county which is in the southeast of Wuxi, Jiangsu province. Seven tombs dating to the early Warring States period were excavated by the Archaeological Institute of Nanjing Museum and Cultural Relics Management Committee of Xishan district, Wuxi from 2001 to 2004. Over 2300 cultural relics were found, so this which is considered as one of the most important findings for the archeology of Yue State.

The largest and highest grade discoveries are from the tombs of Yue State at Hongshan, in the noble tombs at Qiuchengdun (邱承墩). A large number of celadon, other high fired pottery, ordinary lower fired pottery, jade and a small number of glass objects were excavated from the Qiuchengdun (邱承墩) noble tombs. Besides the relics mentioned above, four colourful glazed pots were also found.

Fig. 20.1 The fragments and repaired colorful glazed pot found in the tombs of Yue State at Hongshan

The colorful glazed pots were placed together with jade and glass objects in same area of the main coffin chamber.

All the colourful glazed pots unearthed were broken when they were found. Two of four were badly broken and cannot be repaired. The sizes of the two repaired colorful glazed pots are 6.7 cm (the diameter of the foot) by 5.8 cm (the height), and 6.6 cm (the diameter of the foot) and 6.4 cm (the height), respectively. These two colourful glazed pots are very rare because of their large size, unique shape and their early date. One of the two colorful glazed pots is shown in Fig. 20.1.

The colourful glazed pot is described as "exquisite coiled snake spherical device with glaze" in the archaeological report. The body of the pottery is grey argillaceous clay with a spherical hollow shape. The spherical body is composed by eight snakes with the body or tail of one snake gripped by the mouth of the other snake. The head and body of the snakes are decorated with alternate blue and white glaze. The archaeological report presumes that the master of the tombs of Qiuchengdun (邱承墩) is the *Dafu* (大夫, a high-rankingperson in ancient China of the Warring States Period) of Yue State in the early Warring States, and the "exquisite coiled snake spherical device with glaze" is the highest level of ritual for the Yue State and is a symbol of royal power or divine right.[1]

In order to identify the raw materials and chemical composition of the colourful glazed pottery the Hongshan Museum and the centre of Sci-tech Archaeology, Shanghai Institute of Optics and Fine Mechanics cooperated with each other, and a non-destructive chemical analysis of the colourful glazed pot was performed using a portable XRF.

20.2 Experimental

A portable X-ray fluorescence spectrometer (pXRF, OURSTEX 100FA) provided by Shanghai Institute of Optics and Fine Mechanics was used for the chemical analysis. The equipment consisted of four parts: the detection unit (low vacuum detection unit and air detection unit), power unit, control unit and data processing unit. More operating parameters used can be found in Reference [2]. The colorful glazed pottery analyzed is shown in Fig. 20.1.

20.3 Results

The air detection unit was used because of the large size of the pot. Seven areas with different colours were selected and analysed. The results of the semi-quantitative analysis are listed in Table 20.1, and the qualitative XRF spectra are shown in Fig. 20.2.

As shown in Table 20.1, a certain amount of PbO and BaO have been detected in the blue parts of area 1 and area 2 and the contents of Al_2O_3 and K_2O are relative high. The blue hue is probably caused by Cu ions (cobalt could not be detected). The results for brown pigments in area 3 showed that the levels of Fe_2O_3 are very high, and that limonite may have been used as a colorant producing the reddish brown color. The main chemical components of area 4 are CaO SO_2 and TiO_2, which indicates that the area was repaired after the excavation, and that gypsum and a titanium pigment may have been used during the repair process. The results for area 5 show that the contents of PbO and BaO are relative high, while K_2O is low. Area 6 is also repaired according to the chemical results obtained because the chemical composition is similar to that of area 4. The results obtained of area 7 show that the contents of PbO and BaO are relative high, while that of K_2O is low. This is similar to that for area 5.

Table 20.1 The semi-quantitative results of the colorful pottery (air detection unit)

No.	Area	Al_2O_3	SiO_2	SO_3	Cl	P(ppm)	K_2O	CaO	TiO_2	Cr_2O_3	Fe_2O_3	CuO	ZnO	BaO	PbO
1	Blue	7.35	63.58	—	0.17	n.d.	6.21	4.57	—	0.09	0.64	4.26	n.d.	6.30	6.78
2	Blue	11.56	61.79	—	0.12	n.d.	5.43	1.84	—	0.03	3.29	1.70	n.d.	3.84	10.37
3	Brown, bottom	n.d.	36.86	7.44	n.d.	239	1.23	13.41	3.78	0.01	35.07	n.d.	n.d.	1.47	0.74
4	White	n.d.	2.33	29.83	n.d.	1754	n.d.	41.29	25.74	n.d.	0.62	n.d.	0.04	0.77	0.04
5	Grey1, bottom	5.52	52.34	—	n.d.	n.d.	0.44	8.79	—	0.21	2.03	n.d.	0.05	5.86	24.59
6	Grey2, bottom	n.d.	8.43	24.36	n.d.	1596	n.d.	34.39	32.82	n.d.	1.07	n.d.	n.d.	0.34	88ppm
7	White+brown, head	7.73	46.91	—	0.18	n.d.	0.22	4.63	—	n.d.	8.20	0.07	0.03	5.56	26.49

Note: "—"denotes overlapping peaks so the element can't be analyzed semi-quantitatively.

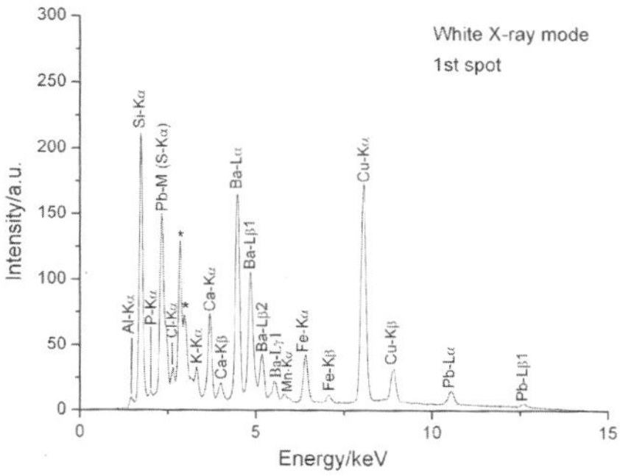

(a) spot 1 (circle blue area, direct model)

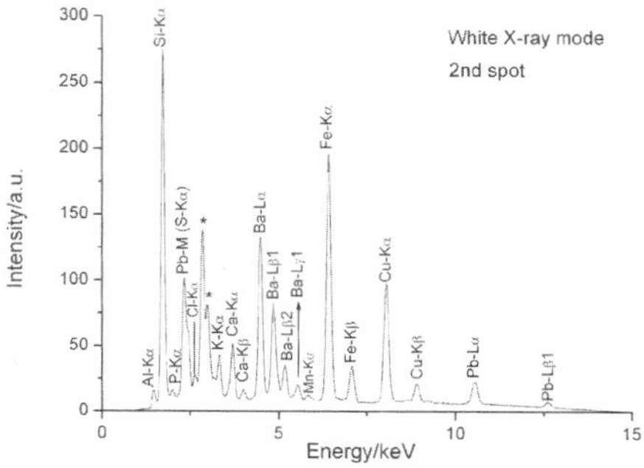

(b) spot 2 (circle blue area, direct model)

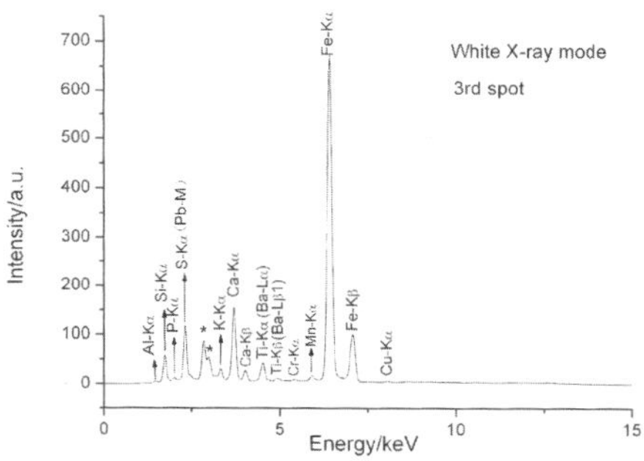

(c) spot 3 (brown area at bottom, direct model)

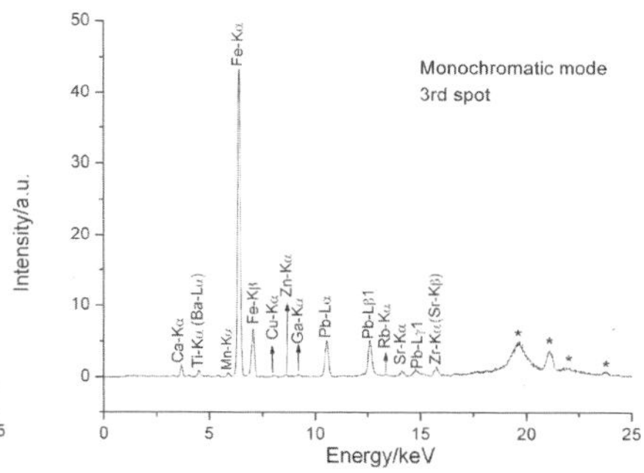

(d) spot 3 (brown area at bottom, monochrome model)

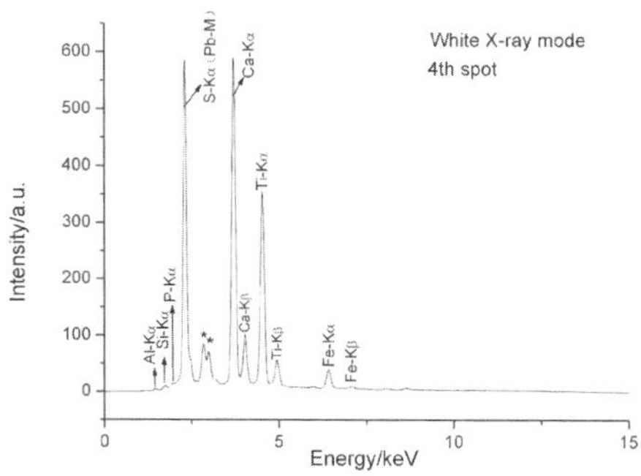

(e) spot 4 (white band at middle, direct model)

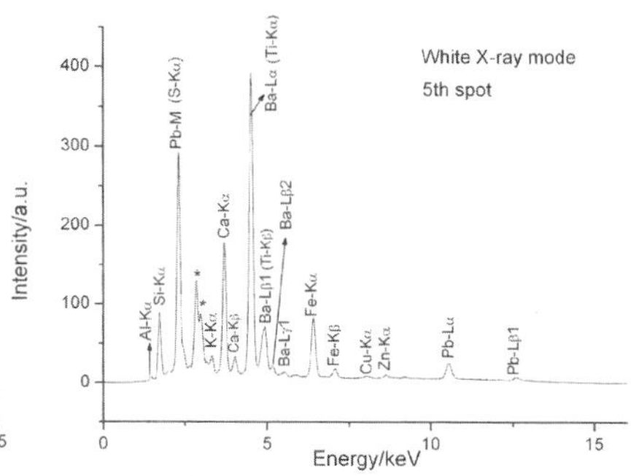

(f) spot 5 (grey area 1 at bottom, direct model)

Fig. 20.2 (*Continued*)

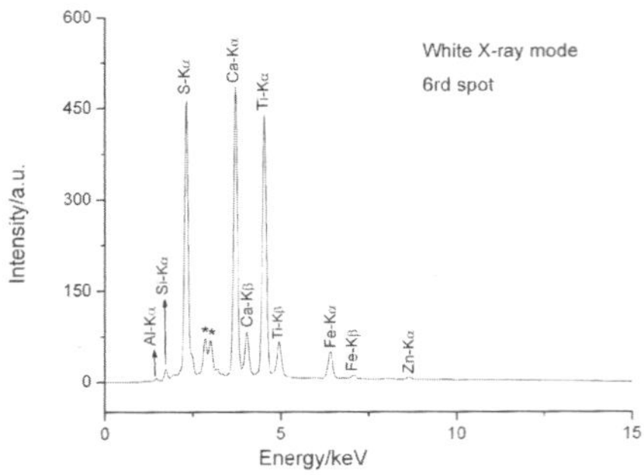

(g) spot 6 (grey area 2 at bottom, direct model)

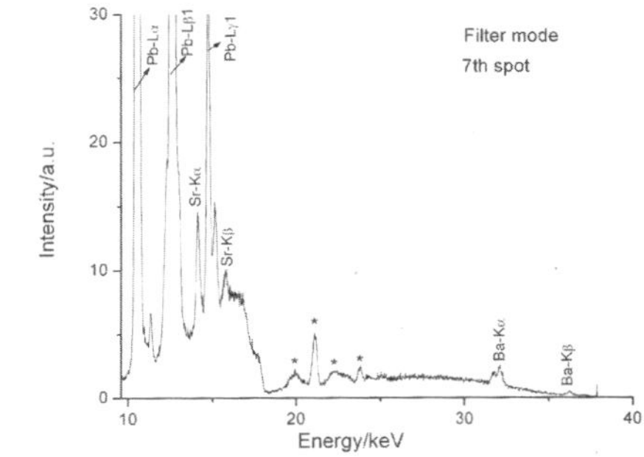

(h) spot 7 (wihte+brown area, head of snake, filter model)

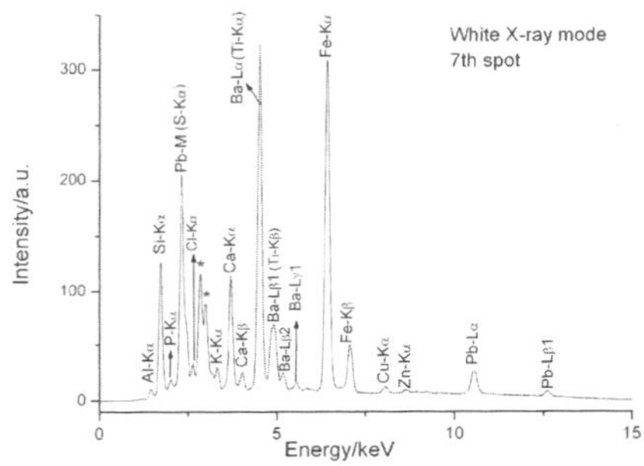

(i) spot 7 (white+brown area, head of snake, direct model

Fig. 20.2 The qualitative XRF spectra for the seven areas obtained by different work models

Therefore, the white surface material of the colourful glaze pottery belongs to the lead-barium silicate (PbO-BaO-SiO_2) glass type and the blue pigments which belong to lead-barium silicate materials, may be glass or possibly Chinese blue. The raw material used in the reddish brown pigment may be limonite.

20.4 Discussion

20.4.1 *Manufacturing techniques*

The chemical results show that the compositions of areas 1 and 2 (white) are similar to that of areas 5 and 7 (blue), and are lead-barium silicate. The white materials are vitreous, while the blue may not be.

The difference between the two is in the CuO colorant content. But the contents of PbO, SiO_2 and K_2O are also different, especially for the PbO. The content of PbO in the white material is higher than that detected in the blue material. Combined with results for the eye beads found in the Tombs of Yue State at Hongshan, the characteristics in common between the colourful pottery and eye beads are that white materials are used as a base or matrix, and blue ones for eye-ball or pottery decoration.

The inside of the colourful glazed pottery has no colour; the white glaze on the reddish brown pigments appears only in local areas on the pottery surfaces. This indicates that the dripping glaze technique was used. Differences also exist among the white glazes. In some areas it has been better sintered than in others, is a flat surface, has a hard texture there are no bubbles and the glaze and the body fit closely. But in some other areas this is not the case. The different appearances of the glaze on the same pot shows that the dripping glaze technique is not very advanced.

The decorative style of the white vitreous mater on the snake body is similar to the blue “S” patterns on the surface of the eye bead found in M48 tomb, Er-li-gang (二里岗), Zhengzhou, Henan

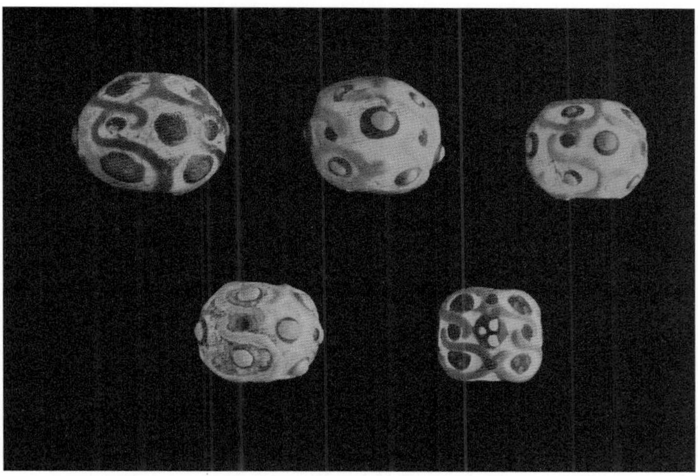

Fig. 20.3 Glazed pottery beads with “S” pattern dated to the Warring States period

province,[3] and dated to the Warring States period. The manufacturing techniques for the two decorations may be also similar.

According to the chemical analyses and shapes of the colourful glazed pot, its production process may include four steps: shape formation using hand sculpting techniques, surface painting with limonite, local painting with white materials, and then it was placed in the kiln and fired. Two different ways may have existed for making the blue decoration on the colourful glazed pot and the blue eyeball for the eye beads. One is that blue decorative glass was added using the dripping or casting method after the pottery was fired. The other is that blue decorative glass was added before the pottery was fired. The author believes that the first way may have been used for the production of the colourful glazed pottery.

20.4.2 *The significance of the present study for the origin of Chinese low fired lead glazed pottery*

The development process of pottery and porcelain mainly includes four steps: argillaceous clay, fired pottery under high tempertaure, glazed pottery, and celadon. The low temperature lead glazed pottery is an important invention and it was produced successfully in the Han Dynasty and thereafter proliferated widely. The manufacturing techniques of low temperature lead glazed pottery may be compared to that of lead-barium silicate glass in the Pre-Qin period.[4] The chemical compositions of colourful glazed pottery found in tombs of Yue State at Hongshan validate the conjecture mentioned above. As shown by chemical analysis, manufacture techniques, shape and function are different in Han Dynasty lead glazed pottery; the lead-barium glaze used as decoration may therefore result from experiments in making lead glazed pottery.

A lot of archaeological information shows that the pottery decoration techniques in ancient China had developed since the Neolithic period. The manufacturing techniques of colourful pottery, painted pottery and black pottery were mature and glorious. A large number of painted potteries are also derive from the pre-Qin period and are common in the tombs of East Zhou. Therefore, the decoration technology of colourful glazed pottery found in the tombs of Yue State at Hongshan are a continuation and development of the traditional ones, and innovation lies in attempts at pottery glazing.

According to the developmental levels of craftsmanship in East Zhou, the appearance of pottery glazing may be inspired by two aspects: one is in the manufacturing techniques for eye beads and the other is celadon glazing techniques.

There are many ways for in which eye beads can be made, but the most popular one is that the body of the eye beads is composed of clay and quartz sand. The surface of the body is covered by a thin layer of vitreous material with a variable thickness. The characters of eye beads mentioned above are very similar to that of colourful glazed pottery. The blue and white eye on the surface of colourful glazed pottery is in imitation of eye beads. The chemical compositions of the vitreous material of colourful glazed pottery and the eye beads are consistently similar to each other. All the eye beads found in the tombs of Yues tate at Hongshan belong to lead-barium silicate glass type. The biggest differences in the appearance of the two lie in the shape, the size of the body and the thickness of the glaze. Eye beads are small their glaze is thick and inhomogeneous; the pottery is relatively large and the glaze is thin and variable in thickness. Some glazed beads were unearthed from tombs dating to the Warring States period in Jiudian, Sui County, Hubei province, Changsha, Hunan province, Er-li-gang, Henan province and Ta-er-po (塔尔坡), Shaanxi province. The diameters of the beads range from 1 to 3 cm, the sizes are therefore between that of eye beads and the colourful pottery. These beads may be in imitation of the eye beads and may have been precursors for the colourful glazed pottery.

Three samples, an eye bead (Fig. 20.4) collected by cultural relics shop of Luoyang, Henan province[5]; a glazed pottery collected by the Museum of Chinese History[6]; and a lead-barium silicate ring (Fig. 20.5) unearthed from the tombs of Warring States in yuanshi county, Hebei province,[7] must now be considered. The sizes of the eye bead and glazed pottery are close to that of the colourful glazed pottery. The outside diameter of the glass ring is 4.45 cm. The white, brown, blue and yellow pigments on the surface of the glass ring

Fig. 20.4 The eye bead of Warring States collected in Luoyang, Henan province

Fig. 20.5 The glass ring of Warring States founded in Hebei province

found in Hebei province were confirmed to be lead-barium silicate glaze which is similar to that of the colourful glazed pottery found in tombs of Hongshan. The decoration on the glazed pottery found in Luoyang, Henan province, is also consistent with that of colourful glazed pottery.

A large number of high quality celadons were found in the tombs of Yue State at Hongshan. It is possible that colourful glazed pottery glazing techniques are the imitation of that of celadon. The difference is that the celadon is a high temperature glaze while the colourful glazed pottery is a low temperature glaze (the firing temperature of celadon found in tombs of Yue State at Hongshan reached above 1283°C[8]). The contemporaneous potteries with lead-barium silicate glaze exist in other regions: the *Huang* (璜) found in tombs of Chu in Jiangling, Hubei province, for example.[9]

It is strange that the colourful glazed pottery found in tombs of Yue State at Hongshan, used as ritual vessels and a symbol of royal power or divine right, was made with clay and not from hard pottery or celadon given that manufacturing techniques for celadon and other hard pottery was already mature. The reasons can be ascribed to the physical and chemical properties of the raw materials used. The design, manufacturing techniques and decorations show that the pottery is not an experimental product, but a 'final' product following hundreds of tests. This kind of colourful pottery found in tombs of Yue State at Hongshan must have been made from successive processes including origin, development and maturation. More colourful glazed pottery like the one found in tombs of Yue State at Hongshan may be found in the future.

20.5 Conclusion

Colourful glazed pottery shows a close relationship with glass making and ceramic making. It could eventually provide useful clues about where early Chinese glassmaking workshops were located. The workshops for early glassmaking in ancient China have not been found yet. The production of colourful glazed pottery gives us some inspiration. The workshops for glassmaking in pre-Qin periods may exist independently or together with ceramic workshops. If the latter is true, seeking the clues for glassmaking in the workshops of ceramics in pre-Qin period is one important aspect for glass archeology.

References

1. Nanjing Museum, Jiangsu Provincial Institute of Arhaeology and Cultural Relics Management Committee of Xishan District, Wuxi (Eds.), *Excavation Report on the Burials of the Yue State at Honghan* (Cultural Relics Press, Beijing, 2007), pp. 335–342.
2. S. Liu, Q. H. Li, F. X. Gan, *et al.*, *X-Ray Spectrometry*, **40**, 364 (2011).
3. S. Kwan (eds.), *Early Chinese Glass* (Art Museum, The Chinese University of Hong Kong, Hong Kong, 2001), pp. 17–54.
4. The Chinese Ceramic Society. *History of Chinese Porcelain*, (Cultural Relics Press, Beijing, 1982), pp. 114–116.
5. W. X. Zhao, *Relics from Central Plain*, **2**, 112 (1993).
6. J. Y. An, *A Brief History of Glasswares in China*, (Encyclopedia of China Publishing House, Beijing, 2000), pp. 20–27.
7. Q. L. Li, X. Y. Yu *et al.*, *Spectroscopy and Spectral Analysis*, **31**, 3395 (2011).
8. J. Wu, *et al.*, The studies of Celadon unearthed from the tombs of Hongshan, in *Excavation Report on the Burials of the Yue State at Honghan* (Cultural Relics Press, Beijing, 2007), pp. 354–375.
9. F. X. Gan, H. X. Zhao, Q. H. Li, *et al.*, Jiang Han Kaogu (江汉考古), **2**, 108 (2010).

Chapter 21

第二十一章

Characterization of the Ancient Glaze Used in Different Types of Ceramics from the Huangye Kiln Site in Gongyi County Using OCT Technique

Junqing Dong, Xin Yan,* Qinghui Li,**
Musen Guo† and Yongqing Hu†
*Shanghai Institute of Optics and Fine Mechanics,
Chinese Academy of Sciences,
Shanghai, 201800, China
†Institute of Cultural Relics in Henan Province,
Zhengzhou, Henan 450000, China

21.1 Introduction

Chinese ceramic glazes can be divided into two types: one is the high-temperature porcelain glaze such as calcium glaze, lime-alkali glaze and alkali-lime glaze, and the other is the low-temperature pottery glaze such as lead glaze and lead-barium glaze. In principle, the ceramic glazes from different kilns, pottery types and colours, made using different production systems from different districts and used to make different shapes should have various characteristics. There are also diverse features of the glaze developed in different historical periods or applied for different purposes, even from the same kiln. However, for the history and technology transfer of firing technology, as well as the physical and chemical changes in product processing, many ceramics appear similar, so there may be difficulties in identifying them visually.

During the past decades, the results of trace element and other microanalysis have provided valuable information about ceramic production, production processes and coloration characteristics. However, as most of the research objects were fragments, and there was a lack of representativeness, these results are not applicable to precious intact relics. Being elegant and rare, the products fired by the famous kilns were fascinating to ancient peoples and to modern populations. For this reason, lots of imitations have been made not only in modern but also in ancient times. It is a focus and a challenge to find out how to distinguish between the excellent ceramics produced by famous kilns and imitated products using nondestructive analytical techniques. Features of the shape and ornamentation are important avenues for research on the ancient ceramic glazes.[1] But their assessments are often influenced by subjective factors such as the ranges of experience and knowledge of the researcher. In terms of technology, most analyses on the chemical composition of ancient porcelain and pottery are obtained mainly by nondestructive analytical techniques, for example PIXE and XRF. [2–4] Recently, combined with other methods, important advances have been made in discrimination using micro-Raman spectroscopy (LRS),[5,6] providing the identification of the ceramic raw materials and the relationship between the index of polymerization (I_p) and firing temperatures. However, the results of qualitative and quantitative analysis by PIXE and XRF, and phase analysis by LRS are possibly influenced by the uniformity of the ceramic glaze and the status of the body/glaze. Optical coherence tomography (OCT) is a fundamentally new type of optical sensing technology based on the principle of Michelson interferometer and confocal microscopy that can produce cross sectional scans of the internal structure of materials and biological samples. Since its invention, the advantages of non-contact, high-resolution, high-sensitivity,[7] OCT systems have been attracting the attention of researchers all around the world, [8–10] and have been widely applied in medical, biomedical, material science, cultural heritage,[11–15] and so on. OCT technology can obtain two dimensional sectional images and three dimensional tomography images in the subsurface of the glaze layers and the body/glaze boundary. This provides a new important nondestructive analytical method for researching the glaze structure, crackle, defects, crystal grains, bubbles, layering and split-phase in glazes, the fit of body to glaze, as well as the relationship between the color, lustre and transparency with the chemical compositions of glazes.

In this work, taking the Huangye kiln at Gongyi, Henan province as an example, we propose to characterize different types of ceramic glazes and the body/glaze boundaries unearthed from this typical kiln dated from Sui Dynasty to the late Tang Dynasty (581 A.D.–907 A.D.) using OCT technique combined with the portable energy dispersive X-ray fluorescence spectrometer. The size and distribution characteristics of bubbles and strong scattering particles are described. The perspectives for future applications of statistical analysis of OCT images in different ceramics are explored in a preliminary way.

21.2 The archaeological background and samples

Known for producing the tri-color glazed pottery of Tang Dynasty, the Huangye kiln located north of Gongyi city, Henan province, is the earliest kiln which fired tri-color glazed pottery, and of the largest area producing the most excellent quality. The discovery of the Huangye kiln is a milestone in ceramics history, unveiling the answer to the historical question about which kilns producing tri-color glazed pottery in the Tang Dynasty. The Huangye kiln, originated in the Han Dynasty, developed during the Southern and Northern dynasties, and entered its mature phase in the Sui Dynasty. It began to fire tri-color glazed pottery as well as twisted body and twisted glazed (marble-texture) ceramics at the early stage of the Tang Dynasty in Huangye kiln. It is the earliest known application of the technique to the production of twisted bodies in China. The kiln culminated at the heyday of the Tang Dynasty. After a continuous development during the middle and late periods of Tang Dynasty, its scale gradually began to decline during the Five Dynasties. The history of the Huangye kiln spanned up to 500 years until it came to an end in the Song and Jin Dynasties. The type of products fired in the Huangye kilns was quite wide, including not only various colored porcelains but also single-color, double-color and tri-color glazed pottery.[16]

In the present work, 29 ceramic fragments (Table 21.1) dated from the Sui Dynasty to Jin Dynasty were nondestructively analyzed.

Table 21.1 List of ceramic samples unearthed from the Huangye kiln in Gongyi city, Henan province, China

Sample no.	type	color	amount	date
HY-I-2	porcelain	celadon	3	Sui Dynasty
HY-II-1-A	porcelain	white	1	The Early Tang Dynasty
HY-II-2	porcelain	bluish-white	3	The Early Tang Dynasty
HY-III(Q)-1	porcelain	white	3	The Glorious Tang Dynasty
HY-III(Q)-7-A	pottery	white	1	The Glorious Tang Dynasty
HY-III(Q)-12-B	pottery	tri-coloured glaze	1	The Glorious Tang Dynasty
HY-III(H)-3	porcelain	black	3	The Middle Tang Dynasty
HY-III(H)-4	porcelain	tea-dust glaze	3	The Middle Tang Dynasty
HY-III(H)-5-A	pottery	light yellow	1	The Middle Tang Dynasty
HY-III(H)-6-B	pottery	yellow	1	The Middle Tang Dynasty
HY-III(H)-8	pottery	blue	2	The Middle Tang Dynasty
HY-III(H)-9-A	pottery	green	1	The Middle Tang Dynasty
HY-IV-5-C	pottery	dark reddish brown glazed pottery	1	The Late Tang Dynasty
HY-IV-9-B	pottery	white and blue	1	The Late Tang Dynasty
HY-IV-10-A	pottery	white and green	1	The Late Tang Dynasty
HY-IV-11-A	pottery	tri-coloured glaze	1	The Late Tang Dynasty
HY-IV-13	pottery	yellow glaze with the twistable body	1	The Late Tang Dynasty
HY-IV-14	pottery	yellow glaze with the twistable body	1	The Late Tang Dynasty

There are 16 porcelain chips and 13 glazed pottery chips including celadon, white, bluish-white, black and tea-dust glazed porcelains, as well as white, yellow, green, dark reddish-brown, white and blue, white and green glazed potteries, yellow glazed pottery with a twisted body and tri-colored glazed pottery. The white glazed porcelain contains no apparent bubbles, and is thin, homogeneous and transparent. The glaze of bluish-white glazed porcelain is of a light green and turquoise color: the glaze thickness of different samples varies. It also apparently contains no bubbles. The character of the glaze is common among the bluish-white glazed samples: the samples were glazed inside and outside but not on the bottom. The tea-dust glaze and black glaze with low transparency have big differences. The transparency of dark tea-dust glaze is slightly higher, yet some are hardly transparent. The black glaze layer is relatively thick, with higher transparency than the tea-dust glaze. Compared with calcium glazes, the lead glaze layer is thinner, more transparent, and almost bubble-free.

21.3 Experimental

21.3.1 *The scanning optical coherence tomography (OCT) imaging technology*

The samples were analyzed using the optical coherence tomography (OCT) imaging technology illustrated in Fig. 21.1 in the Center of Sci-tech Archaeology, Shanghai Institute of Optics and Fine Mechanics (ISOM), Chinese Academy of Sciences (CAS). The OCT system consists of a low-coherence near-infrared light source, interferometer (Santec, IVS-2000, Japan), scanning probe and computer. The near-infrared light is high-speed scanning laser (HSL-2000), of which the probe power is less than 20 mW, with wavelengths ranging from 1260 nm to 1360 nm (centered at 1320 nm) at 20 kHz scanning rate. The axial resolution of this system is 10 μm in air, equivalent to approximately 5.3μm in ordinary ceramic material (assuming a refractive index of around $n = 1.5$), and the lateral resolution is about 17 μm centered at 1310 nm spectra at 20 kHz scan rate. The laser emitted by HSL was divided into two beams via the interferometer: one is

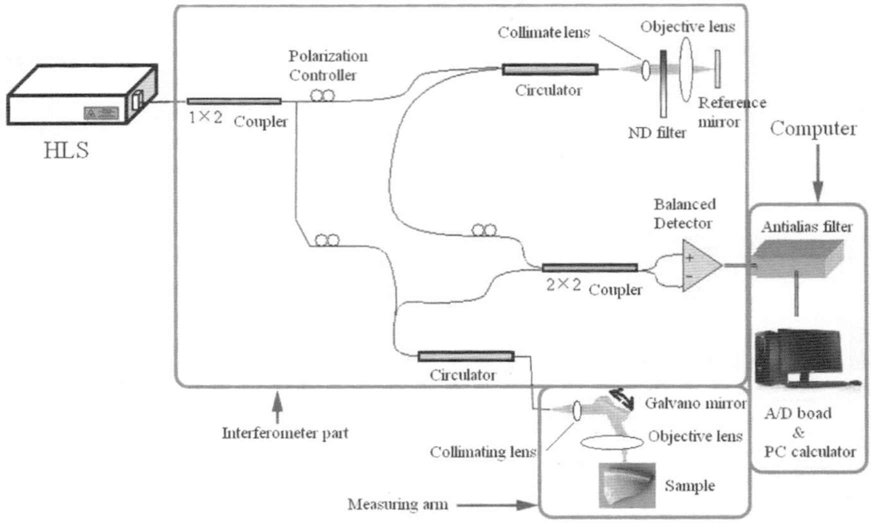

Fig. 21.1 Schematic diagram of OCT system

directed onto the mirror, the other is directed onto the sample. Both of them are reflected as reference beams. The backscattered light from the sample is coupled back in the system with the reference beam, digitized within a time scale and then analyzed in the Fourier domain to reveal the depth-resolved reflectivity profile (A-scan) at each point along the scanned section. By combining the series of A-scans in a linear fashion across the sample, a cross-section (B-scan) is obtained. Finally, cross-sectional images can be created by converting the B-scan raw data into gray-scale images.[15,17]

21.3.2 *The portable energy dispersive X-ray fluorescence spectrometer (pXRF)*

The pXRF spectrometer used in this analysis is an OURSTEX 100FA portable energy dispersive XRF spectrometer made by OURSTEX Co., Ltd. Japan, and modified by the Center of Sci-tech Archaeology, Shanghai Institute of Optics and Fine Mechanics. The pXRF spectrometer uses a palladium (Pd) excitation tube with a voltage of 40 kV, and a current of 0.5 or 1.0 mA. The maximum power is 50W. The X-ray beam spot focused on the sample is about 2.5 mm in diameter. The instrument is composed of four separate parts: an X-ray spectrometer, a compact digital processing unit controlled by a microcomputer, a compact power unit including high-voltage transformer, a power supply and a vacuum sample chamber. The X-ray spectrometer mainly provides the high voltage for the X ray source. The compact digital processing unit consists of control software and qualitative and quantitative analysis software. A silicon drifted detector (SDD) is specially designed for light element detection. The spectral resolution is 145 eV at Mn Kα. The detector can produce accurate quantitative analyses in air for elements heavier than Al, but it is not good for determining light elements (Na and Mg). However, its advantage is that it can be used to analyse any sample size, which gives us more freedom to select the measured area. In order to reduce the effect of atmospheric absorption on the characteristic spectral lines of light elements, a vacuum pump is present. The lowest air pressure is 400–600 Pa.[18]

21.4 Results

21.4.1 *The bulk chemical composition analyses using pXRF*

The bulk chemical compositions of samples (Table 21.2) were analyzed by pXRF before OCT technique analysis was performed. They could be divided into two types: the high-temperature calcium glazed porcelain and the low-temperature lead glazed pottery. The former consists mainly of SiO_2, CaO and Al_2O_3, with the average values of 61.62%, 11.77% and 15.36% respectively. The latter mainly contains SiO_2, PbO, and Al_2O_3, with the average values of 40.15%, 46.84% and 5.30%, respectively. There is a close relationship between the glaze color and levels of Fe_2O_3, the content of which decreases successively in black glaze, tea-leaf glaze, celadon glaze, bluish-white glaze and white glaze. For the glazed pottery, the coloring elements of yellow and dark reddish-brown glaze, blue glaze and yellow glaze are Fe, Co and Cu, respectively.

21.4.2 *Analysis of two-dimensional OCT data*

21.4.2.1 *Information about glazes conveyed in the two dimensional*

Table 21.2 Compositional results for glazes from the Huangye kiln site (wt. %)

Sample no.	Colour	Na_2O	MgO	Al_2O_3	SiO_2	P_2O_5	K_2O	CaO	TiO_2	Fe_2O_3	CoO	CuO	PbO	Type
HY-I-2-A	Light green	0.85	2.62	12.99	53.37	0.46	4.08	22.20	0.53	2.90	n.d.	tr	tr	Lime
HY-I-2-B	greenish-yellow	0.96	2.64	12.53	57.14	0.74	1.94	20.76	0.56	2.73	n.d.	tr	tr	glaze
HY-I-2-C	greenish-yellow	0.75	2.45	14.76	54.35	0.34	1.92	21.98	0.65	2.80	n.d.	tr	tr	
HY-II-1-A	white	1.52	2.34	17.74	63.49	0.56	2.35	10.99	0.14	0.88	n.d.	n.d.	n.d.	
HY-II-2-A	greenish-white	1.64	2.07	15.36	60.94	0.68	1.98	16.31	0.05	0.96	n.d.	tr	tr	
HY-II-2-B	greenish-white	2.10	1.63	15.74	64.76	0.58	2.30	11.93	0.05	0.92	n.d.	tr	tr	
HY-II-2-C	greenish-white	1.56	2.26	18.45	63.10	0.47	2.40	10.2	0.29	1.27	n.d.	tr	tr	
HY-III(Q)-1-A	white	1.35	3.00	16.30	58.37	1.01	2.65	16.28	0.25	0.79	n.d.	tr	tr	
HY-III(Q)-1-B	white	1.04	3.00	20.01	60.92	1.02	2.60	10.23	0.36	0.82	n.d.	tr	tr	
HY-III(Q)-1-C	white	2.07	2.31	25.10	54.68	0.56	2.92	10.72	0.80	0.83	n.d.	tr	tr	
HY-III(H)-3-A	black	1.01	3.72	13.72	64.83	0.32	3.13	6.96	0.86	5.45	n.d.	tr	tr	
HY-III(H)-3-B	black	1.19	2.36	12.77	66.55	0.34	3.96	7.23	0.67	4.93	n.d.	tr	tr	
HY-III(H)-3-B	black	1.14	2.78	13.57	67.27	0.39	2.79	6.15	0.76	5.15	n.d.	tr	tr	
HY-III(H)-3-C	black	1.14	3.83	13.28	65.93	0.38	3.67	5.76	1.02	4.98	n.d.	tr	tr	
HY-III(H)-4-A	tea-dust	1.68	3.02	12.76	63.22	n.d.	3.05	9.98	1.01	5.28	n.d.	tr	tr	
HY-III(H)-4-B	tea-dust	1.19	5.42	11.05	62.96	n.d.	3.86	8.85	0.84	5.82	n.d.	tr	tr	
HY-III(H)-4-B	tea-dust	1.28	3.64	14.09	65.45	n.d.	2.98	6.23	0.97	5.35	n.d.	tr	tr	
HY-III(H)-4-C	tea-dust	1.31	2.43	14.62	66.82	0.08	2.93	5.74	1.01	4.65	n.d.	tr	0.41	
HY-III(H)-4-C	tea-dust	1.53	2.26	14.57	67.91	0.16	2.85	5.26	1.07	3.99	n.d.	tr	0.4	
average value	tea-dust	1.34	2.82	15.45	61.34	0.54	2.86	12.26	0.57	2.88				

HY-III(Q)-7-A	tea-dust	0.86	0.72	3.85	41.76	0.35	0.58	3.91	0.16	0.29	n.d.	n.d.	47.53	Lead glaze
HY-III(Q)-12-B-b	Blue	0.97	0.34	4.90	37.47	0.29	0.38	0.54	0.19	0.97	0.48	0.46	53.01	
HY-III(Q)-12-B-y	yellow	1.02	0.40	7.51	34.50	0.27	0.65	1.40	0.20	4.38	0.00	0.08	49.59	
HY-III(H)-5-A-w	white	0.95	0.21	3.19	42.86	0.24	0.55	0.42	0.17	0.20	n.d.	n.d.	51.21	
HY-III(H)-5-A-y	yellow	1.19	0.73	8.00	43.99	0.10	0.93	1.64	0.24	2.81	n.d.	n.d.	40.37	
HY-III(H)-6-B	yellow	0.93	0.34	4.14	40.87	0.25	0.94	0.65	0.17	4.30	n.d.	n.d.	47.39	
HY-III(H)-8-B	bluish green	0.98	1.03	4.02	34.62	0.29	0.42	2.17	0.18	1.39	0.40	0.71	53.78	
HY-III(H)-8-C	dark blue	0.94	0.48	3.91	41.27	0.24	0.62	1.31	0.17	0.97	0.61	0.48	49.00	
HY-III(H)-9-A	green	0.94	0.64	5.44	37.53	0.16	1.32	1.30	0.17	1.10	n.d.	3.09	48.32	
HY-IV-5-C	yellow	0.92	0.31	8.27	38.03	0.15	1.30	1.59	0.19	6.52	n.d.	0.17	42.55	
HY-IV-9-B-w	white	0.92	0.44	3.82	47.75	0.11	0.67	3.14	0.17	0.51	n.d.	n.d.	42.48	
HY-IV-9-B-b	blue	0.97	0.66	4.20	39.43	0.17	0.65	3.37	0.17	1.20	0.68	0.34	48.17	
HY-IV-10-A-g	green	0.86	0.59	6.19	39.50	0.23	0.63	1.78	0.17	0.73	n.d.	3.79	45.52	
HY-IV-10-A-w	white	0.87	0.28	3.57	45.72	0.30	0.55	1.39	0.15	0.41	n.d.	0.33	46.44	
HY-IV-11-A-b	blue	1.02	0.31	6.39	41.21	0.30	0.84	0.74	0.18	1.44	0.15	5.05	42.36	
HY-IV-11-A-y	yellow	0.89	0.22	5.78	31.80	0.40	0.38	2.97	0.18	9.90	n.d.	0.17	47.32	
HY-IV-11-A-w	white	0.86	0.15	4.91	52.50	0.32	0.53	0.88	0.16	0.36	n.d.	0.15	39.17	
HY-IV-13-y1	yellow	0.98	0.32	6.02	33.10	0.44	0.72	1.30	0.20	4.41	n.d.	0.3	52.22	
HY-IV-13-y2	yellow	1.07	0.39	5.84	33.83	0.44	0.77	1.31	0.20	5.75	n.d.	0.18	50.21	
HY-IV-14-y	yellow	0.93	0.39	9.62	41.01	0.19	1.53	1.41	0.20	4.96	n.d.	0.21	39.53	
HY-IV-14-dy	dark yellow above the twistable body	0.86	0.59	9.52	37.8	0.12	1.78	1.68	0.20	8.52	n.d.	0.22	38.71	
average value	**N = 21**	**0.95**	**0.45**	**5.67**	**39.84**	**0.26**	**0.80**	**1.66**	**0.18**	**2.91**	**0.39**	**0.98**	**46.42**	

Note: b, blue; w, white; g, green; y, yellow; bk, black; l, light; d, dark; n.d., not detected; tr, trace.

cross-sectional images

A uniformity of glazes in most of the samples is observable except for the presence of unmelted grains. The glazes are usually thick on the vessel rims due to the firing process. Glazes are often weathered, especially of low-temperature glazed wares. Homogeneous glaze areas, with little or no apparent weathering were chosen: the results are shown in Figs. 21.2–21.8, in which the approximate tested positions are indicated by a red line. It has been found that the bright dot in the OCT grayscale image, which indicates that the scattered light from the object is higher is due to the relative refractive index. For the glaze layer, the relative refractive index appears in the interfaces between air and glaze as well as body and glaze. The bright line is the representation of the air/glaze boundary and this is different from the body/glaze boundary. Because the glaze infiltrated the body during the firing process and the obvious refractive index differences between glaze and body, the body is full of strong light scattering and the boundary is broadened producing a bright band. In addition, the air bubbles and crystal grains are very common in the glaze layer. There is a great refractive index difference between the inner air and the outer glaze due to the scattered light from the air bubble surfaces, forming a brilliant circle. For the strong scattered light from the lateral bubble beyond the detection range, only the scattered light of the upper and lower bubble surfaces are detected. Therefore we can see

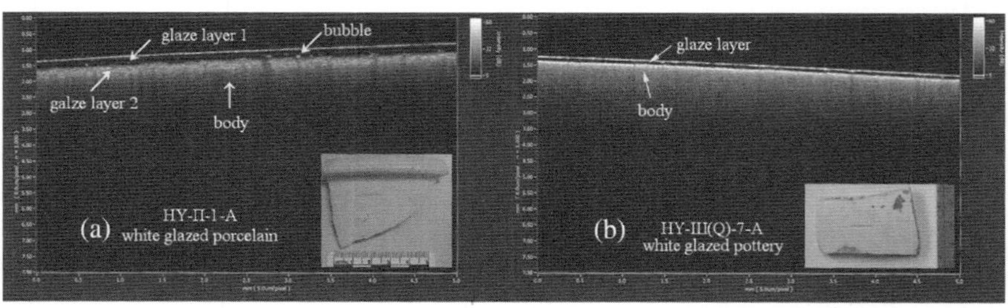

Fig. 21.2 2D gray-scale images of white glazed porcelain and white glazed pottery

two light lines in the grayscale image. The crystal grain in the glaze layer will form bright points or lumps for the similar reason.

There are obviously differences in the OCT gray-scale images of the high-temperature calcium glazed porcelain and low-temperature lead glazed pottery due to the different recipes and manufacturing processes. The 2D images of white glazed porcelain and white glazed pottery are illustrated in Fig. 21.2. The OCT image of the opaque body is uniformly grayish-white because of the strong intensity light scattering. An obvious stratification phenomenon is observed in the calcium glaze layer, commonly thicker than the lead glaze layer. Lots of little bubbles and little scattering particles are distributed in the glaze layer 1 in the top surface of the calcium glaze. The middle layer lying between the glaze layer 1 and the body is very thin — a thick line in the gray-scale image due to its low light absorption capacity and high light scattering capacity (Fig. 21.2a). We provisionally named it as glaze layer 2; it maybe due to glaze seeping into the body, whereas the situation of the lead glaze is clearly different from the calcium glaze. There are scarcely any bubbles or scattering particles in the thin lead glaze which has a 'high quality' light absorption. Furthermore, there is no obvious difference amongst the different colors of lead glazes in the OCT image. Compared to the calcium glaze, there are only two bright lines in the lead glaze gray-scale image. The upper line is the representation of the air/glaze boundary, whereas the lower line relates to the body/glaze boundary and is thinner and more compact than the one in the calcium glaze, which is a sharp contrast to the body (Fig. 21.2b).

21.4.2.2 *Two dimensional results for the high-temperature calcium glaze*

Figs. 21.3 and 21.4 present the 2-dimensional gray-scale images of five types of calcium glazes: white glaze, celadon glaze, greenish-white glaze, tea-leaf glaze and black glaze. Compared with tea-dust glaze and black glaze, the white glaze, celadon glaze and greenish-white glaze with a high grade of transparency (which equates to a weak absorption and scattering of light), with a few little air bubbles, have a clear body/

glaze boundary. As shown in Figs. 21.3a, 21.3b and 21.3c, for the white glaze sample HY-III (Q)-1, glaze layer 1 looks almost black with a rare occurrence of small scattering particles and bubbles; glaze layer 2 is like an uneven coarse line, which indicates that glaze has seeped to a varying depth into the body perhaps due to the degree of roughness and porosity of the body. The glaze/body boundary is very clear equating to strong scattering. Figs. 21.3d, 21.3e and 21.3f present the OCT images of the celadon sample HY-I-2. From the images, we can easily find the smooth boundaries among glaze layer 1, glaze layer 2 and the body. There are a few bubbles in glaze layer 1. The bubbles in samples of HY-I-2-A and HY-I-2-C are bigger than those in HY-I-2-B. Relatively large bubbles are distributed in bluish-white porcelain sample HY-II-2 (Figs. 21.3g, 21.3h and 21.3i). For sample HY-II-2-B, the glaze is more transparent, with a few larger strongly scattering particles. The layer of glaze that has seeped into the body is less clear and neat than into the celadon sample. The larger quantity of scattering particles in the glaze layer 1 of samples HY-II-2-A and HY-II-2-C reduces its transparency and have prevented light passing through the body.

The characteristics of tea-dust glaze and black glaze are widely different from white glaze, celadon glaze and bluish-white glaze. The

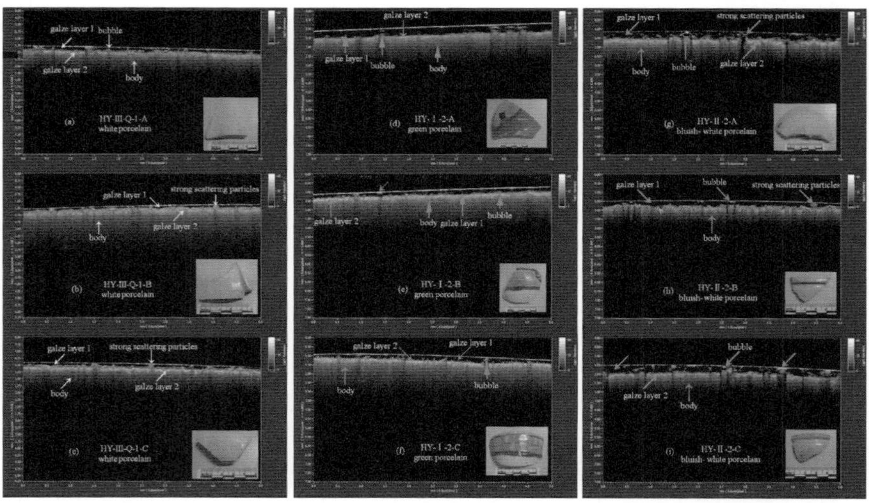

White glaze sample [HY-III(Q)-1] | Celadon sample(HY-I-2) | Bluish-white glaze sample (HY-II-2)

Fig. 21.3 2D gray-scale images of white, celadon, and bluish-white glazed porcelain

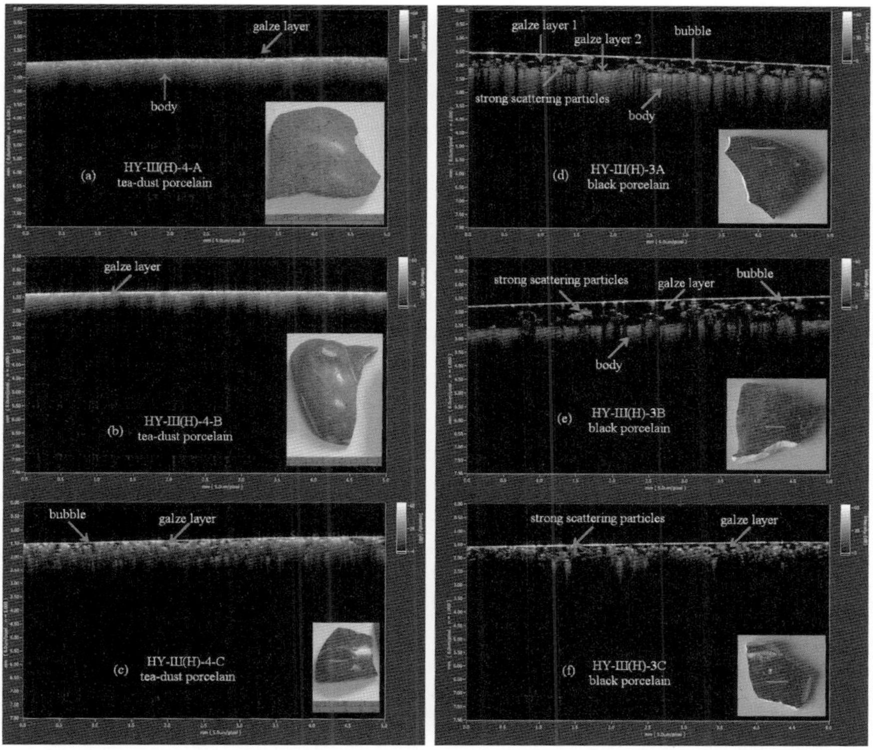

tea-dust glaze sample [HY-III(H)-4] Black glaze sample [HY-III(H)-3]

Fig. 21.4 2D gray-scale images of tea-dust and black glazed porcelain

glaze/body boundaries of tea-dust glaze and black glazes are obscure, since the light is not easily transmitted into the body from the glaze. The tea-dust glaze is almost opaque, for example HY-III (H)-4-A and HY-III (H)-4-B (Figs. 21.4a and 21.4b). The scattering in the glaze layer without bubbles is so strong that the light could not penetrate into the body and the glaze/body boundary can not be observed. Although the scattering intensity is also strong in the dark sample HY-III (H)-4-C, relatively more light is transmitted through the glaze layer (Fig. 21.4c).

Compared to tea-dust glazed samples, the light is more easily transmitted into the body of the black glazed samples, and the glaze/body boundary can be observed. There are more strong scattering particles in the glaze. The glazes of HY-III (H)-3-A and HY-III (H)-3-B are more transparent and the glaze/body boundaries are clearer than in HY-Ш (H)-3-C; large bubbles are distributed through them (Figs. 21.4d and 21.4e). The scattering intensity is strong in sample HY-Ш (H)-3-C with more strong scattering particles in the glaze (Fig. 21.4f).

21.4.2.3 *Two dimensional results for the low-temperature lead glazes*

Fig. 21.5 shows the 2D images of two single-color glaze samples (yellow glaze and blue glaze). It can be seen that there are no obvious differences between different colors of lead glazes. A relatively thick glaze layer, a wide crack and a few little bubbles are present in the dark reddish-brown glazed sample (Fig. 21.5a). The scattering particles of different sizes and tiny bubbles are distributed through the glaze of yellow glazed pottery sample HY-Ш (H)-6-B (Fig. 21.5b).

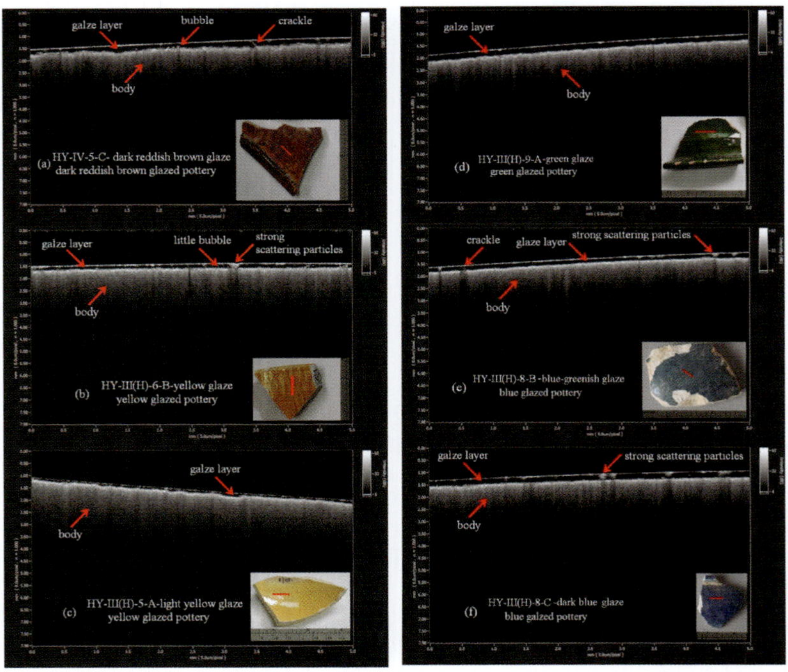

Dark reddish-brown, yellow and light yellow glazes

Green, blue-greenish and dark blue glazes

Fig. 21.5 2D gray-scale images of singe-color lead glazed pottery

The light yellow glaze layer from the inner of sample HY-Ш (H)-5-A is very thin, this is different from the other samples. The scattering particles and bubbles can hardly be observed in the light yellow glaze (Fig. 21.5c)- this is the same as in the green glaze sample HY-Ш (H)-9-A (Fig. 21.5d). The green, blue-greenish and dark blue glazed samples are relatively thick. Large scattering particles and wide cracks are distributed through the blue-greenish glaze (Fig. 21.5e), and many large scattering particles in dark blue glaze (Fig. 21.5f).

Fig. 21.6 shows the OCT images of the dual-colored glazed pottery. From the images, it can be seen that the distinction between the white glaze and blue or green glazes are not obvious, but the areas of blue and green glazes are thicker than the white glaze areas. There are many strong scattering particles with varying sizes in the glaze of white and blue glazed samples HY-IV-9-C (Figs. 21.6a and 21.6b). For the green and white glaze sample HY-IV-10-A, less strong scattering particles are distributed in the green glaze (Fig. 21.6c), while more strong scattering particles and wide cracks in the white glaze (Fig. 21.6d). The 2D grayscale images of Tang Dynasty tri-color glazed

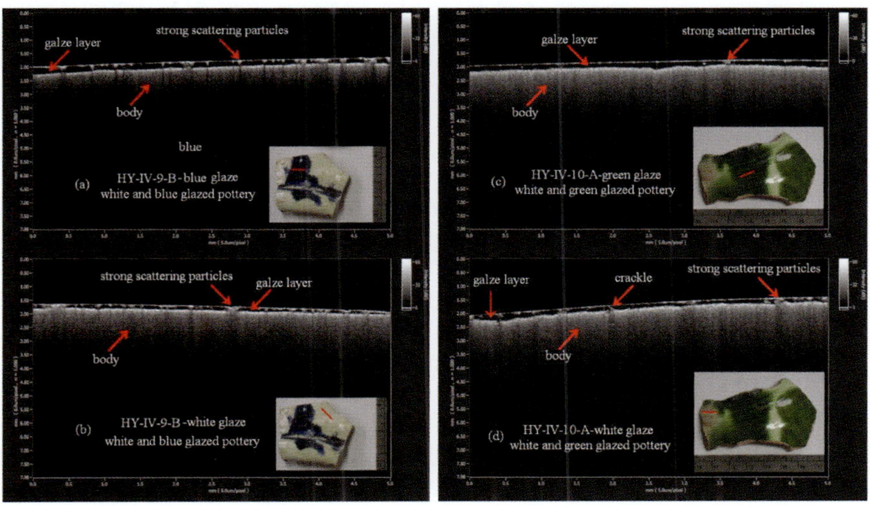

White and blue glazes [HY-IV-9-B] White and green glaze (HY-IV-10-A)

Fig. 21.6 2D gray-scale images of dual-colored lead glazed pottery

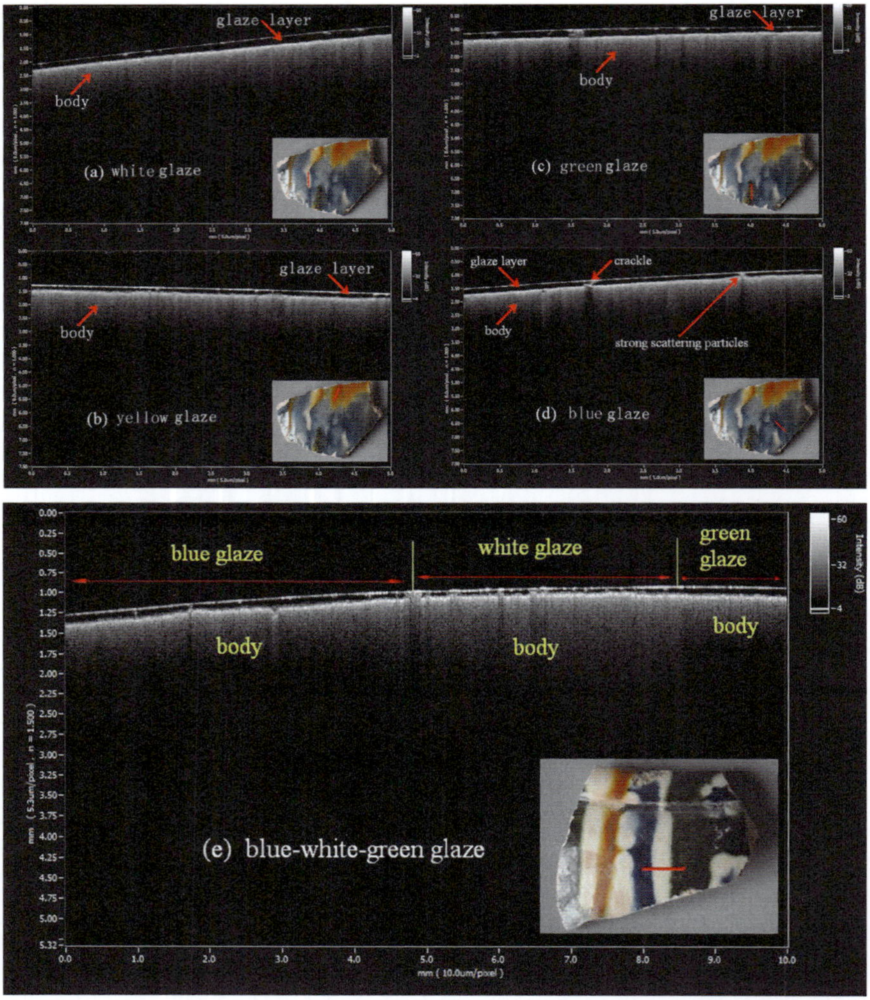

Fig. 21.7 2D gray-scale images of Tang Dynasty tri-color lead glazed pottery (a-d:HY-III(Q)-12-B; e:HY-IV-11-A)

pottery, sample HY-Ш (Q)-12-B, are shown in Fig. 21.7. There are no bubbles or strong scattering particles in the thick white glaze or in the thin yellow glaze (Figs. 21.7a and 21.7b), while a few scattering particles can be found in the green and blue glazed parts (Figs. 21.7c and 21.7d). In addition, there are clear cracks in the blue glaze. From Fig. 21.7e, we can find that the differences among the blue, white and

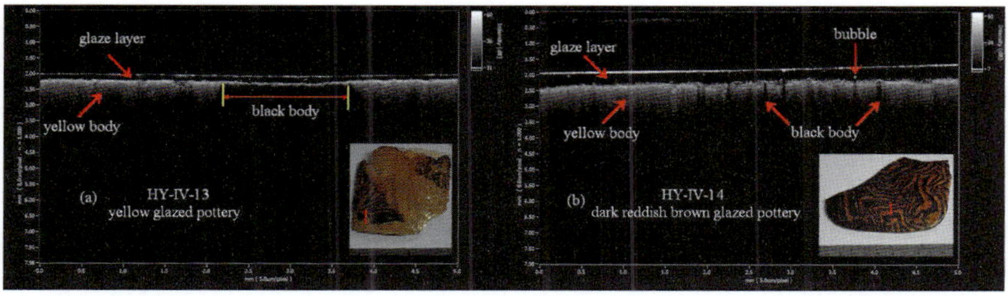

Fig. 21.8 2D gray-scale images of the twisted body lead glazed pottery

green areas of sample HY-IV-11-A is very small, but the scattering intensity of white glaze and the absorption intensity of green and blue glazes are strong.

The 2D image of the twisted body of lead glazed pottery appears the same as the other lead glazed pottery, but the body is obviously different. The black body area has bright contrasts with the greyish-white main-body area, which makes the mixed area like grain of wood. The black area between the two white bright lines is the glaze layer (Fig. 21.8). The light transmittance of the lead glaze is so high that the light could easily pass through glaze layer into the body. The absorbing and reflecting light levels of the dark area and the greyish-white main-body area shows a striking difference: the black body area has low scattering intensity and a high light transmittance, which is just the opposite of the greyish-white main-body area (Fig. 21.8). From the results of sample HY-IV-14 determined by pXRF, we found that the Fe_2O_3 contents in the surface of the glaze area on the black body is up to 8.52%, sharply higher than the glaze area on the greyish-white body with Fe_2O_3 content 4.95%. However, from the 2D OCT images we can see that the whole glaze layer is relatively homogeneous, and the OCT character is basically the same, while the body has sharp differences. The differences in Fe_2O_3 contents in the glaze mentioned above might be caused by the fact that the glaze layer is so thin that it is penetrated by X-rays and the chemical component of the body was determined by pXRF.

In summary, both the calcium glazed porcelain and lead glazed pottery are decorated with various colored glazes. The 2D OCT image characters of the glazes with the same colour appear similar, the characters of ceramics with several glaze colours present variable differences. For the calcium glazed porcelain, the white glaze, celadon glaze and bluish-white glaze have similar characteristicss; with obvious differences between the black glaze and tea-dust glaze. The former glaze has a high transparency, relatively few bubbles scatting particles and clear glaze/body boundary, while the light transmittance of glaze is low, and glaze/body boundary is not obvious in the latter. The tea-dust glaze is hardly transparent. Weathering can effect glazes so the shapes from which images were takes varied: the OCT images of different wares varied even though they are from glazes of the same color such as white, black and tea-dust glazes. In contrast the lead glazed pottery with the highly transparency, almost without bubbles and scattering particles, is usually very thin. Moreover, the OCT image characters among single-color glazes, dual-colored glazes, tri-colored glazes and the glaze with twisted bodes present great similarity. Comparatively speaking, the scattering intensity of yellow, green and blue glazes is lower than that of white glaze. But the lead glaze is easily weathered. The unweathered glaze area is relatively pure with few scattering particles, while predictably lots of scattering particles are distributed in the weathered area.

21.4.3 *Analysis of the 3D OCT data stop*

To further acquire the distinction between the high-temperature calcium glaze and the low-temperature lead glaze and their formation mechanisms, one bluish-white glazed porcelain sample HY-II-2-C and one yellow glazed pottery sample with twisted body HY-IV-13 were chosen for 3D scanning. The 3D images are slices along the planes perpendicular to the *z*-axis, *y*-axis or *x*-axis to acquire different sections. The test parameters are set as follow: the measurement units of *x*-axis and *y*-axis both are 3 mm; the pixels of *x*-axis, *y*-axis and *z*-axis are 700, 700, and 512, respectively; the scale of the *z*-axis is 5.3 μm/pixel.

Fig. 21.9a shows the 3D images of bluish-white glaze sample HY-II-2-C with pseudo-color added. It has been found that the

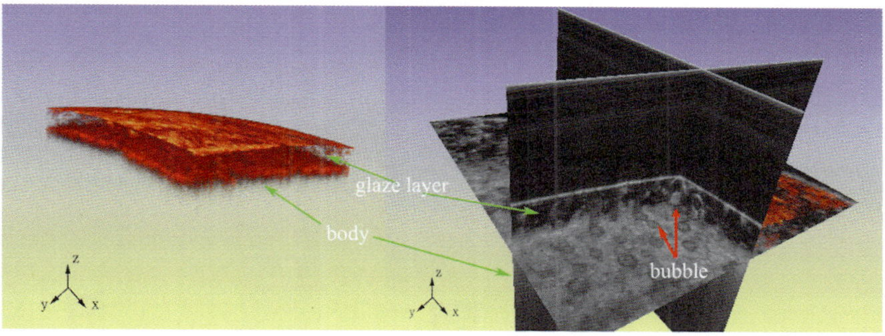

(a) (b)

Fig. 21.9 3D image of the bluish-white glazed porcelain sample HY-II-2-C (a: 3D pseudo-color image; b:3D sliced diagrammatic sketch)

boundaries of surface, glaze and body are very clear. The 3D diagrammatic sketch of the slices is shown as in Fig. 21.9b that is created from Fig. 21.9a and slices along both the planes perpendicular to the *z*-axis and *y*-axis. The chosen section position is shown in Fig. 21.9b, where the positions of bubbles are indicated as red arrows.

The conform strong scattering area is below the bubbles; around this strong area of scattering area is the black area with no scattering or with low scattering. The diameter of the bottom of the conform area is similar to the diameter of the bubble. The phenomenon is caused by a process whereby the spaces are formed after gases such as N_2, H_2O, CO, O_2, SO_2, H_2 spread from the body and into the glaze, and when the liquid glaze could seep into the body during firing. This suggests that most of the bubbles of glaze come from the body.

In order to explore the bubble characters in ancient glazes, six sections sliced along the plane perpendicular to the *z*-axis were analyzed, as shown in Fig. 21.10. The slices appear to be tilting as the result of the uneven surface of samples. Slice 1 lies in the glaze (Fig. 21.10a) positioned at a distance of 220 pixels from the top of the image in Fig. 21.9b (shorter form of H = 220); slices 2 to 5 lie in positions that correspond to different stages of the glaze seeping into the body (Figs. 21.10b-10e, H = 215, 235, 276 and 310, respectively); slice 6 lies in the body (Fig. 21.10f, H = 380). The white and black

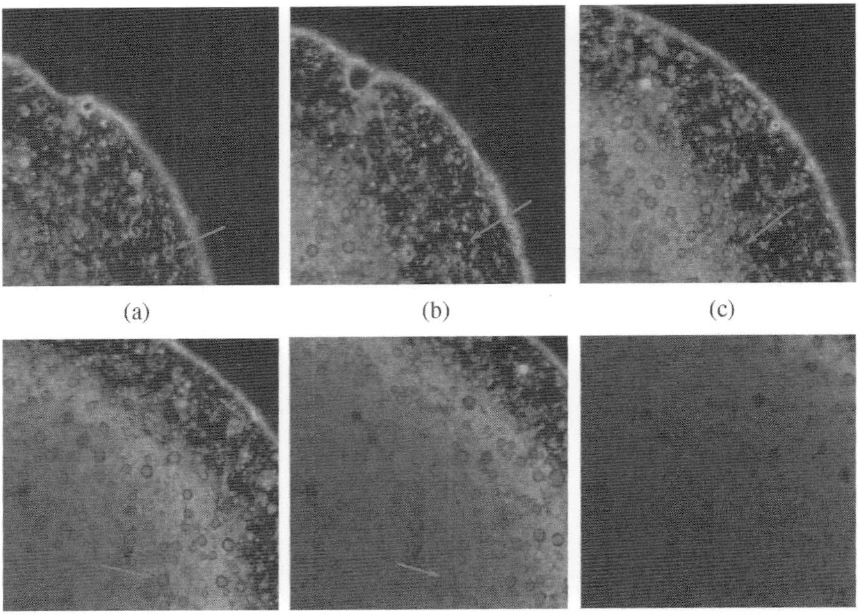

(a) (b) (c)

(d) (e) (f)

Fig. 21.10 Slices perpendicular to *z*-axis of the bluish-white glazed porcelain sample HY-II-2-C (a: gray-scale slice 1, H = 201; b: gray-scale slice 2, H = 215; c: gray-scale slice 3, H = 235; d: gray-scale slice 4, H = 276; e: gray-scale slice 5, H = 310; f: gray-scale slices 6, H = 380)

rings respectively indicate the route the bubble took when escaping from the glaze and from the body. In addition, the two types of rings correspond to each other. The red arrows represent the different positions and shapes of the same bubble in the different slices. The size and distribution of the bubbles are different based on the sections sliced from different positions, which suggests that the bubbles are mainly formed due to the gas spreading from the body and into the glaze during firing.

Fig. 21.11a is the 3D image of lead glazed pottery with twisted body, sample HY-IV-13 with pseudo-color added. The 3D sliced diagrammatic sketch of the sample in Fig. 21.11a is cuts along both the planes perpendicular to the *z*-axis and *y*-axis as shown in Fig. 21.11b,

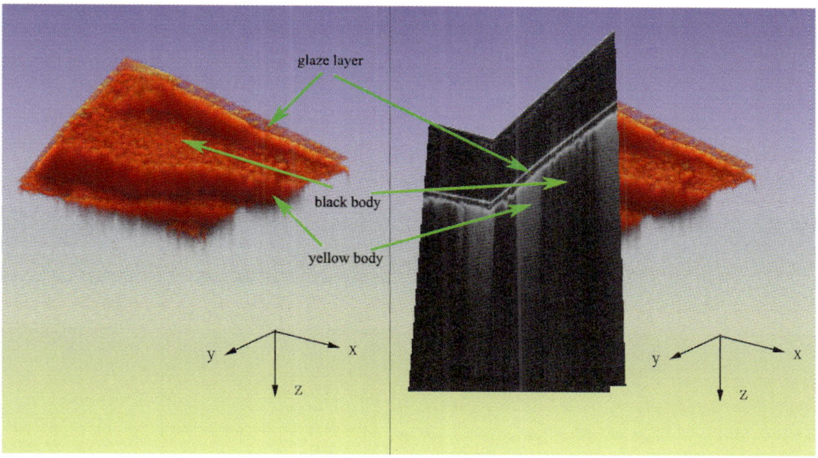

(a) (b)

Fig. 21.11 3D OCT image of lead glazed pottery with twisted body sample HY-IV-13 (a: 3D pseudo-color image; b: 3D sliced diagrammatic sketch)

from which the surface, glaze layer and greyish-white and black body are clear. To acquire the details of the lead glaze and twisted body, six slices are taken along the plane perpendicular to the *z*-axis and analyzed, as shown in Fig. 21.12. The slices appear to be tilting as the result of the uneven surface of samples. Slices 1 and 2 lie in the surface and middle of the glaze layer respectively (Figs. 21.12a and 21.12b), which are positioned at the distance of 200 and 220 pixels from the top of the image in Fig. 21.11b (H = 200 and 220); slice 3 lies in glaze/body boundary (Fig. 21.12c, H = 230); slices 4 to 6 lie in the different depths of the body (Figs. 21.12d and 21.12f, H = 255, 350 and 450, respectively). This suggests that the lead glaze layer is almost without bubbles but has lots of fine crackles, is thin and pure, and that the grayish-white and black bodies in the lead glazed pottery with marble-texture body alternate with each other, and that they have sharply different properties for absorbing and scatting light.

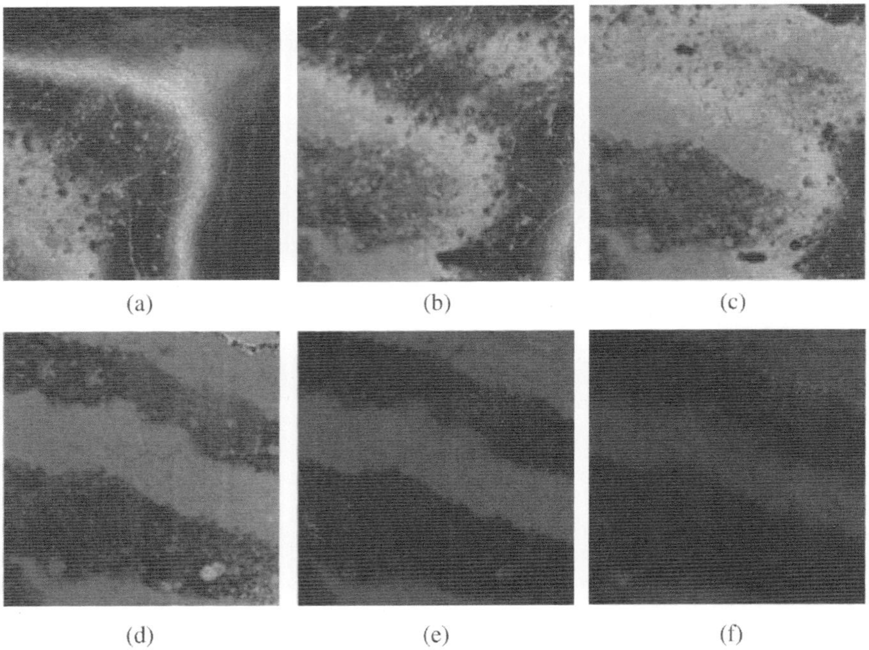

Fig. 21.12 Slices perpendicular to the z-axis in the lead glazed pottery with twisted body HY-IV-13 (a: gray-scale slice 1, H = 200; b: gray-scale slice 2, H = 220; c: gray-scale slice 3, H = 230; d: gray-scale slice 4, H = 255; e: gray-scale slice 5, H = 350; f: gray-scale slices 6, H = 450)

21.5 Discussion and conclusions

The results of the above 2D cross-section and 3D tomography analyses suggest that shigh-temperature calcium glaze can be easily distinguished from the low-temperature lead glazes by OCT imaging. The bubbles and strong scattering particles are common in various calcium glazes, while the bubbles and scattering particles are hardly found in the relatively pure and thin lead glaze. There is a middle layer between the main glaze (glaze layer 1) and the body, which we called 'glaze layer 2'. In contrast to the calcium glaze, there is not an obvious layering phenomenon either in the single-colored l, dual-colored or tri-colored lead glazes. But two light lines can clearly be observed in the lead glazes, the upper one representing the interface of glaze and air, the lower one is the glaze/body boundary; the latter has a sharp contrast to the body which is a thinner and denser than in a calcium glaze. The differences between the calcium glaze and lead glaze are due to the different production processes. With the high porosity of the body of the once-fired calcium glazed porcelain, a little glaze seeps into the body and then forms the middle layer (glaze layer 2). But in the lead glazed pottery, most of air has escaped from the body during the first firing the body after which the body was glazed; in the second firing process, the liquid glaze would hardly seep into the body, so there is no middle layer.

The difference in the characteristics of bubbles between high-temperature glazed porcelain and low-temperature glazed pottery is realted to their different production processes as mentioned above. The structure of the unfired porcelain body is relatively porous. During the process of firng porcelain glaze and body together, the N_2, H_2O, CO, O_2, SO_2, H_2 from the body would enter into the glaze and form the bubbles. In addition, the CO_2 can also from from the decomposition of lime ($CaCO_3$), one of the major raw materials in the calcium glaze at the high temperature. A part of these gases in the body and glaze will escape from the glaze, creating small pits in the glaze surface, while the other part will be distributed at different depths in glazes as differenty the shaped bubbles. The low-temperature lead glazed pottery including the single-colored, dual-colored and tri-colored glazed pottery have two firings; the body has been sintered during biscuit firing and most of the air has escaped. Moreover, being lime-free, the lead glaze has little gas. Therefore the lead glaze is almost without bubbles and appear to be relatively pure.

In summary, a comparison has been made between the chemical compositions determined by pXRF and the 2D cross-sectional and 3D tomography image characters of high-temperature porcelain and low-temperature pottery with various color glazes have been analyzed using the OCT imaging technology. This work suggests that the OCT imaging technology, as a nondestructive tomography analysis technology, could effectively be applied to the characterization of the subsurface of ancient ceramic glazes. The results intuitively describe the characters of bubbles in the glazes, the scattering particle distributions, the glaze uniformities, the glaze/body boundaries and the middle layer of different types of porcelain and pottery. It furthermore allows the a means of distinguishing and classifying the characteristics of glazes from different lineages, glazes of different colors from the same lineage and glazes from different kilns with the same shade of color.

Acknowledgments

This research was supported by the Major State Basic Research Development Program of China (973 Program, Grant No. 2012CB720901, 2012CB720906), the Shanghai R&D Public Service Platform Construction Project (Grant No. 13DZ2295800) and the National Natural Science Foundation of China (Grant No. 51402326)

References

1. D. Huang, E. A. Swanson, C. P. Lin *et al.*, *Science* **254**, 1178 (1991).
2. J. Swerts, A. Aerts, N. De Biscop *et al.*, *Chemometrics and intelligent laboratory systems* **22**, 97 (1994).
3. J. L. Dong, W. J. Zhao, G. D. Liu *et al.*, *Nuclear Physics Review* **25**, 380 (2008).
4. P. L. Leung, M. J. Stokes, C. Tiemei *et al.*, *Archaeometry* **42**, 129 (2000).
5. L. D. Kock, *Journal of Raman Spectroscopy* **44**, 97 (2013).
6. N. Q. Liem, N. T. Thanh and P. Colomban, *Journal of Raman Spectroscopy* **33**, 287 (2002).
7. B. E. Bouma and G. J. Tearney, *Handbook of optical coherence tomography*, Marcel Dekker, Inc, New York, USA, 2002.
8. A. F. Fercher, W. Drexler, C. K. Hitzenberger *et al.*, *Reports on Progress in Physics* 66, 239–303 (2003).
9. W. Drexler and J. G. Fujimoto, Optical Coherence Tomography: Technology and Applications, Springer-Verlag, Berlin, Germany, 2008.
10. G. Smolka, *Optical Coherence Tomography: Technology, Markets, and Applications 2008–2012*, BioOptics World, Penn Well Corporation, New York, USA, 2008.
11. P. Targowski, M. I. Wanicka, M. Sylwestrzak *et al.*, *SPIE* **8790**, 87900N (2013).
12. H. D. Liang, B. Peric, M. Hughes *et al.*, *SPIE* **6618**, 661805 (2007).
13. P. Targowski, M. I. Wanicka, *Applied Physics A* **106**, 265 (2012).
14. X. Yan, J. Q. Dong, Q. H. Li *et al.*, *Chinese Journal of Lasers* **41**, 0908001 (2014).

15. X. Yan, J. Q. Dong, Q. H. Li *et al.*, *Spectroscopy and Spectral Analysis* **35**, Accepted, (2015).
16. Institute of Cultural Relics in Henan Province, China Institute of Cultural Property, Nara National Research Institute for Cultural Properties, *The new archaeological discover at Huangye kiln site*, Da Xiang press, Zhengzhou, China, 2005.
17. S. Alireza, N. Syozi, S. Yasushi *et al.*, *SPIE* **8208**, 82080H (2012).
18. S. Liu, Q. H. Li, F. X. Gan *et al.*, *X-Ray Spectrometry* **40**, 364 (2011).

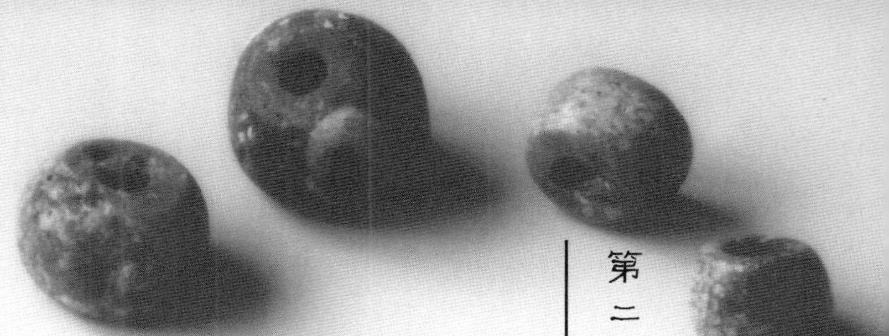

Chapter 22

第二十二章

Chemical Analysis of Ceramics Unearthed from the Huangye Kiln Site in Henan Province Using a Portable XRF Spectrometer

*Song Liu**, *Yongqing Hu*†, *Musen Guo*†, *Dong Junqing*† *and Qinghui Li**

*Shanghai Institute of Optics and Fine Mechanics,
Chinese Academy of Sciences, Shanghai 201800, China
†Henan Provincial Institute of Cultural Relics and Archaeology, Zhengzhou 450000, China

22.1 Introduction

The Huangye kiln site is renowned for the production of three-color pottery of the Tang Dynasty and is located in the northeast of Gongyi city, Henan Province. Its discovery confirmed the kiln sites, used for the production of three-color pottery (a milestone in Chinese ceramic history). The production of three-color pottery in the Huangye kiln began in the early Tang Dynasty, matured in the prosperous Tang Dynasty, and continued in the middle and the late Tang Dynasty. The scale of production became smaller in the Five Dynasties Periods, and stopped in the Song and Jin Dynasties. The history of the use of the Huangye kilns is over 500 years long, and the Huangye kiln took an important role in the history of Chinese ceramic production. Besides the three-color pottery, high temperature porcelains were also produced.

Three-color glazed potteries have always been the focus of archaeological scientists because of the importance of the Huangye kiln. Scientific researchers have studied the three-color pottery from the Huangye kiln site using many techniques including proton induced X-Ray emission (PIXE) techniques, neutron activation analysis technique (NAA), scanning electron microscopes (SEM-EDS) and so on.[1–3]

In this paper, a batch of ceramic fragments including three-color pottery, celadon, white glaze porcelains, and black glaze porcelains, unearthed from the Huangye kiln site and dated to Tang Dynasty, were analyzed using a portable XRF. Based on the chemical compositions obtained, the ceramic fragments are classified, the differences in the chemical compositions of glazes with different colors in each group are given. In Sec. 3 and the evolution of glaze recipes is also discussed.

22.2 Experimental

22.2.1 *Equipment and methods*

An OURSTEX 100FA portable energy dispersive XRF (*p*XRF) spectrometer provided by OURSTEX Co. Ltd. was used for the chemical analysis. A palladium (Pd) excitation tube was employed, with a voltage of 40 kV in monochromatic and filter modes, and 15 kV for direct mode, and with a current of 0.5–1.0 mA. The detector is a silicon drifted detector (SDD) with an active area of 5 mm^2 equipped with a polymer window (MOXTEK AP 3.3 film), which can provide superior X-ray transmission in the low-energy range down to Be-Ka. The X-ray beam spot focused on the sample is about 2.5 mm in diameter. Together with the introduction of a vacuum sample chamber, this *p*XRF spectrometer allowed us to quantify the light elements such as Na and Mg. The *p*XRF spectrometer has already been employed for the quantitative analysis of ancient glass samples found in Xinjiang and Guangxi Provinces, and more experimental parameters associated with its use can be found elsewhere.[4–6]

22.2.2 *Quantitative analysis method*

The calibration curve method is used to analyze the ceramics quantitatively.[7] The glaze and the body of the ceramics have different matrices so the calibration curves of detected elements in glazes and bodies needed to be created separately.

Glaze is a vitreous coating to a ceramic material and can be divided into two groups: high temperature glaze and low temperature glaze. The difference between the two glazes lies to the contents of PbO and CaO. The high temperature glaze contains low PbO, while the low temperature glaze contains high levels of PbO, and that's why the low temperature glaze is also called lead glaze. The two glazes have largely different compositions, so the calibration curves for the two glazes also needed to be created separately.

The calibration curves for ancient glasses are also used for glazes, because the composition of the high temperature glaze is close to that of ancient glass. The calibration curves have already been used in quantitative analyses of ancient glasses found in Xinjiang, Guangxi, successfully, and the precision and accuracy of the calibration curves have been validated.[4–6]

For the low temperature glaze, six optical glasses containing PbO, such as barium flint glass (BaF), dense barium flint glass (ZBaF) and dense flint glass (ZF), have been selected as reference samples. The reference values for the components of the optical glass are obtained by liquid Inductively Coupled Plasma-Atomic Emission spectroscopy (ICP-AES). Meanwhile, three glass standard samples produced by Breitlander Eichproben und Labomaterial GmbH, Germany, and belonging to the $PbO\text{-}SiO_2/PbO\text{-}BaO\text{-}SiO_2$ silicate glass type were also used as reference samples. Calibration curves are created using the samples mentioned above, and calibration curves for some typical components (PbO, BaO, Na_2O, CaO, SiO_2 and K_2O) are shown in Fig. 22.1. The standard sample Corning glass C was measured 12 times to estimate the precision and accuracy of the calibration curves, and the results are listed in Table 22.1.

As shown in Table 22.1, the errors for most components are less than 7%, and that for BaO reaches 15.62% which is a little higher but acceptable; the errors for P_2O_5, SnO_2 and Sb_2O_3 are 16.37%, 21.71% and 31.39%, respectively. The low level of the three components is the main reason for causing large errors. The results obtained in Table 22.1 indicate that the calibration curves for lead-barium glass have good precision and accuracy.

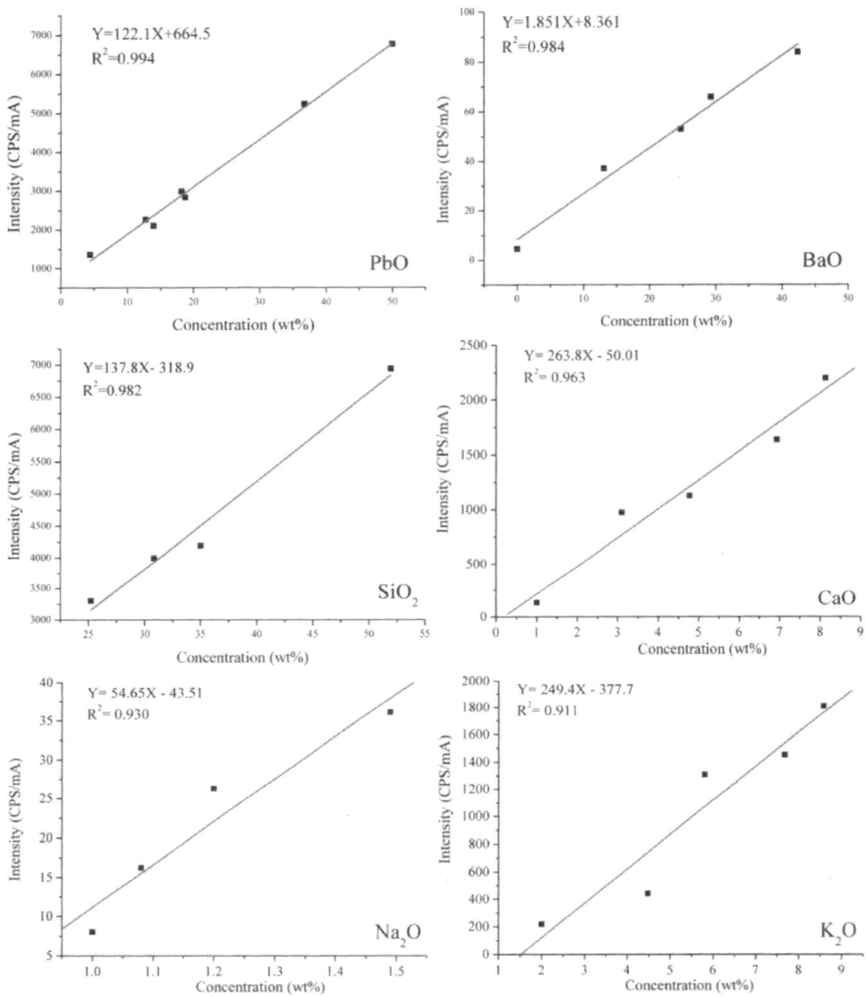

Fig. 22.1 Typical component calibration curves for low temperature glaze

As for the bodies, a set of ceramic bodies (13 pieces) provided by Jingdezhen Ceramic Institute was used as reference samples. The main components of Na_2O, MgO, Al_2O_3, SiO_2, K_2O, CaO, TiO_2, Fe_2O_3 and the elements P, Cr, Mn, Cu, Zn, Pb, Rb, Sr for the ceramic bodies can be analyzed quantitatively using the calibration curves created. The typical components and elements calibration curves for the body are shown in Fig. 22.2.

Table 22.1 Evaluation of precision and accuracy of calibration curves for low temperature glaze

Corning C	Average (wt%)	STD (wt%)	RSD (%)	Certified Values (wt%)	Error (%)	LLD (μg/g)
Na_2O	1.07	0.02	1.76	1.07	0.02	467
MgO	2.93	0.13	4.42	2.76	6.31	275
Al_2O_3	0.87	0.04	4.70	0.87	0.16	136
SiO_2	34.56	0.23	0.67	34.87	0.90	120
P_2O_5	0.12	0.02	16.04	0.14	16.37	488
K_2O	2.99	0.03	0.86	2.84	5.12	125
CaO	5.18	0.04	0.76	5.07	2.08	87
TiO_2	——	——	——	0.79	——	——
MnO	0.76	0.09	11.69	0.82	6.78	305
Fe_2O_3	0.34	0.01	4.36	0.34	0.27	44
CoO	0.19	0.01	3.50	0.18	3.70	34
CuO	1.17	0.01	1.27	1.13	3.42	37
ZnO	0.05	0.01	10.05	0.052	3.04	41
SrO	0.30	0.00	1.47	0.29	3.56	20
SnO_2	0.15	0.01	5.35	0.19	21.71	47
Sb_2O_3	0.04	0.01	23.39	0.03	31.39	39
PbO	38.19	0.27	0.71	36.7	4.06	124
BaO	9.62	0.47	4.87	11.4	15.62	1879

†: Error (%)=$|C_{Measured} - C_{Certified}| / C_{Certified} \times 100\%$; "LLD" denotes the detection limit. "——" represents the components can not be analyzed quantitatively because of the overlap.

22.2.3 *Samples*

136 pieces of ceramic fragments were analyzed in this work. The ceramic fragments can be divided into two groups: porcelains with high temperature glaze and three-color pottery. The porcelain has four colors: white, greenish white, black and tea-leaf color, while the pottery with low temperature glazes belong to three-color pottery group, and have yellow, blue, green, brown and white colors. It is

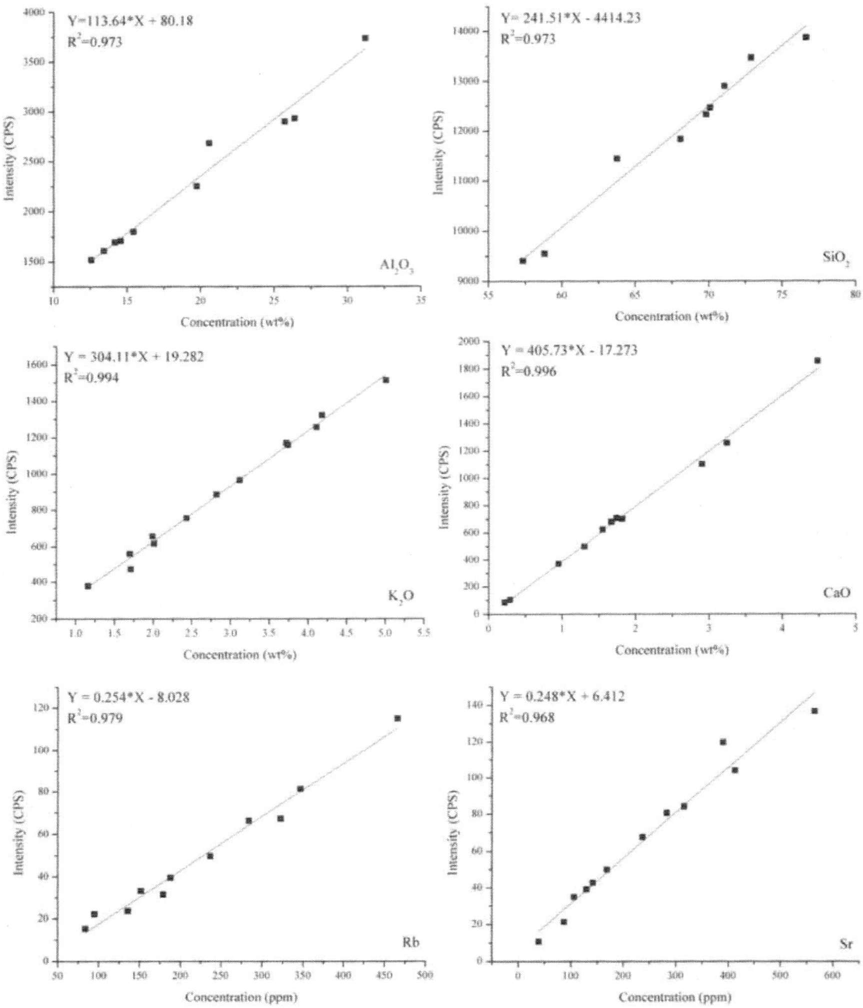

Fig. 22.2 Typical components and elements calibration curves for ceramic body

worthy of note that "three-color" doesnot mean three colors. Some glazes of three-color pottery samples only have one color, such as yellow, blue or green, while some have two or three colors. All the samples are dated to the Tang Dynasty, but belong to different sub-periods, which are listed in Table 22.2.

Table 22.2 Sample information list

No.	Type*	Color	Quantity	Site	Period	Time†
HY-II-1	H	white	2	IIT4H16	II	Tang-E
HY-II-2	H	greenish white	3	IIT4H16	II	Tang-E
HY-II-3	H	white	3	IT1L4	II	Tang-E
HY-II-4	H	black	3	IIT4C1	II	Tang-E
HY-II-5	H	black	4	IIT4H16	II	Tang-E
HY-II-6	H	Tea-flake	3	T3C2	II	Tang-E
HY-II-7	H	Tea-flake	3	IT116	II	Tang-E
HY-II-8	H	Tea-flake	1	IIT4H16	II	Tang-E
HY-II-9	L	yellow	3	IT116	II	Tang-E
HY-II-10	L	three-color	3	IIT4C1	II	Tang-E
HY-III(Q)-1	H	white	3	IIT4H15	III-E	Tang-P
HY-III(Q)-2	H	greenish white	3	IIT4H15	III-E	Tang-P
HY-III(Q)-3	H	black	3	IIT4H15	III-E	Tang-P
HY-III(Q)-4	H	Tea-flake	3	IT1⑩	III-E	Tang-P
HY-III(Q)-5	L	blue	3	IIT4H15	III-E	Tang-P
HY-III(Q)-6	L	blue	3	IIT4H15	III-E	Tang-P
HY-III(Q)-7	L	white	3	IIT4H15	III-E	Tang-P
HY-III(Q)-8	L	white glaze with blue pattern	3	IIT4H15	III-E	Tang-P
HY-III(Q)-9	L	white glaze with green pattern	3	IIT8⑩	III-E	Tang-P
HY-III(Q)-10	L	yellow	3	IIT4H15	III-E	Tang-P
HY-III(Q)-11	L	greenish white	3	IIT8⑩	III-E	Tang-P
HY-III(Q)-12	L	three-color	3	IIT4H15	III-E	Tang-P
HY-III(H)-1	H	white	3	IIT4⑥	III-L	Tang-M
HY-III(H)-2	H	inside white glaze with black outside glaze	3	IT1⑤	III-L	Tang-M
HY-III(H)-3	H	black	3	IT1⑤	III-L	Tang-M
HY-III(H)-4	H	Tea-flake	3	IIT4⑥	III-L	Tang-M
HY-III(H)-5	L	white	3	IT1⑤	III-L	Tang-M

(*Continued*)

Table 22.2 (*Continued*)

No.	Type*	Color	Quantity	Site	Period	Time†
HY-III(H)-6	L	yellow	3	IT1⑤	III-L	Tang-M
HY-III(H)-7	L	brown	3	IT1⑤	III-L	Tang-M
HY-III(H)-8	L	blue	3	IT1⑤	III-L	Tang-M
HY-III(H)-9	L	green	3	IT1⑤	III-L	Tang-M
HY-III(H)-10	L	white glaze with blue pattern	3	IT1⑤	III-L	Tang-M
HY-III(H)-11	L	white glaze with green pattern	3	IT1⑤	III-L	Tang-M
HY-III(H)-12	L	three-color	3	IT1⑤	III-L	Tang-M
HY-IV-1	H	white	3	IIG2	IV	Tang-L
HY-IV-2	H	black	3	IIT35⑨	IV	Tang-L
HY-IV-3	H	Tea-flake	3	IIT4H7	IV	Tang-L
HY-IV-4	L	yellow	3	IIT35⑨	IV	Tang-L
HY-IV-5	L	brown	3	IIG2	IV	Tang-L
HY-IV-6	L	blue	3	IIG2	IV	Tang-L
HY-IV-7	L	green	3	IIG2	IV	Tang-L
HY-IV-8	L	green	3	IIG2	IV	Tang-L
HY-IV-9	L	white glaze with blue pattern	3	IIG2	IV	Tang-L
HY-IV-10	L	white glaze with green pattern	2	IIT35⑨	IV	Tang-L
HY-IV-11	L	three-color	4	IIG2	IV	Tang-L
HY-IV-12	H	white	1	IIT35⑨	IV	Tang-L
HY-IV-13	L	yellow	1	IIT35⑨	IV	Tang-L
HY-IV-14	L	brown	1	IIT35⑨	IV	Tang-L
HY-IV-15	L	green	1	IIT35⑨	IV	Tang-L

*:"H" represents porcelains with high temperature glaze; "L" represents pottery with low temperature glaze.

†: "Tang-E" represents early Tang Dynasty, "Tang-P" represents prosperous Tang Dynasty, "Tang-M" represents middle Tang Dynasty and "Tang-L" represents later Tang Dynasty.

22.3 Results

The chemical results obtained by pXRF for glazes and bodies are listed in Tables 22.3 to 22.5. The chemical compositions for high temperature glazes, low temperature glazes and bodies are discussed sequentially below.

22.3.1 *Chemical compositions of high temperature glazes*

The chemical results obtained by pXRF for high temperature glazes are listed in Table 22.3. As shown in Table 22.3, the prominent characteristic of the high temperature glaze is the high level of CaO which varies from 4.63 wt% to 16.50 wt%. The high level of K_2O is another characteristic of high temperature glaze, which ranges from 1.98 wt% to 6.45 wt%. The contents of MgO range from 1.63 wt% to 6.99 wt%, which indicates that plant ash was used for the production of high temperature glazes. The Al_2O_3 and SiO_2 contents vary from 10.98wt% to 25.10 wt%, and 54.68 wt% to 68.41 wt%, respectively. The Fe ions are the main colorant for the glaze, and the levels vary according to the colors of glaze. The black and tea leaf coloured glazes have higher levels than that of greenish white and white glaze.

22.3.2 *Chemical compositions of low temperature glazes*

The chemical results obtained by pXRF for low temperature glazes are listed in Table 22.4. Low temperature glazes contains large amounts of PbO which was used as a flux and makes the melting point of low temperature glazes much lower than high temperature glazes. The PbO contents in low temperature glazes range from 7.12 wt% to 57.90 wt%, with a mean value of 48.24 wt%; the CaO contents vary from 0.42 to 5.12 wt% with a mean value of 1.75, which is much lower than that of high temperature glazes; the MgO and K_2O contents range from 0.06 to 4.28 wt% with a mean value of 0.53 wt%, and from 0.00 to 4.22 wt% with a mean value of 0.70 wt%; the levels of SiO_2 and Al_2O_3 vary from 29.46 to 56.63 wt% and from 0.62 to 10.17 wt%, respectively. The differences (in the chemical

Table 22.3 Chemical results of glaze for high temperature porcelain unearthed from the Huangye kiln site

No.	Na_2O	MgO	Al_2O_3	SiO_2	P_2O_5	K_2O	CaO	TiO_2	Fe_2O_3	PbO
HY-II-1-A	1.52	2.34	17.74	63.49	0.56	2.35	10.99	0.14	0.88	n.d.
HY-II-1-B	1.22	2.3	18.61	61.76	0.77	2.36	11.79	0.25	0.94	n.d.
HY-II-2-A	1.64	2.07	15.36	60.94	0.68	1.98	16.31	0.05	0.96	n.d.
HY-II-2-B	2.1	1.63	15.74	64.76	0.58	2.3	11.93	0.05	0.92	n.d.
HY-II-2-C	1.56	2.26	18.45	63.1	0.47	2.4	10.2	0.29	1.27	n.d.
HY-II-3-A	1.41	2.22	18.05	58.15	1.03	3	14.03	0.27	1.84	n.d.
HY-II-3-B	1.52	2.64	15.9	58.77	0.43	3.1	16.5	0.1	1.04	n.d.
HY-II-3-C	2.21	2.5	16.99	58.48	0.63	4.06	13.82	0.19	1.12	n.d.
HY-II-4-A	1.14	3.1	14.63	64.43	n.d.	3.38	6.02	0.98	6.31	n.d.
HY-II-4-B	1.27	3.71	14.26	64.39	n.d.	4.04	4.66	0.98	6.69	n.d.
HY-II-4-C	n.d.	2.64	14.45	66.09	0.14	5.35	5.7	0.97	4.65	n.d.
HY-II-5-A	1.2	3.21	14.65	65.35	0.23	3.24	5.54	1.16	5.42	n.d.
HY-II-5-B	1.15	2.49	13.97	67.29	0.23	3.79	5.3	0.83	4.96	n.d.
HY-II-5-C	1.28	3.2	13.18	65.28	n.d.	3.85	7.57	0.79	4.85	n.d.
HY-II-6-A1	1.21	4.56	12.16	62.79	n.d.	3.13	10.76	0.72	4.68	n.d.
HY-II-6-A2	1.48	3.26	12.12	66.36	n.d.	2.9	7.47	0.87	5.54	n.d.
HY-II-6-B	1.46	2.87	15.14	64.6	0.08	3.19	6.49	1.04	5.14	n.d.
HY-II-6-C1	1.64	4.28	12.67	61.28	n.d.	3	11.26	0.71	5.17	n.d.
HY-II-6-C2	1.49	2.98	12.5	65.64	n.d.	3.07	7.87	0.83	5.62	n.d.
HY-II-7-A1	1.31	4.55	12.47	60.81	n.d.	3.48	10.63	0.8	5.94	n.d.
HY-II-7-A2	1.28	3.55	13.48	62.46	0.24	3.16	8.85	0.84	6.15	n.d.
HY-II-7-B	1.02	4.93	13.62	57.33	n.d.	6.45	10.44	0.73	5.47	n.d.
HY-II-8	1.16	5.43	12.88	62.13	n.d.	3.86	8.65	0.87	5.01	n.d.
HY-III(Q)-1-A	1.35	3	16.3	58.37	1.01	2.65	16.28	0.25	0.79	n.d.
HY-III(Q)-1-B	1.04	3	20.01	60.92	1.02	2.6	10.23	0.36	0.82	n.d.
HY-III(Q)-1-C	2.07	2.31	25.1	54.68	0.56	2.92	10.72	0.8	0.83	n.d.
HY-III(Q)-2-A	1.43	2.78	16.15	58.3	0.95	4.05	15.09	0.09	1.17	n.d.
HY-III(Q)-2-B	1.08	2.4	19.01	60.41	0.54	2.07	13.32	0.2	0.98	n.d.
HY-III(Q)-2-C	1.25	2.2	16.81	62.05	0.4	2.06	14.08	0.08	1.06	n.d.

(*Continued*)

Table 22.3 (*Continued*)

No.	Na_2O	MgO	Al_2O_3	SiO_2	P_2O_5	K_2O	CaO	TiO_2	Fe_2O_3	PbO
HY-III(Q)-3-A	1.36	2.95	13.49	65.16	0.26	4.26	6.92	0.79	4.82	n.d.
HY-III(Q)-3-B	1.15	3	14.03	66.6	0.3	3.85	5.06	1.06	4.94	n.d.
HY-III(Q)-3-C	1.14	3.05	13.58	64.65	0.26	3.72	6.41	0.92	6.28	n.d.
HY-III(Q)-4-A1	1.48	4.49	12.92	61.94	n.d.	3.24	9.39	0.93	5.6	n.d.
HY-III(Q)-4-A2	1.5	3.51	12.09	67.64	0.04	2.98	6.28	0.95	5.02	n.d.
HY-III(Q)-4-B	0.79	6.65	11.29	60.07	0.05	3.53	11.23	0.85	5.55	n.d.
HY-III(Q)-4-C	0.83	4.35	12.92	63.23	0.15	4	7.91	0.78	5.83	n.d.
HY-III(H)-1-A	1.41	3.1	18.31	56.67	0.53	4.88	13.92	0.3	0.88	n.d.
HY-III(H)-1-B	1.08	2.69	18.19	60.2	0.96	3.18	12.6	0.31	0.8	n.d.
HY-III(H)-1-C	1.43	3.14	20.01	56.5	0.51	3.79	13.33	0.34	0.95	n.d.
HY-III(H)-2-A-bk	1.16	3.28	13.14	66.21	0.16	4.06	5.4	0.81	5.78	n.d.
HY-III(H)-2-A-w	1.2	1.7	19.57	66.34	0.41	2.99	6.46	0.49	0.84	n.d.
HY-III(H)-2-B-bk	0.98	3.25	13.5	64.97	0.15	4.87	5.89	0.77	5.62	n.d.
HY-III(H)-2-B-w	0.93	2.02	19.64	63.4	0.53	3.16	8.84	0.53	0.95	n.d.
HY-III(H)-2-C-bk	0.94	2.8	14.79	66.81	0.42	3.47	4.63	0.9	5.25	n.d.
HY-III(H)-2-C-w	0.83	2.32	17.76	62.61	0.79	3.04	11.61	0.29	0.75	n.d.
HY-III(H)-3-A	1.01	3.72	13.72	64.83	0.32	3.13	6.96	0.86	5.45	n.d.
HY-III(H)-3-B	1.19	2.36	12.77	66.55	0.34	3.96	7.23	0.67	4.93	n.d.
HY-III(H)-3-B	1.14	2.78	13.57	67.27	0.39	2.79	6.15	0.76	5.15	n.d.
HY-III(H)-3-C	1.14	3.83	13.28	65.93	0.38	3.67	5.76	1.02	4.98	n.d.
HY-III(H)-4-A	1.68	3.02	12.76	63.22	n.d.	3.05	9.98	1.01	5.28	n.d.
HY-III(H)-4-B1	1.19	5.42	11.05	62.96	n.d.	3.86	8.85	0.84	5.82	n.d.
HY-III(H)-4-B2	1.28	3.64	14.09	65.45	n.d.	2.98	6.23	0.97	5.35	n.d.
HY-III(H)-4-C	1.31	2.43	14.62	66.82	0.08	2.93	5.74	1.01	4.65	0.41
HY-IV-1-A	0.86	2.46	14.1	68.41	0.81	2.8	9.79	0.08	0.7	n.d.
HY-IV-1-B	0.92	2.48	17.02	61.41	0.61	2.24	14.07	0.4	0.87	n.d.
HY-IV-2-A	0.64	3.08	14.93	63.59	0.12	3.94	6.61	0.95	6.15	n.d.
HY-IV-2-B	0.92	3.16	15.49	63.67	0.05	3.88	5.85	0.96	6	n.d.
HY-IV-3-A	0.79	6.39	12.19	60.83	n.d.	3.54	9.86	0.86	5.54	n.d.
HY-IV-3-B	0.84	6.99	10.98	59.43	n.d.	4.2	11.07	0.88	5.61	n.d.
HY-IV-12	0.72	2.11	13.8	64.31	0.91	2.48	14.85	0.07	0.75	n.d.

Table 22.4 Chemical results of glaze for low temperature glazed potteries unearthed from the Huangye kiln site

No.	color	Na_2O	MgO	Al_2O_3	SiO_2	P_2O_5	K_2O	CaO	TiO_2	Fe_2O_3	CoO	CuO	PbO
HY-II-9-A	yellow	0.9	0.16	7.69	36.25	0.19	0.56	1.49	0.18	4.84	n.d.	n.d.	47.72
HY-II-9-B	yellow	1.04	0.29	4.87	29.91	0.32	0.31	0.96	0.2	9.41	n.d.	n.d.	52.69
HY-II-9-C1	yellow	0.98	0.38	6.25	32.3	0.28	0.5	1.14	0.2	5.1	n.d.	n.d.	52.87
HY-II-9-C2	yellow	0.96	0.32	8.6	32.9	0.19	0.75	1.64	0.2	3.57	n.d.	n.d.	50.87
HY-II-10-A	blue	1.11	0.34	4.76	33.91	0.38	0.3	0.59	0.21	0.72	0.5	0.43	56.77
HY-II-10-B	yellow	1.05	0.33	5.98	29.46	0.32	0.34	1.06	0.21	4.37	n.d.	n.d.	56.89
HY-II-10-C	yellow	1.05	0.31	4.84	29.86	0.41	0.34	0.75	0.21	4.32	n.d.	n.d.	57.9
HY-III(Q)-5-A	blue	0.92	0.44	4.29	41.52	0.24	0.4	0.57	0.17	1.01	0.87	0.45	49.13
HY-III(Q)-5-B	blue	0.98	0.6	3.85	37.69	0.34	0.49	1.14	0.17	0.97	0.52	0.55	52.7
HY-III(Q)-5-C	blue	0.97	0.51	5.65	40.07	0.23	1.25	0.86	0.18	1.33	0.76	0.48	47.71
HY-III(Q)-6-A1	blue	1.01	0.11	8.32	33.29	0.33	0.61	0.6	0.2	0.71	0.26	0.24	54.33
HY-III(Q)-6-A2	yellow	1.13	0.43	6.99	33.99	0.25	0.46	1.24	0.22	1.89	n.d.	0.38	53
HY-III(Q)-6-B1	blue	0.85	0.71	5.04	46.35	0.12	0.68	1.18	0.15	1.55	0.59	0.42	42.35
HY-III(Q)-6-B2	blue	1.06	0.75	5.48	41.27	0.03	0.47	1.72	0.18	1.31	0.46	0.33	46.94
HY-III(Q)-6-C	blue	0.76	0.91	7.43	44.41	0.25	1.6	3.67	0.14	1.87	0.3	0.43	38.24
HY-III(Q)-7-A1	white	0.86	0.72	3.85	41.76	0.35	0.58	3.91	0.16	0.29	n.d.	n.d.	47.53
HY-III(Q)-7-A2	white	1.34	1.23	9.54	42.21	0.93	0.66	3.67	0.3	2.27	n.d.	n.d.	37.85
HY-III(Q)-7-B	white	0.74	0.39	4.67	48.18	0.12	0.5	1.65	0.14	0.28	n.d.	n.d.	43.33
HY-III(Q)-7-C	white	0.76	0.36	4.36	49.83	0.09	0.43	0.93	0.13	0.22	n.d.	n.d.	42.88

HY-III(Q)-8-A-b	blue	0.83	4.22	3.49	42.02	0.04	0.37	4.33	0.14	1.26	0.79	0.45	42.07
HY-III(Q)-8-A-w	white	0.96	4.28	5.34	51.72	0.98	0.69	4.61	0.16	0.82	n.d.	n.d.	30.43
HY-III(Q)-8-B-b	blue	0.84	0.71	4.43	43.67	0.18	1.71	1.74	0.17	0.82	0.16	0.13	45.58
HY-III(Q)-8-B-w	white	0.84	0.51	6.9	44.17	0.2	1.36	1.03	0.17	0.35	n.d.	n.d.	44.49
HY-III(Q)-8-C-b	blue	0.84	0.57	6.08	42.53	0.14	0.86	0.78	0.17	1.64	0.4	0.23	45.76
HY-III(Q)-8-C-w	white	0.87	0.56	5.08	44.31	0.1	0.66	1.64	0.16	0.34	n.d.	n.d.	46.27
HY-III(Q)-9-A-g	green	1.11	0.32	5.41	34.1	0.26	0.41	0.66	0.2	0.61	n.d.	2.53	54.4
HY-III(Q)-9-A-w	white	1.04	0.12	5.86	41.43	0.26	0.68	1.07	0.19	0.31	n.d.	0.18	48.86
HY-III(Q)-9-B-g	green	1.01	0.44	3.63	35.37	0.26	0.43	1.66	0.18	0.65	n.d.	2.13	54.24
HY-III(Q)-9-B-w	white	0.96	0.94	7.39	42.91	0.07	0.67	2.49	0.18	0.43	n.d.	0.08	43.87
HY-III(Q)-9-C-g	green	1.05	0.34	2.93	37.43	0.27	0.32	1.71	0.19	0.66	n.d.	2	53.1
HY-III(Q)-9-C-w	white	0.96	0.42	4.03	41.5	0.11	0.4	3.22	0.18	0.44	n.d.	0.11	48.64
HY-III(Q)-10-A	yellow	0.91	0.62	10.17	34.66	0.11	0.73	1.79	0.21	4.24	n.d.	0.12	46.44
HY-III(Q)-10-B	yellow	0.96	0.5	6.77	31.43	0.3	0.71	1.68	0.2	5.23	n.d.	0.07	52.15
HY-III(Q)-10-C	yellow	0.97	0.2	5.91	46.91	0.15	0.68	0.46	0.18	4.6	n.d.	0.18	39.76
HY-III(Q)-11-A	green	1.05	0.31	7.05	35.74	0.17	0.8	0.83	0.2	0.94	n.d.	3.99	48.93
HY-III(Q)-11-B	green	0.86	0.59	5.13	41.73	0.11	0.25	1.42	0.15	1.43	n.d.	5.04	43.29
HY-III(Q)-11-C	green	1.05	0.59	5.15	32.87	0.27	0.37	2.01	0.19	0.87	n.d.	4.13	52.51
HY-III(Q)-12-A-lg	green	1.08	0.55	5.48	39.58	0.17	0.61	1.73	0.19	1.86	n.d.	5.87	42.88
HY-III(Q)-12-A-dg	green	0.84	1.64	5.77	41.35	0.04	1.08	2.63	0.15	2.31	n.d.	5.78	38.44

(Continued)

Table 22.4 (*Continued*)

No.	color	Na_2O	MgO	Al_2O_3	SiO_2	P_2O_5	K_2O	CaO	TiO_2	Fe_2O_3	CoO	CuO	PbO
HY-III(Q)-12-A-b	blue	1.13	0.48	6.54	39.98	0.21	0.46	0.85	0.2	1.36	1.02	0.69	47.07
HY-III(Q)-12-A-y	yellow	0.96	0.8	4.81	35.16	0.29	0.27	1.33	0.19	5.37	n.d.	0.38	50.43
HY-III(Q)-12-A-w	white	0.97	0.12	4.03	49.27	0.24	0.34	0.94	0.16	0.28	n.d.	0.21	43.43
HY-III(Q)-12-B-b	blue	0.97	0.34	4.9	37.47	0.29	0.38	0.54	0.19	0.97	0.48	0.46	53.01
HY-III(Q)-12-B-y	yellow	1.02	0.4	7.51	34.5	0.27	0.65	1.4	0.2	4.38	0	0.08	49.59
HY-III(Q)-12-C-g	green	1.04	0.47	5.09	38.59	0.22	0.51	1.68	0.18	2.04	0	6.86	43.32
HY-III(Q)-12-C-b	blue	1.04	0.45	4.35	36.67	0.22	0.36	1.03	0.19	1.47	1.11	1.98	51.14
HY-III(Q)-12-C-y	yellow	1.17	0.72	3.92	37.82	0.29	0.27	1.39	0.21	4.49	n.d.	0.36	49.35
HY-III(H)-5-A	white	0.95	0.21	3.19	42.86	0.24	0.55	0.42	0.17	0.2	n.d.	n.d.	51.21
HY-III(H)-5-A	yellow	1.19	0.73	8	43.99	0.1	0.93	1.64	0.24	2.81	n.d.	n.d.	40.37
HY-III(H)-5-B	white	0.98	0.23	4.32	41.83	0.25	0.92	1.1	0.17	0.34	n.d.	n.d.	49.85
HY-III(H)-6-A	yellow	1.08	0.5	3.25	35.45	0.16	1.14	0.92	0.18	3.73	n.d.	n.d.	53.58
HY-III(H)-6-B1	yellow	0.93	0.34	4.14	40.87	0.25	0.94	0.65	0.17	4.3	n.d.	n.d.	47.39
HY-III(H)-6-B2	yellow	0.96	0.21	3.93	39.81	0.2	0.9	0.94	0.17	4.17	n.d.	n.d.	48.71
HY-III(H)-6-C-y	yellow	0.98	0.5	6.85	32.47	0.19	0.75	2.96	0.19	4.75	n.d.	0.1	50.26
HY-III(H)-6-C-g	green	1.05	0.69	4.81	34.74	0.3	0.45	1.62	0.21	2.57	n.d.	1.76	51.8
HY-III(H)-7-A-y	yellow	1.05	0.33	2.12	34.07	0.34	0.35	1.63	0.21	4.92	n.d.	0.39	54.59
HY-III(H)-7-A-g	green	1.09	0.65	4.09	34.94	0.32	0.4	2.2	0.21	2.53	n.d.	1.74	51.83
HY-III(H)-7-B	yellow	0.98	0.13	7.02	38.88	0.19	1.37	0.96	0.2	6.35	n.d.	0.17	43.75
HY-III(H)-7-C-y	yellow	0.96	0.5	7.53	31.61	0.29	0.52	1.59	0.2	4.29	n.d.	n.d.	52.51

HY-III(H)-7-C-dy	yellow	1.01	0.54	6.06	32.17	0.18	0.48	2.41	0.23	13.66	n.d.	n.d.	43.25
HY-III(H)-8-A	blue	1.07	0.64	4.84	35.18	0.08	1.13	1.87	1.31	0.73	0.57	0.55	52.02
HY-III(H)-8-B	blue	0.98	1.03	4.02	34.62	0.29	0.42	2.17	0.18	1.39	0.4	0.71	53.78
HY-III(H)-8-C	blue	0.94	0.48	3.91	41.27	0.24	0.62	1.31	0.17	0.97	0.61	0.48	49
HY-III(H)-9-A	green	0.94	0.64	5.44	37.53	0.16	1.32	1.3	0.17	1.1	n.d.	3.09	48.32
HY-III(H)-9-B	green	1.02	0.45	4.73	32.25	0.33	0.61	3.36	0.19	1.36	n.d.	3.14	52.57
HY-III(H)-9-C	green	1.03	0.51	5.15	34.51	0.22	0.78	2.11	0.19	0.78	n.d.	4.62	50.11
HY-III(H)-10-A-w	white	0.92	0.42	3.63	37.96	0.31	0.45	3.84	0.17	0.47	n.d.	n.d.	51.82
HY-III(H)-10-A-b	blue	0.98	0.42	2.83	35.34	0.39	0.37	4.57	0.18	0.6	0.2	0.43	53.68
HY-III(H)-10-B-w	white	1.02	0.27	4.66	35.01	0.37	0.29	0.64	0.21	0.41	n.d.	n.d.	57.13
HY-III(H)-10-B-b	blue	0.98	0.46	5.23	34.66	0.25	0.55	1.28	0.19	0.72	0.21	0.23	55.24
HY-III(H)-10-C-w	white	0.6	0.2	0.88	56.63	n.d.	n.d.	0.88	n.d.	0.08	n.d.	n.d.	40.73
HY-III(H)-10-C-b	blue	0.91	0.57	3.18	40.45	0.3	0.39	1.23	0.17	0.89	0.3	0.4	51.23
HY-III(H)-11-A-w	white	1.07	0.62	1.96	38.61	0.11	0.67	2.69	0.19	0.35	n.d.	n.d.	53.74
HY-III(H)-11-A-g	green	1.09	0.83	1.61	31.73	0.33	0.77	3.28	0.19	1.03	n.d.	3.19	55.94
HY-III(H)-11-B-w	white	1.21	0.46	5.73	46.74	0.33	1.62	1.29	0.22	0.63	n.d.	n.d.	41.75
HY-III(H)-11-B-g	green	1	0.49	3.21	36.92	0.07	0.74	1.21	0.18	0.77	n.d.	2.53	52.88
HY-III(H)-11-C-w	white	1.07	0.77	1.81	31.95	0.57	0.76	4.65	0.21	0.41	n.d.	n.d.	57.8
HY-III(H)-11-C-g	green	1.01	0.85	2.08	31.94	0.97	0.74	3.79	0.18	0.99	n.d.	2.83	54.61
HY-III(H)-12-A-g	green	1.02	0.37	5.63	32.25	0.27	0.78	1.37	0.19	0.95	n.d.	3.58	53.6

(Continued)

Table 22.4 (*Continued*)

No.	color	Na_2O	MgO	Al_2O_3	SiO_2	P_2O_5	K_2O	CaO	TiO_2	Fe_2O_3	CoO	CuO	PbO
HY-III(H)-12-A-y	yellow	1.05	0.29	5.8	33.26	0.3	0.4	0.89	0.22	2.98	n.d.	n.d.	54.83
HY-III(H)-12-B	white	0.86	0.4	5.93	44.22	0.01	4.22	2.58	0.15	0.45	n.d.	n.d.	41.17
HY-III(H)-12-B	green	1.22	0.52	5.35	31.29	0.18	0.67	1.72	0.22	1.47	n.d.	6.66	50.7
HY-III(H)-12-B	yellow	0.94	0.39	6.04	31.08	0.24	0.69	2.14	0.19	6.99	n.d.	n.d.	51.3
HY-III(H)-12-C-g	green	1.01	0.52	5.11	34.26	0.13	1.09	1.27	0.18	0.85	n.d.	4.41	51.17
HY-III(H)-12-C-y	yellow	1.04	0.39	6.13	32.72	0.29	0.56	1.74	0.21	3.63	n.d.	0.27	53.02
HY-III(H)-12-C-w	white	1.03	0.23	4.6	40.03	0.25	0.62	0.79	0.19	0.26	n.d.	n.d.	52
HY-IV-4-A	yellow	0.91	0.49	9.19	39.06	0.18	2.04	2.05	0.2	5.96	n.d.	n.d.	39.92
HY-IV-4-B	yellow	0.92	0.44	7.21	35.13	0.22	0.61	1.22	0.19	5.01	n.d.	n.d.	49.05
HY-IV-4-C	yellow	0.97	0.41	7.79	35.59	0.26	0.79	1.78	0.2	5.57	n.d.	n.d.	46.63
HY-IV-5-A	yellow	0.99	0.56	4.41	32.27	0.31	0.55	1.88	0.2	6.5	n.d.	n.d.	52.33
HY-IV-5-B	yellow	0.92	0.16	5.91	36.59	0.25	0.39	1.15	0.18	6.83	n.d.	n.d.	47.61
HY-IV-5-C	yellow	0.92	0.31	8.27	38.03	0.15	1.3	1.59	0.19	6.52	n.d.	0.17	42.55
HY-IV-6-A	blue	0.97	0.56	3.99	39.03	0.31	0.58	0.63	0.18	1.11	1.11	0.33	51.18
HY-IV-6-B	blue	1.01	0.63	4.56	32.29	0.4	0.61	2.31	0.19	0.85	0.41	0.33	56.42
HY-IV-7-A	green	1.02	0.3	4.73	35.3	0.28	0.37	0.86	0.19	0.91	n.d.	4.23	51.8
HY-IV-7-B1	green	1.01	0.33	5.97	36.66	0.25	0.66	1.55	0.18	1.08	n.d.	4.3	48.02
HY-IV-7-B2	green	1.14	0.22	4.15	36	0.42	0.32	0.72	0.2	1.07	n.d.	4.5	51.26
HY-IV-7-C-g	green	1.08	0.34	6.5	31.93	0.54	0.71	1.17	0.2	0.93	n.d.	3.9	52.69

HY-IV-7-C-w	white	1.08	0.31	3.85	45.19	0.44	0.38	1.23	0.2	0.41	n.d.	0.14	46.77
HY-IV-8-A1	green	0.95	0.75	6.06	35.26	0.35	0.89	1.89	0.18	1.53	n.d.	4.6	47.54
HY-IV-8-A2	green	1.37	0.47	4.97	34.24	0.4	0.45	1.79	0.24	1.23	n.d.	3.97	50.86
HY-IV-8-B1	green	0.99	0.42	9.12	35.06	0.32	1.4	1.72	0.2	1.78	n.d.	3.55	45.45
HY-IV-8-B2	green	1.17	0.35	7.49	35.39	0.31	0.62	1.03	0.22	1.95	n.d.	3.53	47.92
HY-IV-8-C1	green	0.99	0.66	3.74	34.03	0.34	0.48	1.85	0.17	2.11	n.d.	7.35	48.29
HY-IV-8-C2	green	1.01	0.77	5.59	38.96	0.17	0.78	2.84	0.18	2.17	n.d.	6.09	41.44
HY-IV-9-A-w	white	1.01	0.3	3.07	37.32	0.25	0.25	0.84	0.18	0.29	n.d.	n.d.	56.48
HY-IV-9-A-b	blue	1.13	0.92	4.73	35.58	0.19	0.38	3.36	0.2	0.73	0.24	0.2	52.33
HY-IV-9-B-w	white	0.92	0.44	3.82	47.75	0.11	0.67	3.14	0.17	0.51	n.d.	n.d.	42.48
HY-IV-9-B-b	blue	0.97	0.66	4.2	39.43	0.17	0.65	3.37	0.17	1.2	0.68	0.34	48.17
HY-IV-10-A-g	green	0.86	0.59	6.19	39.5	0.23	0.63	1.78	0.17	0.73	n.d.	3.79	45.52
HY-IV-10-A-w	white	0.87	0.28	3.57	45.72	0.3	0.55	1.39	0.15	0.41	n.d.	0.33	46.44
HY-IV-10-B-g	green	0.78	0.85	8.33	41.63	0	0.39	3.46	0.15	0.6	n.d.	2.24	41.56
HY-IV-10-B-w	white	0.75	0.48	8.7	45.29	2.11	0.47	5.12	0.16	0.74	n.d.	0.13	36.05
HY-IV-11-A-b	blue	1.02	0.31	6.39	41.21	0.3	0.84	0.74	0.18	1.44	0.15	5.05	42.36
HY-IV-11-A-y	yellow	0.89	0.22	5.78	31.8	0.4	0.38	2.97	0.18	9.9	n.d.	0.17	47.32
HY-IV-11-A-w	white	0.86	0.15	4.91	52.5	0.32	0.53	0.88	0.16	0.36	n.d.	0.15	39.17
HY-IV-11-B-g	green	1.06	0.27	5.36	36.07	0.39	1.1	1.11	0.2	1.19	n.d.	5.18	48.07
HY-IV-11-B-w	white	1.01	0.6	5.91	51.89	0.37	1.67	2.96	0.18	0.98	n.d.	0.25	34.18

(Continued)

Table 22.4 (*Continued*)

No.	color	Na_2O	MgO	Al_2O_3	SiO_2	P_2O_5	K_2O	CaO	TiO_2	Fe_2O_3	CoO	CuO	PbO
HY-IV-11-B-y	yellow	1.09	0.28	8.24	31.52	0.34	1.46	1.89	0.23	8.01	n.d.	0.16	46.77
HY-IV-11-C-g	green	1.09	0.41	5.79	34.43	0.41	0.43	1.66	0.2	2.17	n.d.	4.03	49.39
HY-IV-11-C-y	yellow	0.99	0.41	6.42	30.97	0.38	0.49	2.54	0.21	5.59	n.d.	0.21	51.78
HY-IV-11-D-y	yellow	1.01	0.17	3.92	33.46	0.48	0.39	1.12	0.2	3.3	n.d.	0.54	55.41
HY-IV-11-D-g	green	1.01	0.2	3.63	37.82	0.51	0.43	0.93	0.19	0.68	n.d.	1.58	53.02
HY-IV-11-D-w	white	0.98	0.23	3.58	38.43	0.48	0.37	3.01	0.18	0.35	n.d.	0.16	52.24
HY-IV-13-y1	yellow	0.98	0.32	6.02	33.1	0.44	0.72	1.3	0.2	4.41	n.d.	0.3	52.22
HY-IV-13-y2	yellow	1.07	0.39	5.84	33.83	0.44	0.77	1.31	0.2	5.75	n.d.	0.18	50.21
HY-IV-14-y	yellow	0.93	0.39	9.62	41.01	0.19	1.53	1.41	0.2	4.96	n.d.	0.21	39.53
HY-IV-14-dy	yellow	0.86	0.59	9.52	37.8	0.12	1.78	1.68	0.2	8.52	n.d.	0.22	38.71
HY-IV-15-g	green	1	0.51	2.97	36.6	0.27	0.32	1.33	0.18	1.5	n.d.	1.86	53.45
HY-IV-15-dg	green	1.01	0.46	2.99	37.47	0.52	0.48	1.98	0.18	1.79	n.d.	1.63	51.5

Table 22.5 Chemical results for ceramic bodies unearthed from the Huangye kiln site

No.	Na_2O	MgO	Al_2O_3	SiO_2	K_2O	CaO	TiO_2	Fe_2O_3
HY-II-1A	1.1	0.43	26.59	65.9	1.97	0.77	0.57	2.66
HY-II-1B	1.01	0.78	28.59	64.64	2.14	0.37	0.69	1.78
HY-II-2A	1.05	0.98	27.54	64.38	2.33	1.06	0.67	1.99
HY-II-2B	1.1	0.65	26.61	66.32	2.09	0.52	0.68	2.03
HY-II-2C	1.11	0.81	27.72	65.38	1.93	0.48	0.56	2.01
HY-II-3A	0.86	0.91	27.4	65.29	1.94	0.6	0.61	2.39
HY-II-3B	0.98	0.79	24.08	67.75	1.84	1.45	0.6	2.51
HY-II-3C	0.89	1.49	26.87	64.29	2.54	1.06	0.73	2.13
HY-II-4A	1.11	1.17	31.36	56.58	2.04	4.44	0.6	2.71
HY-II-4B	0.94	1.2	29.34	60.67	2.22	2.34	0.62	2.67
HY-II-4C	0.7	0.91	29.09	63.33	2.56	0.44	0.66	2.3
HY-II-5A	0.97	1.16	28.76	62.78	2.29	0.96	0.66	2.41
HY-II-5B	1.05	1.53	24.62	64.83	2.4	2.2	0.65	2.72
HY-II-5C	0.94	0.9	27.19	65.39	2.06	0.56	0.63	2.33
HY-II-5D	1.02	0.85	28.34	64.34	2.09	0.51	0.65	2.2
HY-II-6A	1	1.04	29.04	62.7	1.8	1.59	0.55	2.28
HY-II-6B	0.99	0.71	28.43	64.07	2.1	0.57	0.64	2.48
HY-II-6C	1.04	1.21	29	62.06	1.85	1.84	0.59	2.4
HY-II-7A	0.93	0.79	27.11	64.97	1.95	1.23	0.6	2.41
HY-II-7B	1.03	0.96	26.84	65.6	2.1	0.5	0.62	2.35
HY-II-7C	0.94	0.87	26.66	65.51	2.1	0.85	0.58	2.48
HY-II-8	0.99	0.46	27.13	62.91	2.24	3.2	0.61	2.44
HY-II-9A	0.89	1.24	29.28	61.72	2.25	1.86	0.66	2.08
HY-II-9B	1.1	0.66	28.33	63.8	2.11	1.09	0.6	2.31
HY-II-9C	0.9	0.77	27.75	63.24	2.13	2.33	0.64	2.25
HY-II-10A	0.93	0.71	28.69	62.88	2.24	1.31	0.64	2.6
HY-II-10B	1.02	0.68	29.95	63.48	1.87	0.4	0.61	1.99
HY-II-10C	1.02	0.73	29.89	63.25	1.91	0.57	0.63	2
HY-III(Q)-1A	1.9	2.71	28.76	58.41	4.46	1.23	0.72	1.82

(*Continued*)

Table 22.5 (*Continued*)

No.	Na_2O	MgO	Al_2O_3	SiO_2	K_2O	CaO	TiO_2	Fe_2O_3
HY-III(Q)-1B	0.99	1.85	29.19	62.51	2.43	0.4	0.74	1.88
HY-III(Q)-1C	0.96	1.67	29.43	62.68	2.19	0.36	0.71	2
HY-III(Q)-2A	0.85	2.38	30.67	60.36	2.31	0.63	0.68	2.12
HY-III(Q)-2B	1.07	2.4	29.59	59.36	3.33	1.04	0.66	2.56
HY-III(Q)-2C	1.15	1.85	25.98	64.06	2.27	0.75	0.6	3.33
HY-III(Q)-3A	1.06	2.14	27.85	61.73	2.85	0.84	0.67	2.86
HY-III(Q)-3B	0.92	1.71	28.43	63.53	1.98	0.6	0.64	2.19
HY-III(Q)-3C	0.99	1.12	26.99	65.3	1.92	0.42	0.69	2.58
HY-III(Q)-4A	1.12	2.02	26.41	62.42	2.56	1.88	0.63	2.96
HY-III(Q)-4B	0.95	2.01	28.7	61.53	2.43	0.98	0.7	2.7
HY-III(Q)-4C	1.07	2.55	25.18	62.03	3.18	2.15	0.73	3.11
HY-III(Q)-5A	1	1.85	27.63	62.83	2.28	1.48	0.69	2.23
HY-III(Q)-5B	0.85	1.16	28.17	63.21	2.13	1.24	0.76	2.47
HY-III(Q)-6A	0.92	1.29	28.52	63	2.04	1.03	0.69	2.51
HY-III(Q)-6B	0.9	1.35	27.73	64.34	1.84	0.64	0.72	2.47
HY-III(Q)-6C	0.97	2.17	24.63	64.83	2.59	1.69	0.71	2.41
HY-III(Q)-7A	1.08	1.44	25.46	64.37	2.53	1.46	0.69	2.97
HY-III(Q)-7B	1.09	2.27	23.61	64.4	2.62	2.53	0.6	2.87
HY-III(Q)-7C	0.99	2.58	23.11	64.94	2.74	1.99	0.63	3.01
HY-III(Q)-8A	1.05	2.05	25.23	63.62	2.5	2.24	0.7	2.61
HY-III(Q)-8B	1.12	1.62	23.37	66.1	2.27	1.95	0.65	2.93
HY-III(Q)-8C	0.94	1.18	26.69	64.27	2.19	1.48	0.66	2.61
HY-III(Q)-9A	0.84	1.33	28.44	62.95	2.06	0.7	0.64	3.05
HY-III(Q)-9B	0.83	1.05	28.76	63.12	1.91	1.12	0.65	2.56
HY-III(Q)-9C	1	1.41	28.85	61.17	2.58	0.99	0.65	3.35
HY-III(Q)-10A	0.81	1.06	27.65	64.29	2.21	0.86	0.67	2.44
HY-III(Q)-10B	0.99	1.58	24.83	61.69	2.67	4.92	0.61	2.71
HY-III(Q)-10C	1.01	1.26	25.88	63.54	2.26	1.95	0.66	3.43
HY-III(Q)-11A	0.73	1	29.03	62.79	2.14	0.89	0.61	2.81
HY-III(Q)-11B	0.86	0.9	27.48	64.6	1.94	0.83	0.65	2.74

(*Continued*)

Table 22.5 (*Continued*)

No.	Na_2O	MgO	Al_2O_3	SiO_2	K_2O	CaO	TiO_2	Fe_2O_3
HY-III(Q)-11C	0.81	1.14	31	62.02	2.01	0.38	0.73	1.92
HY-III(Q)-12A	0.87	1.19	27.54	63.65	2.43	0.73	0.68	2.92
HY-III(Q)-12B	1.09	1.22	26.06	64.51	1.93	0.97	0.55	3.66
HY-III(Q)-12C	0.93	1.14	26.71	63.21	2.57	1.73	0.64	3.08
HY-III(H)-1A	1.08	1.46	28.22	63.49	2.45	0.6	0.69	2.02
HY-III(H)-1B	1.38	0.44	27.85	63.52	1.97	2.23	0.62	1.98
HY-III(H)-1C	0.78	1.39	27.69	65.07	1.98	0.56	0.65	1.88
HY-III(H)-2A	0.86	1.16	28.78	64.59	1.59	0.61	0.68	1.73
HY-III(H)-2B	0.88	0.83	28.1	65.64	1.42	0.62	0.67	1.82
HY-III(H)-2C	0.87	0.87	27.09	65.98	2.03	0.73	0.65	1.79
HY-III(H)-3A	0.85	0.97	26.68	65.78	1.78	0.95	0.7	2.3
HY-III(H)-3B	1.1	0.84	27.65	64.71	2.07	0.71	0.63	2.28
HY-III(H)-3C	0.92	1.19	26.95	64.31	2.28	1.16	0.7	2.5
HY-III(H)-4A	1.02	0.9	28.1	63.85	1.83	1.13	0.62	2.55
HY-III(H)-4B	1.05	2.15	27.85	62.24	2.18	1.05	0.64	2.85
HY-III(H)-4C	0.85	2.13	26.02	61.62	2.12	3.37	0.67	3.23
HY-III(H)-5A	0.91	0.95	29.23	62.34	2	1.25	0.7	2.61
HY-III(H)-5B	0.92	1.35	28.39	62.8	2.22	0.9	0.71	2.71
HY-III(H)-5C	0.95	1.31	25.84	64.82	1.89	2.02	0.67	2.5
HY-III(H)-6A	0.94	0.92	28.56	63.36	2.24	0.67	0.65	2.66
HY-III(H)-6B	0.83	1.24	28.89	63.16	2.32	0.6	0.72	2.25
HY-III(H)-6C	0.98	1.14	28.61	64.28	1.71	0.47	0.7	2.12
HY-III(H)-7A	0.87	1.06	27.6	64.36	1.79	1.37	0.7	2.26
HY-III(H)-7B	0.8	1.01	28.63	64.08	1.74	0.94	0.68	2.1
HY-III(H)-7C	0.77	0.84	28.3	62.98	2.39	1.76	0.71	2.25
HY-III(H)-8A	1.09	1.08	27.54	63.15	3.21	0.82	0.62	2.48
HY-III(H)-8B	1.06	1.04	30.34	61.81	2.39	0.43	0.69	2.24
HY-III(H)-8C	1.03	0.77	26.07	64.21	2.13	2.6	0.75	2.45
HY-III(H)-9A	0.82	1.1	27.13	63.16	2.28	1.99	0.64	2.88
HY-III(H)-9B	1.01	1.15	28.23	64.04	2.08	0.55	0.66	2.28

(*Continued*)

Table 22.5 (*Continued*)

No.	Na_2O	MgO	Al_2O_3	SiO_2	K_2O	CaO	TiO_2	Fe_2O_3
HY-III(H)-9C	0.82	1.09	29.12	62.8	2.17	0.71	0.69	2.6
HY-III(H)-10A	0.87	1.46	27.29	62.09	2.36	2.34	0.67	2.92
HY-III(H)-10B	0.8	1.04	28.09	63.86	1.98	0.89	0.72	2.64
HY-III(H)-10C	0.87	1.02	27.15	65.05	2.13	0.82	0.68	2.27
HY-III(H)-11A	0.88	1.09	26.79	64.09	2.07	1.57	0.64	2.88
HY-III(H)-11B	1	0.83	28	64.18	1.77	0.75	0.63	2.85
HY-III(H)-11C	1	1.01	29.8	62.85	1.74	0.46	0.59	2.55
HY-III(H)-12A	0.78	0.96	28.63	62.38	2.51	1.67	0.71	2.35
HY-III(H)-12B	0.93	1.42	28.35	62.82	2.12	0.95	0.68	2.73
HY-III(H)-12C	0.83	0.76	27.15	63.73	2.12	1.88	0.73	2.81
HY-IV-1A	0.91	1.02	28.54	64.86	1.82	0.33	0.75	1.78
HY-IV-1B	0.81	1.03	28.19	64.68	2.05	0.61	0.65	1.98
HY-IV-1C	0.79	0.98	28.41	64.6	2.04	0.62	0.65	1.93
HY-IV-2A	1.02	1.09	28.86	61.64	2.17	0.67	0.67	3.9
HY-IV-2B	0.72	0.97	27.56	62.55	2.26	1.49	0.63	3.83
HY-IV-2C	0.96	0.67	28.09	64.74	2.12	0.32	0.61	2.49
HY-IV-3A	0.8	0.88	29.39	63	2.12	0.87	0.64	2.3
HY-IV-3B	1.14	0.73	28.1	64.23	1.94	0.77	0.6	2.49
HY-IV-3C	0.93	0.98	29.1	63.03	2.04	0.88	0.63	2.42
HY-IV-4A	0.85	1.4	27.84	62.25	2.39	1.48	0.68	3.11
HY-IV-4B	0.89	1.56	28.01	62.22	1.7	2.53	0.61	2.49
HY-IV-4C	1.06	0.85	26.63	63.96	2.09	1.27	0.59	3.54
HY-IV-5A	1.03	1.19	27.96	62.74	2.31	1.11	0.6	3.05
HY-IV-5B	0.88	0.9	28.11	63.92	1.91	0.98	0.68	2.6
HY-IV-5C	0.98	1.49	28.62	62.41	2.18	0.95	0.65	2.72
HY-IV-6A	0.96	1.14	29.36	60.45	2.05	2.47	0.68	2.9
HY-IV-6B	0.88	0.75	29.65	62.53	2.23	0.61	0.74	2.6
HY-IV-6C	0.9	0.99	29.55	63.24	1.94	0.59	0.69	2.11
HY-IV-7A	0.93	1.18	29.95	61.74	1.98	0.58	0.68	2.97
HY-IV-7B	0.9	0.99	28.51	63.56	1.97	0.66	0.67	2.74

(*Continued*)

Table 22.5 (*Continued*)

No.	Na_2O	MgO	Al_2O_3	SiO_2	K_2O	CaO	TiO_2	Fe_2O_3
HY-IV-7C	0.69	1.45	29.42	62.23	2.06	0.97	0.67	2.51
HY-IV-8A	1.01	1.65	29.56	59.54	2.08	1.93	0.68	3.55
HY-IV-8B	1.05	0.87	29	62.58	2.05	0.5	0.69	3.26
HY-IV-8C	0.88	1.1	27.8	62.01	1.81	2.64	0.62	3.14
HY-IV-9A	1.09	1.4	28.2	62.41	1.9	1.36	0.59	3.04
HY-IV-9B	0.79	0.85	29.17	62.57	2.41	0.72	0.66	2.83
HY-IV-9C	0.96	0.64	26.13	65.05	1.85	1.61	0.59	3.17
HY-IV-10A	1.01	1.93	24	63.21	3.35	1.85	0.57	4.08
HY-IV-10B	1.09	1.26	27.77	61.81	2.33	1.81	0.58	3.35
HY-IV-11C	0.94	1.39	29.17	60.06	2.27	2.11	0.66	3.41
HY-IV-11A	1.07	1.45	24.12	62.71	2.23	3.73	0.63	4.06
HY-IV-11B	0.77	0.8	28.16	63.8	2	0.77	0.69	3
HY-IV-11D	0.87	0.89	27.97	64.32	1.93	0.77	0.64	2.59
HY-IV-12	0.79	0.81	30.79	61.91	2	0.6	0.68	2.4
HY-IV-13	0.83	1.22	30.12	60.86	2.65	1.15	0.66	2.5
HY-IV-14	0.75	1.07	29.09	64.17	1.61	0.44	0.66	2.22
HY-IV-15	1.06	1.22	27.62	62.12	3.18	1.41	0.58	2.81

compositions) between low temperature glaze and high temperature glaze are (chemical). Another difference is that the low temperature glazes are more colorful than high temperature ones, and usually have blue, green, yellow and white colors, which can be attributed to Co ions, Cu ions, Fe ions, respectively.

22.3.3 *Bodies*

The chemical compositions obtained by pXRF for the ceramic bodies, including high temperature glaze porcelain and low temperature glaze pottery, are listed in Table 22.5. As shown in Table 22.5, the SiO_2 contents vary from 29.46 wt% to 56.63 wt%, and the PbO contents from 30.43 wt% to 57.90 wt%; the range of Al_2O_3 contents is

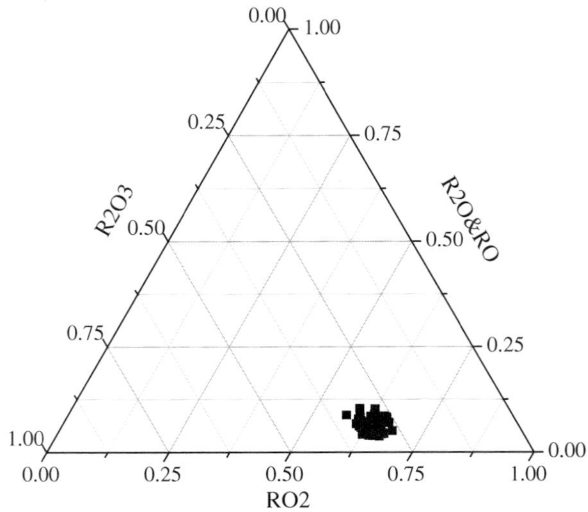

Fig. 22.3 A ternary plot of ceramic bodies unearthed from the Huangye kiln site

from 0.88 wt% to 10.17 wt%. A ternary plot of RO_2 (SiO_2+TiO_2), R_2O(Na_2O+K_2O) +RO(MgO+CaO), R_2O_3 (Fe_2O_3) is shown in Fig. 22.3. As shown in Fig. 22.3, although the samples are composed by porcelain and pottery, the recipes of the bodies are very similar.

22.4 Discussion

22.4.1 *Differences among the high temperature glaze with different colors*

Fe ions are the main colorant for the high temperature glaze. The Fe contents vary among the different glaze colors. A plot of CaO versus Fe_2O_3 for high temperature glazes is shown in Fig. 22.4(a). As shown in Fig. 22.4(a), the black and tea leaf glazes have higher levels of Fe_2O_3 than greenish white glaze and white glaze. Although the Fe_2O_3 contents of black glazes and tea leaf glazes are comparable, the CaO contents are different. The levels of CaO in black glaze are lower than that of tea flake glaze. The contents of Al_2O_3 and TiO_2 can also distinguish the black and the tea leaf glaze from the greenish white and white glaze shown in Fig. 22.4(b). The black and tea leaf glaze have

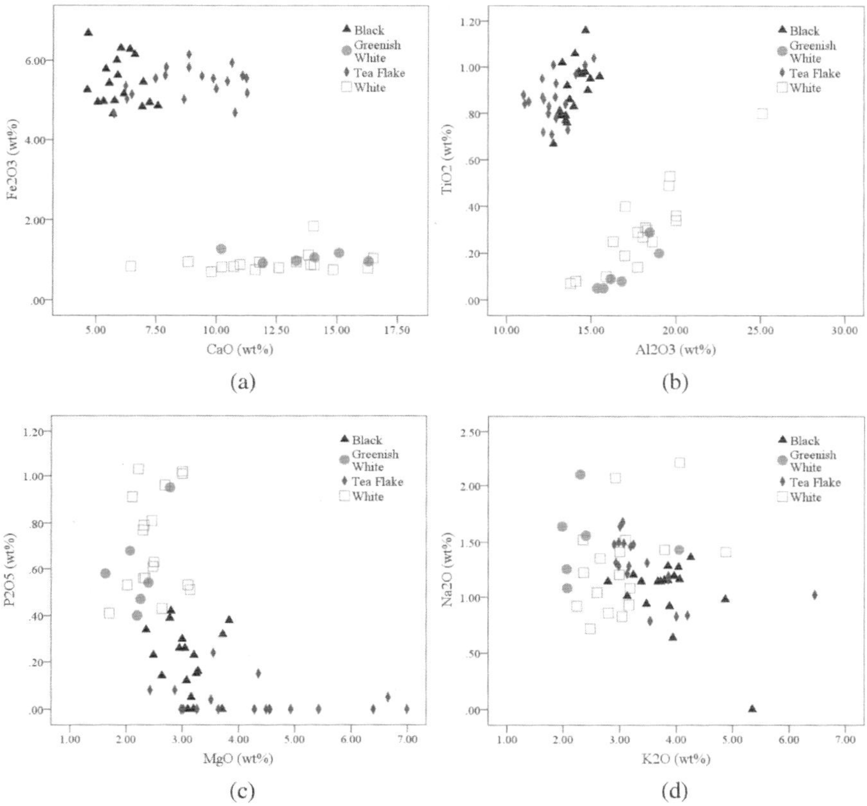

Fig. 22.4 Plots for high temperature glaze unearthed from Huangye kiln site (a) CaO *vs* Fe_2O_3 (b) Al_2O_3 vs TiO_2 (c) MgO vs P_2O_5 (d) K_2O vs Na_2O

relative high levels of TiO_2 and low levels of Al_2O_3. Positive correlations between Al_2O_3 and TiO_2 are also shown in Fig. 22.4(b), especially for the greenish white and white glaze. A plot of MgO versus P_2O_5 is shown in Fig. 22.4(c). As shown in Fig. 22.4(c), the tea leaf glaze has the highest levels of MgO and lowest levels of P_2O_5 which makes it different from the black glaze. Figure 22.4(d) is a plot of K_2O *vs* Na_2O and indicates that the K_2O levels of greenish white glaze are lower than that of the white glaze except for one sample.

In summary, the high temperature glaze can be divided into two groups, one group includes black glaze and tea leaf glaze, the other includes greenish white glaze and white glaze. The chemical compositions of black glaze and tea leaf glaze are similar, but still distinguishable. They can be discriminated by the contents of CaO, and P_2O_5. As for white glaze and greenish white glaze, they are also similar in chemical composition, but hardly distinguishable from each other.

22.4.2 *Evolution of the high temperature glaze recipes*

The box-plots of components for the high temperature glazes with different colors are shown in Fig. 22.5. The time sequence for the four periods is as follows: early Tang Dynasty (Tang-E), prosperous Tang Dynasty (Tang-P), middle Tang Dynasty (Tang-M) and later Tang Dynasty (Tang-L). We focus on black, tea leaf and white glazes because the greenish white glaze only appears in the early and prosperous Tang periods and only 6 pieces, 3 pieces for each period fragments in total.

The variation in chemical compositions for the glazes reflects the evolution of the glaze recipes. The chemical results obtained reveal some clues about recipe evolution. As shown in Fig. 22.5, the Na_2O contents of black glaze have a remarkably consistent decrease, while SiO_2 contents have a slight decrease and Al_2O_3 contents have a slight rise in the late Tang period. For tea leaf glaze, the contents of SiO_2 and Na_2O have relatively low levels, while the CaO, MgO and K_2O contents show a significant rise in the late Tang period compared to that of the other three periods. For the white glaze, the turning point appears not only in the late Tang period, but also in the prosperous Tang period, which is clearly shown in Figs. 22.5(a), (b) and (h). The SiO_2 contents of white glaze have the lowest levels in the prosperous Tang period. After a decrease from the early Tang period to the prosperous Tang period, the SiO_2 contents saw a continuous growth from the prosperous Tang to the late Tang. The trends in Al_2O_3 and TiO_2 contents are exactly the opposite.

In summary, the recipes of black and tea leaf glazes have similar trends, and the turning points appear to be in the late Tang period. But for the white glaze, the turning point appeared to be in the prosperous Tang period.

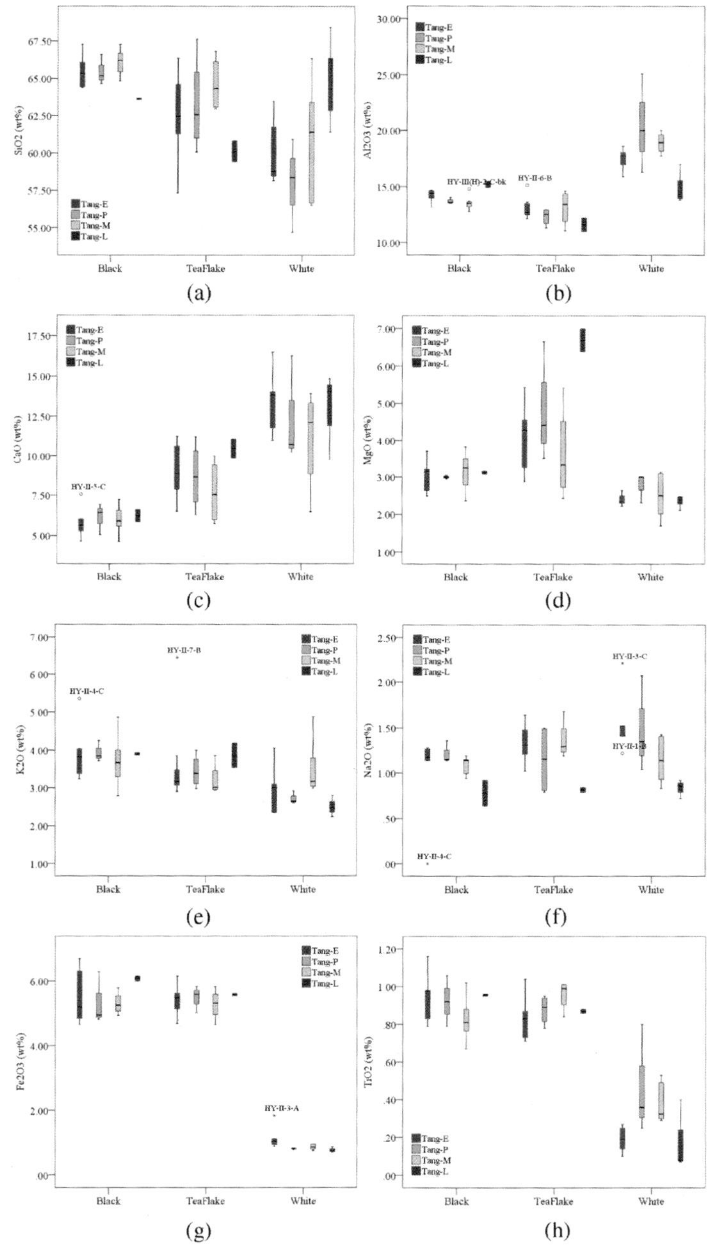

Fig. 22.5 Box-plots for high temperature glaze with different colors unearthed from the Huangye kiln site: (a) SiO_2 (b) Al_2O_3 (c) CaO (d) MgO (e) K_2O (f) Na_2O (g) Fe_2O_3 (h) TiO_2

22.4.3 *Differences and evolution of recipes for the low temperature glaze*

One difference between the low temperature glaze and high temperature glaze is in the colorants employed. The colorants for low temperature glaze include Co, Cu, Fe ions. Thus, the low temperature glaze with different colors can be distinguished by the colorants, and only by the colorants, which is shown in Fig. 22.6. As shown in Fig. 22.7, the low temperature glaze cannot be distinguished by other components except for the colorants. PbO, SiO_2, Al_2O_3 and CaO are the main components of the low temperature glaze matrices but not the colorants. As shown in Fig. 22.8, the contents of PbO, SiO_2, Al_2O_3 and CaO vary in a limited way — this indicates that the recipes used for the low temperature glazes in the four periods are similar.

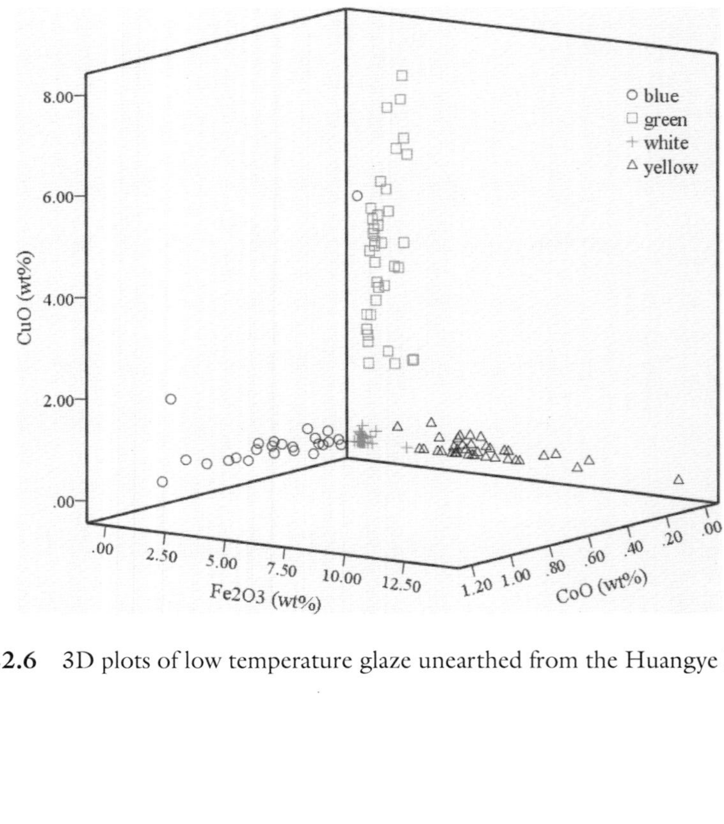

Fig. 22.6 3D plots of low temperature glaze unearthed from the Huangye kiln site

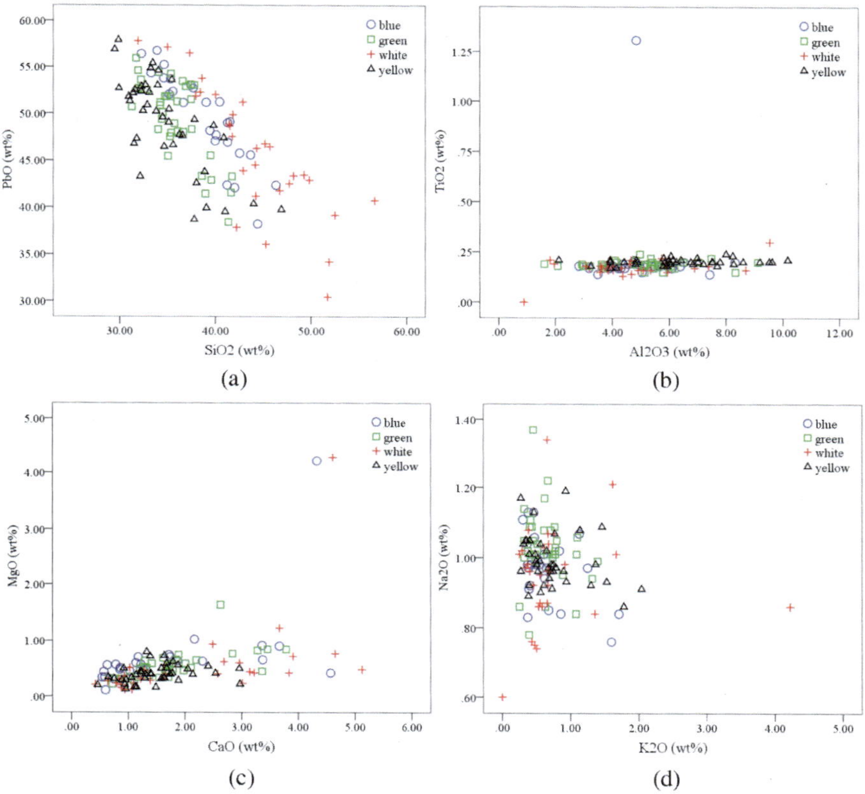

Fig.22.7 Plots for low temperature glaze unearthed from the Huangye kiln site (a) SiO_2 *vs* PbO (b) Al_2O_3 *vs* TiO_2 (c) CaO *vs* MgO (d) K_2O *vs* Na_2O

22.5 Conclusion

The chemical compositions for the glazes and bodies of 136 ceramic fragments unearthed from the Huangye kiln site were obtained by a portable XRF. The results show that the glaze can be divided into two groups: high temperature glaze and low temperature glazes. The high temperature glazes contain high levels of CaO, while the low temperature glaze contains high levels of PbO. They also have many other differences, such as colorants used and concentration levels of components. Although the chemical compositions of glazes are different, the recipes for the bodies are similar in the two glaze types.

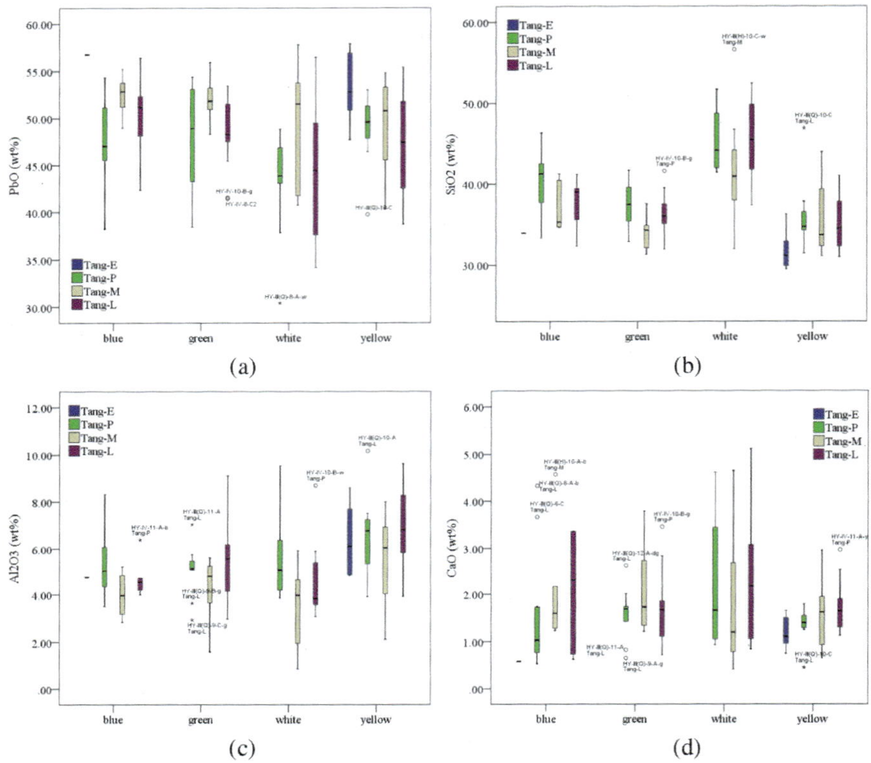

Fig. 22.8 Box-plots for high temperature glaze with different colors unearthed from the Huangye kiln site: (a) PbO (b) SiO_2 (c) Al_2O_3 (d) CaO

The high temperature glazes can also be divided into two groups: one group includes black glaze and tea leaf glazes, the other includes greenish white glazes and white glazes. Although the chemical compositions of black glazes and tea leaf glazes are similar, they are still distinguishable. As for white glaze and greenish white glaze, they too have similar chemical composition, but can hardly be distinguished apart. The recipes of black and tea leaf glazes have similar trends in the four periods, and the turning point appears to be in the late Tang period. But for the white glaze, the turning point appears to be in prosperous Tang period.

The low temperature glazes with different colors can be distinguished only by the colorants: except for the colorants, the recipes for the main matrices of the low temperature glazes are mainly consistent not only for different glaze colors, but also for glazes used in different periods.

Acknowledgements

The research was supported by the 973 Program (Grant Nos. 2012CB720901 and 2012CB720906), the Shanghai R&D Public Service Platform Construction Project (Grant No. 13DZ2295800) and the National Natural Science Foundation of China (Grant No. 51402326).

References

1. J. L. Dong, W. J. Zhao, G. D. Liu *et al.*, *Nuclear Physics Review* **25**, 380 (2008).
2. Y. Yang, Y. Ji, G. X. Li *et al.*, *Journal of Xinyang Normal University: (Science Version)* **21**, 190 (2008).
3. X. Qiu, The studies of origins and firing technology of ceramics unearthed from Huangye and Baihe kilns using modern analytic technique, PhD thesis, Zhengzhou University (2007).
4. S. Liu, Q. H. Li, F. X. Gan *et al.*, *Journal of Archaeological Science* **39**, 2128 (2012).
5. S. Liu, Q. H. Li, Q. Fu *et al.*, *X-Ray Spectrometry* **42**, 470 (2013).
6. S. Liu, Q. H. Li, F. X. Gan *et al.*, *X-Ray Spectrometry* **40**, 364 (2011).
7. R. M. Rousseau, M. Bouchard, *X-Ray Spectrometry* **15**, 207 (1986).

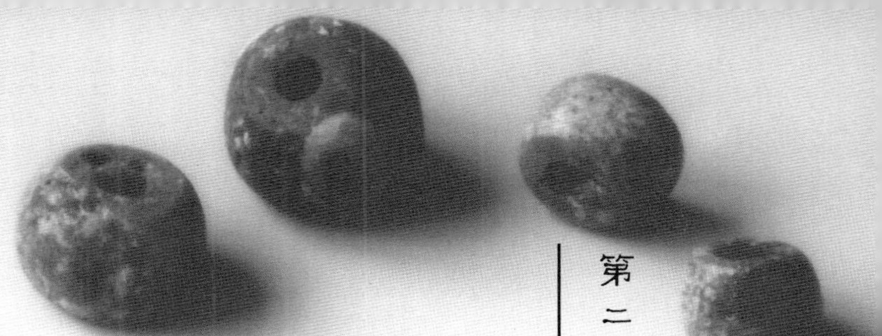

Chapter 23

第二十三章

Microstructure and Composition of Ancient Chinese Octagonal Sticks from the Warring States Period (475–221 B.C.)

Qinglin Ma, Zhiguo Zhang,* David A. Scott,*† *Heinz Berke*‡ *and Detlef Günther*§

*Chinese Academy of Cultural Heritage, Beijing, 100029

†Cotsen Institute of Archaeology, UCLA, Los Angeles, CA 90095-1510

‡Inorganic Chemistry at the University of Zurich, Zurich, Switzerland

§Inorganic Chemistry at the University of ETH, Zurich, Switzerland

23.1 Introduction

Egyptian Blue ($CaCuSi_4O_{10}$) was the first synthetic pigment and was the most extensively used blue pigment from the early dynasties in Egypt until the end of the Roman period in Europe. Egyptian Blue was used as a painting pigment on stone, gypsum plaster, pottery, wooden objects, papyrus, canvas and wall paintings, for the production of faience and also for the manufacturing of compact body objects.[1] Egyptian Blue was synthetically produced by chemical processes. Cuprorivaite, the mineral corresponding to Egyptian Blue in chemical composition was found as very small deposits, for instance, in the volcano at Vesuvius, Italy. It is so scarce in nature that it could not be collected in large enough quantities to be used as a pigment.

Chinese Blue (or Han Blue $BaCuSi_4O_{10}$) and Chinese Purple (or Han Purple ($BaCuSi_2O_6$) are two man-made copper barium silicate pigment materials used to manufacture compact body objects, such as faience beads. These pigments were first discovered by scientists of the Freer Gallery on objects in the Chinese collections, such as painted bronze vessels, painted pottery and octagonal sticks from the Warring States period (475–221 B.C.) to Han dynasty (206 B.C.–220 A.D.) in 1983 and 1992, respectively.[2] As the barium analogue of the calcium containing Egyptian Blue, Chinese Blue was found in minute quantities as a mineral in a South African mine in 1994 named Effenbergerite.[3] Until now, no mineral was found corresponding to Chinese Purple.

The onset of the production of Chinese Purple has been dated to the early part of the Spring and Autumn period (770–476 B.C.). In contrast, Chinese Blue seemed to have emerged significantly later in the middle of the Warring States period (475–221 B.C.). During these times, these compounds were not used as pure purple or blue pigments, but were instead found as minor constituents in the outer layers of the barium and lead containing faience beads. During the late Warring States Period and Western Han Dynasty (206 B.C.–8 A.D.), Chinese Purple became a more frequently used pigment, while Chinese Blue was then only rarely found as a unique pigment constituent; rather it emerged as an admixture to Chinese Purple.[4] All these man-made Chinese blue and purple pigments were used till the end of Eastern Han dynasty (25–220 A.D.).[5] In this context it is very interesting to note that besides the finds from ancient China, cylindrical beads and one glass vessel object bearing Chinese Blue as the colorant were unearthed in Japan.[6]

From the archaeological faience objects made from Chinese Blue or Chinese Purple, which comprise spherical and cylindrical beads, rings, and cup forms etc. one type of object sticks out, since it is special in shape. These are the octagonal sticks. The outer layers of these heavy and dense sticks appear in most cases to be yellow or earthen in colour owing to a leaching process by weathering, which occurred during burial. The inner or centre parts of these objects are a purple or dark blue colour. Such artifacts were occasionally found in tombs of ancient noblemen.

For instance in 1992 archaeologists discovered one tomb, C1M3943, located in the north-eastern part of the ancient capital city of Luoyang,[7] which was dated to the later part of the Warring States period belonging to a person that accompanied one unnamed ruler of the Zhou kingdom. Sixty-three pieces of precious artifacts were discovered in the coffin, such as gold, bronze, jade, agate, bone and stone objects. Among these objects were also six octagonal sticks (collection of the Luoyang Municipal Museum, Luoyang, China), which had a bluish-purple colour and their appearance indicates, in conjunction with many precious objects, that these sticks were objects of high social esteem. These sticks had different sizes: two of them (C1M3943: 27) had a length of 1.5 cm and a diameter of 0.7 cm; two others (C1M3943:28, 32) a length of 7.5 and a diameter of 0.5 cm and still two others (C1M3943:29, 31) a length of 6.8 cm and a diameter of 0.8 cm. They were placed on a bronze mirror at the Southeastern corner of the coffin room. One of the bronze mirrors found in this tomb was inlayed with eye-beads, the other mirror had a special decoration with the tips covered by golden leaves. Archaeologists considered that this tomb was occupied by an aristocratic lady. [8] Two octagonal sticks were studied analytically and their compositions were consistent with Chinese Purple with minor admixtures of Chinese Blue. [9]

In 1992, 17 octagonal sticks were discovered a tomb located at the Shang Wang village of Linzi, Shandong province, dated to the Warring States period,. These sticks could be divided into two classes, the hollow and the compact body class, and they are all conical shape with one side smaller in diameter than the other side.[10] The BaO content of one stick was found by EDXRF to be 0.69%. [11]

In 1996, four octagonal sticks were excavated from tomb No. 20 at You Jia Zhuang in northern Xi'an, Shaanxi province, dated to the Warring States period. Three of the sticks were compact body type sticks and one was a hollow stick.[12] It was estimated that these two sticks belonged to a class of vitreous materials. They have efflorescent surfaces.

The purpose and use of the octagonal sticks in ancient China is presently not clear. Various octagonal sticks, which often have weathered surfaces, were considered as a substitute for jade artifacts except for some glass sticks. Our groups started to study octagonal sticks in 1998, but based on archaeological and chemical information a distinct and unifying meaning for their use could as yet not be determined. One proposal is that they were used to cover the so-called seven holes of dead bodies in the tombs (nose, eyes, ears and mouth) and sometimes the sticks were seen to have been put into the hands of the corpses ('grasped') or sometimes judged from their positions in the tombs, they were meant to be decoration objects.[13] According to analyses from abroad these faience octagonal sticks were found to have in most cases high Pb and Ba contents with major phase contents of Chinese Purple and minor contents of Chinese Blue.[14]

In this work, samples were obtained from a private collector. Although they are not the archaeological finds these samples were found in the Southeastern corner of the Gansu province of China and are dated to the later period of the Warring States period. By various analyses these antiques could be put into a firmer context and in addition we obtained clues about their manufacturing techniques.

23.2 Sampling and analytical instrumentation

23.2.1 *Sample preparation*

The samples **MAHP08** and **MAHP10** were degreased using acetone, ultrasonic cleaning in water for 10 minutes, then washing by acetone and drying in air before being subjected to XRF or XRD analysis.

From the other octagonal sticks 1–2 mm pieces were cut off using a corundum blade. They were cleaned as described above, then embedded in epoxy resin and polished for use in an Electron Microprobe (EMPA) and a Scanning Electron Microscope (SEM).[15]

23.2.2 *Analytical instrumentation and measuring conditions*

The analytical instruments used include a stereoscopic microscope (Leica microscope of Getty Conservation Institute, ambient temperature); a Nicolet (Nic-Plan) FT — IR microscopy used to obtain the infrared spectra (FTIR) (Getty Conservation Institute, infrared transmission light focusing spot diameter of 50 μm, wave number range of 4000–400 cm^{-1}, spectral resolution of 4 cm^{-1}, sweep accumulation: 32 times, transmission mode); a Bruker (Siemens) D5005 X-ray diffractometer (Getty Conservation Institute, 40 kV voltage, electric current 40 mA, exposure times 54 minutes); an Electron microprobe (EPMA) (Cameca SX100, wavelength dispersion (WDS), 20 kV and 10 mA, dwell time of 50 msec/pixel.,Getty Conservation Institute); and a Renishaw 1000 type laser Raman spectrometer equipped with Leica DMLM microscope (excitation wavelength of 514 nm, Deparment of Chemistry at the University of Zurich, Switzerland).

The laser ablation inductively coupled plasma mass spectrometry (LA-ICP-MS) used a system of the Swiss Federal Institute of Technology Zurich, which was composed of an ArF excimer laser ablation system (GeoLas Q, MicroLas Lasersysteme GmbH, Germany) and one quadrupole-ICPMS Elan DRC plus (Perkin-Elmer Sciex, Thornhill, Canada). The sample analysis data are the average of the three different points of the laser beam. The laser wavelength used was 193 nm. Denudation chamber sampling was adjustable in the X, Y direction. The ICP-MS Eland DRC working parameters are: an RF power of 1400 W; sample gas flow of 0.8 l/min Ar; auxiliary gas flow of 0.7 l/min Ar; cooling gas flow of 16.0 l/min Ar; carrier gas flow of 1.0 l/min He. The NIST SRM 610 sample was used as an internal standard.

Some details of the scanning electron microscope (SEM) systems are litsted below. (1) Environmental scanning electron microscope (ESEM) (Getty Conservation Institute): INCA ESEM, voltage 20 kV, vacuum degree 3 torr, working distance of 8 to 10 mm.

(2) Philips SEM 515 (PW 6703) (Institute for inorganic chemistry at the University of Zurich, Switzerland): voltage 26 kV, Tracor Northern 5400 EDS X-ray spectrometer, Si/Li detector. (3) Hitachi S-3600N type SEM (Chinese Academy of Cultural Heritage): voltage 20 kV, The EDAX company Genesis type 2000 XMS X-ray energy spectrometer.

23.3 Results and discussion

23.3.1 *Observations by optical microscopy*

The sizes and appearance of the octagonal sticks **MAHP08** and **MAHP10** are shown in Figs. 23.1a and 23.1b. Their surfaces are yellow or earthen yellow. One side of **MAHP10** has a bronze looking corrosion layer, which was presumably formed during burial. A part of stick **MAHP08** was detached, which revealed that there are three sub-layers: its outer yellow crust has been softened (Fig. 23.1c) and detached areas of the stick have an appearance of sandstone (Fig. 23.1d), the middle layer is purple and degraded and the inner area is dark blue (Fig. 23.1e). For subsequent investigations cross-sections of samples were prepared in agreement with the sample owner.

The cross-sectioned stick **MAHP08** can be divided into three sub-sections, which probably extend deeper into the cylinders along its main axis.

(1) The yellow outer section is at a first glance assumed to be formed by corrosion and chemical changes in contact with the surrounding earth material.
(2) The purple middle section is presumed to consist a great part of original material, which was exposed less to wheathering influences and therefore underwent less chemical changes.
(3) The dark blue inner section had seemingly not suffered from chemical changes at all.

The outer and the middle sections sar degradation, while the inner section was probably conserved — the original state of the stick

(a) Optical microscopy of the octagonal stick **MAHP08** showing degradation, length 2.3 cm, diameter 1.2 cm.

(b) Optical microscopy of the octagonal stick **MAHP10** showing degradation, length 2.2cm, diameter 1.3cm.

(c) Optical microscopy of the octagonal stick **MAHP08,** part of the outer section became detached showing the purple section underneath.

(d) Optical microscopy of the octagonal stick **MAHP08,** light greenish-blue crystals formed where the outer layer became detached.

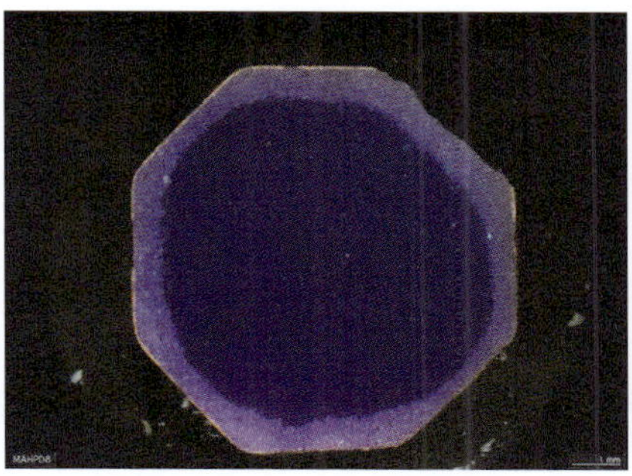

(e) Optical microscopy of the cross-section of the octagonal stick **MAHP08**. Inner section shows glazed dark blue, middle section purple and the thin outer section shows yellow or earthen color.

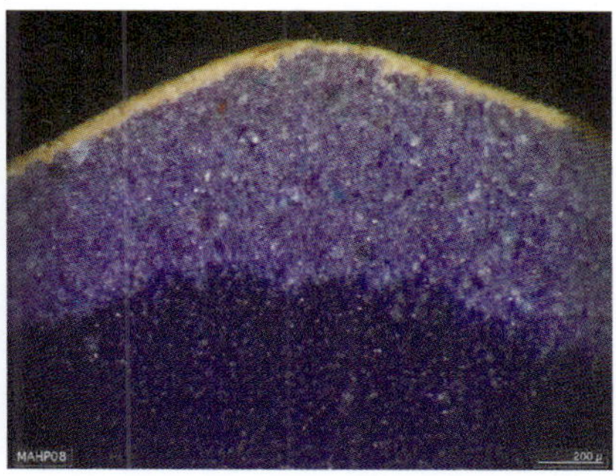

(f) Optical microscopy of a cross-section of the octagonal stick **MAHP08**.

Fig. 23.1 Views of the **MAHP08** and **MAHP10** octagonal sticks by optical microscopy.

during burial (Fig. 23.1e). Driven by chemical processes various elemental constituents of **MAHP08** are anticipated to have diffused into the earthen environment of the stick, while other elements have travelled in the opposite direction diffusing from the environment into the stick. The chemical processes are thus also dependent on the given composition of the earthen environment of the burial site. They are generally attributed to the stick's degradation or decomposition. Decomposition of the vitreous and crystalline barium copper silicate materials included also that of the lead containing phases, which had been added in the synthetic procedure of **MAHP08**, and cannot be fully excluded. These 'self-decomposition' processes seem however to be very slow relative to the timescale of the externally influenced decomposition processes of the outer and middle sections. Produced in ancient times, the original color of this stick on the whole has to be assumed to have been once dark blue. In the decomposition sections there is reddish to purplish inclusions. They are assumed to be remains of cuprite (Cu_2O) — formed in the synthetic process of the barium copper silicates– after the barium copper silicates and the Pb-Ba glass had decomposed and undergone decoloration. (Fig. 23.1f).

23.3.2 *FTIR analysis*

The FTIR spectrum of the powdered samples of the purple middle section and the dark blue inner section of stick **MAHP08** were compared to the spectrum of a sample of an octagonal stick (Han color stick #8) from the Freer gallery of Washington D.C., U.S. and to a Chinese purple sample prepared by Dr. Brill of the Corning Glass Museum (Fig. 23.2). The spectra turned out to be very similar with the main constituent being Chinese Purple.

The phase composition of the outer section of the **MAHP08** stick could be determined by Micro-FTIR only in conjunction with XRD and ICDD analyses. Strong FTIR evidence was indicative that this layer was mainly composed of $BaSO_4$ and $CaCO_3$.

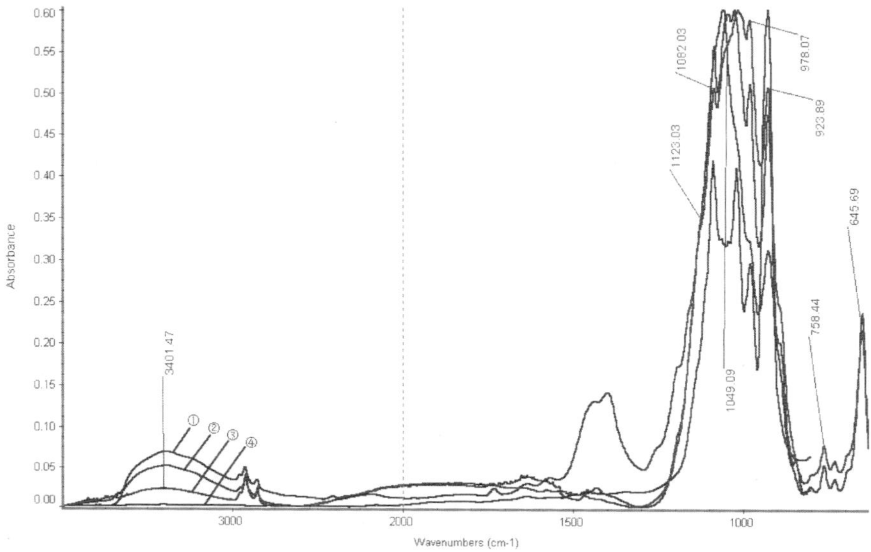

(a) FTIR spectra of stick **MAHP08** referenced to a stick of the Freer Gallery prepared according to a modern times procedure by Dr. Brill of the Corning Glass Museum.

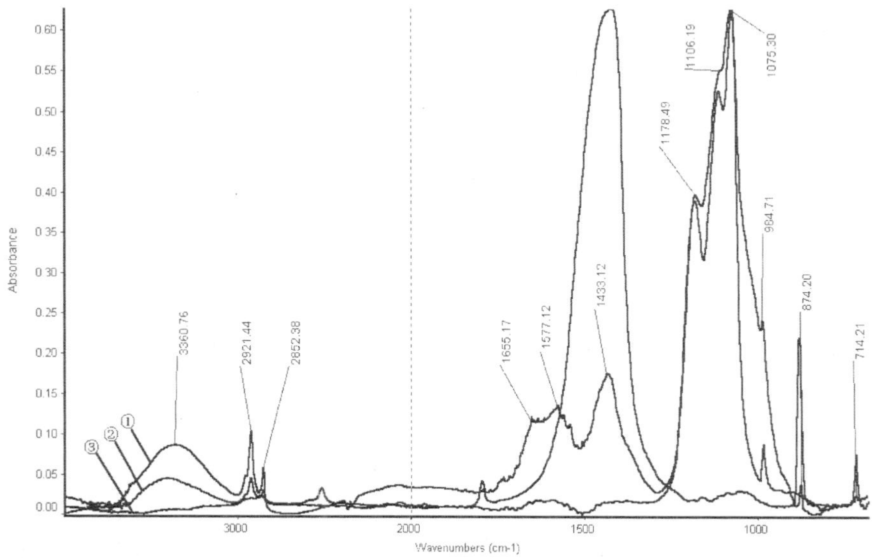

(b) IR spectroscopic comparison between the outer section of stick **MAHP08** and reference samples of $BaSO_4$ and $CaCO_3$.

Fig. 23.2 Comparison of the FTIR spectra of stick **MAHP08** and reference samples of $BaSO_4$ and $CaCO_3$

23.3.3 *XRD analysis*

The powder XRD results also revealed that the dark blue section of the stick **MAHP08** consists of Chinese Purple (Table 23.1) and minor amounts of β-$BaSi_2O_5$ and $BaSiO_3$. The yellow section is composed of $BaSO_4$ and minor amounts of $PbCO_3$ (Table 23.2).

23.3.4 *LA-ICPMS analysis*

An attempt was made to compare the semi-quantitative analyses of the purple middle part and the inner dark blue part **MAHP08** by XRF. Part of the experimental problem was to clearly confine the given sections of the stick. LA-ICP-MS was applied as a micro-analytical method to determine the major and trace elemental contents of the various sections (Tables 23.3 and 23.4).

The partly decomposed purple middle sections were additionally analyzed by XRD using samples from several other octagonal sticks. Tables 23.3 and 23.4 show the concentrations of the inner blue section of the **MAHP08** and **MAHP10** in the range of 32.57–29.07 wt% for SiO_2 (**MAHP08ip**, (ip = inner part)), 4.93–6.38 wt% for CuO, 25.31–31.99 wt% for BaO, 24.45–34.95 wt% for PbO. While the compositions of the outer purple parts were 24.58–24.85 wt% for SiO_2 (MAHP08op, op = outer part)), 3.03–3.29 wt% for CuO, 18.83–21.59 wt% for BaO and 35.16–36.82 wt% for PbO. Compared with the inner blue section, the purple middle sections have lower SiO_2, CuO, BaO and higher PbO contents, which indicates that leaching of SiO_2, CuO and BaO compensated by an increase in the PbO content has occurred.

The P and S contents of the inner blue sections of the **MAHP08** and **MAHP10** were determined to be 0.04–0.30 wt% P_2O_5 and 0.28–0.63 wt% SO_2, which are lower than 3.86–5.70 wt% P_2O_5 and 3.17–3.49 wt% SO_2 in the outer purple sections. This clearly indicates that the elemental compositions of octagonal sticks changed, perhaps also as a result of proximity to human remains in the burial site being part of the material exchange with the environment. Some of the components in the octagonal sticks decomposed and respective elements were leached into the environment, while other ingredients penetrated into the stick from the burial environment.

Table 23.1 XRD result of inner dark blue part of stick **MAHP08**

MAHP08 Blue part		$BaCuSi_2O_6$ 43-300		β-$BaSi_2O_5$ 26-176		$BaSiO_3$ 26-1402	
d-spacing	I/I_0	d-spacing	I/I_0	d-spacing	I/I_0	d-spacing	I/I_0
5.99	9	5.965	25				
5.21	10	5.581	8				
4.99	9	4.988	20				
3.97	10			3.973	85		
3.71	41	3.719	90				
3.53	15	3.527	18				
3.41	22			3.424	50	3.41	100
3.33	60			3.343	70	3.33	65
3.25	18	3.291	30	3.234	30		
3.09	12	3.0363	25	3.097	100	3.10	55
2.97	100	2.9830	100				
2.72	20	2.79	5	2.732	35	2.80	25
2.49	33	2.4935	50				
2.43	36	2.4344	65			2.35	17
2.27	13	2.2766	14	2.227	35	2.29	17
2.22	14	2.2306	10				
2.18	23	2.1880	5	2.164	25		
2.13	29	2.1279	8			2.07	20
2.04	39	2.0712	30			2.03	30
1.92	10	1.9272	7			1.89	20
1.85	10	1.8596	5				
1.81	9	1.8217	2				
1.78	12	1.7637	20				
1.76	20	1.7427	40				
1.67	13	1.6919	2			1.69	15
1.64	22	1.6820	2				
1.62	8	1.6456	12				
1.59	8	1.5939	8	1.5812	10		
1.57	10	1.5778	8				
1.51	28	1.5182	16				

Table 23.2 XRD result of yellow outer part of stick **MAHP08**

MAHP08 Surface		$BaSO_4$ 02-1199		$PbCO_3.2H_2O$ 41-677	
d-spacing	I/I_0	d-spacing	I/I_0	d-spacing	I/I_0
4.82	12				
4.53	12			4.45	27
4.38	23	4.27	40	4.27	13
3.89	29	3.82	40		
3.76	10				
3.67	10				
3.55	20	3.55	40	3.59	100
3.43	58	3.42	80	3.5	77
3.31	39	3.3	60		
3.09	56	3.08	80	3.07	22
2.82	36	2.84	50		
2.72	33	2.72	50		
2.51	8			2.52	25
2.47	12	2.47	20	2.49	60
2.32	12	2.33	60		
2.20	20	2.21	40		
2.10	100	2.11	100	2.08	50
2.04	23	2.06	40	2.01	23
1.92	9	1.93	20	1.98	20
1.84	16	1.86	50	1.93	37
1.66	16				
1.63	11	1.64	40	1.63	14
1.59	12	1.59	40	1.58	13
1.52	19	1.53	60	1.5	7

Table 23.3 LA-ICP-MS results on octagonal sticks (main elements) (wt%)

	MAHP08ip	MAHP08op	MAHP10ip	MAHP10op	HB1	HB2	HB3
Na_2O	1.55	0.13	0.22	0.18	0.18	0.07	0.02
MgO	0.18	0.20	0.44	0.29	0.45	0.13	0.05
Al_2O_3	0.82	1.57	1.88	2.23	1.42	1.10	0.90
SiO_2	32.57	24.58	29.07	24.85	45.47	29.87	25.44
P_2O_5	0.04	3.86	0.30	5.70	2.96	1.23	1.45
SO_2	0.28	3.17	0.63	3.49	1.26	2.13	0.51
K_2O	0.26	0.41	0.27	0.47	0.05	0.06	0.03
CaO	0.55	3.42	0.87	4.60	1.79	1.01	1.23
FeO	0.52	0.75	0.67	0.77	0.94	0.46	0.48
CuO	6.38	3.29	4.93	3.03	5.17	4.94	7.14
SrO	0.41	0.21	0.44	0.39	0.16	0.27	0.27
BaO	31.99	21.59	25.31	18.83	18.02	27.83	24.64
PbO	24.45	36.82	34.95	35.16	22.12	30.90	37.84

Most of the Al_2O_3 and CaO in the purple middle degradation layers could be related to the burial environment and particularly the presence of barite and lime based compounds in the outer yellow crust of the octagonal sticks could be explained on the basis of this specific surrounding. The analyses of the purple middle sections of other octagonal sticks however also contained considerable amounts of P_2O_5, Al_2O_3, SO_2 and CaO, which are very close to those of the purple sections of **MAHP08** and **MAHP10** giving rise to the alternative interpretation that these constituents were not formed by diffusion of the respective ions from the environment but rather have been present from the very beginning of the preparation process of the sticks.

23.3.5 *EPMA analysis*

In order to further detail the analyses of changes of the elemental distributions in the inner unweathered dark blue sections, weathered

Table 23.4 LA-ICP-MS results on octagonal sticks (Main and minor elements) (wt%) (Internal Standard Si = 100,000ppm)

	MAHP08ip	MAHP08op	MAHP10ip	MAHP10op	HB1	HB2	HB3
Na 23	7556.8	821.5	1215.8	1141.0	641.4	381.7	107.4
Mg 25	712.8	1029.8	1962.9	1530.2	1272.7	545.9	262.7
Al 27	2862.9	7223.8	7328.7	10168.6	3536.1	4174.4	3986.7
P 31	122.7	14671.7	966.2	21426.0	6084.9	3858.8	5326.3
S 34	908.5	13812.9	2325.5	15045.8	2971.2	7638.1	2147.9
K 39	1423.0	2959.4	1654.4	3369.8	187.0	349.6	220.6
Ca 42	2569.2	21259.5	4575.5	28324.3	6027.9	5187.2	7376.8
Sc 45	0.7	1.3	1.0	2.1	1.0	1.2	1.6
Ti 47	95.4	182.6	272.0	353.0	73.7	109.4	125.4
V 51	3.2	249.7	15.6	149.0	15.4	13.1	50.7
Cr 53	2.5	321.9	23.6	106.4	7.8	5.6	11.1
Mn 55	99.7	67.2	19.6	111.9	14.8	27.6	13.4
Fe 57	2670.1	5053.6	3826.0	5122.8	3441.4	2587.6	3115.7
Co 59	42.8	8.2	6.1	5.7	6.4	13.0	11.5
Ni 60	22.2	16.8	9.7	8.4	10.3	15.0	20.3
Cu 65	33771.0	23215.8	30250.8	23550.0	20170.5	29167.4	49333.5
Zn 66	45.1	164.5	109.3	303.4	52.4	79.9	132.7
As 75	88.7	119.4	618.9	802.2	40.9	158.2	36.3
Rb 85	2.8	3.0	2.3	2.7	0.3	0.6	1.1
Sr 88	2272.2	1514.2	2741.4	2822.2	646.1	1620.4	1903.3
Y 89	5.4	8.2	5.6	17.8	2.1	4.3	6.4
Zr 90	4.9	8.2	11.2	14.2	8.5	4.6	11.7

Nb 93	0.3	0.7	0.8	1.1	0.3	0.4	0.4
Mo 95	0.4	5.3	1.6	3.6	0.4	1.4	1.1
Ag 107	97.4	111.3	10.1	23.3	52.5	21.5	21.0
Cd 111	1.6	3.0	0.4	0.8	0.2	0.2	0.3
Sn 118	2.7	16.5	2.6	3.3	2.7	0.6	5.4
Sb 121	257.2	200.6	139.4	198.6	113.4	114.5	281.1
Ba 137	188214.1	168300.2	166822.6	145201.8	75901.7	178500.5	185567.8
La 139	4.1	8.9	6.9	13.9	2.1	4.6	6.6
Ce 140	8.9	17.3	12.3	28.1	4.7	8.6	12.9
Pr 141	1.0	1.9	1.5	3.3	0.6	1.1	1.7
Nd 146	4.4	8.3	5.9	14.5	2.7	4.7	7.2
Sm 147	1.0	1.7	1.2	3.1	0.6	1.0	1.6
Eu 153	2.2	2.5	2.0	2.3	2.2	3.6	3.5
Tb 159	0.1	0.2	0.1	0.4	0.1	0.1	0.2
Dy 164	0.6	1.2	0.7	2.7	0.4	0.6	0.9
Yb 173	0.4	0.7	0.4	1.4	0.2	0.4	0.7
Hf 180	0.1	0.3	0.3	0.4	0.3	0.2	0.4
Ta 181	0.0	0.0	0.1	0.1	0.0	0.1	0.0
W 182	0.1	1.6	0.1	0.3	0.2	0.4	0.1
Pb 207	143542.9	273764.9	227710.7	268863.6	93711.6	197949.3	286716.8
Pb 208	149063.6	297440.8	238707.3	280934.4	96573.3	205428.9	295363.0
Bi 209	3.6	5.4	1.3	1.6	94.3	43.6	18.8
Th 232	0.5	1.0	1.1	1.7	0.6	1.8	0.7
U 238	1.1	4.1	6.4	8.3	1.7	2.6	3.0

purple middle sections and the outer crusts, electron microprobe was employed to scan the areas for elemental distributions in the **MAHP08** stick. Fig. 23.3 shows the EPMA results of the inner dark blue section and the purple middle section. In Fig. 23.4 the purple section and the outer crust are compared.

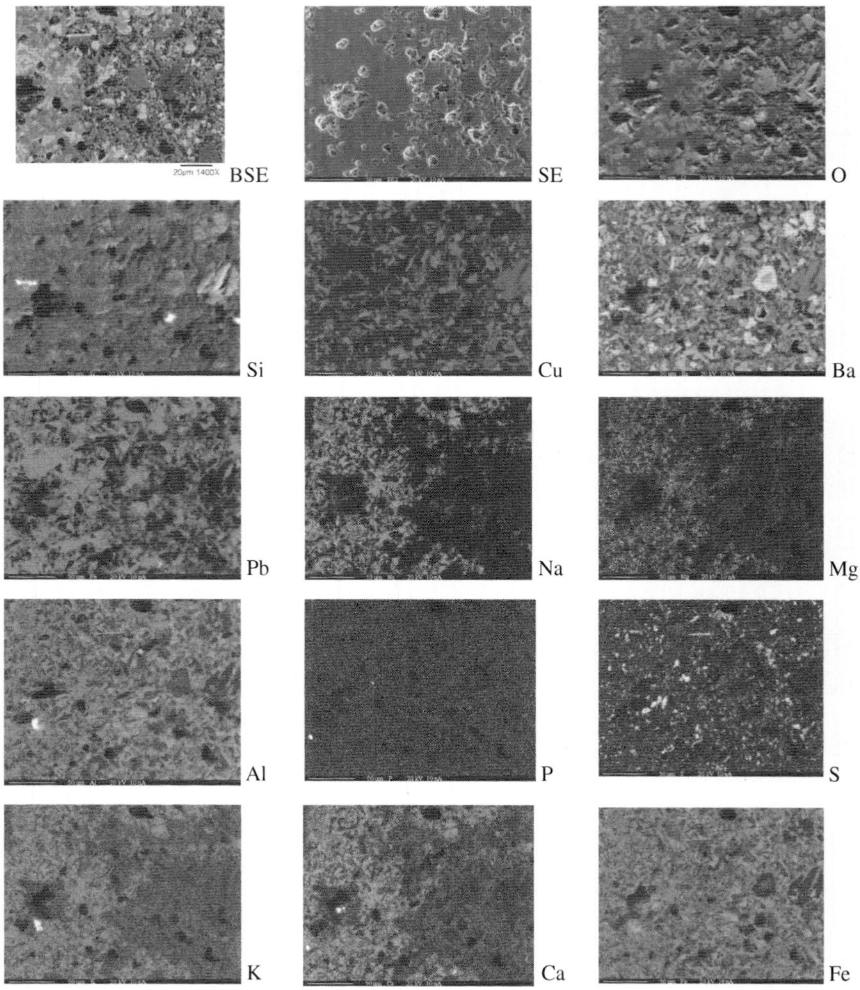

Fig. 23.3 EPMA elemental distributions of **MAHP08** in the dark blue section and purple section

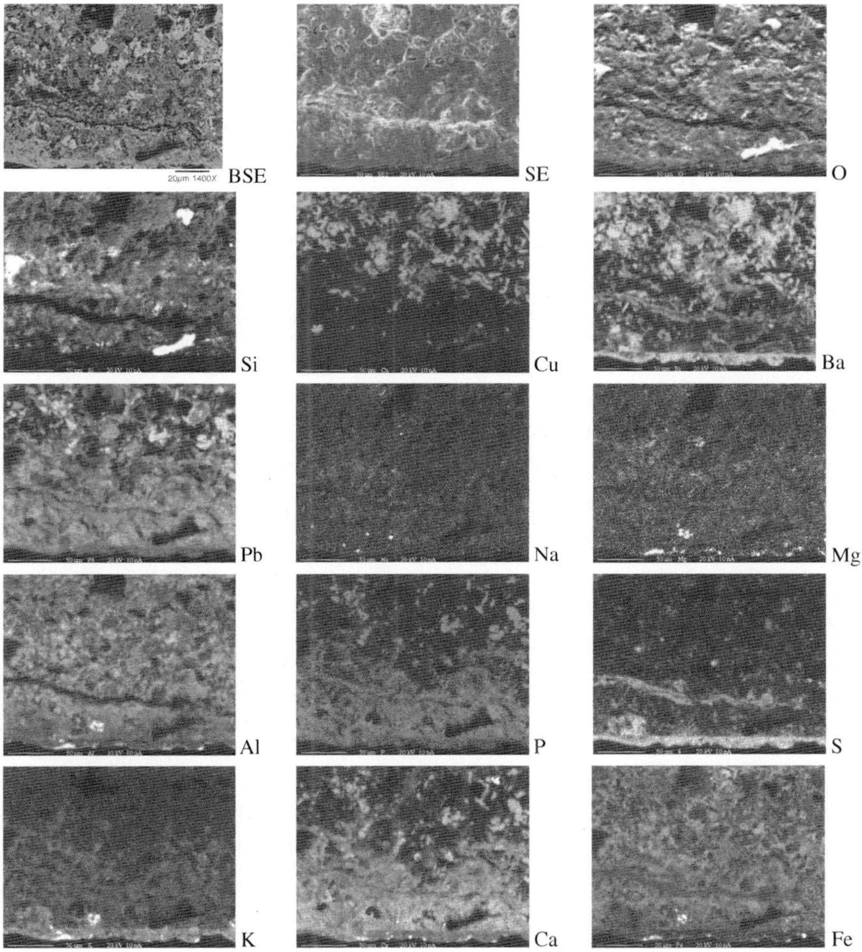

Fig. 23.4 EPMA element distributions in a **MAHP08** section between the purple part and outer crust.

Fig. 23.3 demonstrates the deterioration of the glass phase in the purple section and the presence of various remaining particles present in various shapes, colours and sizes. The contents of Na, Mg, K and Pb decline from the blue inner section to the purple middle section and sulphur has a scattered distribution, which may correlate with the distribution of the barium source ($BaSO_4$). The Cu elemental distribution shows no obvious variation, which may point to the fact that the basic forms of Cu in this section are Chinese Purple and Chinese Blue. The losses of elemental concentrations of Pb and Ba in the purple section are attributed to a leaching process from the Pb glass or Pb — Ba glass phases.

In Fig. 23.4 one can clearly identify boundaries between the purple section and the yellow section, which have been formed during the time of burial. The Ba and S distributions show a special sandwich structure in the outer crust, which mirrors the layer formation by the sedimentary process or the changes originating from the glass weathering morphology. Moreover, the high level of Ca and P contents in the crust indicates that they are caused by the influence of the environment. In addition, there are low Cu concentrations in this section meaning there is lack of Cu in the barium copper silicates, which may point to the decomposition or transformation of large parts of the Chinese Purple in this area, inducing Cu deficiency by means such as leaching copper into the burial environment.

Fig. 23.3 and Fig. 23.4 also reflect the fact that Pb or Pb-Ba glass phases of the blue octagonal sticks have deteriorated and became porous and softened during the long period of burial. The purple areas have been transformed with a decrease in the Pb and Ba concentrations. In the yellow section, Chinese Purple weathered and reacted with sulphates, calcium and phosphates, which have at least partly diffused into the stick from the environment to form insoluble $BaSO_4$ and other highly insoluble products.

23.3.6 *SEM-EDX analysis*

The dark blue section, the purple section and the yellow section of **MAHP08** were examined by SEM, in order to determine their microstructures and surface morphology (Fig. 23.5).

A rough surface of the dark blue section is seen at a natural break in the stick and is depicted in the SEM micrograph of Fig. 23.5a. The particles in this section possess a compact and very dense structure except for a few pores, while the natural break in the purple section has a lot of pores and little particular coherence (Fig. 23.5b).

(a) Cross section of the dense blue inner section.

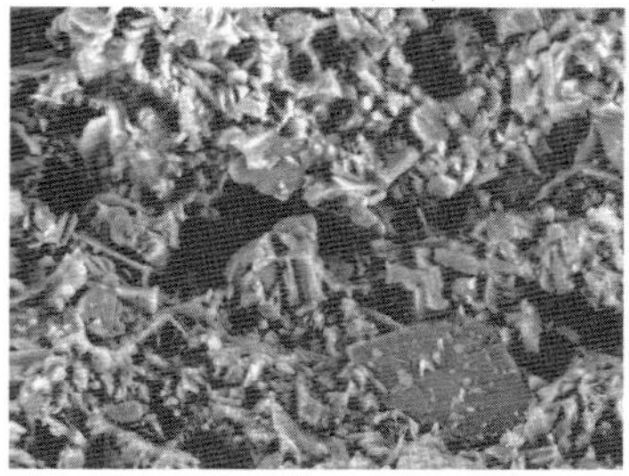

(b) Cross section of the soft and porous purple middle section.

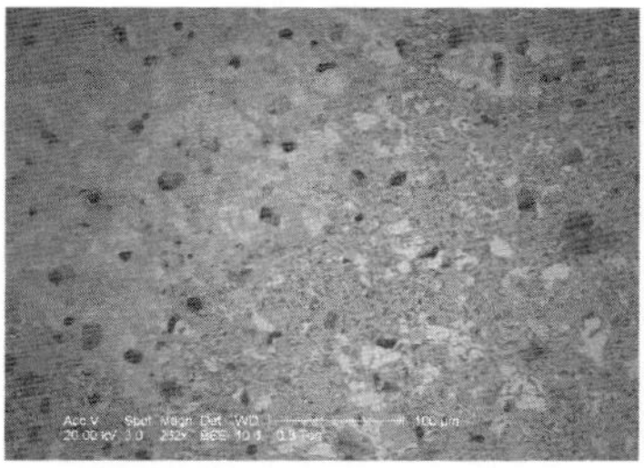

(c) Polished section of the transition between the inner blue area and purple arae.

(d) Polished section of the purple loose area.

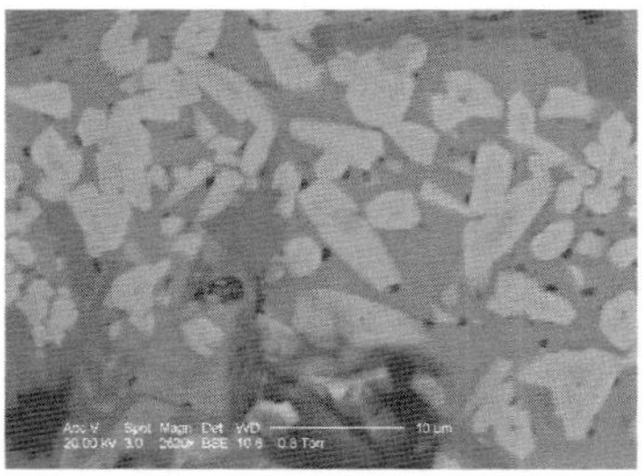

(e) Polished area of the blue inner section showing large amounts ofbarium silicate embedded in Pb or Pb-Ba glass phases.

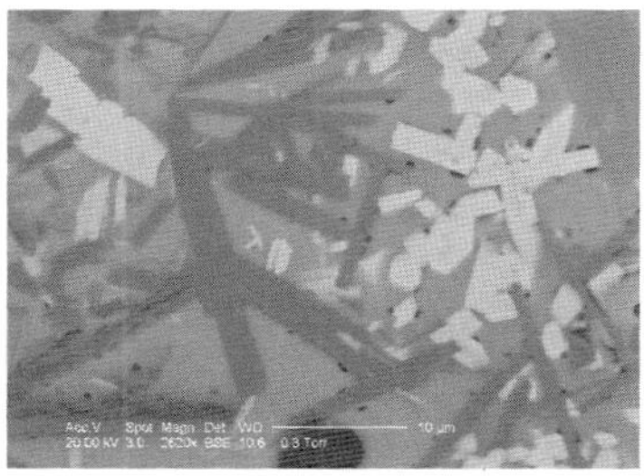

(f) Polished area of the blue inner section showing barium-copper-silicate compounds embedded in Pb or Pb-Ba glass phases.

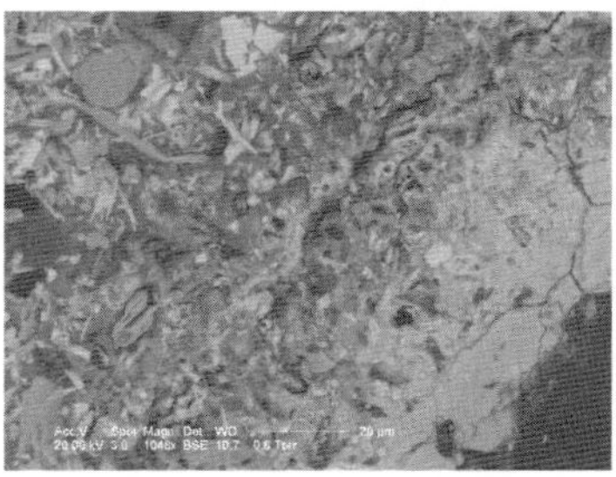

(g) Polished transition area between the purple middle sectionand the outer crust.

(h) Magnified sectionof the lower right corner of Fig.23.5g.

Fig. 23.5 SEM analysis of the various sections of the octagonal stick **MAHP08**

Fig. 23.5c shows the polished section of the transition area between the dark blue and purple sections, which clearly illustrates the deterioration of the glass phase in the changed purple middle section (at the right side of the picture). Furthermore the modified loose purple middle section is shown in Fig. 23.5d, which mainly contains silicates based on the elemental distribution in the centre of the picture.

The dark blue section consists of large amounts of barium silicates and barium-copper-silicate compounds embedded in a "matrix" of Pb or Pb-Ba glass (Fig. 23.5e–23.5f).

The morphology of the polished surface of the cross section of the stick is shown in Figs. 23.5g and 23.5h in the transitional area between the purple section and the yellow section. In comparison, the crystalline particles in the outer crust layer are fewer than in the purple middle section, and most material here seems to mainly exist in an amorphous form.

SEM-EDX data revealed the presence of several compounds in the inner blue section, purple middle section and the outer crust of the octagonal stick **MAHP08**. They are displayed in Fig. 23.6 and listed in Table 23.5.

Fig. 23.6a details the structural composition of the dark blue section and purple section of the octagonal stick **MAHP08** with analytical points plotted. The spots EDX1-EDX2 provide an elemental analysis for the composition of zones changed by weathering, while EDX3-EDX4 show the compositions of unweathered areas. By contrast, the Pb elemental content in the yellow section is less than the Pb content in unweathered zones and this suggests that some of the Pb has leached out into the environment.

Fig. 23.6b–23.6e show the structure of the dark blue section of **MAHP08**. The gray particles (Fig.23.6b) in area EDX1 and EDX3-EDX5 have the same compositions of Chinese purple including two relative pure areas (EDX4 and EDX5). The dark grey group in EDX6–EDX8 areas are identified as being relatively pure Chinese blue without Pb. The white particles in positions of EDX2, EDX5 and EDX10 are $BaSi_2O_5$, and EDX9 also mainly contains $BaSi_2O_5$ with a slightly higher content of Cu. The white area EDX11 is a mixture of Chinese purple and Pb-Ba glass phase.

Table 23.5 SEM-EDX analysis of the cross-section of the octagonal stick **MAHP08** (See Fig. 23.6)

No.	Si	Cu	Ba	Pb	Na	Al	Cl	Ca	Fe	Si/Cu	Si/Ba	Cu/Ba	Possible phase
a EDX1	24.92	8.21	37.48	26.82		1.44			1.14	6.88	3.26	0.47	Degradation layer
EDX2	25.02	8.87	38.64	25.35		1.27			0.85	6.40	3.18	0.50	Degradation layer
EDX3	22.86	7.52	36.77	30.36		1.07		0.66	0.77	6.89	3.05	0.44	Blue layer
EDX4	22.40	6.82	35.13	33.50		0.97		0.61	0.57	7.45	3.13	0.42	Blue layer
b EDX1	26.67	22.43	50.90							2.70	2.57	0.95	$BaCuSi_2O_6$
EDX2	33.81		66.19							–	2.50	0.00	$BaSi_2O_5$
EDX3	26.81	22.42	50.77							2.71	2.59	0.95	$BaCuSi_2O_6$
EDX4	26.84	21.79	51.37							2.79	2.56	0.92	$BaCuSi_2O_6$
EDX5	26.48	22.47	51.05							2.67	2.54	0.95	$BaCuSi_2O_6$
EDX6	39.31	18.45	42.24							4.83	4.56	0.94	$BaCuSi_4O_{10}$
EDX7	40.51	18.01	41.48							5.10	4.79	0.94	$BaCuSi_4O_{10}$
EDX8	41.81	17.03	41.16							5.57	4.98	0.89	$BaCuSi_4O_{10}$
EDX9	23.41	9.48	36.98	30.13						5.60	3.10	0.55	$BaSi_2O_5$
EDX10	19.14	2.05	30.82	48.00						21.17	3.05	0.14	$BaSi_2O_5$
EDX11	24.23	14.39	38.47	20.93			1.46	0.52		3.82	3.09	0.81	Glass+$BaCuSi_2O_6$
c EDX1	15.71	2.04	39.34	39.06	1.95	0.83		0.47	0.60	17.46	1.96	0.11	$BaSi_2O_5$

(*Continued*)

Table 23.5 (*Continued*)

No.	Si	Cu	Ba	Pb	Na	Al	Cl	Ca	Fe	Si/Cu	Si/Ba	Cu/Ba	Possible phase
EDX2	12.55	1.72	50.15	33.77	0.97	0.35		0.14	0.34	16.55	1.23	0.07	$BaSiO_3$
EDX3	12.84	1.80	52.30	33.06						16.18	1.20	0.07	$BaSiO_3$
EDX4	26.03	23.05	50.92							2.56	2.51	0.98	$BaCuSi_2O_6$
d EDX1	16.78	2.35	31.92	48.46		0.49				16.19	2.58	0.16	$BaSi_2O_5$
EDX2	20.13	4.32	32.48	42.13		0.95				10.57	3.04	0.29	$BaSi_2O_5$
EDX3	13.37	1.92	49.70	35.01						15.79	1.32	0.08	$BaSiO_3$
EDX4	13.87	2.65	50.92	32.57						11.87	1.34	0.11	$BaSiO_3$
EDX5	19.25	3.07	26.65	47.04	2.02	0.99			0.98	14.22	3.54	0.25	$BaSi_2O_5$
e EDX1	30.07	8.90	28.09	32.93						7.66	5.25	0.69	Glass + $BaCuSi_4O_{10}$
EDX2	35.10	13.07	33.69	18.14						6.09	5.11	0.84	$BaCuSi_4O_{10}$
f EDX1	39.03	14.84	36.94	9.20						5.96	5.18	0.87	$BaCuSi_4O_{10}$
EDX2	25.07	21.52	47.84	5.57						2.64	2.57	0.97	$BaCuSi_2O_6$
EDX3	24.09	19.72	47.79	8.04						2.77	2.47	0.89	$BaCuSi_2O_6$
EDX4	15.13	2.47	30.65	51.75						13.89	2.42	0.17	$BaSi_2O_5$
EDX5	19.66	1.42	15.11	54.23	3.18	1.02	3.15	1.36	0.87	31.40	6.38	0.20	Glass

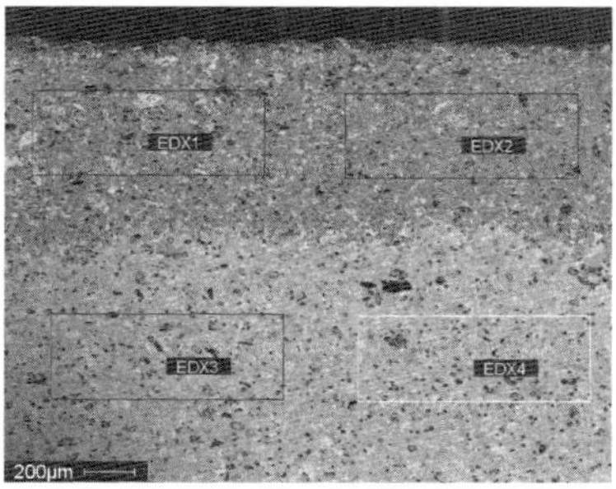

(a) Polished area of the transition between inner bluesection and purple middle section.

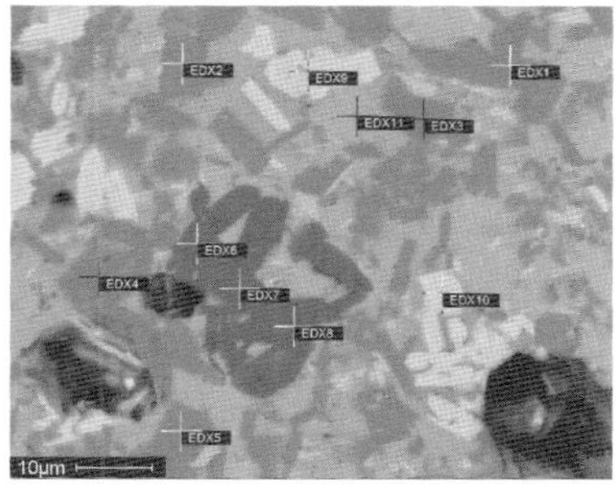

(b) Polished area of the denser part of the middle section, appearing dark purple. The group of crystals which appeared dark correspond to $BaCuSi_4O_{10}$.

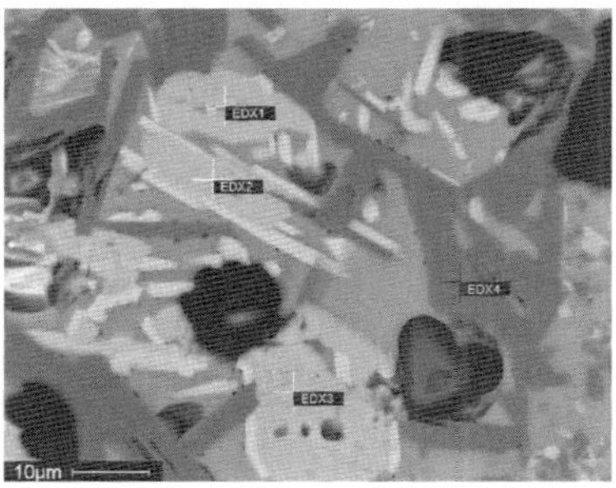

(c) Polished area of the blue inner section showing crystals ofsilicates embedded in Pb-Ba glass phases.

(d) Magnification of Fig.23.6c next to a dark spot somewhat lower and left to the center of the SEM micrograph.

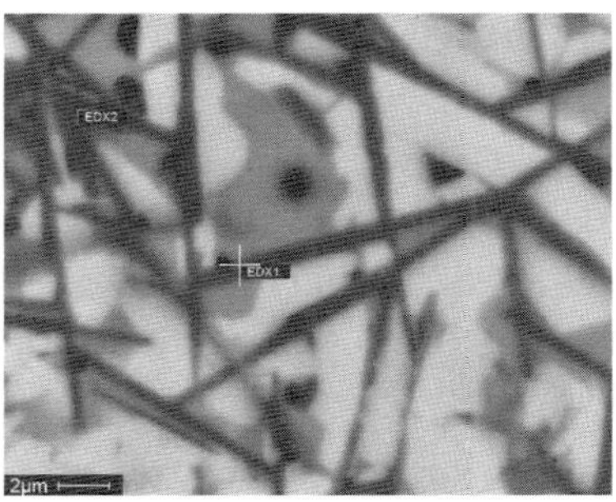

(e) Polished area of blue inner section, which consists of largeamounts of dark needle-like $BaCuSi_4O_{10}$.

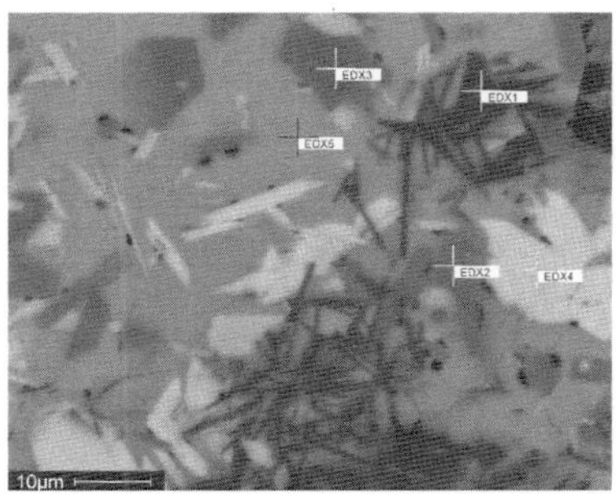

(f) Polished section of the blue inner area, which contains great amounts of dark needle-like $BaCuSi_4O_{10}$.

Fig. 23.6 SEM-EDX analysis of the cross-sectioned octagonal stick **MAHP08**

Based on the EDX analysis of Fig. 23.6c, the light gray particles in the position of spot EDX1 correspond to $BaSi_2O_5$, the light gray EDX2 and EDX3 areas are interpreted to be $BaSiO_3$, and the gray EDX4 area probably corresponds to Chinese Purple.

The enlargement of the laminated particle of Fig. 23.6c is shown in Fig. 23.6d. The off-white areas of EDX1, EDX2 and EDX5 of this particle are identified as $BaSi_2O_5$, while the inner light gray areas of EDX3 and EDX4 correspond to $BaSiO_3$. The barium metasilicates are assumed to be reaction products between barium minerals and the surrounding quartz. This situation can only happen during synthesis. Therefore we find $BaSiO_3$ to be generated in the inner layer with low Si content and $BaSi_2O_5$ in the outer layer with higher Si content. However, Fig. 23.6d shows an opposite situation. The bright area is an area with a high Pb content in the outer area (white) with an admixture of barium metasilicate particles. This also implies that $BaSO_4$ or $BaCO_3$ and SiO_2 particles have gradually transformed into barium silicates under the influence of lead salt catalysis and flux.

Fig. 23.6e shows a dense blue area, which contains large amounts of dark needle-like $BaCuSi_4O_{10}$ (EDX1and EDX2) surrounded by the white $BaSi_2O_5$.

The dark needle-like particle in the position of Fig. 23.6f (EDX1) corresponds to Chinese Blue and the dark block compounds in EDX2 and EDX3 are identifited as Chinese Purple. Moreover, the white block of EDX4 corresponds to the presence of $BaSi_2O_5$ and the large light grey area in the mixture in EDX5 is identified as Pb-Ba glass.

In the above analysis, two areas contain high concentrations of chloride, which may originate from copper chloride sources. Perhaps ancient craftsmen may have used copper compounds containing copper green, vice-atacamite and sodium chloride as the copper colorant.

23.3.6 *Micro-Raman spectroscopy of MAHP08*

In order to confirm the SEM-EDX analyses in terms of a phase analysis, micro-Raman spectroscopy was used to analyze the colorants of the octagonal stick MAHP08. A powdered sample was scraped from the blue inner section. Selected Raman spectra are shown in Fig. 23.7.

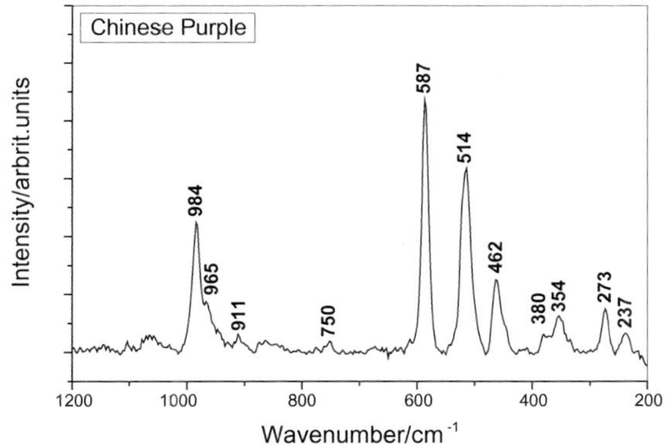

(a) Raman spectrum of Chinese purple

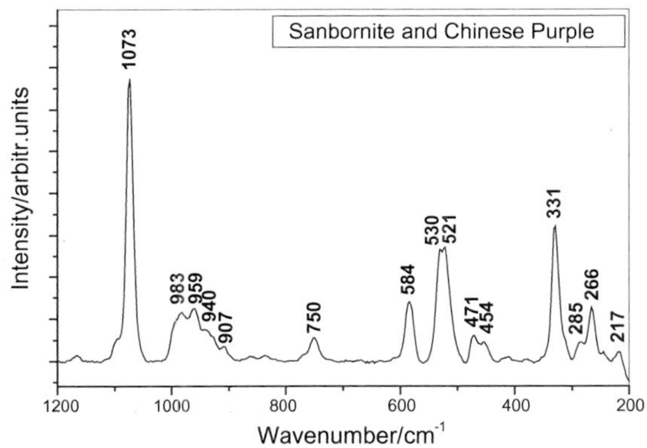

(b) Sanbornite and Chinese purple

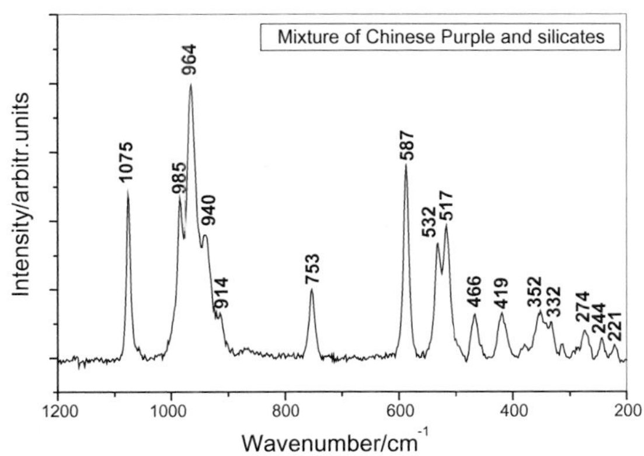

(c) Mixture of Chinese purple and various silicates

Fig. 23.7 Micro-Raman spectra of the octagonal stick **MAHP08**

Under the microscope, the main part of the colorant consists of Chinese Purple phases identified by their spectral bands at 984, 587, 514, 462, 354 and 273 cm^{-1} (Fig. 23.7a). In general, Chinese Purple and barium silicates always appear as mixtures together with sanbornite (α-$BaSi_2O_5$) particles identified from bands at 1073, 530, 471, 331, 285 and 266 cm^{-1} (Fig. 23.7b).

Fig. 23.7c shows a spectrum of a mixture of Chinese Purple, barium silicates and other silicates. A band at 964 cm^{-1} indicates the presence of an orthosilicate compound containing SiO_4^{4-} ions, bands at around 940 and 420 cm^{-1} accounting for glassy phases with poly-silicate units (SiO_3^{2-}) and the band present at 914 cm^{-1} accounting for the presence of disilicates ($Si_2O_7^{6-}$)[16]. Chinese Purple can typically be identified by the broad bands of a bridging oxygen vibration of the $(SiO)_4$ local units of the eight-membered ring structures in the range of 513–521 cm^{-1}[17].

Although Chinese Blue particles were found in the SEM-EDX analysis, it was difficult to locate them by micro-Raman spectroscopy. This suggests that the Chinese Blue phase content is very low in the octagonal stick **MAHP08**. The "third" copper barium silicate compound (Chinese dark blue, $BaCu_2Si_2O_7$), which has been detected in ancient samples, could not be detected in the micro-Raman analysis of the **MAHP08** stick.

23.3.7 *The four barium copper silicate compounds*

Four kinds of barium copper silicate compounds exist in the ternary BaO-CuO-SiO_2 phase diagram and can be identified by XRD as constituents: $BaCuSi_2O_6$ (Chinese Purple, JCPDS 43-0300, 81-0199, 82-0600), $BaCuSi_4O_{10}$ (Chinese Blue, JCPDS 81-0510, 81-1240, 82-1868; natural mineral Effenbergerite, JCPDS 12-0510, 47-1779), $BaCu_2Si_2O_7$ (Chinese Dark Blue, JCPDS 39-0253, 80-0691) and light Blue $Ba_2CuSi_2O_7$ (JCPDS 40-0084, 0084-73), respectively.[18] At present, the first three kinds of compounds could be identified by Raman spectroscopy as constituents of coloured ancient Chinese antiques.[19] $Ba_2CuSi_2O_7$ seems to be a rare constituent appearing only as a mixture with the others.

Regarding the formation of these compounds in preparations of the barium copper silicates, a model of progressive silicate condensation has been developed. Starting from the basic structural element SiO_4^{4-} (orthosilicate), the $Si_2O_7^{6-}$ ion (disilicate) is generated at first. It is the silicate base for the stable $BaCu_2Si_2O_7$ phase. In the next step, the above-mentioned four-membered ring unit $Si_4O_{12}^{8-}$ (cyclotetra-silicate) is generated by advancing condensation: this constitutes the basic silicate unit of Chinese Purple. The $Si_4O_{12}^{8-}$ units may in turn condense further building the infinitely connected layers of $Si_8O_{20}^{8-}$ eight-membered ring units and planar four-membered rings of Egyptian and Chinese Blue. The reaction scheme of Fig. 23.8 stresses the fact that, along the pathway of silicate condensations Chinese Blue, Chinese Purple are generated as intermediate products and Chinese Dark Blue Chinese light Blue, are expected to be seen as initial kinetic products formed after short reaction times.

Of course, the phase changes in the polycondensation process are not only dependent on reaction temperature and time, but via diffusion processes also related to the amounts of the starting materials present. Furthermore, the surface colour and compositional analyses of a large number of archaeological faience beads and of octagonal sticks have indicated very clearly that ancient people knew how to

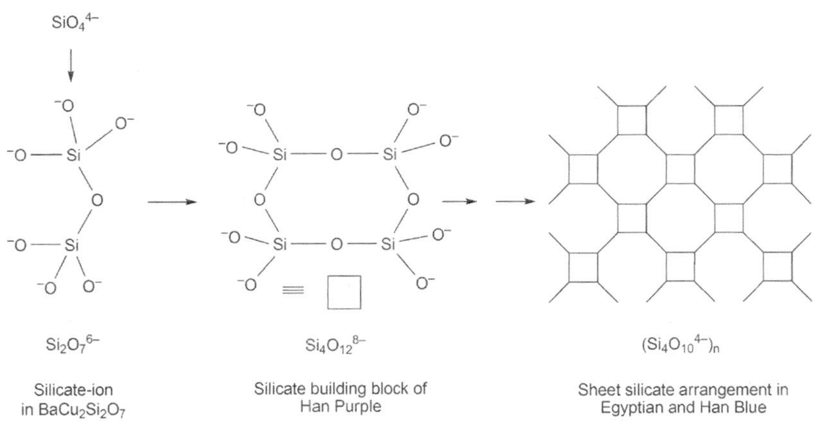

Fig. 23.8 Schematic sketch of the polycondensation process of silicate compounds

generate the different colour shadings, such as light green, blue and purple by mixing the starting materials in different ratios. As Chinese Dark Blue was up to now found mixed with Chinese Purple in the pigments layer of wall paintings, colour painted pottery and faience, it supports the notion of a progressing polycondensation process according to Fig. 23.8. Going on during prolonged heating procedures and for the disilicates Chinese Dark Blue and Chinese Light Blue it seems plausible that they are absent as products in those preparations of $BaCuSi_2O_6$ and $BaCuSi_4O_{10}$ requiring prolonged heating processes. Involving also greater amounts of SiO_2 in extended heating processes, Chinese Dark Blue will gradually be changed into Chinese Purple and Chinese Blue, and will therefore also disappear within the multi-step syntheses of the refining processes of Chinese Purple and Chinese Blue.

Chinese Purple features a layered structure, but its framework differs greatly from the layered structure of Chinese Blue. Its basic units are isolated $(SiO)_4$ four-ring units, where terminal oxygen atoms bind with two connected copper atoms. This results in the formation of an infinite arrangement of Cu_2 units. From a chemical point of view, this copper–copper bond is a very uncommon feature. Metal–metal bonds are in general chemical curiosities, when they appear in materials other than metals. Physicists hope to be able to eventually address the electronic spins of the Cu_2 units constructing 'spin quantum computers' to increase in this way the storage capacity of such devices greatly.[21] The crystals of Chinese Purple are also dichroic. As a consequence the substance's appearance gets lighter when ground as do the other layered copper silicate, Chinese Blue and Egyptian Blue, as well.

23.3.8 *Manufacture and moulding of the octagonal sticks*

The condensation mechanism of Fig. 23.8 is a suitable model helping us to understand the manufacturing process of octagonal sticks. According to the results of SEM-EDX analysis, the barium copper silicate content in the **MAHP08** sample is mainly due to Chinese Purple and occasionally there is an admixture of Chinese Blue. Most of the particles in **MAHP08** are between 5 to 15 microns in dimension and only minorities of particles are between 15 to 25 microns, which indicates that the raw materials have been carefully ground.

According to a synthetic simulation tests [22] at least two or three synthesis steps to manufacture octagonal sticks from the powdered materials are required. At first, depending on the ratio of the starting materials, if Chinese Purple was to be made, the materials were mixed and ground sufficiently and then calcined at high temperature to get a primary raw product. According to the mechanism of the silicate polycondensation not too short reaction times and low stoichiometry in quartz will lead to formation of large amounts of Chinese Purple, maybe in rare cases also some small amounts of Chinese Dark Blue, and small amounts of Chinese Blue to be present in the product mixture of octagonal sticks obtained along a 'not refined' synthesis procedure. To the crushed and finely ground primary product, honey or similar viscous media were added in the ancient synthesis treatments, and the material would be shaped into the desired shape pressing it directly or in a mould, then calcined continuously till the desired colour and quality was achieved. Since another heating stage got involved the product is expected to mainly contain now only Chinese Purple and Chinese Blue. This second step could have been repeated particularly so as to get still further refined object shapes and maybe, also in appropriate cases of applied stoichiometries of the starting materials, almost phase pure products, as for instance seen for some octagonal sticks. It is hard then to trace any Chinese Dark Blue in these objects, because the total heating process is long. In the case of the octagonal stick **MAHP08** we expect that it was produced on total by a three- or even four-stage procedure.

Considering the shapes of octagonal sticks, like **MAHP08**, we suppose that ancient Chinese craftsmen have indeed applied another stage of preparation after the synthesis of the barium copper silicate. First, a finely ground colored barium copper silicate product was prepared to enable shaping in moulds, then calcination followed at an appropriate sintering temperature. The vitreous state needed to be avoided to keep the macroscopic shapes. Secondly, the octagonal stick was fired in the mould. Thirdly, these raw compact bodies were again calcined to refine their cylindrical shapes and phase compositions, and were then polished. Most octagonal sticks are presumed to be manufactured via one or two stages of the compact body manufacturing procedure; only a few such objects may have reached a third refining stage. Meanwhile, it has also been taken into consideration whether there was a simplified one stage process that would have included the main manufacturing technique, because there was apparently mass production of octagonal sticks in ancient Chinese times as reflected in the large numbers of archaeological finds.

23.3.9 *The use of the octagonal sticks*

The historical significance of octagonal sticks is presently not clear. From the presently known archaeological discoveries it can be concluded that the purple octagonal sticks were in some cases used as plugs for the seven orifices of dead bodies. Some octagonal sticks in jade and glass were found in orifices or in the hands of the corpses as gripped objects. One example is the octagonal sticks excavated from a Qin tomb in the south of Xi'an, one white mouth jade plug in tomb No.19, three purple-blue stone plugs in tomb No.106 and 12 coal plugs in another eight tombs.[23] Chinese Purple and Chinese Blue were detected as main phases in the purple-blue octagonal stick in tomb No.106 according to Raman analysis.[24] From the finds of the octagonal sticks in Xi'an Qin tombs, they were used as the seven orifices plugs.

There were six octagonal sticks excavated from tomb C1M3943 as mentioned above. They were placed above one bronze mirror. These sticks can be divided into three pairs according to their size. They were presumed to be earrings for aristocratic ladies as judged from the literature of in ancient Chinese anthology collections (Shijing) or from the ancient Chinese game Six Bo.

In the late Warring States to Western Han dynasty periods, the five elements theory and Zhouyi was popular, thus was concluded that octagonal sticks had relationship with these religious theories or mythologies myths. It was also forward that the sticks could have symbolized the linga (penis) totem in Hinduism, or were symbols of the buddhist pagoda with an octagonal basal shape. For a more definite explanation of the archaeological significance and usage of these sticks, we are looking forward to more discoveries from archaeological excavations.

23.4 Conclusions

Octagonal stick **MAHP08** belongs to the class of barium copper silicate faience made from refined ingredients with complete reaction, expected to be produced in three- or even four-stage procedure. It is expected to appear as a shiny or glazed blue object in the final state. Barium copper silicate compounds in the sample **MAHP08** are mainly Chinese Purple and occasionally Chinese Blue accompanied by abundant barium silicates. Most of the particles in stick **MAHP08** are between 5–15 microns and only minorities of the particles are between 15 to 25 microns, which indicates that raw materials were carefully ground.

The outer layer of the octagonal stick **MAHP08** appears yellow or earthy yellow because of weathering during burial. The purple middle section of the octagonal sticks were formed, because Pb and Ba elemental contents were lost from Pb glass or Pb — Ba glass phases, and this moderated the original dark blue tone. At the same time, when Chinese Purple decomposes the decomposition products react with S, Ca and P compounds that penetrated from the environment to form great amounts of insoluble $BaSO_4$ and other insoluble products.

According to the archaeological discoveries up to now, the cylindrical or cone objects may have several usage, such as the plugs of seven orifices of dead bodies, or as the as gripped objects in the hands of the corpses, or as the ancient Chinese game Six Bo, or as the symbolism of the linga (penis) totem in Hinduism, and or as the symbols of the buddhist pagoda with an octagonal basal shape. The stick MAHP08 is similar. The exact archeological significance and other uses are still unclear. To establish a more comprehensive historical picture of these octagonal sticks, we are looking forward to more unusual archaeological discoveries in the future.

Acknowledgements

We would like to thank Dr. Neville Agnew, the Getty Conservation Institute, for supporting Qinglin Ma as a visiting scientist who started the research on Chinese (Han) Blue and Chinese (Han) Purple in 2001. Scientists Eric Doehne, David Carson and Herant Khanjian camed out the IR microscopic analysis, environmental SEM and electron probe analyses. Thanks to Dr. Robert H. Brill (The Corning Museum of Glass), Dr. Hans-Georg Wiedemann (Switzerland) and Elisabeth West FitzHugh (Freer Gallery of Art, Smithsonian Institution, Washington D.C), for their encouragement given to Qinglin Ma. This work was carried out in 2003 under the auspices of a postdoctoral scholarship from the Institute of Inorganic Chemistry (ACI), University of Zurich, Switzerland. The study benefited from the postdoctoral work in the Institute of Inorganic Chemistry of the University of Zurich, and the visiting professor plan at the University of Zurich and the Swiss Federal Institute of Technology Zurich in 2009 and 2011. Dr. Armin Portmann, Dr. Ferdinand Wild, Dr. Zhou Yin and Dr. Reto Glaus helped in Raman microscopy, SEM-EDX and ICP-MS measurements. Thanks are also due to Professor Wang Julin and Ms. Li Mo, Beijing Chemistry and Engineering University, for their suggestions for the manufacturing of replica of octagonal sticks.

References

1. J. Riederer, Egyptian blue, in 'Artists Pigment. A Handbook of their History and Characteristics', ed. E. W. FitzHugh (Oxford University Press, Oxford, 1997), pp. 23–45.
2. E. W. FitzHugh and L. A. Zycherman, An early man-made blue pigments from China — barium copper silicate, *Studies in Conservation* **28**, 15–23 (1983). E. W. FitzHugh and L. A. Zycherman, A purple barium copper silicate pigment from early China, *Studies in Conservation* **37**, 145–154 (1992).
3. G. Giester and B. Rieck, Effenbergerite, $BaCu[Si_4O_{10}]$, a new mineral from the Kalahari Manganese Field, south Africa: description and crystal structure, *Mineralogical Magazine* **58**, 663–670 (1994).
4. C. Thieme, E. Emmerling and C. Herm, Research on paint materials, paint techniques and conservation experiments on the polychrome terracotta army of the First Emperor Qin Shi Huang, Techna. Paper presented at the CIMTEC Conference, Florence, Italy, July, 1994. pp. 591–601. Ma Qinglin. Brief

introduction of man-made Chinese blue and Chinese purple in Warring States Period and Han Dynasties. *Scientific and technological reasearch on cultural relics* **4**, Science Press, Beijing, 2006, 77–80. Ma Qinglin, Armin Portmann, Ferdinand Wild and Heinz Berke, Raman and SEM Studies of Man-made Barium Copper Silicate Pigments in Ancient Chinese Artifacts, *Studies in Conservation*, 81–98 (2006).

5. Ma Qinglin, Zhang Zhiguo, Xia Yin. Overview of the research of Chinese blue and Chinese purple. *Scientific and technological reasearch on cultural relics* **6**, Science Press, Beijing, 2009, 64–76. S. Greiff, S. P. Yin. Das Grab des Bin Wang-Wandmalereien der Ostlichen Han Zeit in China. Mainz: Harrassowitz Verlag, Wiesbaden. 2002, pp. 95–98.
6. K. Takayasu, Investigation of ancient glass found in Japan — Historical evolution of ancient glass systems. Paper presented at the Independent administrative Institution-Nara National Research Institute for Cultural Properties, Japen, 2002. pp. 147–157.
7. Li Defang. Research on the Area of Eastern Zhou King Mausoleums. *Cultural Relics of Central China*. special issue. 45 (1987).
8. The Archaeological Team of Luoyang City. Excavation of Some Warring States Tombs in Luoyang. *Cultural Relics* **8**, 4–13 (1999).
9. Qinglin Ma, Zhiguo Zhang, Xisheng Gao. Studies on the Chinese Blue and Chinese Purple in the Octagonal Sticks of Warring State Excavated from Luoyang, China. *Cultural Relics* **8**, 83–88 (2008).
10. Zi Bo Museum. Excavation of the Warring States Tomb 1 at Shangwangcun, Linzi, Shandong. *Cultural Relics* **6**, 14–26 (1997).
11. Guan Shanming. China Ancient Glass. Art Museum of The Chinese University of Hong Kong, (2001).
12. Xi'an Institute of Cultural Relics and Archaeology. Excavation of a Warring States Tomb in the Northern Suburbs of Xi'an. *Cultural Relics* **1**, 4–16 (2004).
13. Wang LeLe. Investigation of the octagonal sticks objects in the Warrior States and Qin Han Dynasty. unpublished paper.
14. R. H. Brill, S. S. C. Tong and D. Dohrenwend, Chemical analyses of some early Chinese glasses (Chapter 4) and P. M. Fenn, R. H. Brill, M. G. Shi, 'Addendum to Chapter 4' in *Scientific Research in Early Chinese Glass*. Ed. R. H. Brill and J. H. Martin (the Corning Museum of Glass, Corning NY, 1991), pp. 31–64; H. Berke, Chemistry in ancient times: The development of blue and purple pigments, *Angew Chem Int* **14**, 2483–2487 (2002).
15. D. A. Scott, *Metallography and microstructure of ancient history metals*, The Getty Conservation Institute/ The J. Paul Getty Museum: Malibu, 1991, pp. 72–73.
16. S. Bouherour, H. Berke and H. G. Wiedemann, Ancient man-made copper silicate pigments studied by Raman Microscopy, *Chimia* **1**, 942–951 (2001). J. D. Frantz, B. O. Mysen, Raman spectra and structure of BaO-SiO_2, SrO-SiO_2 and CaO-SiO_2 melts to 1600°C, *Chemical Geology* **121**, 155–176 (1995).

17. D. A. McKeown and M. Bell, Four-membered silicate rings: Vibrational analysis of $BaCuSi_2O_6$ and implications for glass structure, *Physical Review B* **6**, 3114–3121 (1997). D. A. McKeown and M. I. Bell, Linked four-membered silicate rings: vibrational analysis of Gillespite $BaFeSi_4O_{10}$ and implications for glass structure, *Phys. Chem. Minerals* **25**, 273–281 (1998). D. A. McKeown, M. I. Bell and C. C. Kim, Raman spectroscopy of silicate rings: Benitoite and the three-membered ring, *Physical Review B* **22**, 16357–16365 (1993). L. W. Finger, R. M. Hazen and R. J. Hemly, $BaCuSi_2O_6$: a new cyclosilicate with four membered tetrahedral rings. *American Mineralogist* **74**, 952–955 (1989).
18. J. M. Du, H. Y. Zeng, L. J. Song and Z. C. Dong, Synthesis and structure of a new polymorph $Ba_2CuSi_2O_7$, *Jiegou Huaxue* **1**, 33–36 (2003).
19. H. Berke. The invention of blue and purple pigments in ancient times, *Chem. Soc. Rev* **36**, 15–30 (2007). Ma Qinglin, Zhang Zhiguo, Xia Yin. Overview of the research of Chinese blue and Chinese purple. *Scientific and technological reasearch on cultural relics* **6**. Science Press, Beijing, 2009, pp. 97–116. Yin Xia, Shuangcheng Wu, Shengkuan Cui and Yufu Lan. Study on the pigments in polychrome potteries of West Han Dynasty tomb in Weishan, Shandong province. *Sciences of Conservation and Archaeology* **2**, 13–19 (2008).
20. H. G. B. Wiedemann, Formation and stability of Chinese barium copper-silicate pigments [A]. In: *International Conference on the Conservation of Grotto Sites: Proceedings Conservation of Ancient on the Silk Road* [C]. Dunhuang, China. 1993, pp. 379–387. H. Berke. The invention of blue and purple pigments in ancient times, The Royal Society of Chemistry, *Chem Soc Rev* **36**, 15–30 (2007).
21. M. Jaime,V. F. Correa and N. Harrison, Magnetic-field-induced condensation of triplons in Han Purple pigment $BaCuSi_2O_6$, The American Physical Society, *Physical Review Letters* **8**, 087203-1-4 (2004).
22. Tristan Corbiere. Preparation and methods of characterisation of the barium copper silicates $BaCuSi_4O_{10}$ (Chinese Blue), $BaCuSi_2O_6$ (Chinese Purple), $Ba_2CuSi_2O_7$, $BaCu_2Si_2O_7$ used as blue and purple pigments, Ph.D Thesis, University of Zürich, Switzerland. 2009.
23. Xi'an Institute of Cultural Relics and Archaeology. *Qin tomb in South Xi'an.* Xi'an, Shanxi people's publishing press. 2004, pp. 345–346.
24. Xia Yin, Ma Qinglin, Zhang Zhiguo. Research report of Chinese blue and Chinese purple. State Administration of Cultural Heritage. Research subject of Cultural relics protection technology. 20050113. Feng Jian. Primer analysis of two ancient objects contain Chinese blue and Chinese purple from Qin Tomb in South Xi'an. *New technology of Cultural relics protection.* Cultural Relics Press, Beijing, 2008, pp. 296–299.

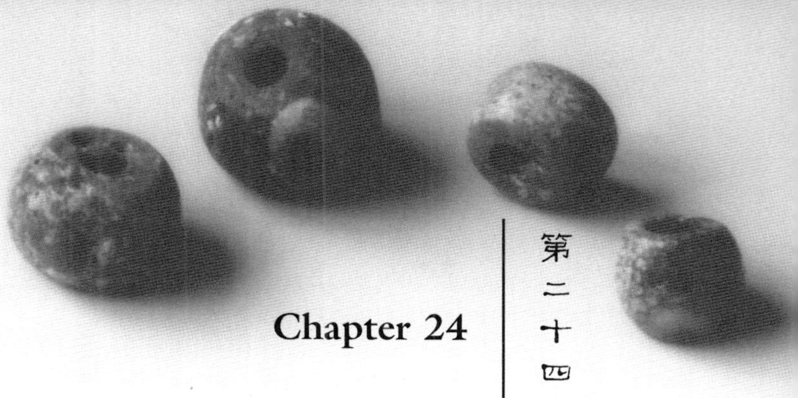

Chapter 24 第二十四章

Chinese Blue and Chinese Purple in Copper Tinted Lead Barium Glass Produced During Firing Experiments

Ying Qin, Shicai Li,* Xi Chen† and Huang Huang**

*Department of History of Science and Technology and Archaeometry, University of Science and Technology of China, Hefei, 230026, China

†Lanxi Museum, Lanxi, 321100, China

24.1 Introduction

The recognition by foreign scholars of the mysterious blue and purple pigments found in Chinese cultural relics occurred in the1980s and 1990s. Since then research into Chinese blue and purple became popular.[1–6] More researchers found Chinese blue and purple, dating to the Warring States Period and Han dynasty. On the other hand, during the predynastic period in ancient Egypt, the Egyptians invented another kind of silicate pigment — Egyptian blue. Its raw materials were easy to obtain and its manufacture was not complex. In the Mediterranean area, including Egypt, the Middle East, ancient Greece and the Roman Empire, ancient people use this blue pigment for thousands years. Its composition is consistent, which suggests that the techniques used to make it were mature. According to the available archaeological material, the technique for the production of Egyptian blue did not spread to China. The Chinese craftsman synthesised a homologue of the silicate pigments — Chinese blue and purple. Now, the questions arise about the origin of the artificial barium-silicate pigment; and why it only appeared in ancient China. Lead barium glass and barium silicate pigments existed at the same time in ancient China, the copper tinted glaze of Liaozhu appeared in the Western Zhou Dynasty, and the copper tinted lead barium faience appeared at least as early as the Spring and Autumn period.[7] In the simulated experiments to produced ancient glazes[8] and lead barium glass, Chinese blue and purple was produced. This may provide a key to its invention and how the technology emerged over time.

24.2 Experiments

The main materials are (natural minerals and the artificial materials which could be obtained in the pre–Qin period): quartz, malachite, lead monoxide, barite, witherite, feldspar. Except for witherite which is extremely rare, the other materials are widely distributed. Because of specific metallogenic conditions, witherite deposits are extremely rare. At present, there are only three countries in which witherite deposits have been found: Settlingstones in Britain (found in 1960s, small and well explored), Karakala in Turkmenistan and the Qinling-Daba region in China. More than 40 independent witherite deposits are located in the northern Daba Mountains in the border areas connecting Sichuan, Hubei and Shanxi,[9] forming the only large witherite metallogenic belt in the world, and they are found on the surface.

The recipes for the experiments were based on lead barium Bi glass (Fig. 24.1) unearthed from Chu Tombs M271 of the Warring States Period at Yiyang in Hunan province. [10] The mass ratio used for PbO, BaO and SiO2 was 30:15:50 and the amount of flux used (such as Na_2O, K_2O, CaO) was about 3.5%, added in the form of feldspar. The content of CuO was about 1.5%. We used copper oxide (calcined malachite). Based on melted Bi glass mentioned above the melting temperature is under 1100°C, the working temperature of lead barium glass cited in R. H. Brill's experiments,[11] we fired glass at 1000°C–1050°C in our experiments. The recipe used our experiments is in Table 24.1.

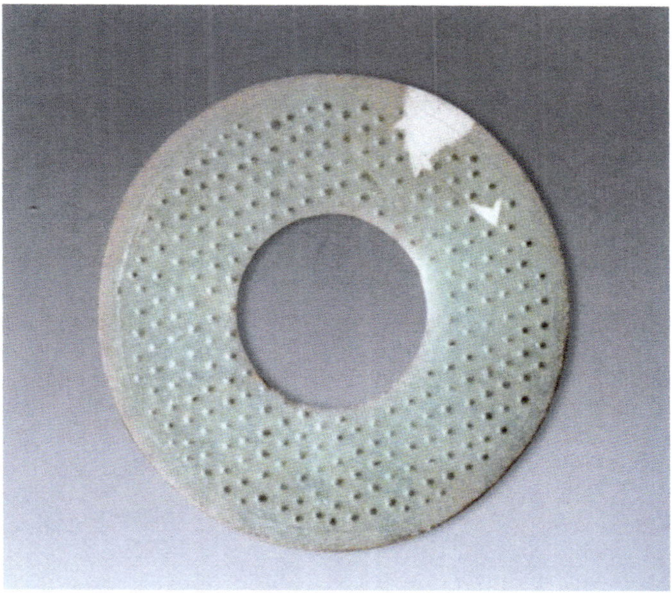

Fig. 24.1 The Bi glass from Chu Tomb, Yiyang, Hunan

Table 24.1 The recipe of the simulated experiments

Sample no.	The recipes used in our experiments
H_1	quartz, copper oxide , lead monoxide, barite, sodium-lime flux
H_2	quartz, copper oxide, lead monoxide, witherite, sodium-lime flux
H_3	quartz, copper oxide, lead monoxide, barium oxide, sodium-lime flux

These ingredients were ground and well blended in our experiments.

H_1, H_2, H_3 was placed in porcelain vessels separately, and heated to 1000 Centigrade in a muffle furnace, for 2 hours and allowed to cool.

H_1 was not vitrified completely, and was covered with particles of white barite, in the surface there were a lot of tiny holes but no blue-violet spots. When heated to 1050 centigrade, and held for six hours at this temperature, no obvious changes occurred.

H_2 and H_3 were blue, glassy and translucent; under the stereomicroscope blue and purple spots (Figs. 24.3 and 24.4) were observed.

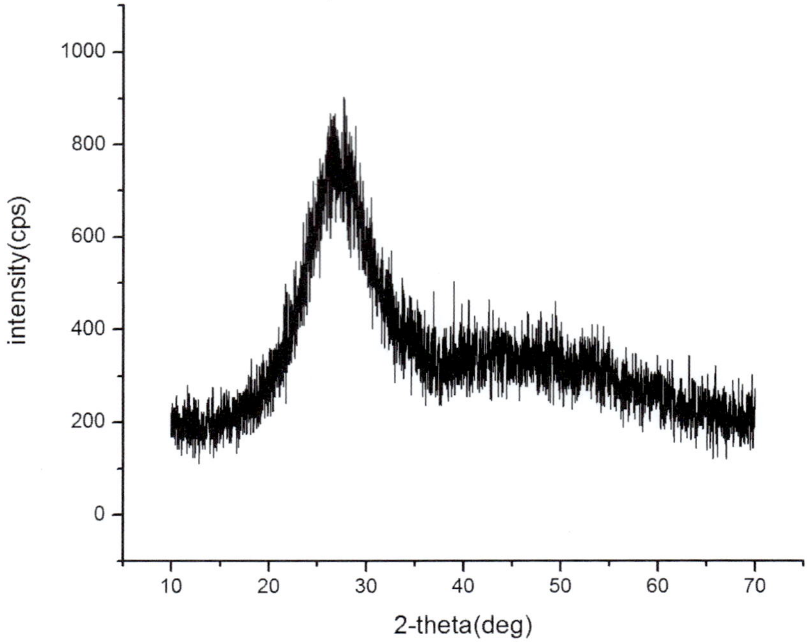

Fig. 24.2 The XRD spectrum of H_2

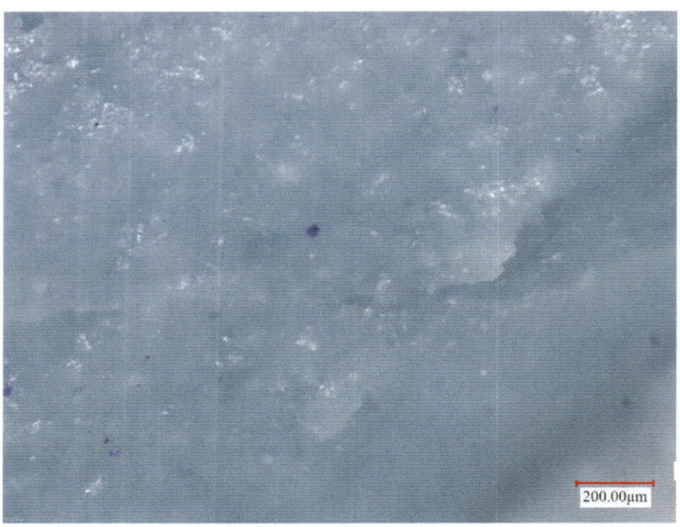

Fig. 24.3 The tiny purple grains in H_2

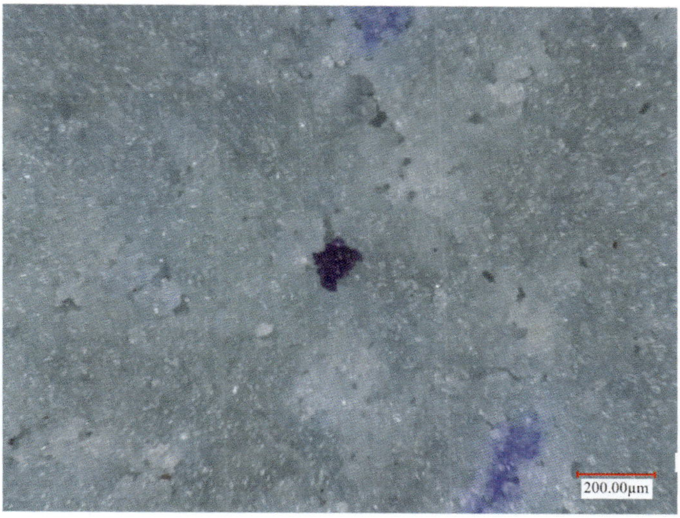

Fig. 24.4 The tiny purple and blue-purple grains in H_3

The purple grains were very fine, and the blue grains consist of fine particle and agglomerations.

For understanding better the relationship between the grains and the composition of the recipe, we regulated the composition by increasing the copper oxide level. The molar ratio of witherite, copper oxide, quartz, lead monoxide is 1:1:5:0.25 in sample H_4. This was heated to 1000 centigrade in a muffle furnace for two hours, then allowed to cool off. It became a blue-violet sintered material (Fig. 24.8).

24.3 Results

After H_2, H_3 and H_4 were calcined they were analysed using X-ray diffraction for phase identification. H_2 and H_3 were analysed by Micro-Raman, in the public experimentation centre at USTC.

LABRAM-HR Micro-Raman was used for structural analysis. The experimental conditions were argon ion laser, excitation wavelength = 514.5 nm, 600 mW, wavlength number = 200–1200cm^{-1} , the precision of the wavelength number is ± 1cm^{-1}, objective lens × 50, spot size = 1μm.

TRR- A high power X-ray diffract meter was used for phase identification, CuKa at 1.5418Å, 40 kV, 200 mA, the range of the scanning angle was (2θ) 5°~70°.

The XRD spectrum shows H_2 and H_3 are vitrified. The XRD spectra of H_2 and H_3 are the same, so we take spectrum of H_2 for example (Fig. 24.2). In Fig. 24.5 the Raman spectrum of tiny purple grains in H_3, shows characteristic peaks for Chinese purple (Figs. 24.6 and 24.7) are the Raman spectra of tiny blue grains in H_2 and H_3. They show the characteristic peaks for Chinese blue. The phase characteristic for Chinese blue and purple is found in the sintered material in H_4 (Fig. 24.9), and no surplus CuO was left. A part of the barium-rich was synthetised with Chinese blue and purple, the other part was with $BaSi_2O_5$.

24.4 Discussion

Lead-barium glaze's and glass' material was not mixed well in ancient times, though there was not enough CuO and BaO, the experimental temperature achieved the requirement for sintering, especially for

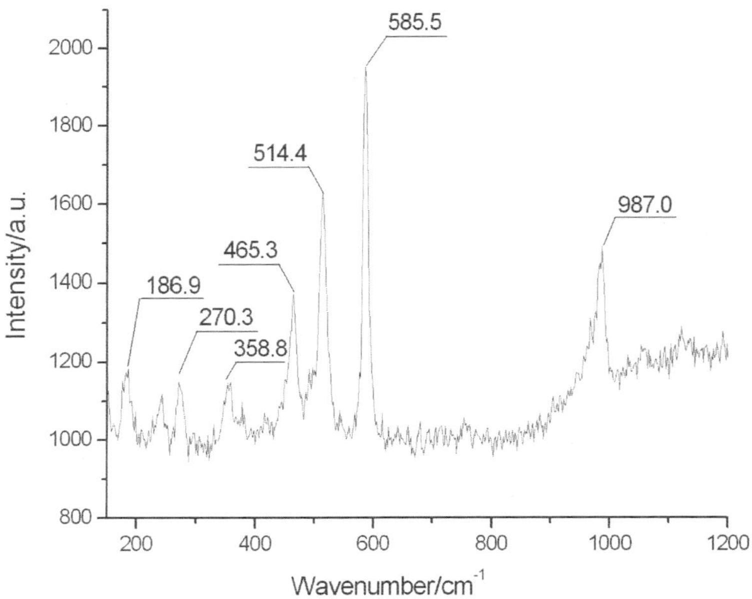

Fig. 24.5 The Raman spectrum of the tiny grains in H_3

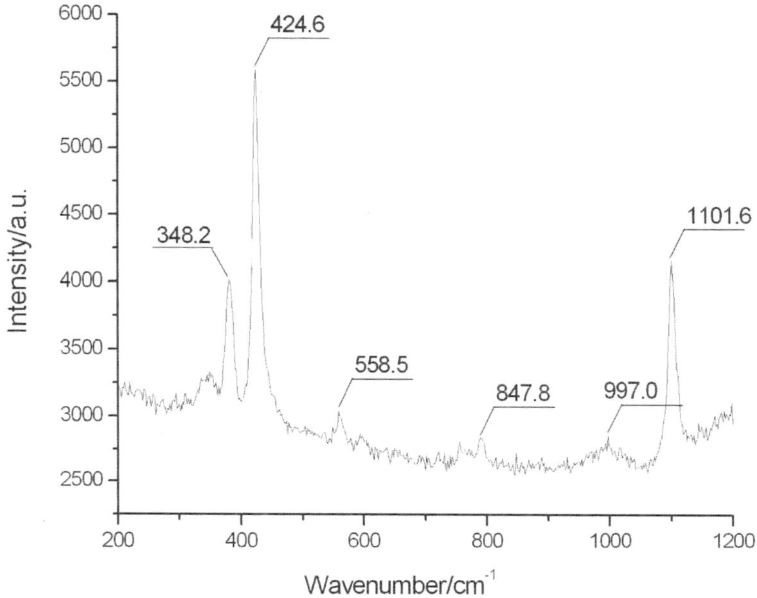

Fig. 24.6 The Raman spectrum of the tiny blue purple grains in H_2

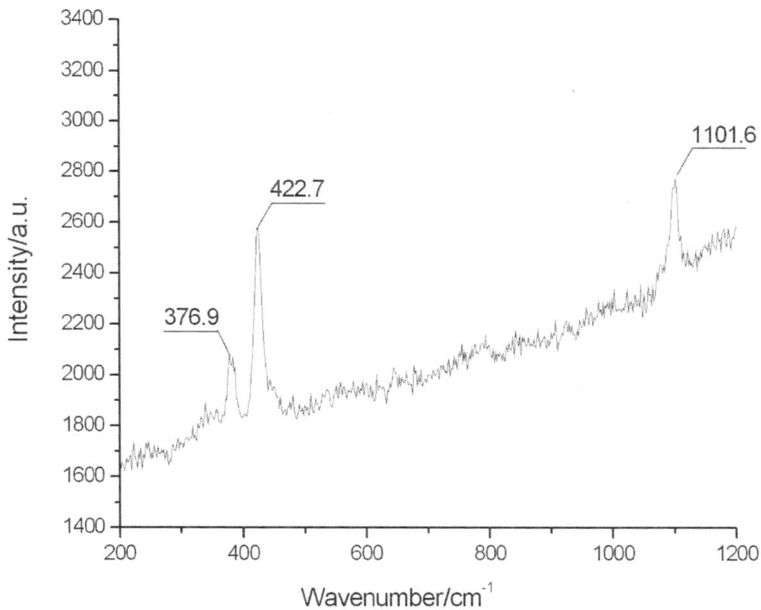

Fig. 24.7 The Raman spectrum of the tiny blue grains in H_3

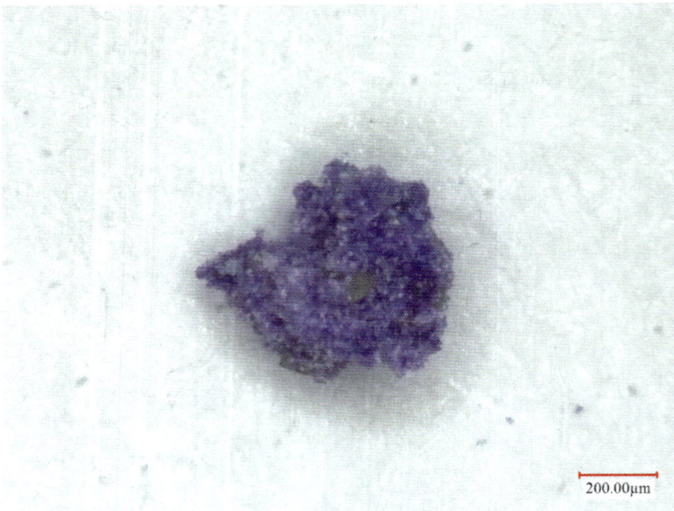

Fig. 24.8 The blue-purple sintered material

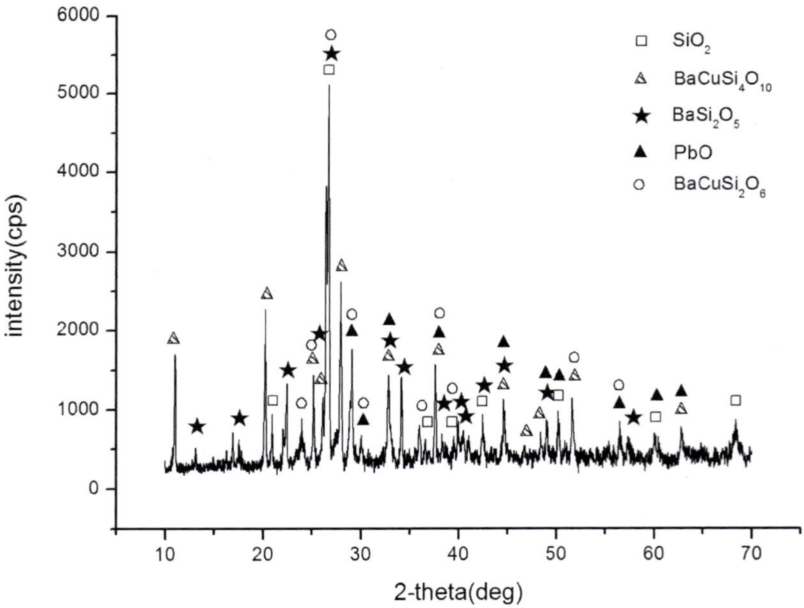

Fig. 24.9 The XRD spectrum of H_4

PbO (catalyst and flux) and a part of the material may meet the conditions for synthesis of Chinese blue and Chinese purple so that tiny blue and purple grains were generated. A longer heating time or a higher temperature produced bigger grains because of ion diffusion. Chinese blue and purple found in faience excavated in Dabaozi mountain, Li county, was analysed using micro-Raman by Qing-Lin Ma etc.[12] When the percent of copper oxide and witherite was increased in the synthetic raw materials, the colour of the products changed from blue to purple.

No Chinese blue and purple developed when barite were used (a common mineral) in the raw materials, while when witherite was used, Chinese blue and purple appeared. In the modern industry, the process converting barite into BaO or $BaCO_3$ is very complex. In China there are witherite deposits: these are rare in other areas of world. The unique artificial barium-silicate materials of ancient China made using the special barium mineral is worthy of deep consideration by the archaeological community and within the context of the history of science and technology.

Comparing the composition of Chinese blue and purple, the silica in the copper-tinted lead barium glass is excessive Chinese blue is more stable than the Chinese purple perhaps because there is the following reaction: $BaCuSi_2O_6+SiO_2 \rightarrow BaCuSi_4O_{10}$ in the firing process which leads to the development of more frequent blue grains that are bigger than in Chinese purple.

In nature, a purple mineral pigment is rare. The accidental products of the process of firing copper tinted lead–barium glaze or lead-barium glass may have been a great technological invention–the synthesis of Chinese blue and purple pigments.

24.5 Conclusion

When using witherite as a raw material that contains barium and firing the copper tinted lead-barium glaze or lead-barium glass, Chinese blue and purple crystals may be generates in the glassy materials, and the higher precent of CuO and $BaCO_3$ in the raw materials, the higher proportions of crystals that are generates.

Unique artificial barium-silicate products could possibly have a compositional relationship with special barium-rich mineral deposits (whitherite).

Acknowledgments

The authors acknowledge the financial support of state administration of cultural heritage in China (No. 20110108).

References

1. E. W. Fitzhugh and L. A. Zycherman, *Studies in Conservation* **28**, 15 (1983).
2. E. W. Fitzhugh and L. A. Zycherman, *Studies in Conservation* **37**, 145 (1992).
3. H. G. Wiedemann and H. Berke, (1999) Chemical and physical investigations of Egyptian Blue and Chinese Blue and Purple, in *Proceeding of the Conference The Polychromy of Antique Sculpture and the Terracotta Army of the First Chinese Emperor: Studies on Materials, Painting Techniques and Conservatio, eds.* Wu Yongqi, Zhang Tinghao, Michael Petzet *et al* (Karl M. Lipp Verlag, Munchen, Germany, 2001), Monuments and Sites III, pp. 154–157.
4. H. Berke, T. Corbiere, A. Portmann, *et al.*, *Relics and Museology* **6**, 251 (2009).
5. Q. L. Ma, Z. G. Zhang and Y. Xia, The Research Review of the Chinese Blue and Chinese Purple, in *Scientific and Technological Research on Cultural Relics*, ed. Chinese academy of cultural heritage (Science Press, Beijing, China, 2009), Vol. 6, pp. 64–76.
6. H. Berke, *Royal Society of Chemistry* **36**, 15 (2007).
7. Q. L. Ma, A. Portmann, F. Wild *et al.*, *Studies in Conservation* **51**, 81 (2006).
8. Y. Qin, X. Chen, X. L. Li *et al.*, *Journal of the Chinese Ceramic Society* **4**, 567 (2012).
9. S. H. Wu, J. J. Liu, N. Zhang *et al.*, *Journal of Geoscience* **2**, 237 (2010).
10. Yiyang Municipal Administrative of Cultural Relics and Yiyang Municipal Museum, *Yiyang Chu Tombs*, Cultural Relics Publishing House, China, 2008, p. 393.
11. R. H. Brill, J. F. Wosinski, Chemical analyses of some early Chinese gasses, in *Scientific Research in Early Chinese glass — Proceeding of the Archaeometry of Glass Session of the International Symposium on Glass*, ed. F. X. Gan(China Architecture & Building Press, Beijing, China, 1984), pp. 15–35.
12. S. Bouherour, H. Berke and H. G. Wiedemann, Ancient Man-Made Copper Silicate Pigments Studied by Raman Microscopy, in *Scientific and Technological Research on Cultural Relics*, (Science Press, Beijing, China, 2006), Vol. 4, pp. 81–95.

Chapter 25

第二十五章

A Study of Simulation of the Production Technology of Ancient Chinese Blue and Purple Faience During the Warring States, Qin and Han Dynasties

Mo Li, Julin Wang* and Qinglin Ma*[†]

*Beijing Key Laboratory of Electrochemical Process and Technology for materials, Beijing University of Chemical Technology, Beijing 100029, China

[†]Chinese Academy of Cultural Heritage, Beijing, 100029

25.1 Introduction

Faience is noted as the first of the 'high-tech ceramics' of human history, and some scholars consider it as an early stage in glass development.[1,2] Excavated Chinese faience mainly dates back to between the Warring States and Western Han Dynasties. These include beads, eye beads, ear spools and octagonal sticks. Numerous chemical analyses have established that the colorants used in these most remarkable purple and blue faience are mainly Chinese-purple and Chinese-blue.[2–8]

In the colour deficient ancient times, colorants mainly came from natural minerals and plant extract. lack of resources led the craftsmen to invent purple and blue materials, which could meet their pursuit of beauty. Thus, Chinese-blue and Chinese-purple emerged and became unique achievements during the history of ancient Chinese science and technology.[9]

Chinese-blue or purple pigments so far identified in beads or in paint layers date from the late Western Zhou period (1207–771 BC), after the booming development in the Qin and Western Han dynasties (221 BC–8 AD). It faded from history after the late Eastern Han dynasty.[10] At *Majiayuan* in *Zhangjiachuan* county, *Gansu* province, archeologists found that Chinese-purple and Chinese-blue were widely used in bead decoration during the late Warring States.[5] During the Qin to Han dynasties the Chinese-purple pigment was widely used in murals and pottery figurines.[11]

When it comes to the *Sanmenxia* tombs in Henan province of the Han Dynasty, blue-purple frit was found in a bottle of No. M17 tomb. The excavation bulletin describes a small amount of blue-purple mineral in an unglazed porous bottle.[12] The blue-purple mineral was finally identified as blue-purple barium copper silicate frit which had been fired at a low temperature or for a short time. With regards to the special manufacturing process, a wealth of information about the origin of ancient faience making has been discovered.[5]

Chinese-purple was widely used to make beads and pigment and it was also the main component of the purple faience octagonal sticks. Due to their unique shape and profound significance, octagonal sticks are worth studying. By using XRD and SEM, E. W. Fitz Hugh analyzed six octagonal sticks and identified the artificially produced material barium copper silicate in these objects probably dating to the Warring States period (475–221 BC) or to the Han Dynasty.[3,4]

Many octagonal sticks were discovered in the collection of Guan Shanming.[1] Octagonal sticks of various sizes are very hard and have a smooth surface covered by brown weathering. The purple colour showed through in some places on the flat facets and (inevitably) on the sharp edges (Fig. 25.1).

In Guan's opinion, the octagonal sticks were mainly used for the plugs for the dead (generally ear, nose and anal plugs).[1] The use of a jade plug, which can keep the spirit in the body eternally, was very well known during the Han Dynasty (especially in the Western Han Dynasty). It is known that jade held a special status in Taoism. Taoist people believed that jade, which they considered to be a magical material, not only held the power to preserve a human body and

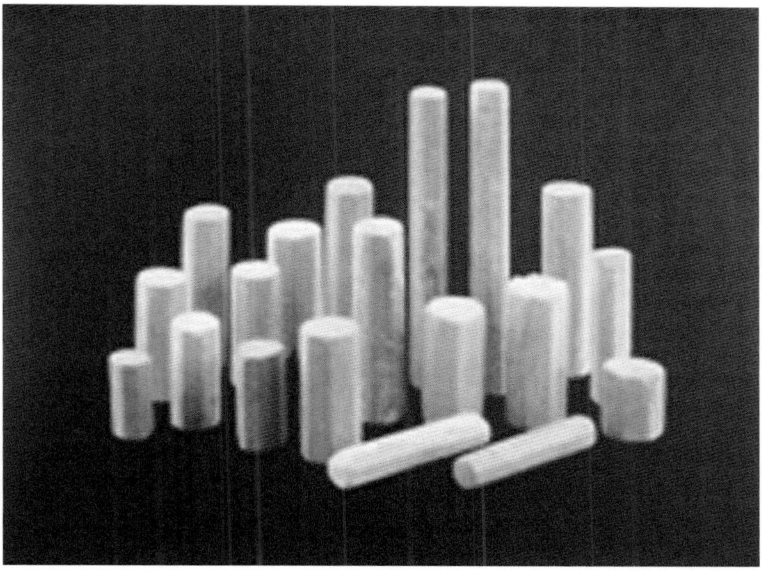

Fig. 25.1 Octagonal faience sticks, Han dynasty (206–220 BC)

spirit, but also was an elixir for achieving physical immortality. In the pursuit to understand and obtain such a precious material, the Taoist monks started to synthesize it themselves.[11] And to some extent, the octagonal sticks could well be the substitutes of the jade in ancient China. For this reason, the study of the simulation of the production technology of faience is not only an exploration of ancient glass technology development but also can provide further understanding of the pursuit of immortality in early China.

So far the study of the production technology of purple and blue faience has mainly focused on the chemical analysis and glazing method of Egyptian Blue.[13–19] As to our unique blue and purple faience, the research has stopped at compositional analysis and theoretical inferences.[20–24] A lack of simulation may potentially lead to research stagnation. Therefore, this study aims at a preliminary simulation of faience products so as to investigate the production technology of ancient Chinese-purple and blue faience during the Warring States to Qin and Han dynasties. The simulated powder used in this study was based on the test results of ancient bead compositions.

Through the moulded or pressed forms, four different shaped faience products were obtained. These synthetic beads or sticks were compared with four objects of faience from China spanning the Warring States to Qin and Han dynasties using macroscopic and microscopic structures. On the basis of these results, it is possible to sketch out the general forming processes of Chinese blue-purple faience during the Warring States, the Qin, and the Han dynasties.

25.2 Samples and experimental

25.2.1 *Ancient beads survey and simulation recipes*

In order to increase the accuracy of simulation, three faience beads (Fig. 25.2) were selected for measurement and chemical analysis. All ancient samples came from personal collections, the first is a tubular blue bead fragment (A-1), the second is a hollow spindle-shaped blue bead (A-2) and the last is a hollow spherical blue bead (A-3). The chemical compositions of the ancient beads are listed in Table 25.1. They were measured by energy dispersive X-ray spectrometry (EDX-800HS): the simulation recipe (Table 25.2) was based on the average of these test results.

25.2.2 *Preparation and analysis of Chinese-purple frit*

The Chinese-purple frit was prepared from $Cu_2(OH)_2CO_3$, $BaCO_3$ and SiO_2 according to the stoichiometry 1:2:4.[8] After adding 5% (wt%) $PbCO_3$, the mixture was finely ground and put into clay

Fig. 25.2 The photos of ancient blue faience beads

Table 25.1 Chemical compositions of ancient Chinese-purple/blue faience beads (wt%)

No.	Description	SiO_2	Al_2O_3	P_2O_5	Fe_2O_3	PbO	BaO	CuO	CaO	K_2O	SrO
A1	Faience bead; tubular; bule; opaque (fragment)	54.690	—	2.520	0.472	27.912	8.489	4.271	1.364	—	0.282
A2	Faience bead; hollow spindly; bule; opaque	50.602	—	—	0.305	21.104	16.008	9.031	1.621	0.968	0.357
A3	Faience bead; hollow spherical; bule; opaque	46.542	—	1.613	0.782	28.955	14.945	3.905	1.885	0.939	—

Table 25.2 Chemical compositions of the simulative beads formula and corresponding sources

	SiO_2	PbO	BaO	CuO	CaO
Wt%	40	25	15	10	10
Source	SiO_2	$PbCO_3$	$BaCO_3$	$Cu_2(OH)_2CO_3$	$CaCO_3$

crucibles. Then the powder was fired at 900°C for 18 h in a muffle furnace and allowed to cool down in the furnace. Finally, the purple frit product CP-18 (Chinese-purple 18 h) was obtained. The reaction equation is as follows:

$$Cu_2(OH)_2CO_3+2BaCO_3+4SiO_2\rightarrow 2BaCuSi_2O_6+H_2O+2CO_2+O_2$$

After the fully grinding in an agate mortar, the chemical composition of CP-18 was determined using the RigakuD/max2200 X-ray diffractometer (with a Cu target). The tube voltage and tube current of the diffractometer were 40 kV and 40 mA respectively.

25.2.3 *The simulation of bead making, and analysis*

The laboratory replication of three kinds of small faience beads have been carried out: a tubular bead, a hollow spindle-shaped bead and hollow spherical bead.

First, a lump of paste was prepared from 10g of faience simulation recipe powder (see Table 25.2) and 1 ml deionized water. Then the paste was moulded by hand and perforated with a small bamboo stick to achieve the hole through the bead. The chemical composition of 5 g of simulation powder was fired as well as the beads.

After one hour drying at room temperature, the beads were fired in a muffle furnace at 900°C for 10 h before being cooled inside the furnace to room temperature. CB-1 (the tubular blue bead), CB-2 (the hollow spindle-shaped blue bead), CB-3 (the hollow spherical blue bead) and blue frit CB-10 (Chinese-blue 10h) were finally obtained.

After being fully ground in an agate mortar, the chemical composition of CB-10 was conducted using a RigakuD/max2200 X-ray diffractometer. By using an Anyty 3R-type video microscope, the microstructure of the replicated and ancient beads were observed.

25.2.4 *The production technology of octagonal stick*

25.2.4.1 *Measurement*

The height and the length of the ancient octagonal stick are 3.0 cm and 0.5 cm.

25.2.4.2 *Clay mould manufacture*

A clay mould was made in two symmetrical parts; each part has three inner edges; two outer edges are parts that fit together. During the experiment, clay was used to manufacture the clay mould; after drying it for three days at room temperature, the mould was fired at 700°C for 6 h followed by furnace cooling. The fired clay moulds are shown as Fig. 25.3.

25.2.4.3 *Moulding and mould release*

Petroleum jelly was evenly applied to the inner wall of the clay moulds as a mould release agent. Pulverized homemade Chinese-purple raw

(a) (b)

Fig. 25.3 (a) Fired clay moulds; (b) the simulation of the closed mould process

Table 25.3 Chemical compositions of the simulative octagonal stick formula and corresponding sources

	SiO_2	PbO	BaO	CuO
Wt%	30	10	40	20
Source	SiO_2	$PbCO_3$	$BaCO_3$	$Cu_2(OH)_2CO_3$

materials (Table 25.3) were mixed with deionized water; a lump of paste was pressed into a cylindrical shape, placed into the mould and then covered by the other half of the clay mould. After being tied attached with tape the mould was arranged vertically. After compaction and removing from the mould the raw block became a pre-octagonal stick.

25.2.4.4 *Firing*

The pre-octagonal stick was placed in a clay crucible, fired at 900°C for 10 h and cooled down in the furnace. Then the product CPOS (Chinese Purple Octagonal Stick) was produced.

25.3 Results and discussion

25.3.1 *Simulated results of Chinese-purple frit*

Figures 25.4–25.5 show the purple frit excavated from *Henan Sanmenxia* tomb and the simulated Chinese-purple frit.

When it comes to colour, ancient frits are purple or purple-black (with an earthy yellow weathered crust). The frits produced in the laboratory are also purple or purple-black so are similar to ancient ones. Ancient pieces have a very irregular shape and no sign of being shaped by hand. So we have deduced that they were not sintered after being moulded but, crushed after the frit was sintered. To confirm this idea, Chinese-purple frit CP-18, and the clay crucible were broken and crushed. After being collected, some purple frit pieces were obtained: these have very similar colour and shape to purple frit excavated from M17 in *Henan Sanmenxia*, as shown in Fig. 25.5. XRD analysis shows that the main component of the purple powder is $BaCuSi_2O_6$ (Chinese-purple), as shown in Fig. 25.6.

Fig. 25.4 Purple frit excavated from M17 in *Henan Sanmenxia*[13]

(a) (b)

Fig. 25.5 (a) Purple frit stuck together in the clay crucible; (b) the crushed simulated purple frit

25.3.2 *Results of forming clue faience beads*

25.3.2.1 *The analysis of blue powder*

By the XRD analysis of the blue powder CB-10 shows it is mainly composed of $BaCuSi_4O_{10}$ (Chinese blue). A little $BaCuSi_2O_6$ (Chinese purple) is also presented in the spectrum (Fig. 25.7).

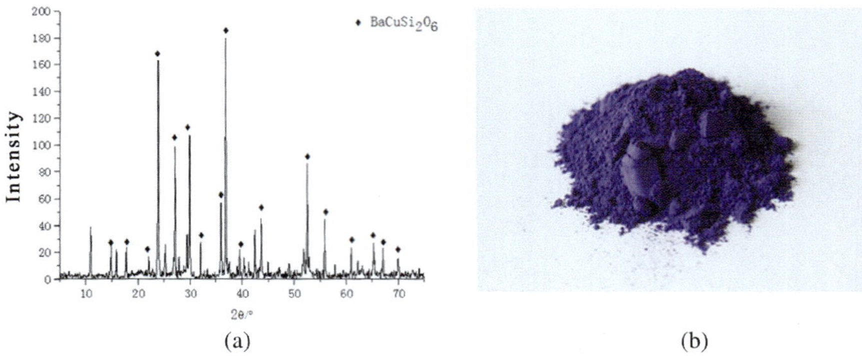

Fig. 25.6 XRD analysis and photo of the purple product CP-18 (a) CP-18 XRD spectrum (b) Chinese-purple pigment simulated in laboratory

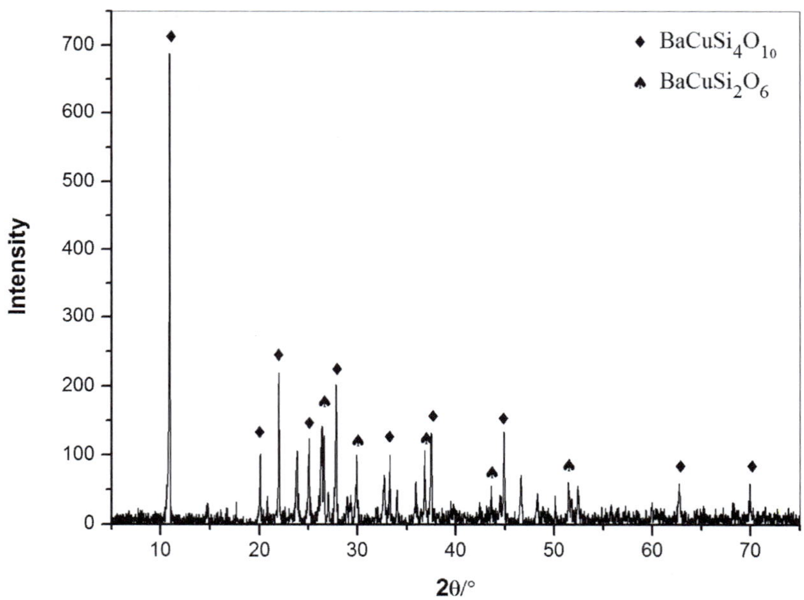

Fig. 25.7 XRD spectrum of the blue powder CB-10

25.3.2.2 *Simulated blue beads and microstructure results*

As shown in Fig. 25.8, the simulated faience tubular bead is quite similar to the ancient one. They both have uniform particle size, a bright blue colour and a dense surface structure. In the ancient bead,

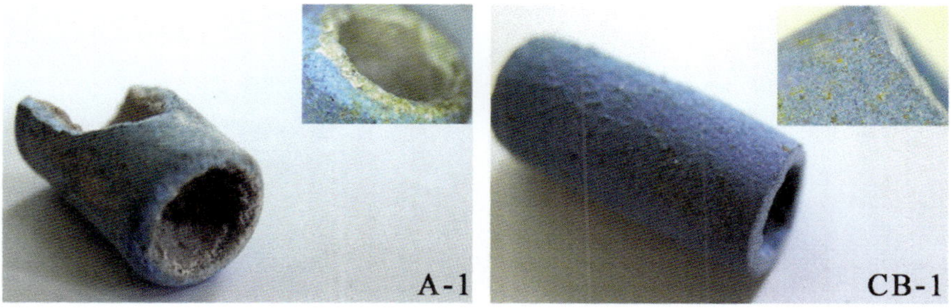

Fig. 25.8 The macroscopic and microscopic images of the ancient tubular bead (A-1, microphotograph: 60x) and the simulated tubular bead (CB-1, microphotograph: 60x)

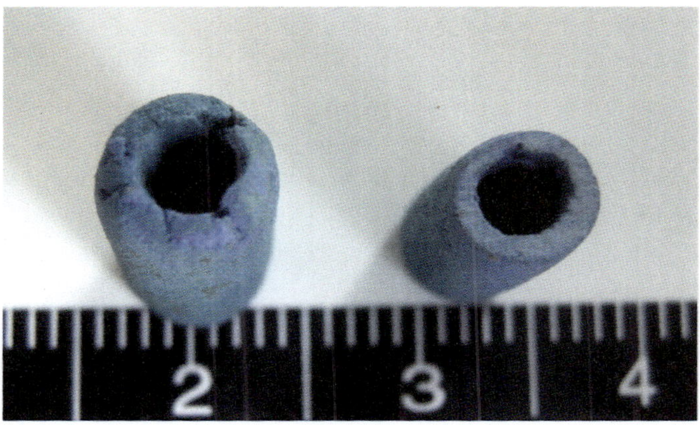

Fig. 25.9 The edges comparison before and after the edges polishing process (Left: unpolished tubular bead; right: polished tubular bead)

the edge left after the perforating tools is very smooth. During the experimental simulation, only by polishing the edges after firing could the smooth edges of the faience beads be obtained. It is not possible to get smooth edges by just using a mould. Fig. 25.9 shows a comparison of the simulated tubular bead edges before and after the polishing process.

The spindle-shaped bead consists of two symmetric cones which can be obtained by inward extrusion of the two halves simultaneously. After the firing process, the simulated bead has a similar shape to the

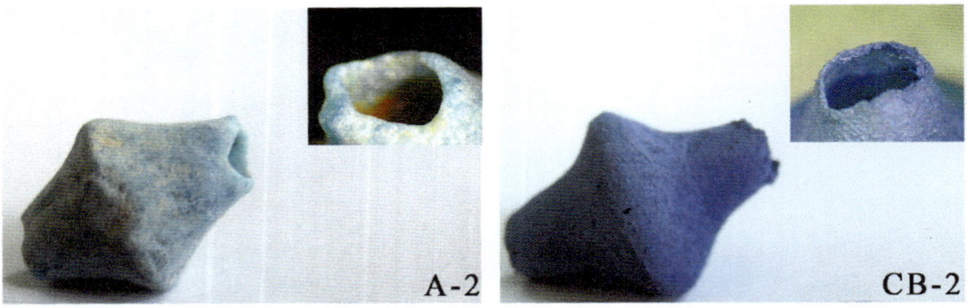

Fig. 25.10 A macroscopic and microscopic structural comparisons between the ancient spindle-shaped bead (A-2, microphotograph: 80x) and the simulated spindle-shaped bead (CB-2, microphotograph: 80x)

Fig. 25.11 A macroscopic and microscopic structural comparison between the ancient spherical bead (A-3, microphotograph: 100x) and the simulated spherical bead (CB-3, microphotograph: 100x)

ancient one. Compared with the tubular bead, the edges of the ancient hollow spindle-shaped bead are quite irregular, which can be seen in the photomicrograph of A-2, as shown in Fig. 25.10. This phenomenon indicates that the spindle-shaped bead was not polished.

The simulated spherical blue bead CB-3 shows a uniform blue color and has a compact surface without any holes. However, the ancient spherical bead A-3 is a blue-grey colour and has a lot of holes in the surface due to serious weathering. So the simulated bead was not very similar to the ancient one, as shown in Fig. 25.11. In addition, the uneven edges indicate that this ancient spherical bead did not go through the polishing process.

In conclusion, the forming process of the hollow blue faience bead generally requires three steps. Firstly, the powdered raw materials were mixed and some water was mixed in to produce a paste. Then the bead shape was moulded by hand and perforated with a small stick or other perforating tool. Finally, the bead was sintered in a kiln after drying for several hours. In order to increase the regularity of the faience beads, ancient craftsmen would sometimes polish the edges after sintering.

25.3.3 *The result of mouding the octagonal stick*

The ancient octagonal stick and the synthetic one are shown in Fig. 25.12.

In the Fig. 25.12, the laboratory simulated Chinese purple octagonal stick CPOS has eight clear edges and an octagonal cross section. These are very similar to the ancient one. Based on the simulation experiment, the production method for the octagonal stick is a one-step moulding process using a polished clay or stone mould. After the mould formation, the faience paste is pressed into a cylindrical shape and placed it on the mould. After drying, removal from the mould and firing, the octagonal stick can be obtained.

(a) (b)

Fig. 25.12 The shape comparison between ancient octagonal stick and the laboratory simulative stick (a) The ancient octagonal stick dated to the Warring States period (b) laboratory simulation of a Chinese-purple octagonal stick

In summary, using the simulation test as a foundation, some inferences about ancient Chinese faience production technology can be drawn. Firstly, our ancestors accidentally discovered blue-purple remains in the bottom of the crucible in the process of smelting of ores. The frit was removed with difficulty, so the craftsmen broke and crushed the frit and discovered the unique bright purple or blue powder. Then the formula of the blue-purple powder was defined and used to make faience beads. By the mid and late stages of the Warring States period, the excellent bronze casting technology was combined with faience moulding technology. So the wide use of octagonal sticks for decoration and burials appeared. This compact solid may have been the replacement for jade which it was believed could not only hold the power to preserve a human body and spirit but also was an elixir for achieving physical immortality.

With the development of the faience production technology, manual forming was followed by mould forming. The microstructure of the objects also changed from loose to dense. Those made during the Warring States, the Qin, and the Han Dynasties using Chinese-blue and purple faience represent a big breakthrough.

25.4 Conclusion

This article has discussed the production technology of ancient Chinese- blue and purple faience during the period from Warring States to the Qin and the Han dynasties. The conclusions are as follows:

(1) After being broken and crushed the purple frit material excavated in the *Henan Sanmenxia* tomb is confirmed as Chinese-purple frit and it was not manually shaped.
(2) The forming of the hollow blue faience bead is mainly achieved by moulding the paste. With the help of perforating tools, compact beads could be obtained after the firing process. Moreover, in order to increase the regularity of the faience beads ancient craftsmen would polish the edges after sintering.

(3) Mould forming technology was used to make ancient faience octagonal sticks, the model having been was made of clay. The wide use of the octagonal sticks reveals the pursuit and the yearning for 'immortality' in ancient China during the Warring States to Qin and Han dynasties.

References

1. S. M. Guan, *Ancient Chinese Glass*, The Chinese University of Hong Kong Art Museum, Hong Kong, 2001, pp. 8–11.
2. R. H. Brill, The Dominick Labino Fund Lecture: Glass and glassmaking in ancient China and some other things from other places, *The Glass Art Society Journal*, 1993, pp. 56–69.
3. E. W. FitzHugh and L. A. Zycherman, *Studies in Conservation* **28**, 15 (1983).
4. E. W. FitzHugh and L. A. Zycherman, *Studies in Conservation* **37**, 145 (1992).
5. Q. L. Ma, R. Glaus, Z. G. Zhang *et al.*, Application of Laser Ablation-Inductively Coupled Plasma-Mass Spectrometry in Ancient Chinese Man-made Copper-Barium-Silicate Faience, in *Scientific and Technological Research on Cultural Heritage*, Chinese academy of cultural heritage, Vol 8 (Science press, Beijing, 2012), pp. 1–15.
6. Q. L. Ma, Z. G. Zhang, D. A. Scott *et al.*, *Journal of National Museum of China* **12**, 116 (2012).
7. Z. G. Zhang, A Study of Artificial Barium Copper Silicate Pigments in Ancient China, PhD thesis, University of Science and Technology Beijing (2011).
8. H. Berke, T. Corbiere and A. Portmann, *WenBo* **6**, 251 (2009).
9. Q. L. Ma, Brief introduction of man-made Chinese blue and Chinese purple in Warring States Period and Han Dynasties, in *Scientific and Technological Research on Cultural Heritage*, Vol. 4 (Science press, Beijing, 2006), pp. 77–79.
10. Z. Liu, A. Mehta, N. Tamura *et al.*, *Journal of Archaeological Science* **34**, 1878 (2007).
11. Y. Xia, S. C. Wu, S. K. Cui *et al.*, *Sciences of conservation and archaeology* **20**, 13 (2008).
12. H. T. Wei, H. Zhao and J. G. Sui, *Cultural Relic (WenWu)* **3**, 4 (2009).
13. E. Kendrick, C. J. Kirk and S. E. Dann, *Dyes and Pigments* **73**, 13 (2007).
14. A. Psbst, *Acta Crystallographica* **12**, 733 (1959).
15. H. G. Wiedemann and G. Bayer, Formation and stability of Chinese barium copper-silicate pigments, in *International Conference on the Conservation of Grotto Sites: Proceedings Conservation of Ancient on the Silk Road*, Dunhuang, China, 3–8 October 1993, ed. N. Agnew (The Getty Conservation Institute, Los Angeles, 1997), pp. 379–387.

16. T. Corbiere, Preparation and Methods of Chracterisation of the Barium Copper Silicates $BaCuSi_4O_{10}$ (Chinese Blue), $BaCuSi_2O_6$ (Chinese Purple), $Ba_2CuSi_2O_7$, $BaCu_2Si_2O_7$ Used as Blue and Purple Pigments, PhD thesis, University of Zurich (2009).
17. G. D. Hatton, A. J. Shortland and M. S. Tite, *Journal of Archaeological Science* **35**, 1591 (2008).
18. L. D. Salvatore, V. Formisano and E. Ciliberto, *Journal of Cultural Heritage* **9**, 113 (2008).
19. M. S. Tite, Y. Maniatis, D. Kavoussanaki *et al.*, *Journal of Archaeological Science* **36**, 370 (2009).
20. Q. L. Ma, A. Portmann, F. Wild *et al.*, *Studies in Conservation* **51**, 81 (2006).
21. S. X. Wang, *WenBo* **2**, 26 (1986).
22. Q. L. Li, *Journal of National Museum of China* **5**, 109 (2012).
23. X. F. Fu, F. X. Gan, *Journal of the Chiness ceramic society* **34**, 427 (2006).
24. F. X. Gan, H. S. Cheng and Q. H. Li, *Science in China Series E Technical science* **37**, 382 (2007).

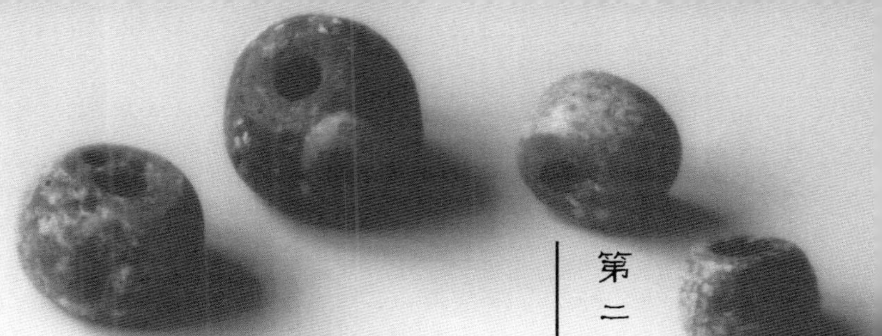

Chapter 26

第二十六章

Production Technology of Faience Beads from Peng State Cemetery, Shanxi Province, China

Zhou Gu,[*,†,¶] *Jian Zhu,*[*,†] *Yaoting Xie,*[‡] *Tiqiao Xiao,*[§] *Changsui Wang*[*,†] *and Yimin Yang*[*,†]

[*]Key Laboratory of Vertebrate Evolution and Human Origins of Chinese Academy of Sciences, Institute of Vertebrate Paleontology and Paleoanthropology, Chinese Academy of Sciences, Beijing 100044, China

[†]Department of Scientific History and Archaeometry, University of Chinese Academy of Sciences, Beijing 100049, China

[‡]Shanxi Provincial Institute of Archaeology, Taiyuan 030001, China

[§]Shanghai Synchrotron Radiation Facility, Shanghai Institute of Applied Physics, Chinese Academy of Sciences, Shanghai 201204, China

[¶]China Printing Museum, Beijing 102600, China

26.1 Introduction

The term faience refers to a ground-quartz body with an alkaline glaze.[1] While it has been discovered in many ancient civilizations, such as ancient Egypt, Mesopotamia, the Indus Valley and China,[2–5] each cultural region developed its own styles of ornaments and production technologies. Compared with the comprehensive study of ancient

Egyptian and Mesopotamia faience, little attention has been paid to Chinese faience. The Chinese faience also called "liaoqi" or "artificial quartz beads" by Chinese archaeologists,[6] suddenly appeared and became popular in the Western Zhou Dynasty (1046BC–771BC). The appearance of Chinese faience is closely related to the prosperity of cultural decoration, and most faience artifacts in the Western Zhou Dynasty are small perforated beads, which were mainly found in Huanghe valley and used for forming fabulous necklaces with different types and shapes of ornaments as a symbol of status and wealth.[7] However, in a necklace the faience beads play a subordinate role. The number of faience beads found in the Central Plain decreased with the great changes in cultural and ritual systems during the Spring and Autumn (770BC–476BC) and Warring States periods (475BC–221BC), and faience disappeared in the Han Dynasty (202BC–221AD). Some scholars think that the production technology of Chinese faience may have been influenced by the West in terms of chemical characteristics[8,9] and the cultural connections between China and the West 3000 years ago.[10–13] Because Chinese faience predates glass, the faience in China has also been considered as the precursor of glass or even protoglass.[14,15] Obviously, studies on early Chinese faience will contribute a lot to further understanding the origin of Chinese glass and the cultural communications between the West and the East.

Previous studies of Chinese faience in the Western Zhou Dynasty have mainly focused on the chemical compositions of glazes. This has contributed a lot to the investigation of the raw materials used in faience production and to the differentiation of faience from other glazed artifacts.[6,16,17] It has been suggested that Chinese faience was formed by the modelling or core forming technologies because of the clay that remained in the perforation.[17] However, it is difficult to judge whether the relic clay has been introduced in the burial environment or not. In fact, the shaping and glazing methods employed in Western Zhou faience production still lack proper scientific investigation. As for the investigation of the production technology of faience, the primary method has been the examination of polished cross-sections of both ancient faience samples and laboratory replicas using a scanning electron microscope (SEM) or electron probe microanalysis with attached analytical facilities: either wavelength dispersive spectrometers (WDS) or an energy dispersive spectrometer (EDS) have therefore been used. Based on the analytical and microstructure data obtained from SEM-EDS or EPMA, raw materials and methods of fabrication employed in ancient western faience production can be deduced, and whether the application, cementation or efflorescence glazing methods were employed in faience production.[1,5,18,19] However, this method required damaging the faience artifacts, and it is very difficult to confirm by SEM which glazing methods were used.[20–23] Recently, Optical Coherence Tomography (OCT) was used to analyze the microstructural features of faience and has provided some useful information which can be used to determine the glazing methods of ancient Egyptian faience.[24] Also, Synchrotron Radiation Micro Computed Tomography (SR-μCT) has been used to distinguish the material types used and to research the detailed manufacturing procedures of faience eye beads in China[25] and as a preliminarily analysis to investigate the forming technology of one Western Zhou faience bead.[26] Compared to SEM, SR-μCT this can provide information about the inner structure of faience beads from different angles in a non-destructive mode.

In this study, SR-μCT was used to infer the manufacturing procedure of faience beads from the Peng State cemetery. Energy dispersive X-ray fluorescence (EDXRF) was used to analyze chemical compositions and determine the glaze recipes. The results obtained from simulation experiments were also produced so as to discuss the production technology of ancient Peng State faience. In the following sections, the chemical compositions of early faience and glass which have been published previously will be combined so as to discuss the relationship between them and to explore the development of Chinese faience.

26.2 Materials and experiment

26.2.1 *Samples*

Sixteen typical faience artifacts were selected from five tombs in the Peng State cemetery located in Hengshui Town, Jiang County, Shanxi Province, China. Peng State cemetery won the honor as one of the top ten archaeological discoveries of 2005 in China. Many faience beads and tubes were discovered at this famous archaeological site. Most of them were found near the tomb occupant's neck or stomach, indicating that the faience beads were parts of necklaces.[27] These beads were firstly observed under an optical stereomicroscope; the exterior blue-green glassy surface was interspersed with many semi-opaque quartz particles. Detailed information about those artifacts is described in Table 26.1 and Fig. 26.1. Their diameters ranges from 4 mm to 12 mm and their lengths vary from 5 mm to 13 mm.

26.2.2 *Determination of structures and chemical compositions*

Since the faience artifacts had to be analyzed in a non-destructive manner, SR-μCT and EDXRF were applied to analyze their structures and chemical compositions.

Table 26.1 Description of some ancient Peng State faience beads

Sample	Description
M2161-1	bead, like a jujube pit; blue outside with a quartz body; length, 8.16 mm; aperture, 2.7 mm; diameter, 5.57 mm.
M2036-1	tube; blue outside; some cinnabar adhered to the inner layer; length, 12.75 mm; aperture, 2.6 mm; diameter, 4.65 mm.
M2002-1	like abacus bead; blue and green outside; length, 11.95 mm; aperture, 2.64 mm; maximum diameter, 11.63 mm.
M2055-2	fragment of a tube with a blue surface; length, 6.59 mm; diameter, 4.71 mm.
M2047-1	like an egg; blue and green outside; aperture, 3.43 mm; diameter, 6.12 mm.
M2047-2	like an egg; aperture, 3.01 mm; diameter, 5.78 mm. It naturally breaks into two parts due to the crack on the body after examination.
M2047-5	short tube; green outside with some remarkable setting marks; some cinnabar adhered to the surface; aperture, 3.67 mm; diameter, 5.1 mm.
M2047-6	like a egg; light green; some cinnabar adhered to the surface; aperture, 3.14 mm; diameter, 5.58 mm.
M2047-8	like an egg; blue outside and adhering some cinnabar; aperture, 3.53 mm; diameter, 5.87 mm.

Fig. 26.1 Photographs of some typical faience objects. (a: M2161-1; b:M2036-1; c: M2002-1; d: M2055-2; e: M2047-1; f:M2047-2; g:M2047-5; h:M2047-6; i: M2055-8. The scale bar is 2 mm.)

All the faience beads were scanned by SR-μCT at the Shanghai Synchrotron Radiation Facility (SSRF), Shanghai, China. During the experiment, the scanned beads were set on an open sample platform. The parallel SR X-ray with the height of 4 mm and the width of 2 cm was directed at the object with the source energy set at 30 keV. The CCD detector has a spacial resolution of 9 μm. The scan time was about 10 minutes. Scan data were imaged and analyzed using Mimics 12 (Materialise, Belgium). Micro-CT produces a slice of an object according to the differences in the absorption of X-rays; the variation in brightness in sectioned sample could is a reflection of the variation of densities and chemical compositions.[28]

Prior to the EDXRF analysis, the beads were cleaned in an ultrasonic tank using deionized water to remove attached soil. Six typical faience artifacts were selected for compositional analysis. Chemical analysis was performed on an Eagle III μ-Probe EDXRF spectrometer (EDAX, USA) with a Mo tube and a 125 μm Be window, which had been calibrated using appropriate ceramic standards. The detector is a liquid-nitrogen-cooled Si (Li) crystal with the resolution of about 160.3 eV at Mn K alpha. The operating conditions of the X-ray tube include an accelerating voltage of 50 kV and a current of 800 μA. The X-ray beam spot was 0.3 mm in diameter. The incident angle of the spectrometer was 65° and the emergent angle was 60°. VISION32 software with all the basic EDX functions was used to analyze the spectra obtained. The concentrations of the analyzed major and some minor elements were above the detection limit of 20 ppm. The relative errors were 1–3% for elements present at 1 wt% or above, and up to 10% for elements present at 0.1 wt% or below. All the samples were tested in a vacuum atmosphere. The results of the chemical analyses of six representative faience beads are listed in Table 26.2 with some previously published data on faience glaze from the Western Zhou Dynasty and glass of the Warring States period.

Table 26.2 The chemical composition of faience glaze and glass (wt %)

Sample	SiO_2	K_2O	Na_2O	CaO	CuO	Fe_2O_3	Al_2O_3	P_2O_5	MgO
M2161-1	82.97	1.37	1.88	4.48	2.63	0.62	1.85	0.31	3.06
M2036-1	94.16	0.43	1.29	0.09	1.84	0.14	1.26	0.21	0.41
M2002-1	91.23	0.86	1.22	0.28	2.59	0.29	1.76	0.31	0.74
M2055-2	90.05	2.02	1.17	0.28	1.61	0.17	2.93	0.35	1.18
M2047-1	88.08	1.07	1.55	0.79	1.47	0.18	5.46	0.29	0.84
M2047-2	90.87	0.6	2.51	0.33	1.76	0.18	2.36	0.29	0.9
G6-Fengxi[6]	94.0	0.3	0.3	0.4	0.8	0.4	0.7	—	0.2
Luoyang[6]	>90	3.4	1.2	0.4	1.6	0.2	0.3	—	0.3
HNWII 67-Ying State[14]	92.98	0.51	1.38	0.30	2.06	0.31	1.26	0.47	0.68
Jiangling[2]	71.63	10.92		2.95		1.99	6.53		

26.2.3 *Laboratory replications*

In this study, the results obtained from laboratory replications of application and efflorescence glazing were used as a supplement for interpreting ancient faience production technology. Different raw materials including finely ground quartz, copper oxide, sodium carbonate, saltpeter and calcium carbonate were used to recreate the ancient faience. In order to explore firing temperature information involved in ancient faience production, the replicates were fired at temperatures ranging from 750°C to 950°C for a range of different amounts of fluxes and colourants. Sodium carbonate and saltpeter were used as fluxes separately so as to investigate the relationship between the flux and glazing methods. The firing simulation results are listed in Table 26.3.

For laboratory replication of the application of glazing, firstly a slurry was prepared which consisted of a mixture of quartz (80%–90% in weight percentage), copper oxide (3%), calcium carbonate (3%) and flux- either sodium carbonate or saltpeter (2%–14%). The glaze was then applied either by dipping the faience body in the slurry, or by pouring the slurry onto the body.[18] For the efflorescence glazing method, the faience was made by a mixture of raw materials including quartz, lime, alkali and a colorant: the compositions are similar to the glaze mixture used for application glazing replicates. All the replicates were shaped by hand, and dried at room temperature for 1 day and

Table 26.3 The simulation experiment results (wt%)

Recipe	Temperature °C	Quartz	Saltpeter	Sodium carbonate	Calcium carbonate	Copper oxide	Efflorescence	Application
1	750	80	14	—	3	3	×	—
2	750	80	—	14	3	3	×	—
3	800	80	14	—	3	3	×	×
4	800	80	—	14	3	3	×	×
5	850	87	7	—	3	3	×	√
6	850	90	—	4	3	3	√	√
7	950	80	14	—	3	3	×	√

√ successful; × failed

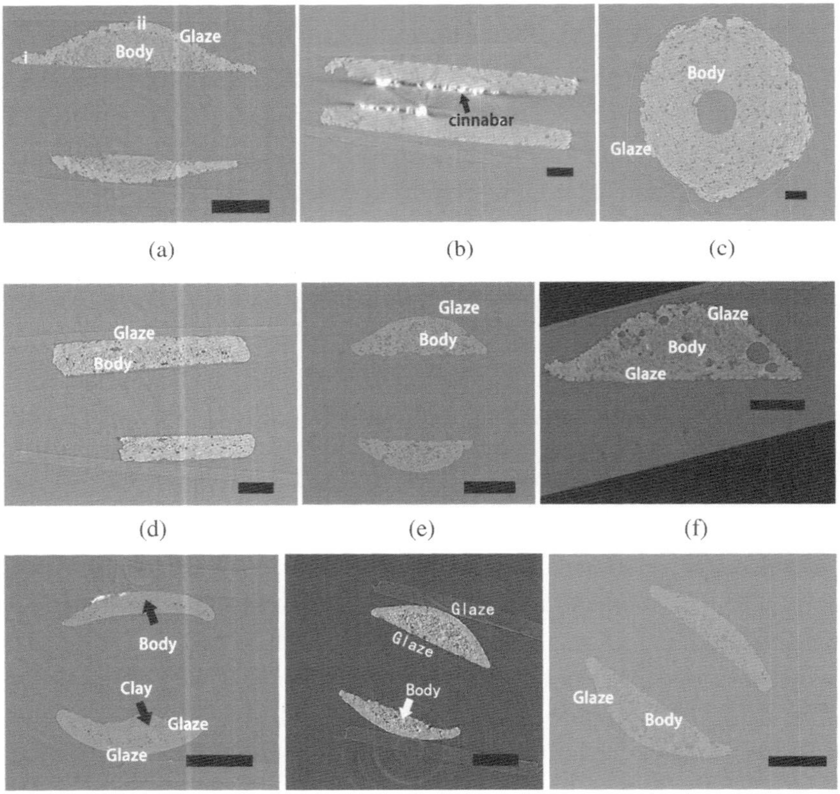

(a) (b) (c) (d) (e) (f) (g) (h) (i)

Fig. 26.2 The CT slices of two different faience types (a: M2161-1; b:M2036-1; c:M2002-1; d: M2055-2; e: M2047-1; f:M2047-2; g:M2047-5; h:M2047-6; i: M2047-8. The scale bar is 2 mm.)

then fired at various temperatures (750°C–950°C) in an electric muffle furnace for up to 5–7 hours.

26.3 Results and discussion

In describing the faience structure, as observed in CT slices, it is convenient to distinguish between the body and the glaze. Because the glaze contained higher amounts of elements such as K, Ca, Fe and Cu (Table 26.2), it has higher X-ray absorption than the quartz body. So glazed areas are brighter than the quartz body in the CT slices and the air pores are black. The clay and cinnabar adhering to the perforation can also be clearly observed in the CT slices. According to structural characteristics revealed by CT slices (Fig. 26.2), the faience beads could be divided into two types. For type A, glazed faience, there is an apparent glaze on the bead surface (Fig. 26.2a, c-i). The vast majority of faience objects belong to type A, and the typical CT slices are shown in Fig. 26.2a, c, d, e, i. Especially, three glazed faience beads (M2047-2, M2047-5 and M2047-6) were glazed on both the outer surface and the inner wall of perforation (Fig. 26.2f, g, h). Type B is glassy faience[1]: the outer layer can not be clearly distinguished from the interior. M2036-1 belongs to type B (Fig. 26.2b).

As shown in Fig. 26.2, the glassy faience bead (Fig. 26.2b) is of a higher density when compared with glazed faience that (air) pores and crevices can be clearly observed in the CT slices. The inner structure of glazed faience artifacts is also distinct. The inner structure of M2161-1, M2002-1, M2055-2, M2047-5 and M2047-8 (Fig. 26.2a, c, d, e, g, h, i) are different from M2047-2 (Fig. 26.2f) in that it contains some big pores. Sample M2047-5, has a body that is much thinner than the glaze (Fig. 26.2g). Micro-CT imaging provides a convenient method to differentiate the glaze, quartz body and pores. Different types of faience can easily be differentiated in a nondestructive manner and the differences of the inner structure can also be clearly shown in CT slices. The structural evidence, obtained by examination of CT slices, provides a very effective approach to investigating the overall inner structure of faience.

26.3.1 *Laboratory replication*

According to Table 26.3, the replicates using recipe 1 do not exhibit an even surface at 750°C and 800°C even though the flux content is more than 14 weight % (Fig. 26.3a). However, at 850°C the replicates using recipe 5 (containing 7% saltpeter) exhibit a shiny blue surface (Fig. 26.3b). The results suggest that the firing temperature of ancient faience is around or above 850°C. More details are listed in Table 26.3.

For objects made with the efflorescence method, the extent of efflorescence was variable for different samples. The replicates which gave the best results were fired using the sodium carbonate (4%) as flux (Fig. 26.4a; Recipe 6 in Table 26.3). However, the replicates of recipe 7 in Table 26.3 using saltpeter (13%) as flux showed no significant efflorescence effect (Fig. 26.4b) and many blue inter-particle glassy phases were found to be interspersed in the body. The different efflorescence effects are related to differences between potassium and sodium oxide levels.[23,29] The results indicate that the efflorescence effect is produced when sodium is used as a flux: potassium could not be used as a flux when the efflorescence glazing method was employed to glaze faience.

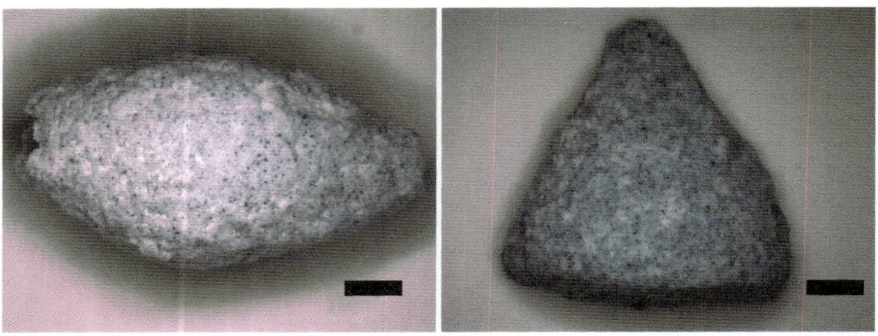

(a) (b)

Fig. 26.3 Photographs of two replicates. (a: the replicates fired at 800°C; b: the replicates fired at 850°C. The scale bar is 2 mm.)

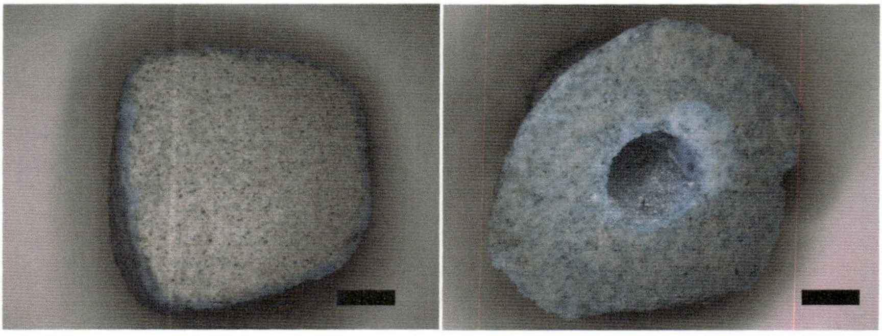

(a) (b)

Fig. 26.4 Photographs of two replicates produced by efflorescence glazing by using different fluxes. (a: sodium carbonate; b: saltpeter. The scale bar is 2 mm.)

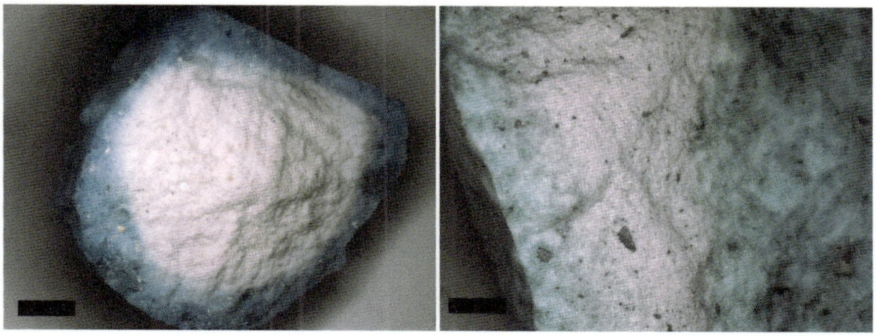

(a) (b)

Fig. 26.5 Photographs of two replicates made using different glazing methods. (a: application glazing method; b: efflorescence glazing method. The scale bar is 2 mm.)

For the application glazing method, the quartz body is formed separately and then coated with a layer of glaze slurry before firing. Hence, the body of replicates is white (Fig. 26.5a). However, for the replicates produced using efflorescence glazing, the body is grey in colour and some black unmelted raw materials can clearly be observed in the body (Fig. 26.5b). The possible explanation for this is that the contents of copper in the body are much higher than the copper contents in the glaze.[23] The copper in the body reacts during firing with flux to form a melt which causes the body to turn grey after firing.

26.3.2 *The manufacturing process*

The analyzed beads in this study have similar shapes and similar glaze recipes (Table 26.2), so it can be deduced that the same manufacturing techniques, including glazing and firing methods were used. As shown in Fig. 26.1 and Fig. 26.2, the longitudinal sections of all faience beads share a similar character and the profile lines of the perforation are straight. The cross-sections of faience beads are also similar since the transverse section of the perforation is like a circle (Fig. 26.2c). On the other hand, the fragments show the perforation is like a cylinder (Fig. 26.1d). Combining the structural information produced by CT slices, the shape of the perforation of unbroken faience artifacts is like a cylinder. Thus, cylindrical inner cores would have been used to support the artifacts during the shaping and firing processes. Furthermore, the inner core of the perforation would have been composed of organic materials; otherwise, it would have been very difficult to remove because the clay would have become sintered together with glaze or quartz particles during firing. The shaping of the quartz body could be described as follows: quartz particles were piled on a cylindrical stick and then glazed.

Although the criteria used to identify the glazing method must be used with caution, the structural evidence, based on the examination of numerous laboratory replications, suggests that the variations in glaze thickness would provide some positive information in determining the possible glazing methods used in the manufacture of ancient faience production.[5,18] The structural information provided by micro-CT can provide possible evidence for the glazing method used in faience bead production.[25] As shown in Fig. 26.2, the glazed faience beads have variable glaze thicknesses (Fig. 26.2a, c–i), and the glaze at the rim (Fig. 26.2a position i) is much thicker than other in other locations (Fig. 26.2a position ii) and this is true of other glazed faience samples. Besides, the glaze on the outer surface of a special glazed faience bead M2047-2 is of a similar thickness to the interior layer. The results suggest that the glazed faience beads would not have been glazed using the efflorescence glazing method because the glaze on the outer surface and at the rims (where drying occurs) are of the same thickness.[18,19] Because the faience was formed on the cylindrical core, either no glaze has covered the interior layer or the glaze thickness of the interior layer is the thinnest: the drying of the interior was much slower than the outer surface during the efflorescence process. As for the application glazing method, the glaze slurry was applied by painting or dipping. The glaze slurry would flow to the rim during the application process, so the glaze was found to be thickest at the rims.[5] During the dipping process, the slurry would enter the perforation. Consequently, the interior surface would be glazed after firing, as M2047-2, M2047-5 and M2047-6 in Fig. 26.2f, g, h. Moreover, the cementation glazing is characterized by a uniform glaze thickness[5] and there is no visible grinding mark on the glaze on those beads using an optical stereomicroscope. As shown in Fig. 26.6, the

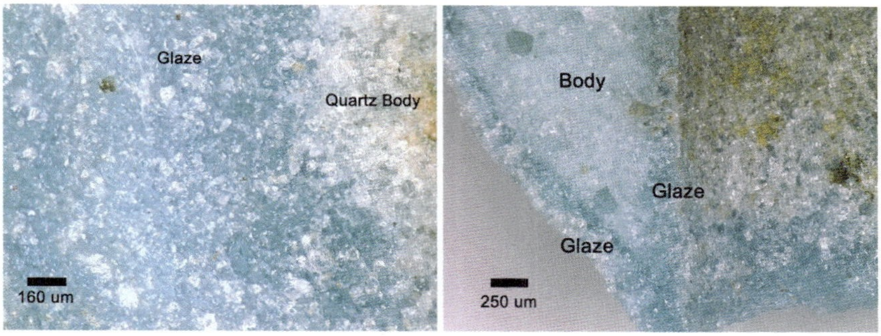

(a) (b)

Fig. 26.6 Magnification of two faience artifacts. (a:M2161-1; b:M2047-2)

body of ancient faience is white — this is consistent with the results obtained from laboratory replications of application glazing. From the above discussion, it is more likely that the faience was glazed using the application method. The possible manufacturing process for these faience beads is that firstly the quartz body was formed on a cylindrical stick and then the glaze was applied by painting the glaze slurry onto the body or dipping the body into the glaze slurry.

The manufacturing process of faience seems to have been influenced by the proto-porcelain production technology. Proto-porcelain production began at least as early as the Xia and Shang Dynasties (c. 2000BC–1046BC).[30] However, there are some differences in the application glazing methods between the analyzed faience beads and proto-porcelain. The proto-porcelain is much larger than faience beads. The edge of porcelain bases is always unglazed because they are in contact with the kiln chamber floor. However, the surfaces of the analyzed faience beads are totally coated with a thin layer of glaze. According to the simulation experiment, when a faience bead was placed directly on a surface before firing, the location where the bead had stood would appear grey because of the lack of contact with air during the firing process. Since the faience beads were shaped on a stick, it is inferred that they were probably fired by suspending them in air: the stick may have been used to support the beads.

26.3.3 *The glaze recipe*

From the chemical data of the faience glaze listed in Table 26.2, the Peng State faience are similar to the glaze recipes of other Chinese early faience beads: they contain high contents of silica (more than 80 wt%) and low contents of flux. The silica derives from quartz. These faience beads are blue-green in color: copper and iron are coloring elements. The copper contents are about 2%, and the iron contents are around 0.2%. When the Fe content is lower than 2 wt% in glaze, it would have no significant influence on the glaze color.[31] Thus, the glaze colour variation is deemed to be caused by different local chemical environments of Cu^{2+} ion in glaze.[32] The faience beads in the Western Zhou Dynasty are the earliest glaze with a copper colorant in China. Some Chinese archaeologists have thought that the source of the copper colorant would be copper slag.[33] It is known that Shang and Western Zhou Dynasty bronzes contain more than 15% Pb and Sn,[34] but very low Pb and Sn contents were detected in these faience glazes. Therefore, bronze scale or copper slag was not used as the source of the colorant. In sum, a copper ore, such as malachite or azurite, and quartz would have formed the glazing mixture.

Scholars have summarized the chemical compositions of faience in the Western Zhou Dynasty (Table 26.4). The chemical compositions of analyzed glazes in this study are not exceptional. The glazes of Egyptian and Near East faience are of the soda-lime-silicate type with soda as the dominant alkali,[5] and the typical glaze compositions are listed in Table 26.4. Compared with the glaze compositions of western

Table 26.4 Typical glaze chemical compositions of Chinese faience and Egyptian faience (wt%)

	Date	SiO_2	Al_2O_3	CaO	MgO	Na_2O	K_2O	CuO	Fe_2O_3
Chinese faience[17]	Western Zhou Dynasty	81–94	1–5			1	1	1–9	
Egyptian faience[5]	2000BC-700BC	65–75			3–7	15–20			

faience, Chinese faience only contained 2–3% soda-plus-potash (Na_2O+K_2O), the potassium usually being the dominant alkali.[17,35] Obviously, the glaze recipe of most of Western Zhou faience in China is distinct from faience in Egypt and Near East. Therefore: Western Zhou Dynasty faience beads were not imported from the West but were indigenous Chinese products.

Western scholars have suggested that faience was a precursor for glass based on the following evidence: (1) the faience predates glass; (2) faience glaze and glass have similar recipes (Na_2O-CaO-SiO_2); (3) Moreover, some faience and early glass were discovered in the same workshop.[3] However, the situation in China is different. Glass production in China started from the late period of the Spring and Autumn and Warring States Period (476BC–221BC). Three types of glass have been found during this period, including Na_2O-CaO-SiO_2 glass (imported from the west), PbO-BaO-SiO_2 glass (Chinese-made, the commonest type), and K_2O-CaO-SiO_2 glass (Chinese-made).[2] Furthermore, as shown in Table 26.2, the K_2O-CaO-SiO_2 glass (The Jiangling sample in Table 26.2 of the, Warring States period) has a much higher flux content than the glaze recipe of Western Zhou faience. Thus, although the faience also predates glass in China, the glaze recipe of Western Zhou faience is distinct from that of PbO-BaO-SiO_2 and K_2O-CaO-SiO_2 glass in the later period. Besides, the early Chinese glass objects in Warring States period were produced by casting in moulds.[2] Therefore, Western Zhou faience has no direct technical relationship with early indigenous glass in China. On the other hand, the Warring States craftsmen also produced faience eye beads using mould casting techniques, and the glaze recipe is similar to K_2O-CaO-SiO_2 glass recipes.[25] Also faience eye beads and K_2O-CaO-SiO_2 glass were mainly discovered in Chu cultural regions, which imply that the K_2O-CaO-SiO_2 glass had a close relationship with faience production during the Warring States period (476BC–221BC). In the future, more emphasis should be placed on the study of faience and glass in the Spring and Autumn Period (770BC–476BC) to understand the development of faience in China.

26.4 Conclusions

The appearance of faience indicates great progress in Western Zhou crafts. It was the first kind of artifact with a glazed quartz body, and the earliest glaze with a copper colorant in China. The analysis of the faience objects derived from the Peng state cemetery, China has suggested that most faience was possibly glazed using the application method. This has enriched our understanding of the manufacturing technology of Chinese faience. The faience can be divided into two types: glazed faience and glassy faience. Chinese glassy faience has been discussed for the first time in this paper. The possible sources of copper colorants used in it were copper ores. By using CT slices, the possible manufacturing process of faience was revealed as a quartz body formed on a cylindrical stick which was then glazed by applying glaze, by painting the glaze slurry onto the body or dipping the body in the glaze slurry. This technology maybe has been influenced by Chinese proto-porcelain production. However, the faience was possibly fired by suspending the bead in the air. The firing temperature would have been around 850°C, or above.

With the start of faience production in the Early Western Zhou Dynasty, most Chinese ancient faience contained a low flux content. This distinguishes it from Western Asian and Egyptian faience both of which commonly have levels of high sodium and low potassium oxides. The research on the production technology of faience shows that most Chinese Western Zhou faience has an indigenous origin in China. Thus, the spread of faience from the West to the East could be as a result of transferring the idea, but not as a result of artifact exchange. Furthermore, Western Zhou faience had no direct technical relationship with the emergence of the earliest Chinese PbO-BaO-SiO_2 glasses.

The structural data obtained by SR-μCT has provided valuable information on the glazing methods used. It has revealed the overall structural features of ancient faience beads in a non-destructive manner, and has clearly shown the differences between inner structures of different objects. The present study has shown that SR-μCT provides a new and effective approach to the study of manufacturing technology of ancient faience.

Acknowledgements

The authors would like to acknowledge Prof. Feng Songling and Dr. Li Li for the experiments with EDXRF. This work was supported by the National Natural Science Foundation of China (40802002) and CAS Strategic Priority Research Program (Grant no. XDA05130303). Also, it was sponsored by the Scientific Research Foundation for the Returned Overseas Chinese Scholars, State Education Ministry and in part by the President Fund of the UCAS. We specially thank Professor J. M. Kenoyer for valuable advice and suggestions.

References

1. M. S. Tite and M. Bimson, *Archaeometry* **28**, 69 (1986).
2. F. X. Gan (eds.), *Technology Development of Chinese Ancient Glass* (Shanghai Science & Technology Press, Shanghai, 2005).
3. P. R. S. Moorey, *Ancient Mesopotamian Materials and Industries: The Archaeological Evidence* (Oxford University Press, Oxford, 1999), pp. 167–189.
4. J. M. Kenoyer, *Ornament* **17**, 36 (1994).
5. M. S. Tite and A. J. Shortland (eds.), *Production technology of faience and related early vitreous materials* (Oxford University School of Archaeology Monograph, Oxford, 2008).
6. F. K. Zhang, Z. H. Cheng and Z. G. Zhang, *Journal of the Chinese Ceramic Society* **11**, 67 (1983).
7. J. Q. Dong, D. H. Gu and F. X. Gan, 2009. The origin and characteristic Chinese ancient faience, in *Proc. 7th International Symposium on Ancient Ceramics*, Shanghai, China, 2009, eds. H. J. Luo and X. M. Zheng (Shanghai Science & Technology Press, Shanghai), pp. 735–743.
8. F. X. Gan, *Journal of Guangxi University for Nationalities (Natural Science Edition)* **15**, 6 (2009).
9. Z. G. Zhang and Q. L. Ma, Faience beads of the Western Zhou Dynasty excavated in Gansu province, China: A technical study, in *Ancient Glass Research along the Silk Road World*, eds. F. X. Gan, R. H. Brill and S. Y. Tian (Scientific Publishing Co. Pte. Ltd., Singapore, 2009), pp. 275–289.
10. K. Chen and F. T. Hiebert, *Journal of World Prehistory* **9**, 243 (1995).
11. S. C. Li and T. Shui, *Cultural Relics* **3**, 36 (2000).
12. M. C. Lin, *The Prevalence of Western Custom in Ancient and Medieval China* (SDX Joint Publishing Co., Beijing, 2000).
13. J. J. Mei, *Bulletin of Archaeology, The University of Kanazawa* **27**, 109 (2004).

14. X. F. Fu and F. X. Gan, *Journal of the Chinese Ceramic Society* **34**, 427 (2006).
15. Q. L. Li, *Journal of National Museum of China* **106**, 109 (2012).
16. F. X. Gan, Y. Q. Hu, J. Q. Dong *et al.*, *Journal of the Chinese Ceramic Society* **37**, 1005 (2009).
17. Q. H. Li, J. Q. Dong, F. X. Gan, *et al.*, *Journal of Guangxi University for Nationalities (Natural Science Edition)* **15**, 31 (2009).
18. A. Kaczmarczyk and R. E. M. Hedges, *Ancient Egyptian faience: an analytical survey of Egyptian faience from Predynastic to Roman times* (Aris and Phillips, Warminster, 1983).
19. M. S. Tite, I. C. Freestone and M. Bimson, *Archaeometry* **25**, 17 (1983).
20. G. D. Hatton, A. J. Shortland and M. S. Tite, *Journal of Archaeological Science* **35**, 1591 (2008).
21. P. T. Nicholson, Egyptian Faience and Glass (Shire Publication, Buckinghamshire, 1993).
22. P. B. Vandiver, A review and proposal of new criteria for production technologies of Egyptian faience, in *La Couleur dans le Peinture et l'Emaillage de 'Egypte Ancienne*, eds. S. Colinart, and M. Menu (Edipuglia, Bari, 1998), pp. 121–139.
23. M. S. Tite, P. Manti and A. J. Shortland, *Journal of Archaeological Science* **34**, 1568 (2007).
24. H. D. Liang, M. Sax, D. Saunders *et al.*, *Journal of Archaeological Science* **39**, 3683 (2012).
25. Y. M. Yang, L. H. Wang, S. Y. Wei *et al.*, *Microscopy and Microanalysis* **19**, 335 (2013).
26. Z. Gu, Y. T. Xie, Y. M. Yang *et al.*, *Nuclear Techniques* **35**, 265 (2012).
27. J. Z. Song, K. Z. Ji, J. W. Tian *et al.*, *Cultural Relics* **8**, 4 (2006).
28. J. Baruchel, J. Y. Buffiere, P. Cloetens *et al.*, *Scripta Materialia* **55**, 41 (2006).
29. M. K. Misra, K. W. Ragland and A. J. Baker, *Biomass and Bioenergy* **4**, 103 (1993).
30. C. S. Wang, W. J. Li and Y. Chen, *Archaeology* **9**, 86 (2014).
31. Y. M. Yang, Archaeometry research about Ru porcelain and imperial blue and white porcelain, PhD thesis, University of Science and Technology of China (2005).
32. W. T. Hao, Y. M. Yang, J. Zhu *et al.*, *Journal of Electron Spectroscopy and Related Phenomena* **196**, 133 (2014).
33. B. D. Yang, *Palace Museum Journal* **2**, 14 (1980).
34. R. F. Han and J. Ke, *A history of science and technology in China: mining and metallurgy* (Science Press, Beijing, 2007), pp. 229–237.
35. R. H. Brill, S. S. C. Tong and F. K. Zhang, *Journal of Glass Studies* **31**, 11 (1989).

Chapter 27

第二十七章

Scientific Analysis of Natural Glasses

Fuxi Gan Hongxia Zhao and Zhenqi Xu
Shanghai Institute of Optics and Fine Mechanics,
Chinese Academy of Sciences, Shanghai 201800, China

27.1 Introduction

Non-artificial glassy materials from nature include mainly the following two kinds of glasses. One is volcanic glass called Obsidian — a natural mineral. It is formed by the quenching of the magma erupted in a volcanic explosion. The other is meteorite glass called Tektite, also a natural mineral. It is formed by the melting and cooling of fire meteors exploded in outer space, and then falling as glassy meteoric rain, or by the melting, falling and impacting of meteoric with the earth and the followed splashing.

Thin slices of natural glasses were used as tools or weapons since Paleolitic period, and as ornaments or sacrificial vessels in the Neolithic time. The discovery of natural glasses, such as the chest ornament of the Pharaoh in ancient Egypt for example, is more than one thousand years earlier than the advent of man-made glasses. In archaeology, this is considered the evidence for the migration of ancient mankind, and the material and cultural exchange among them.[1]

The black natural gemstone called *leigongmo* in southern China is used to ward off evil. It is extremely precious and collected as a kind of very rare gemstone. It is actually Tektite.

With the development in research on the formation of glassy materials under special conditions, such as high-speed shock and nuclear explosion, a high surge in the study of natural glasses rose in the 1980's. These are closely related to the study of the particular substances formed under the condition of nuclear explosion, as well as in the extraterrestrial geological situation.

There are certain special requirements for the study of natural glasses. First, samples used for analysis should be from different parts of the world; then, since the samples are rare and precious, they should not be damaged, i.e., nondestructive analysis should be conducted using certain special equipment; finally, cooperation with experts in the geological and astronomical fields are necessary.

The inner structure of Obsidian and Tektite is all in glassy state, so both of them are called natural glass. They have been regarded as precious stones for appraisal and collection. The most Tektites are in shapes of balls, dumbbells, buttons and irregular forms, while the surface of Obsidian is relatively smooth. Therefore, Obsidian and Tektite can be distinguished from their appearances in certain extent. But it is not accurate to classify them mainly based on their appearance.

Recently, some researchers used modern techniques, such as laser ablation plasma mass spectroscopy, proton induced x-ray emission, neutron activating analysis and Raman spectroscope etc. to measure the composition, trace element and inner-structure of natural glass. They hope to get some information of their inner-structure and provenance, but the work was not systematic, and uses methods of destructive analysis. Because the high quality natural glasses are rare and valuable, the destructive analysis method sometimes is not allowed. In this work, several non-destructive analysis techniques, such as the x-ray diffraction (XRD), x-ray fluorescence (XRF) and laser Raman spectroscope etc. are used to measure the chemical composition, structure and phase of 15 natural glass samples, and to obtain related information. It is a pilot experiment of the non-destructive analysis method applied in natural glass research.

27.2 Experimental

27.2.1 *Methods and equipment*

(1) Determination of the chemical composition (main composition, impurities and trace elements) of the sample by modified portable X-ray excited emission spectrometer (pXRF), OURST EX100FA.[2]
(2) Determination of the mineral phases of the sample by X-ray Diffractometer (XRD), Rigaku ultima IV.
(3) Characterization of the chemical structure of the sample by portable Laser Raman Spectrometer (pLRS), Lab RAM Xplora.
(4) Micro area compositional analysis and surface morphology analysis using SEM-EDS and TEM.
(5) Micro-, trace and isotope analyses using Proton induced X-ray Emission Spectroscope (PIXE) and Atomic Emission spectrometer (AES).

27.2.2 *Samples*

Figure 27.1 shows the photographs and series number of the samples whose origin and appearance characteristic are given in Table 27.1. Tektite and Obsidian are different in physical shapes. The shapes of the Tektites include globular, dumbbell, droplet, button shaped and

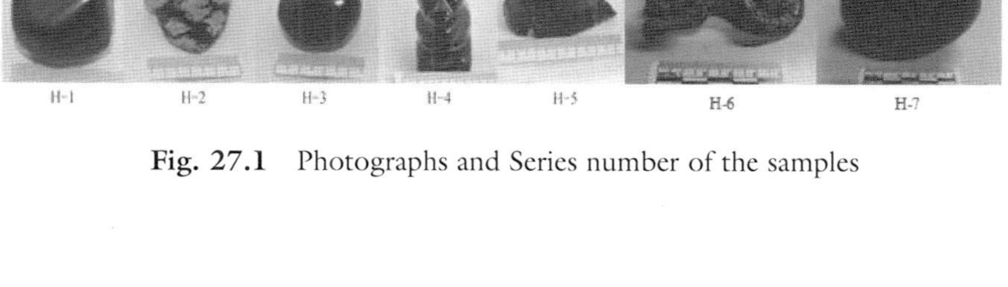

Fig. 27.1 Photographs and Series number of the samples

Table 27.1 Origin and appearance characteristics of the samples

Sample No.	Given name	Locality	Description
B–1	Tektite	Guangxi	4 cm in length and 2 cm in width, have two smooth surfaces, but many shallow pits around
B–2	Tektite	Guangdong	4 cm in length and 2 cm in width, curving and rough surface, many corrosion pits
B–3	Tektite	Australia	5 cm in length and 0.5 cm in diameter, dumbbell like, rough surface with many grooves
B–4	Tektite	Hainan	3 cm in length and 2 cm in width, step shape, rough surface with corrosion pits
B–5	Tektite	Australia	6 cm in length, ellipsoid with rough surface and corrosion pits
B–6	Tektite	Australia	8 cm in length, dumbbell like, rough surface with many grooves and gouges
B-7	Tektite	Hainan	6 cm in length,rough surface with many grooves and gouges
B-8	Tektite	Hainan	8 cm in length,rough surface with many grooves and gouges
B–9	Tektite	Romania	8 cm in length and 5 cm in width, curving and rough surface
B–10	Tektite	Hainan	9 cm in length and 3 cm in diameter, rough surface with many grooves
H–1	Obsidian	Mexico	3 cm in length, 2 cm in width, have smooth surface but with irregular shape
H–2	Obsidian	Mexico	3 cm in length, black matrix with white snowflakes on it, have smooth surface but with irregular shape
H–3	Obsidian	Hawaii	3 cm in length, 2 cm in width, have smooth surface but with irregular shape
H–4	Obsidian	Mexico	10 cm in height, 3.5 cm in width, a handcraft
H–5	Obsidian	Australia	6 cm in length, 2 cm in width, like a arrow
H–6	Obsidian	Mexico	6 cm in height, 3 cm in width, a cat
H–7	Obsidian	Wudalianchi	7 cm in height, 5 cm in width, an egg

some irregular lumps, similar to those of the melting condition, and their sizes are small. The Obsidians have more smooth surfaces and larger dimensions than the Tektites so that they can be processed and carved into works of art relatively easy. The difference in appearance between Tektite and Obsidian is not enough to accurately distinguish them from each other. Therefore it is necessary to distinguish the two natural glasses in terms of chemical composition.

27.3 Results and discussion

27.3.1 *Differentiation of natural glasses by their texture*

Natural glasses are non-crystalline while most of jades are poly-crystalline. There are no sharp peaks in the laser Raman (LR) and X-ray diffraction (XRD) spectra of a non-crystalline substance because of their long-range disorder in the arrangement of atoms and ions, as is shown by the samples B-2 and H-2 in Fig. 27.2. On the contrary, there are obvious and sharp peaks in the spectra of jades as is shown by the samples B-1 and H-3 in Fig. 27.3. In this case, the mineral phases can be determined according to the location of the peaks. The

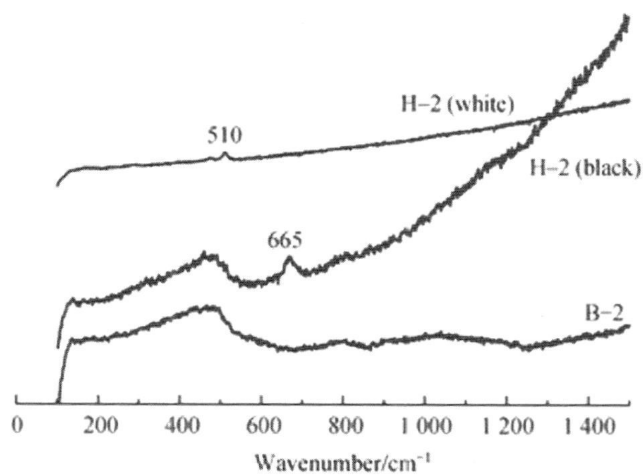

Fig. 27.2 Raman spectra of samples B-2 (Tektite) and H-2 (Obsidian)

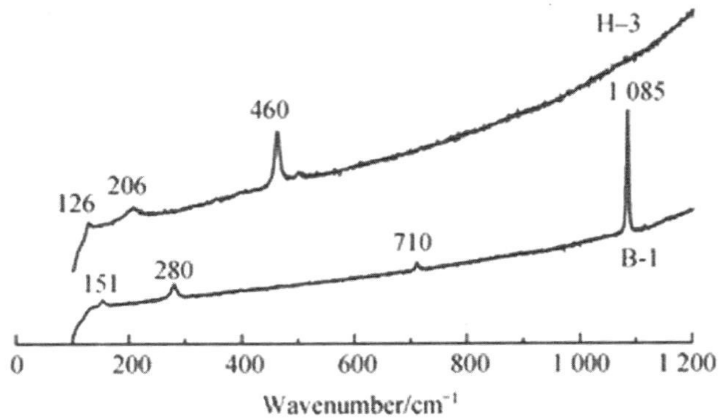

Fig. 27.3 Raman spectra of (B-1) (Calcite) and H-3 (Flint)

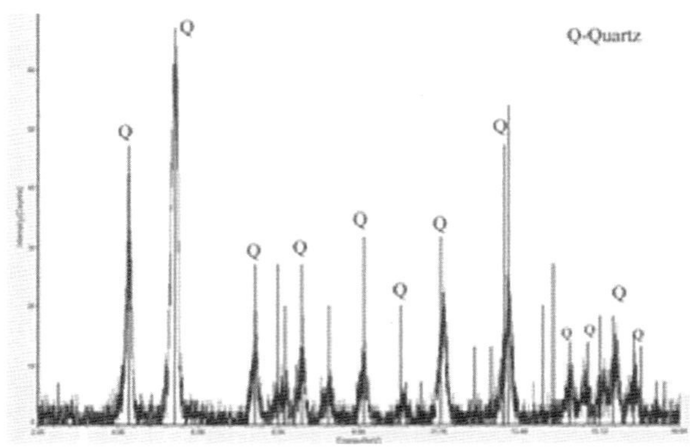

Fig. 27.4 XRD pattern of the H-3 (Flint)

sample B-1, named as Guangxi Tektite, is actually a calcite mineral containing quartz as impurity, while the sample H-3, named as Australia Obsidian, is actually a Flint mineral which is a kind of quartzite microcrystal as is shown in Fig. 27.4.

Natural glasses were formed very rapidly at high temperature in the course of the violent change of geological environment, and their vitreous phase is homogeneous. Figure 27.5 depicts the lattice images of Obsidian and Tektite obtained by high power TEM. Since the red mark-bar in the figure is 2 nm, it can be concluded that the size of the orderly arranged lattice is less than 1 nm , no crystalline

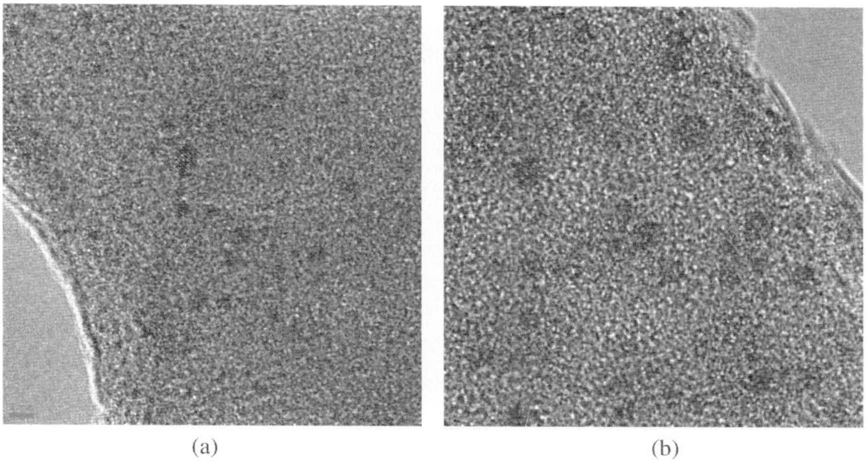

Fig. 27.5 Lattice image of Obsidian and Tektite. (a) Obsidian: H-2, (b) Tektite: B-2

micro-particles exist in the samples and the glassy phase is homogeneous.

27.3.2 *pXRF analytical results*

Tables 27.2 and 27.3 give the chemical composition of the natural glasses obtained using the pXRF method. It can be clearly seen in the two figures that except B-1 (calcite) and H-3 (flint), the other Tektite and Obsidian samples are all aluminosilicate glasses. However, there are obvious difference in chemical composition of the two natural glasses. The Tektites contain less amount of alkaline metal oxides (Na_2O+K_2O) and greater amount of alkaline-earth metal oxides (MgO+CaO) than the Obsidians, and the content of ferric oxide (FeO) in the Tektites is much more greater than in the Obsidians. Thus it is possible to distinguish the two natural glasses from each other in terms of chemical composition.[3] PXRF is a fast and nondestructive analytical method and is especially suitable for the chemical analysis of natural glasses, which have similar structure. Therefore, it possesses incomparable advantage over the other methods used for the analysis of chemical composition of gemstones.

Table 27.2 pXRF analytical results of the Obsidian samples (wt. %)

Sample	Na_2O	K_2O	MgO	CaO	Al_2O_3	FeO	SiO_2	TiO_2	MnO_2	BaO	P_2O_5	R_2O/RO
H-1	4.92	5.25	0.55	0.42	14.08	1.27	72.65	0	0.07	0.32	0.37	10.48
H-2(white)	3.65	6.87	0.49	0.72	13.97	1	72.28	0	0.09	0.31	0.56	8.69
H-2(black)	6.70	2.60	0.40	0.53	13.89	1.14	73.51	0	0.09	0.29	0.77	10.00
H-3	0.95	0.29	0.17	0	1.21	0	95.3	0	0	0.15	1.77	—
H-4	5.29	4.96	0.34	0	11.55	3.55	73.22	0.18	0.11	0.12	0.42	30.15
H-5	5.21	5.26	0.43	0.06	12.58	2.50	72.93	0.14	0.07	0.26	0.32	21.37
H-6(black)	4.16	6.66	0.84	0.74	16.52	0.94	69.43	0.06	0.08	0	0.57	6.85
H-6(white)	5.59	4.15	0.71	0.66	16.83	0.98	70.43	0.04	0.08	0	0.54	7.11
H-7	6.06	4.91	0.63	n.d.	14.30	2.93	70.08	0.27	0.19	0	0.64	17.41

Table 27.3 pXRF analytical results of the Tektite samples (wt. %)

Sample	Na_2O	K_2O	MgO	CaO	Al_2O_3	FeO	SiO_2	TiO_2	MnO_2	BaO	P_2O5	R_2O/RO
B-1	1.35	0	0.82	76.35	0	0.05	21.04	0	0	0.16	0	—
B-2	1.33	2.65	2.29	2.12	16.62	6.51	68.00	0	0.14	0.26	0	0.90
B-3	1.45	2.59	3.42	1.85	11.96	7.18	71.05	0	0.14	0.23	0	0.77
B-4	1.08	2.64	3.44	1.80	12.15	6.90	70.16	1.10	0.14	0.23	0	0.71
B-5	1.48	2.63	2.42	3.00	12.84	6.12	70.89	1.11	0.13	0.20	0	0.76
B-6	1.02	2.41	2.19	2.10	11.24	5.89	73.35	1.07	0.14	0.18	0	0.80
B-7	1.18	2.73	3.05	2.12	13.37	6.31	69.56	1.17	0.12	0.13	0.24	0.76
B-8	1.13	2.55	3.08	2.24	12.94	6.17	70.35	1.09	0.12	0.30	0.03	0.69
B-9	0.86	1.29	1.89	3.89	20.69	19.97	49.80	1.34	0.27	0	n.d.	0.37
B-10	1.78	2.41	2.62	2.20	14.72	5.95	69.11	1.06	0.12	0	0.03	0.87

27.3.3 *Characteristics of volcanic glass (Obsidian)*

All lavas of volcano eruptions could not form glasses. The erupted lavas sometimes are ball-shaped and called volcanic eggs. More than 2000 volcanic eggs have been unearthed in the Tengchong volcano area (China). The volcano eggs in the Yichun geological park (China) are dark red and most of them are porous as is shown in Fig. 27.6.

According to the literature, there are two kinds of volcanic glasses: basalt and rhyolite. The samples in Fig. 27.1 are from Mexico, Australia and Hawaii, respectively. Their chemical compositions are similar, all belonging to basalt.

The author once investigated the craters in Mount Fuji and Sendai in Japan and Wudalianchi in Heilongjiang province, China. No volcanic glasses but volcano eggs were found in most cases (See Fig. 27.6).

Due to the difference in cooling temperature, sometimes snowflake-like polycrystalline materials are formed on the surface of volcanic glasses. They are Albite according to the analytical results shown in Figs. 27.7–27.9.

Fig. 27.6 Porous volcano eggs in Wudalianchi of Heilongjiang province, China

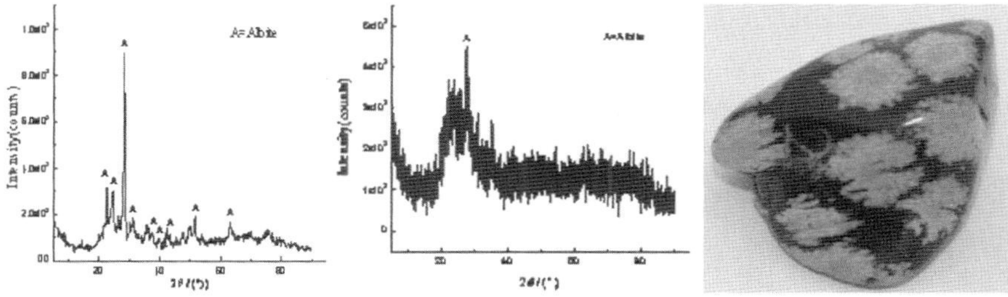

(a) (b)

Fig. 27.7 XRD patterns of sample H-2 (Obsidian). (a) the white area, (b) the black area. The white snowflakes are Albite and the black area is glassy.

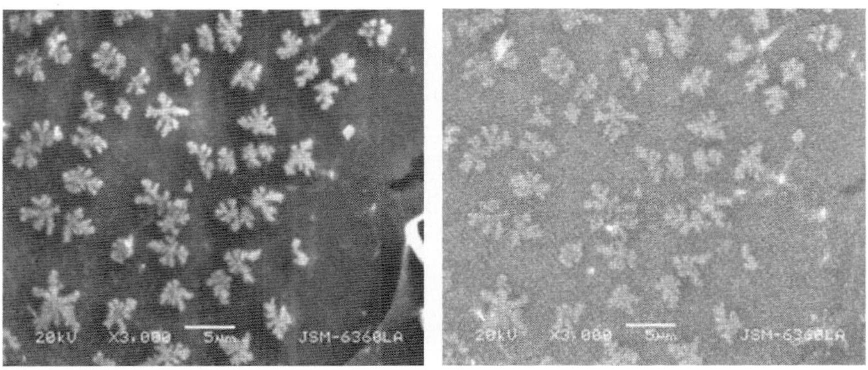

Fig. 27.8 SEM photographs of Sample H-2(Obsidian). (a) secondary electron microscopy, (b) back-scattering electron microscopy.

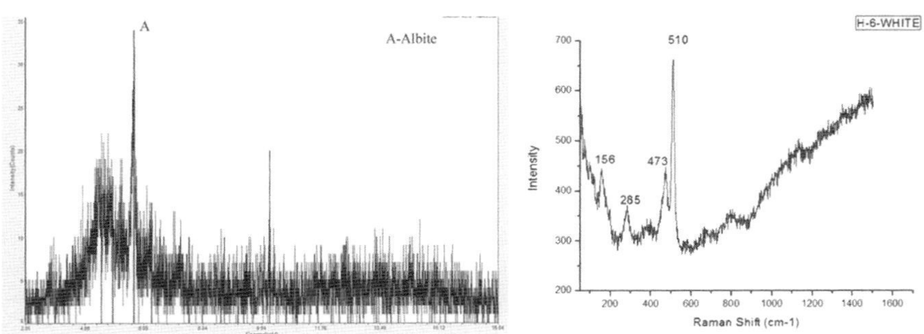

Fig. 27.9 The white area of sample H-6. (a) XRD patterns, (b) Raman spectrum.

The size of the snowflakes on the surface of sample H-2 is about 2.5μm. The analytical results of the snowflakes by EDS are in basic agreement with those by pXRF. The EM-EDS analytical results show that the principal components of the snowflakes are 75.95% SiO_2, 15.18% Al_2O_3 and 5.36% Na_2O, suggesting that the snowflakes are albite.

27.3.4 *Characteristics of Meteorite glass (Tektite)*

Since meteorites rains are rare and most of them felt down into the sea, meteorite glass, as a kind of natural glasses, is very scarce. Figure 27.10 shows the Tektite strewn fields in the world based on the large meteorites rains happened in the past million years. There are North America, Ivory Coast, Czechoslovakian, and Australasian field. China is in the upper side of the Australasian field, meteorites glass have been found mainly in Hainan island, Leizhou peninsular, southern parts of Guangdong and Guangxi provinces.

The author has collected Tektite samples from Mexico, Australia, Hainan island, Guangdong and Guangxi provinces. Their pXRF results are given in Table 27.2.

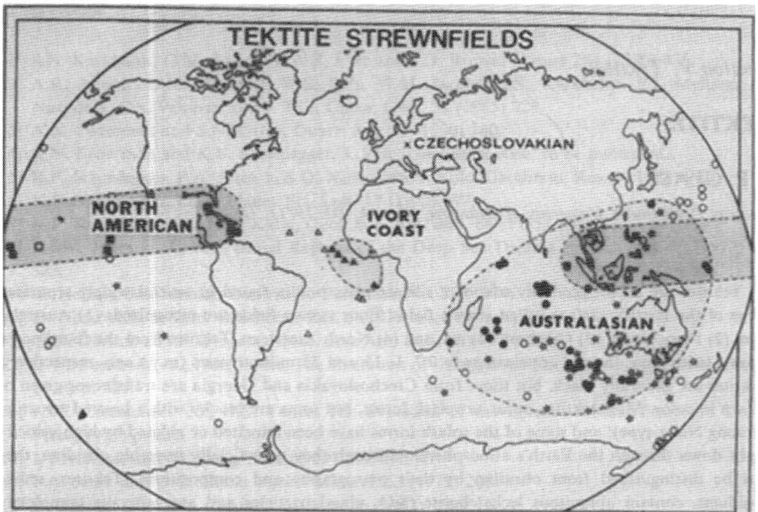

Fig. 27.10 Tektite Strewn fields

Table 27.4 Major compositions of Tektites collected from Guilin, Guangdong and Hainan areas[4]

	Guilin			Hainan			Guangdong		
	1	2	3	1	2	3	1	2	3
SiO_2	74.45	74.88	74.93	75.99	75.62	75.07	73.58	73.16	73.71
Al_2O_3	13.15	12.88	13.20	11.83	12.17	11.83	11.97	11.85	12.07
MgO	1.70	1.69	1.68	1.64	1.65	1.61	2.19	2.18	2.20
FeO	4.54	4.38	4.28	3.84	4.07	4.51	4.74	4.64	4.52
CaO	1.90	1.81	1.72	1.79	1.70	1.88	1.67	1.66	1.65
Na_2O	0.92	0.96	0.91	1.11	1.09	1.13	1.09	1.05	1.03
K_2O	2.69	2.67	2.70	2.95	3.00	2.85	2.61	2.48	2.49
MnO	0.10	0.01	0.06	0.09	0.06	0.05	0.16	0.05	0.04
TiO_2	0.67	0.71	0.68	0.06	0.59	0.57	0.64	0.70	0.54
Cr_2O_3	0.03	0.05	—	—	—	—	0.06	0.03	0.02
Total	100.2	100.0	100.2	99.84	99.95	99.50	98.71	97.8	98.27
$(FeO+MgO)/(Na_2O+K_2O)$	1.73	1.67	1.65	1.35	1.40	1.54	1.87	1.93	1.91

Recently, Lin Su, Guan Yunbin and Hsu Weibiao reported analytical results of the major element compositions of 9 Tektites, as shown in Table 27.4.[4] Their results of the major elements and coloring oxides contents are similar to ours. Based on the analytical results of Tektites from the different strewn fields in the world, as shown in Table 27.5, it is concluded that Tektites found in China belong to the Australasian field.

There have long been debates on the source material of Tektites. One opinion thinks that when an alien entered the atmosphere layer, it exploded and split into small meteorites and melted at a high temperature, and then cooled and fell down onto the different fields of the earth, so the Tektite has the same composition as the meteorite. Another opinion considers that Tektites are the cooled products of melted rocks and soil due to the strong impact by the falling meteorites. There is also a third opinion, but it is actually a mixture of the above two opinions. At present, the first opinion is preferred by most

Table 27.5 Average composition of Tektite samples in different strewn field

Oxide	Australasia (32)	Ivory Coast (7)	Czechoslovakian (33)	North American (21)
SiO_2	73.6	68.02	79.11	76.37
TiO_2	0.68	0.57	0.35	0.76
Al_2O_3	12.23	16.30	10.11	13.78
Fe_2O_3	0.60	0.57	0.19	0.19
FeO	4.14	5.99	1.34	3.81
MgO	2.04	3.32	1.91	0.63
CaO	3.38	1.12	2.64	0.65
Na_2O	1.27	2.06	0.52	1.54
K_2O	2.20	1.89	2.15	2.08
H_2O	(< 0.02)	(< 0.02)	(< 0.02)	(< 0.02)

of the researchers in view of the fact that the chemical compositions of Tektites from the various strewn fields are fairly consistent, and the glass structure of Tektites is reasonably homogeneous. However, further research and experimental results are surely necessary.

27.3.5 *Libyan Desert Glass (LDG)*

Talking about natural glass has no alternative but to involve Libyan Desert Glass (See Fig. 27.11).

The strewn field of LDG is located in the west Egypt desert. Glasses of high SiO_2 content (>90%) have been found in the 100 square kilometers area. They are light green and transparent, and weigh up to 10 kilograms. At present, there are still on sale on the jewelry market.

There have been many investigations on LDG since the end of the 20th century, especially in the 80–90s of the last century. The dominant knowledge about LDG tells that this glass was formed due to the melting, falling and impacting of meteorites with the earth and the splashing followed. Meteorite impact craters were found 150 kilometers away from where the LDG was discovered, but no quartzite mineral has been found.[5]

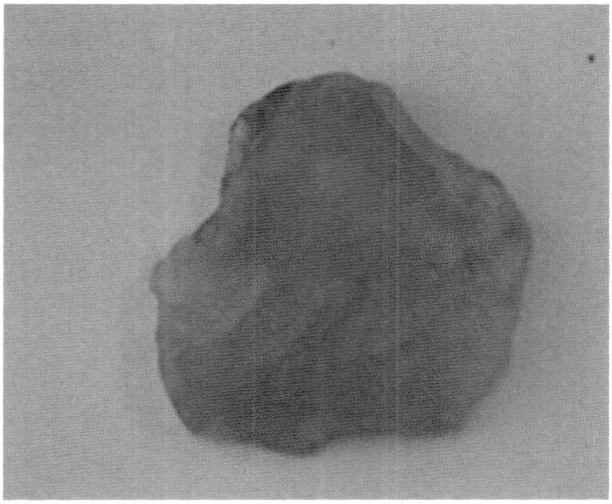

Fig. 27.11 Appearance of a Libyan dessert glass

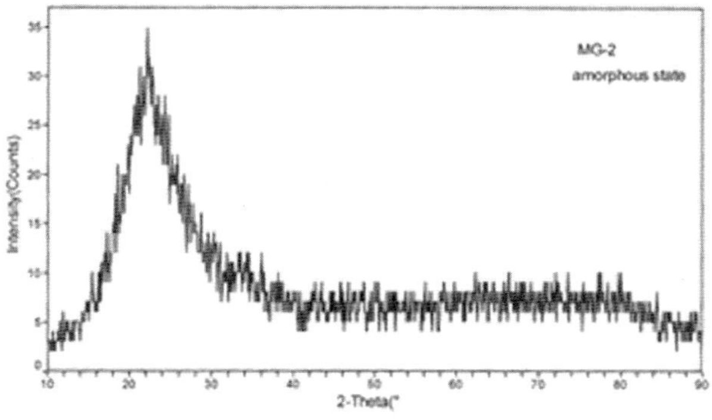

Fig. 27.12 Raman spectrum of a Libyan dessert glass

LDG samples were analyzed in our laboratory. Raman spectrum of the MG-2 sample shown in Fig. 27.12 reveals a typical amorphous state of the sample. The XRF analytical results of chemical composition listed in Table 27.6 indicates a typical aluminosilicate glass containing SiO_2 greater than 90% and Al_2O_3 about 10%. Since there are no minerals with high SiO_2 content found in the surrounding area, It is considered that LDG stemmed from meteorite rains which were formed by the explosion of some little stars with high SiO_2 content in the atmosphere. A higher content of Al_2O_3 is necessary to the formation of the

Table 27.6 Chemical composition of a Libyan dessert glass

Key	Test point	Material	Na_2O wt%	MgO wt%	Al_2O_3 wt%	SiO_2 wt%	P_2O_5 wt%	Cl wt%	K_2O wt%	TiO_2 wt%	Fe_2O_3 wt%	Cr ppm	Ni ppm	Cu ppm	Zn ppm	Sr ppm	Sn ppm	Ba ppm	Pb ppm
MG-2	Natural surface	Quartz glass	0.70	0.74	5.22	90.41	2.03	0.15	0.31	0.12	0.31	85	29	21	80	10	78	57	65

glassy state. Further tracking studies are worthy conducting by means of new methods, especially trace element analysis methods including analysis of trace elements and isotope analysis.

27.4 Summary

The non-natural glass samples, B-1 and H-3, have been successfully distinguished from our experimental natural glasses by the non-destructive analytical methods including XRF, XRD and LRS.

The XRD results show that both the Tektite and Obsidian samples are glassy.

The LRS results make it clear that the diffuse envelope peak around 470 cm^{-1} suggests the presence of cyclic structure units composed of deformed $[SiO_4]$ tetrahedrons in the samples.

The pXRF results indicate that SiO_2 and Al_2O_3 are the major chemical components of natural glasses. It is possible to distinguish Tektite and Obsidian from each other in terms of chemical composition. Tektites contain greater amount of ferric oxide (FeO) and alkaline-earth metal oxides (MgO+CaO) and less amount of alkaline metal oxides (Na_2O+K_2O) than Obsidians.

The SEM-EDS and LRS results reveal that the cooling rate of Tektite is higher than that of Obsidian. The cooling rate is so high that there is no enough time for the melt to crystallize. The glassy phase is homogeneous.

References

1. F. X. Gan, The Silk Road and ancient Chinese glass, in *Ancient Glass Research along the Silk Road*, eds. F. X. Gan, R. H. Brill and S. Y. Tian (World Scientific Publishing Co. Pte. Ltd, Singapore, 2009), pp. 41–109.
2. S. Liu, Q. H. Li, F. X. Gan *et al.*, *Spectroscopy and Spect. Anal* **30**, 2576 (2010).
3. Z. Q. Xu, H. X. Zhao, F. X. Gan, *J. Chin. Ceram. Soc.* **40**, 443 (2010).
4. S. Lin, Y. B. Guan and W. B. Hu, *Science China, (Earth Sciences)* **54**, 349, (2011).
5. R. A. Weeks, J. R. Underwood Jr, and R. Giegengack, Libyan Desert Glass: Areview, in *Natural Glasses*, eds. L. D. Pye, J. A. O'Keefe and V. D. Frechette (North-holland physics publishing, Amsterdam, 1984), pp. 593–621.

GAN Fuxi. Born in Hangzhou (Zhejiang), China, January 1933, a professor of Shanghai Institute of Optics and Fine Mechanics and Fudan University. He received his Ph.D. from the Academy of Science USSR. His research interests cover the fields of optical and laser materials, optoelectronics, and optical storage technology as well as ancient glass. He was elected the Member of Chinese Academy of Sciences in 1980 and Fellow of Third World Academy of Sciences in 1993. An honourable council member of China Association for Science and Technology, and honourable president of Chinese Ceramic Society. He received the Life Time Award of the International Commission on Glass.

LI Qinghui. Born in Xinjiang county, Shanxi, China, October 1973, a professor of Shanghai Institute of Optics and Fine Mechanics, CAS. He received his Master of Science from the China University of Geosciences (Beijing City) and Ph.D. from the Shanghai Institute of Optics and Fine Mechanics, respectively. His research interests cover the fields of mineralogy and crystallography, optical storage, archaeometry of ancient silicate artifacts as well as development of non-invasive techniques. Now he pays more attention on the application of OCT, laser Raman and high-spectral imaging spectroscopy for the scientific research of cultural heritages.

Julian Henderson. Born in London, United Kingdom, December 1953, a Professor of archaeological science at Nottingham University since 1999, a Fellow of the Society of Antiquaries and previously held Fellowships at the Smithsonian Institution, Oxford University and the University of Melbourne. He received his PhD in Archaeological Science from Bradford University. He has been

Head of the Archaeology Department in Nottingham University 4 times and Head of the School of Humanities there. He was editor of the *Journal of Archaeological Science* for 11 years. His research interests focus on ancient vitreous materials and currently on glasses and glazed ceramics along the Silk Road.

Index